WITHDRAWN

BARLEY

BARLEY

D. E. Briggs
British School of Malting and Brewing,
Department of Biochemistry,
The University of Birmingham,
England.

London
CHAPMAN & HALL
A Halsted Press Book, John Wiley & Sons, New York

First published 1978
by Chapman and Hall Ltd.
11 New Fetter Lane, London EC4P 4EE

© 1978 D.E. Briggs

Set by Thomson Press (India) Ltd.
and printed in Great Britain by
Fletcher and Son, Ltd., Norwich

ISBN 0 412 11870 X

All rights reserved. No part of this book
may be reprinted, or reproduced or utilized in
any form or by any electronic, mechanical or
other means, now known or hereafter invented,
including photocopying and recording, or in
any information storage and retrieval system,
without permission in writing from the
Publisher.

Distributed in the U.S.A. by
Halsted Press, a division of
John Wiley & Sons, Inc., New York

Library of Congress Cataloging in Publication Data

Briggs, Dennis Edward.
   Barley.

   " A Halsted Press book."
   Includes bibliographical references and index.
   1. Barley. I. Title.
SB191.B2B75 633'.16 78–7314
ISBN 0–470–26393–8

To Inge

Contents

	Preface	**xiii**
	Units of measurement	**xvii**
1	**The Morphology of Barley; the Vegetative Phase**	**1**
	1.1 Introduction	1
	1.2 The quiescent barley grain	1
	1.3 Changes in the germinating grain	9
	1.4 The growth of the stem and leaves	14
	1.5 The root system	25
	1.6 Plant morphology and lodging	33
	References	35
2	**The Morphology of the Reproductive Parts in Barley**	**39**
	2.1 Introduction	39
	2.2 The development of the ear	39
	2.3 Variations in the form of grains	52
	2.4 The ear	65
	2.5 Some implications of the wide variety of forms of barley	72
	References	73
3	**The Origin and Classification of Barleys**	**76**
	3.1 Introduction	76
	3.2 Classifications of barleys	76
	3.3 The position of barley within the Gramineae	79
	3.4 The origin of cultivated barley	81
	References	85

4 The Biochemistry of Barley — 89
- 4.1 Introduction — 89
- 4.2 Carbohydrates — 89
- 4.3 The glycolytic sequence, the pentose phosphate shunt and the tricarboxylic acid cycle — 102
- 4.4 Barley lipids — 107
- 4.5 Photosynthesis and photorespiration — 121
- 4.6 The formation of porphyrins — 125
- 4.7 Phenolic and aromatic substances — 128
- 4.8 Amino acid metabolism — 136
- 4.9 The metabolism of some amines — 147
- 4.10 Nucleic acids, and some other nitrogenous substances — 155
- 4.11 Barley proteins — 161
- References — 163

5 Grain Quality and Germination — 174
- 5.1 Introduction — 174
- 5.2 Sampling tests with small numbers of grains — 174
- 5.3 Grain evaluation — 176
- 5.4 The penetration of water, and other substances, into grain — 178
- 5.5 Testing for grain germinability — 182
- 5.6 Vigour — 186
- 5.7 Dormancy — 186
- 5.8 The gas exchange of germinating grains — 195
- 5.9 The chemical composition of the quiescent grain — 199
- 5.10 Biochemical changes in germinating grain — 202
- 5.11 Embryo culture *in vitro* — 207
- 5.12 The mobilization of the endosperm reserves — 208
- References — 215

6 The Growth of the Barley Plant — 222
- 6.1 The description of growth — 222
- 6.2 Sequential changes in the growth of the plant — 225
- 6.3 The composition of the growing plant — 231
- 6.4 The composition of the growing grain — 232
- 6.5 Root growth — 240
- 6.6 Water supplies — 245
- 6.7 Water stress — 247
- 6.8 Mineral requirements — 250
- 6.9 The uptake and release of substances by roots — 256
- 6.10 Coleoptile growth and gravity perception — 259
- 6.11 Leaf unrolling and greening — 259
- 6.12 Leaf senescence — 260
- 6.13 Growth regulation — 262
- 6.14 Temperature and growth — 264
- 6.15 Cold hardiness — 265
- 6.16 Vernalization — 266
- 6.17 Some effects of light — 268

	6.18 Some factors that control yield	270
	References	280

7 Agricultural Practices and Yield — 292
7.1	Introduction	292
7.2	Soil preparation	293
7.3	The choice of seed; sowing	294
7.4	Nutrient supply and barley yield	299
7.5	Some chemical treatments	307
7.6	Damaging factors	308
7.7	Water supplies and yield	309
7.8	Barley as forage	311
7.9	Harvesting the grain	313
7.10	Actual and potential yields	314
	References	315

8 Production and Harvesting Machinery — 320
8.1	Introduction	320
8.2	Irrigation and drainage	320
8.3	Tillage	322
8.4	Sowing	327
8.5	Post-sowing treatments	329
8.6	Harvesting and threshing barley	330
8.7	Straw	334
8.8	Harvesting the whole plant	335
8.9	Conclusions	337
	References	337

9 Weeds, Pests and Diseases in the Growing Crop — 339
9.1	Weeds and the need to control them	339
9.2	Weed control	340
9.3	The economics of weed control	343
9.4	Nematode pests	343
9.5	Molluscs	344
9.6	Birds and mammals	345
9.7	Insect and some other pests	345
9.8	Virus diseases of barley	349
9.9	Bacterial diseases	350
9.10	Fungal diseases	351
9.11	Some general considerations	364
	References	366

10 The Reception and Storage of Whole Plants and Grain. The Micro-organisms and Pests of Stored Grain — 369
10.1	Introduction	369
10.2	Barley hay	369
10.3	Straw	372
10.4	Barley silage	372

x   *Contents*

		10.5	Grain reception	375
		10.6	Handling grain	375
		10.7	Weighers	377
		10.8	Cleaning and grading grain	377
		10.9	Drying principles	382
		10.10	Grain drying in practice	385
		10.11	Grain storage facilities	389
		10.12	Seed longevity and grain deterioration	394
		10.13	Micro-organisms in grain	399
		10.14	Insects and mites	404
		10.15	The mites of stored grain	410
		10.16	Insecticides and fumigants	410
		10.17	Rodents and their control	412
			References	414
11	**Barley Genetics**			**419**
		11.1	Introduction	419
		11.2	The inheritance of 'distinct' factors	419
		11.3	Cytology and chromosome behaviour	422
		11.4	Chromosomal abnormalities	427
		11.5	Ploidy levels	430
		11.6	Mutations and mutagenesis	434
		11.7	The expression of some mutant and other genes	437
		11.8	The genetics of complex characters	440
			References	441
12	**Barley Improvement**			**445**
		12.1	Introduction	445
		12.2	Plant introductions, and adapted forms	446
		12.3	Plant selections	449
		12.4	Mutation breeding	450
		12.5	Hybridization	451
		12.6	Crossing barley	453
		12.7	The choice of parents	456
		12.8	Selection sequences applied to hybrid progenies	457
		12.9	Competition and 'natural selection' in barley	460
		12.10	Breeding for quality	463
		12.11	Some other objectives in breeding	466
		12.12	Breeding for higher yields	468
		12.13	The quantitative evaluation of parents	470
		12.14	'Hybrid' barley	471
		12.15	Trial procedures	474
		12.16	The multiplication of seed	475
		12.17	Conclusion	476
			References	476
13	**Some Actual and Potential Uses of Barley**			**481**
		13.1	Introduction	481
		13.2	Barley grain; a source of starch and protein	481

13.3	Minor uses of straw	482
13.4	Straw in building	482
13.5	Animal bedding, litter, farmyard manure and compost	483
13.6	Soil protection, conditioning, or replacement	484
13.7	Some industrial uses of barley	486
13.8	Paper, cardboard and millboard	487
	References	490

14 Barley for Animal and Human Food — 492

14.1	Introduction	492
14.2	The nutritional requirements of animals	492
14.3	Forage and hay	500
14.4	Silage	504
14.5	Barley straw	505
14.6	Barley grain	507
14.7	By-products for animal feed, derived from barley	514
14.8	Non-alcoholic beverages	516
14.9	Other potential feeding stuffs	516
14.10	The technology of preparing grain for food	518
14.11	Future uses of barley as food	521
	References	522

15 Malting — 526

15.1	Introduction	526
15.2	The selection and acceptance of malting barley	527
15.3	Barley handling	533
15.4	Steeping	533
15.5	Germination equipment	538
15.6	Kilns and kilning	541
15.7	Malt analyses	545
15.8	Changes that occur in the malting grain	550
	References	557

16 Some Uses of Barley Malt — 560

16.1	Introduction	560
16.2	Mashing	561
16.3	Some aspects of yeast metabolism	570
16.4	Malt extracts and barley syrups	572
16.5	Brewing beer	573
16.6	Malt vinegar	577
16.7	Distilled 'potable spirits'	580
	References	584

Index — 587

Preface

This book was written to provide an integrated account of barley, including its cultivation, nature and uses. An attempt has been made to cut across the unjustified and obstructive divisions between pure science, applied science, technology, botany, biochemistry, agronomy, and so on. Limitations of space preclude the use of more illustrative material or references, or even complete accounts of various topics. However sufficient information is given to enable the reader to understand the general principles and to find his or her way readily into the literature to obtain further information. Emphasis has been placed on general principles rather than details. In becoming familiar with the literature one becomes acquainted with the effects of the cereal or religion, the English language and the development of agriculture and biochemistry. The comparison between 'parallel literatures' is often stimulating also. For example one is forced to conclude that many of the agricultural problems of poor 'seed vigour' would be overcome if seedsmen used the maltsters techniques for breaking dormancy and speeding 'post-harvest maturation'.

Barley is the world's fourth most important cereal after wheat, rice, and maize. It is the most widely cultivated, being grown from the equator to $70°N$ (Scandinavia), from the humid regions of Europe and Japan to the Saharan and Asiatic Oases, and from below sea level in Palestine to high up mountains in the Himalayas, E. Africa and S. America. Somewhere in the world it is being sown or harvested at every time of the year. The world crop fluctuates year by year, but overall the trend is upwards (Preface Table 1). The highest yields are obtained in Europe. There is an absolute need to improve food supplies as probably $1-1.5 \times 10^9$ people, about half the world's population, are poorly fed or starving. With barley improvements must involve achieving greater yields of materials, probably naked grains with high protein contents, which are suited for direct human consumption.

Preface table 1 Estimates of barley production (after *Mthly Bull. Agric. Econ. Stat.* **25**(2), 1976, p. 13).

Region	Area harvested (10^6 ha)				Yield (tonnes/ha)				Yield (10^6 metric tonnes)			
	1961–1965	1973	1974	1975	1961–1965	1973	1974	1975	1961–1965	1973	1974	1975
World	68.0	87.9	87.9	91.3	1.47	1.93	1.94	1.69	99.7	169.3	170.7	154.0
Africa	5.0	5.4	4.2	3.7	0.71	0.69	0.90	0.76	3.5	3.7	3.8	2.8
N. & Central America	7.0	9.3	8.3	8.2	1.81	2.12	1.89	2.21	12.7	19.8	15.7	18.1
S. America	1.1	1.1	0.9	1.1	1.11	1.22	1.13	1.19	1.3	1.3	1.1	1.3
Asia*	22.6	22.9	23.3	24.0	1.19	1.30	1.36	1.44	27.0	29.9	31.6	34.4
Europe	13.0	17.8	18.2	19.0	2.61	3.20	3.39	3.12	33.8	56.9	61.6	59.1
Oceania	0.9	2.0	1.9	2.4	1.19	1.35	1.45	1.36	1.1	2.7	2.8	3.3
U.S.S.R.	18.3	29.4	31.1	32.9	1.11	1.87	1.74	1.06	20.3	55.0	54.2	35.0

*excluding U.S.S.R

I wish to thank the many people who have helped me in the production of this book, and have guided me away from various pitfalls. I particularly thank Miss B. Ronchetti and her staff in the University of Birmingham library for obtaining some hundreds of 'outside' publications, Dr R.H. Waring for checking the proofs, and Mrs Pauline Hill for her careful production of the artwork. Above all I thank my wife Inge for help with the typing and for unfailing encouragement and support through six difficult years.

Units of measurement

It has proved impossible to convert all the values in this book into standard metric units. To assist the reader, and to allow an understanding of data published in non-metric units, this abbreviated set of common equivalents is given.

Length
    1 mile = 1760 yards (yd) = 1.609 34 kilometres (km) = 1609.34 metres (m)
    1 yard (yd) = 3 feet (ft) = 36 inches (in) = 0.9144 metres (m)
    1 inch (in) = 25.4 millimetres (mm) = 2.54 centimetres (cm)

Area
    1 acre = 10 square chains = 4840 yd^2 = 4046.86 m^2 = 0.404 686 ha
    1 are (a) = 100 m^2 1 hectare (ha) = 10^4 m^2
    640 acres = 1 square mile = 258.999 ha

Mass
    1 ton = 20 hundredweights (cwt) = 2240 pounds (lb)
        = 1016.05 kilograms (kg) = 1.016 05 tonnes (t)
    16 ounces (oz) = 1 pound (lb) = 0.453 592 37 kilograms (kg)
    1 quintal (q) = 100 kg = 1.969 cwt = 220.46 lb
    28 pounds (lb) = 1 quarter (Qr, *Avoir du pois*)
    But 1 quarter barley (Qr; U.K.) = 448 lb
        1 quarter malt (Qr; U.K.) = 336 lb

Volume, Capacity
    1 litre (l) = 1 decimetre cubed (dm^3; approximately) = 0.035 315 ft^3
        = 0.219 97 gal (U.K.) = 0.264 17 gal (U.S.A.)

xviii Units of measurement

1 millilitre (ml) = a centimetre cubed (cm³; cc; approximately)
    = 0.061024 in³
1 hectolitre (hl) = 0.1 metre cubed (m³; approximately) = 100 l
1 cubic foot (cu ft, ft³) = 28.3168 l
1 bushel (bu; U.K.) = 8 gallons (gal U.K.) = 36.368 7 l

But

U.K., S. Africa. { 1 bu (barley, nominal) = 56 lb = 25.401 kg
                  { 1 bu (malt, nominal) = 42 lb = 19.051 kg
Australia,   { 1 bu (barley, nominal) = 50 lb = 22.680 kg
New Zealand   { 1 bu (malt, nominal) = 40 lb = 18.144 kg
U.S.A., Canada   { 1 bu (barley, nominal) = 48 lb = 21.772 kg
                  { 1 bu (malt, nominal) = 34 lb = 15.422 kg
1 gallon (gal, U.K.) = 4 quarts = 8 pints (pt)
    = 1.201 gallon (gal U.S.A.) = 4.546 09 litres
1 barrel (Brl, U.K.) = 2 kilderkins = 4 firkins = 36 gallons (gal U.K.)
    = 1.6365 hl
1 standard barrel (U.S.A.) = 31.5 gal (U.S.A.) = 1.1924 hl
    = 1.016 19 barrel (U.S.A.)

Weights and volumes/units area

1 ton/acre = 20 cwt/acre = 2240 lb/acre = 0.251 071 kg/m²
    = 2510.71 kg/ha
1 gallon (U.K.)/acre = 8 pints (pt)/acre = 11.232 l/ha
1 acre-inch = 254 m³/ha
Fertilizers (U.K.); 1 unit/acre = 1.12 lb/acre = 1.25 kg/ha (approximately)

Other units

1 British thermal unit (B.T.U.) = 1.055 06 kilojoules (kJ)
1 therm = 105.506 megajoules (MJ)
1 calorie (thermochemical) = 4.184 J
1 lb/bu (U.K.) = 1.274 kg/hl
1 lb/bu (U.S.A.) = 1.287 kg/hl

Temperature

The relationship between the Celsius (Centigrade) scale, °C, and the Fahrenheit scale, °F, is
temperature °C = $\frac{5}{9}$ (temperature °F − 32).
Ice point, 0°C = 32°F. 100°C = 212°F.

Chapter 1

The morphology of barley; the vegetative phase

1.1. Introduction

Barley, like other plants, is highly variable. Varieties differ greatly in their morphological and other characters, and a wide variation of characters occurs in 'pure races'. When a character is quantified for a group of plants this number is usually some sort of an average. A wide range of individual values will occur. This is obvious when plants are credited with fractional numbers of roots or stems, but in other cases the 'average' nature of a description may not be apparent.

1.2. The quiescent barley grain

The barley grain is roughly a spindle-shaped body, tapering at each end, with a shallow furrow running along the ventral side. In most commercial barleys the flowering glumes, (husk or chaff), adhere to the grain, but many varieties are known in which the naked or hull-less grain separates from the chaff. Within the husk is the caryopsis, a fruit in which the pericarp (the remains of the ovary wall) is fused to the seed coat (testa, tegmen). Within the testa the largest tissue is the starchy endosperm, bounded at the periphery by the aleurone layer, which is also part of the endosperm. At the basal end of the grain is the embryo, which is complex in structure (Fig. 1.1).

The husk at the apex and base of the grain is usually damaged where the awn, or other 'apical appendage', and basal attachment have been broken in threshing. It is usually pale yellow or buff, and is slightly patterned with wrinkles. The lemma, (palea inferior, flowering glume) invests the dorsal, rounded side of the grain. It carries five longitudinal surface ridges, or 'nerves', below which run vascular bundles. The thinner edges of the lemma overlap the thinned edges of the palea (palea superior,

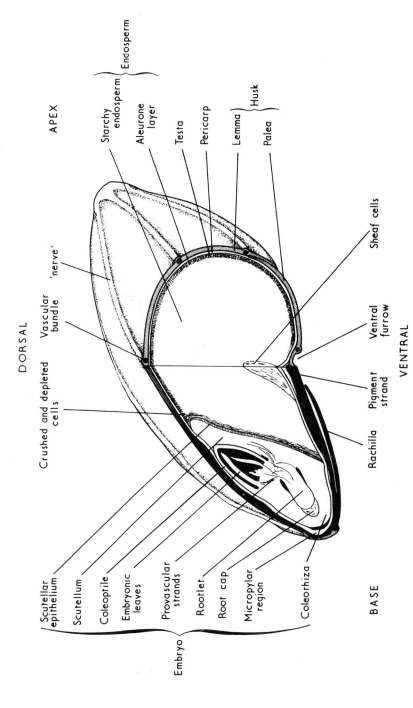

Fig. 1.1 Idealized diagram of a barley grain, with a sector removed, to show the disposition of the tissues (from Briggs [11]: copyright by Academic Press).

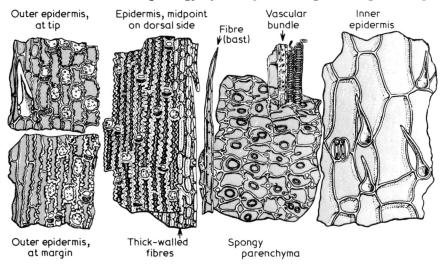

Fig. 1.2 The microscopic appearance of the tissues of the husk (lemma), separated and seen in surface view (after Winton and Winton [100]).

palet). The palea has two 'nerves' and is indented over the ventral furrow. At the base a more or less hairy appendage, the rachilla (basal bristle) may be found. The rachilla, and the two flattened, hairy lodicules that lie over the embryo, within the husk, are very variable in form (Chapter 2). The detailed microscopic structure of the grain is of interest, for example in identifying milled products, as well as for understanding the physiology of the grain [19,54,55,58,59,92,94,100].

The husk is made up of four types of cell which are dead at maturity (Fig. 1.2). The outer, longitudinally elongated epidermal cells have mainly thick, wavy walls except towards the thinner edges. Silica cells and occasional hairs are present but in general stomata are not, except for single rows on the palea, each side of the furrow [65]. The epidermal cells are heavily silicified [33]. The bast layer, below the outer epidermis, is made up of more or less thickened sclerenchyma fibres, which have lateral, tooth-like protrusions which fit into recesses in adjacent cell walls. Below the bast is a thin-walled parenchyma, which was photosynthetic in the immature grain. The vascular bundles are embedded in this layer. The inner face of each husk component is bounded by a thin-walled epidermis containing stomata, and having numerous single-celled hairs.

In husked (hulled) barley the palea and lemma adhere to the pericarp, although air-spaces sometimes occur. The pericarp is crushed during development. In naked grains the pericarp is less compressed and is more robust. However, it is not lignified: in the quiescent grain lignin is practically confined to the husk. The pericarp is closely adherent to the

4 Barley

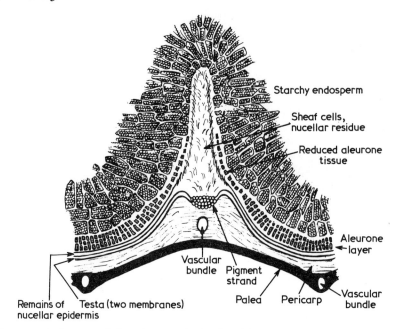

Fig. 1.3 A diagram of the microscopic details of the structures found at the ventral furrow.

testa all over the grain except at the apex, where it may be in folds and carry the vestiges of the ovary tip. It may have a more open structure at the base, in the region of the micropyle, and at the ventral furrow where it becomes thicker, it carries the remains of a vascular bundle, and abuts the pigment strand, the remains of the chalazal tissue (Fig. 1.3). In the pericarp three, rarely four, types of cell occur. The outer, epidermal layer (the epicarp) is of flattened, thin-walled cells, slightly elongated along the longitudinal grain axis, which at the apex may carry single-celled hairs. The next layer, the hypodermis, is of similarly shaped cells, but next comes a double layer of 'cross cells', which are elongated at 90° to the grain axis; these are the residues of the photosynthetic tissue of the ovary. Lastly a few scattered 'tube cells', where visible, represent the remains of the inner epidermis of the ovary wall (Fig. 1.4).

The testa (tegmen, spermoderm, seed-coat) invests the entire grain except where it abuts and merges with the ventral pigment strand, and possibly at the micropyle, at the base of the grain. The testa is so crushed that, although in surface view the outlines of thin-walled cells can be detected, it is usually regarded as a tough membrane of 'cellulose', between two cuticularized layers of which the outer is appreciably thicker than the inner. This layer effectively separates the exterior from the interior of the grain. It is thicker in the flanks of the furrow and at the apex

The morphology of barley; the vegetative phase 5

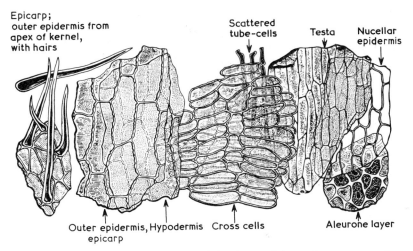

Fig. 1.4 The microscopic appearance of the tissue of the pericarp, testa and aleurone layer, ('bran'), separated and seen in surface view (after Winton and Winton [100]).

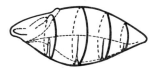

Fig. 1.5 The variable extent of penetration of the furrow (chalazal tissue) and the tip of the nucellar sheaf cells into the grain (after Collins [19]).

of the grain, thins down at the sides and on the dorsal side, is thin over the embryo and may be further reduced, even to one cutin layer, or even be absent, at the micropyle [19,54,64,92]. At the ventral furrow the testa merges with the pigment strand (Fig. 1.3). This strand, which runs the length of the grain, is the chalazal tissue (abutting the funiculus or funicle), which in the mature grain has thickened, brown 'suberized' cell walls, and in which the cell lumena contain brown, waxy inclusions. It provides an effective seal between the edges of the testa. If sections of grain are treated with acid dichromate the cuticles of the husk and testa, and the pigment strand survive.

Within the testa is a crushed, hyaline layer. This can be detected easily in sections swollen with solutions of caustic alkali, when the remains of cellular structures may be detected [19]. This, the perisperm, is the remains of the nucellar tissue. At the ventral furrow it merges with the 'sheaf cells'. The 'sheaf cells' form a crest of tissue projecting into the starchy endosperm to various extents (Fig. 1.3 and 1.5). These consist of empty cells crushed together so that the separate walls are hard to

6 Barley

distinguish. The cells of the starchy endosperm appear to radiate from this tissue.

At the base of the grain is the embryo. The supposed 'homologies' of the parts of this complex body, like those of other parts of the plant, have been the subject of much heated and unprofitable argument and will not be pursued here [13]. The embryonic axis consists of an apical meristem, three or rarely four embryonic leaves, usually two axillary bud primordia, and a tubular coleoptile (leaf-sheath, acrospire) with an apical pore (Figs. 1.1 and 1.6). Usually the coleoptile has two vascular bundles, but in some Chinese varieties three to five are found, and up to six are known [43,81]. There is always a large bud primordium within the coleoptile [66]. Below the shoot, and directed towards the base of the grain are the seminal root initials, each with its root cap (calyptrogen). In

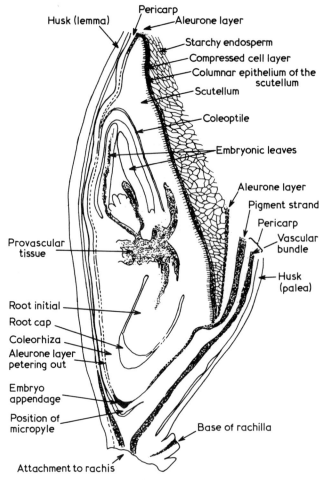

Fig. 1.6 The embryo in longitudinal section (after Lermer and Holzner [59]).

The morphology of barley; the vegetative phase

general there is one main root and between one and ten secondary roots. This root number is partly characteristic of a variety, but within one variety there is a large spread of numbers, smaller grains tending to have fewer rootlets [29,37,77]. Around the roots is the root sheath, or coleorhiza. Between the tip of the coleorhiza and the micropylar region of the testa is the 'embryonic appendage', which is the crushed and empty remains of the suspensor. From the nodal region, between the shoot and rootlet initials, provascular strands extend into the shoot, the roots and into the scutellum. The scutellum (Lat. 'shield') is a flattened expanded organ. On the outer side it is recessed to receive the embryonic axis and on the inner side it fits against the starchy endosperm. The scutellum is made up mainly of thin-walled parenchyma, but at the interface with the endosperm it is covered with a single-celled layer of palisade-type columnar epithelium. The subcellular anatomy of the dry embryo is not easy to study. Few or no starch grains are present, but mitochondria, protein (aleurone?) bodies, lipid-containing spherosomes, Golgi bodies and rough endoplasmic reticulum have been noted, together with large nuclei and thin cell walls, which are traversed by plasmodesmata [1].

The endosperm is divided into four main sections. Around the entire periphery of the starchy endosperm, except where it peters out at the ventral furrow, is the living aleurone layer (Figs. 1.1, 1.3, 1.4 and 1.7).

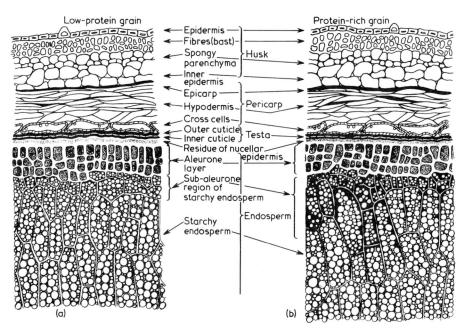

Fig. 1.7 Schematic transverse sections of the dorsal side of a barley grain, over the starchy endosperm, of a protein-poor grain (a) and a protein-rich grain (b) (after Briggs [11]:|copyright by Academic Press).

Normally this is two to four cells (50–110 μm) thick; some African barleys have layers two cells thick. The aleurone layer is reduced to a single layer of reduced, flattened cells overlapping a large part of the embryo, but ceasing towards the base of the grain (Fig. 1.6). The aleurone cells are roughly cuboid, and are separated by thick cell walls crossed by plasmodesmata. They are filled with dense cytoplasm containing prominent nuclei, and a wide range of organelles; rough endoplasmic reticulum, mitochondria, microbodies, proplastids, numerous lipid-containing spherosomes and complex, spherical aleurone grains. No starch is present [16,20,45]. The aleurone grains have protein ground substance, and two types of inclusions; one apparently of protein-polysaccharide and the other the potassium–calcium salt of phytic acid [28,42,60]. Ultracentrifugation stratifies the cell contents and allows their quantification. The thick cell walls contain two structural components which may be separated by selective enzyme digestion during the preparation of protoplasts. The inner, more resistant component, which is immediately adjacent to the plasmalemma, surrounds each cell and lines the pores carrying the plasmodesmata [90].

Three types of dead tissue make up the starchy endosperm. Immediately beneath the scutellum it consists of cell walls, without cellular contents, crushed together and sometimes termed the 'depleted cell layer'. In the central parts of the starchy endosperm the dead cells consist of thin-walled parenchyma packed with starch-grains of widely varying sizes (Chapter 6). Again the cell walls contain two major polysaccharide components. The starch grains are embedded in a proteinaceous matrix. Unless the tissue is immature no intact nuclei are found. The cell walls may carry depressions from the pressure of the starch grains. The residues of the amyloplasts, in which the starch grains were formed, may be detected. In the starchy endosperm adjacent to the aleurone layer, the sub-aleurone region, the cells are smaller and contain substantially more protein and less starch which tends to be small-grained. This difference is particularly noticeable in protein-rich barleys (Fig. 1.7). Protein-containing storage bodies also occur in the protoplasm of the starchy endosperm. By conventional light microscopy they may be mistaken for small starch grains [72,91]. Immature, high-protein grains sometimes contain detectable remains of nuclei in the starchy endosperm [14]. In many cases the endosperms of immature grains, when cut across have a flinty (steely, horny, hard, glassy) appearance. On the other hand mellow, plump, well-matured grains have endosperms that appear chalky (floury, mealy, soft and white), due to the presence of many thousands of minute air-containing cracks around the starch grains and possibly between the cell walls [12]. The endosperms of steely grains are under tension, since attempting to cut sections for microscopy, or soaking them in water and redrying, often makes them mealy. Steely endosperms also have a higher

The morphology of barley; the vegetative phase 9

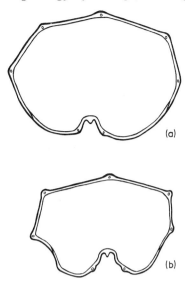

Fig. 1.8 Outlines of transverse sections through the midpoints of (a) a plump, well-filled barley grain, and (b) a thin, poorly-filled grain (after Percival [74]).

specific gravity than mealy endosperm, e.g. 1.345 against 1.305 [12]. In cross-section plump, well-filled grains have a 'kidney' shape. In poorly grown grains the caryopsis does not 'stretch' the husk, so in transverse section it has a more angular appearance (Fig. 1.8).

When kernels that have been peeled, or decorticated with sulphuric acid, are viewed by transmitted light many 'zebroid', mainly transverse, faint shadow-like 'striations' may be visible. Possibly these constitute a useful varietal character [89].

1.3. Changes in the germinating grain

The surface layers of grain readily pick up moisture, which then penetrates into the interior mostly entering through the micropyle (Chapter 5). Fig. 1.9 represents the micropylar region, but accounts do not agree in their description of this structure [54,92]. As water is taken up so the grain swells, the embryo becomes turgid, and noticeably increases in size (Fig. 1.10). The first sign of germination is the emergence of the coleorhiza from the base of the grain (Fig. 1.11) [14]. The main rootlet then emerges, closely followed by the secondary seminal rootlets, which are all nearly indistinguishable. In moist conditions they are supplied with numerous fine root hairs. The coleoptile (blade, spire, acrospire) breaks through the testa and grows up the 'back', or dorsal side of the grain. In husked grain it often presses a groove into the endosperm as it grows; it usually emerges near the apex of the grain, or from beneath the

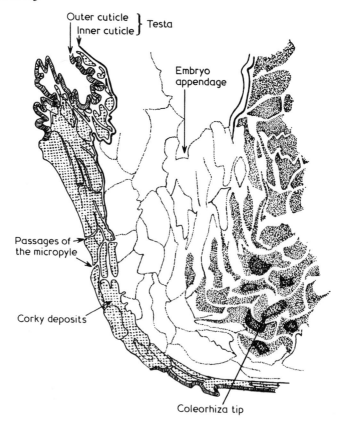

Fig. 1.9 A diagram of the structure of the micropylar region of a husked barley, seen in longitudinal section. In other barleys different structures have been noted (after Krauss [54]).

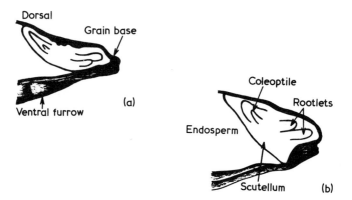

Fig. 1.10 Diagrammatic longitudinal sections of the basal end of (a) a dry grain, and (b) a grain swollen from imbibing water.

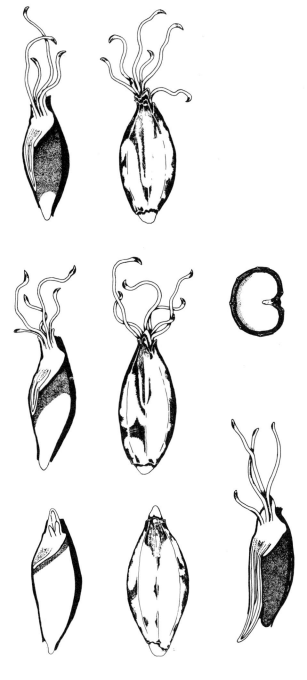

Fig. 1.11 Three early stages of grain germination viewed from the dorsal side and four in longitudinal section, to one side of the furrow structures. The approximate extent of cell wall degradation 'modification', is indicated.

lemma part way along one side. In naked grains the coleoptile often follows a similar course, as it grows beneath the pericarp. In contrast, in decorticated grains from which the husk and pericarp have been removed, the coleoptile grows freely away from the grain surface.

The initial growth of the embryo is supported by its endogenous reserves; subsequently it uses nutrients which are the products of the dissolution of the endosperm. Endosperm breakdown begins immediately below the scutellum, in the layer of compressed cells, and moves towards the apex of the grain. After a period it proceeds faster immediately beneath the aleurone layer [10,63]. The observed asymmetry of breakdown is due to the manner in which the scutellar face is set against the starchy endosperm at an angle to the long axis of the grain, the sequence in which degradative enzymes are formed at various sites, and because the residual walls of the nucellar 'sheaf cells' are relatively resistant to enzymic degradation (Figs. 1.11 and 1.12; Chapter 5). In grain growing on a wet substratum degradation is so complete that the starchy endosperm liquefies beneath the scutellum and beneath the aleurone layer. In grain grown under malting conditions, i.e. adjusted to a chosen moisture level and then germinated in a moist atmosphere, the endosperm becomes progressively softer and more easily 'rubbed out'. On drying the partly degraded endosperm is friable, rather than tough, and is readily crushed. The endosperms of mealy grains undergo this limited breakdown (modification) more readily than those of steely grains. Disputes regarding the extent of cell wall breakdown during 'modification' were due to differences between grains, their moisture contents during germination, and the exact manners in which they were grown [10,11]. As the breakdown of the starchy endosperm begins the cell walls swell, separate into lamellae, lose their ability to take up stains, and apparently disintegrate. A thin insubstantial membrane, not stainable with dyes like Congo Red, survives longer. Following some cell wall degradation, protein degradation begins, and the starch grains lie free in the remains of the cell. Later still, visible degradation of the starch-grains occurs. The grains become pitted, and concentric lamellae separate [14,76]. The degradation pattern of the starch grains is noticeably different to that occurring in living cells [14].

Modification of the starchy endosperm is accompanied by a series of changes in the aleurone layer. These have been studied in whole grains, and in separated but functionally intact aleurone layers treated with gibberellic acid (Chapter 5). As the aleurone hydrates the aleurone grains swell. There is a progressive increase in the cristae of the mitochondria, and the rough endoplasmic reticulum increases in amount, as do the dictyosomes (Golgi apparatus), and microbodies. Gradually the lipid content of the spheroplasts declines, and the aleurone grains are used up, gradually being transformed into vacuoles which occupy progressively

The morphology of barley; the vegetative phase

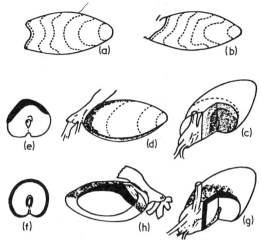

Fig. 1.12 The successive stages of endosperm breakdown in an 'average' malting grain (a-c), and in an 'average' grain germinating on a wet substratum (d-h; from Briggs) [10]. (a-c) The grains were malted in the dark at 14.4°C (about 58°F); the pattern of breakdown was the same whether or not the corns were decorticated. (a) and (b) the approximate extent of modification in 'average' grains successively on days 1–6 indicated by dashed lines in (a) longitudinal horizontal section above the level of the sheaf cells, and (b) in longitudinal, vertical section to one side of the furrow. (1) A schematic view of the extent of modification in grain malted for 2 days. (m-q) The extent of endosperm liquefaction in decorticated grain germinated in the dark, on wet filter paper at 25°C (77°F). (d,g,h) – The shaded areas indicate the extent of the fluid-filled spaces seen in silhouette from outside the grain. (d) The external appearance of the grain after 2 days growth. The dotted lines indicate the position of the leading edge of the fluid-filled space, revealed by dissecting frozen grains, after 2,3 and 4 days. (e) Vertical, transverse section and (f) transverse section cut parallel to the face of the scutellum of grains germinated for 2 days. The black zones indicate the positions of fluid-filled spaces. (g) A schematic view of a grain germinated for 2 days. (h) A view of a grain germinated for 4 days, frozen, and with a piece of the near-side cut away to show how the fluid-filled space (black) extends beneath the scutellum and around a residual, softened 'core' of starchy endosperm, which remains anchored to the sheaf cells.

greater proportions of the cells. Beyond a certain point the subcellular organelles also appear to degenerate. Electron micrographs show that the inner tangential walls of the aleurone layer are selectively degraded, separating into lamellae and being eroded from the aleurone-cell side, inwards towards the starchy endosperm. The channels carrying the plasmodesmata are also enlarged and become irregular [20,28,46–48,95].

The embryo also undergoes striking microscopic changes. Initially it is almost devoid of starch, but on hydration starch appears and is deposited, notably in the coleoptile, the rootlets, and around the provascular bundles. This occurs whether or not the embryo is separated

14   Barley

from the grain (Chapter 5). As endosperm breakdown proceeds starch grains appear in the scutellar parenchyma, apparently being synthesized from the products of endosperm dissolution. Later, when these granules of transitory starch are utilized, they undergo reductions in their radii and, in contrast to the granules of the endosperm, do not show pitting. As germination proceeds the cells of the coleorhiza become microscopically distinct [70], and lignification spreads along the provascular strands. The cells of the columnar epithelium, which are at first isodiametric in cross section, separate along the middle lamellae of their walls and become roughly cylindrical in shape. The cytoplasm alters in appearance and becomes cloudy and remains in this state until the endosperm is depleted. The mitochondria, rough endoplasmic reticulum, dictyosomes, and leucoplasts first increase and later decline in quantity, while the contents of the spherosomes and aleurone grains are utilized. At one stage the separated epithelial cells, which project into the dissolving endosperm like microvilli, elongate by an extent variously estimated as 2–4 times [14,28,62,63,69].

1.4. The growth of the stem and leaves

If the grain is planted at a shallow depth the coleoptile will reach the surface. Its elongation is checked by light. The first leaf emerges from the pore at the coleoptile tip (Fig. 1.13). At the same time the seminal roots are growing, branching and extending into the soil. If the seed is more

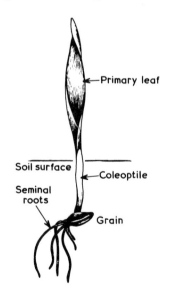

Fig. 1.13   Seedling with primary leaf emerging (compare Chapter 6; after Percival [74]).

The morphology of barley; the vegetative phase

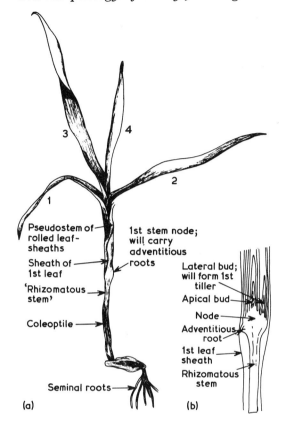

Fig. 1.14 (a) A seedling with four leaves visible, and a swelling indicating the position of the developing crown. (b) The crown region in longitudinal section (after Percival [74]).

deeply planted an elongated 'rhizomatous stem' is formed, from above the coleoptile node, and this throws out leaves when it reaches the surface [3,74]. This 'rhizome' may be one or several internodes in length. The nodes may carry adventitious roots. Apparently the anatomy of the 'rhizome' has never been studied in detail. Gradually leaves, either preformed in the embryo or generated afterwards at the growing point, grow rolled up from the tube formed by the leaf-bases of earlier leaves. When a blade has emerged it unrolls and the auricles and ligule develop. Gradually the upper part of the stem, situated at or near the soil surface and carrying the leaf bases, begins to swell so forming the crown, and adventitious root initials form (Fig. 1.14). At this time the visible main and branch 'stems' or tillers, that are formed from adventitious buds, are really pseudostems composed of rolled leaf bases, the leaf sheaths. The adventitious (coronal, crown, nodal) roots grow out of the crown as the

16 *Barley*

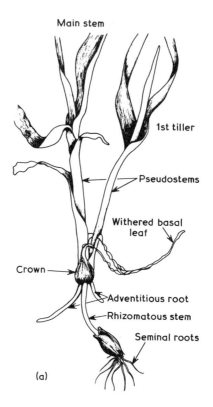

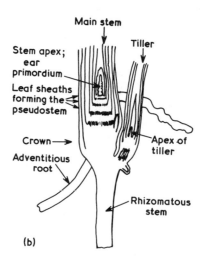

Fig. 1.15 (a,b) A seedling in which the pseudostems of the main axis and first tiller are visible and erect (after Percival [74]).

The morphology of barley; the vegetative phase 17

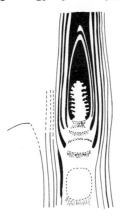

Fig. 1.16 A more detailed view of the stem apex, in longitudinal section, when elongation is beginning, showing the position of the ear rudiment, and the beginning of the separation between the nodal and internodal regions (after Smith [85]).

tillers develop (Fig. 1.15). In time the apical primordia cease producing leaf initials and generate the ear primordia, each of which is situated at the base of a cylindrical sheath of concentric, rolled leaf bases (Chapter 2). Various 'juvenile habits' occur in young plants. In some varieties the pseudostems are more or less prostrate on the ground, in others they are semi-erect, while in yet others they are erect [3,17,79]. When the stem starts to elongate, at the shooting or 'jointing' stage, the stem, or culm, becomes more clearly differentiated into the nodes, which remain solid and from which the leaves arise, and the internodal regions which elongate and become hollow (Fig. 1.16).

By the time the stems have fully elongated the seminal and coronal root systems are at or near their greatest size (Chapter 6). In the field the crown typically carries several culms, a main stem and two or three tillers. Tillering is a varietal character, as well as being strongly influenced by growth conditions. In 'uniculm' varieties generally only the main stem is formed. In other cases about 120 tillers have been noted, or in one record of 1660, 249 stalks [2]. The tillers may form a genetic mosaic. In chimaeras induced by mutagenic treatments nine mutant sectors occur in plants from large seeds having nine meristems [41]. Stem elongation may be sufficient to carry a node from within the sheath of the leaf borne by the node below (Fig. 1.17). The recognition of different growth stages is of importance in agriculture (Chapter 6). The apex of each fertile tiller carries an ear (Chapter 2). As the stem shoots the ear is carried upwards. The last, 'flag' leaf generally contains the ear within its sheath, which swells, and is called the 'boot'. Eventually the terminal internode, the peduncle, lengthens sufficiently to carry the awns into view (Fig. 1.18),

18 *Barley*

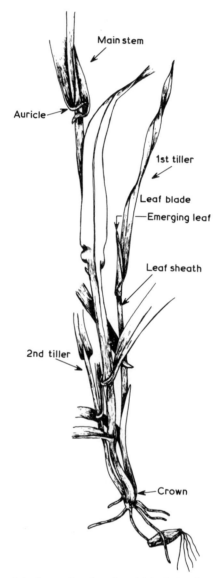

Fig. 1.17 Diagram of the base of a plant in which the main stems and two tillers are shooting, i.e. the stems are elongating. The ear primordium of the main stem is now within the leaf sheath at the top of the shoot (after Percival [74]).

and then in most varieties, to carry the ear clear of the boot. The mature leaves progressively senesce, the older basal leaves first, losing their green colour, becoming brownish, brittle and withered. Gradually the whole plant dries out until full maturity, when the grain is ripe. Flowering or 'anthesis' generally occurs in the newly-emerged ear.

The morphology of barley; the vegetative phase

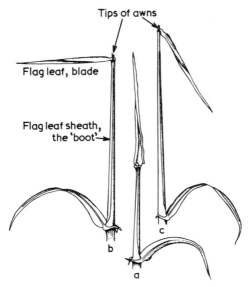

Fig. 1.18 The emergence of the flag leaf, (a) and early stages of the ear in boot (b,c), where the awns are just visible, but the ear is not sufficiently grown to cause the flag leaf sheath to bulge appreciably (from Bergal and Friedberg [4]).

The duration of the different developmental periods varies widely depending on the area of cultivation, the time of sowing, and the variety being grown. In oceanic Europe, including the U.K., spring sown barley requires about 140 days from sowing to maturity, while in western continental Europe the figure is about 120 days, and in N. Europe, e.g. Finland, about 100 days. In contrast autumn sown barley may take about nine months to reach maturity [3].

In most barley varieties each fully grown stem, or culm, consists of a series of cylindrical hollow internodes (usually 5–8) separated at the nodes (joints) by transverse septa. A leaf sheath has its origin at each node, just above the node the sheath is swollen at the meristematic pulvinus. The base of the internode within the leaf sheath, which is not swollen, also remains meristematic, soft and lacking in lignification until after heading. It is a weak zone in the stem which is, however, braced by the ensheathing leaf. If a stem is lodged (laid, displaced from the vertical), asymmetric growth of the internode base and the pulvinus tends to return it to an upright stance (Fig. 1.19). The basal internodes are the shortest and in many varieties each internode is longer than the one below it. However in some forms the peduncle is shorter than the penultimate internode [3,35,52]. The length of the culm differs depending on the variety and the growing conditions. Heights of between 13 cm (5 in) and 153 cm (5 ft) are encountered [17]. The dry stems, together with the shrivelled

20 Barley

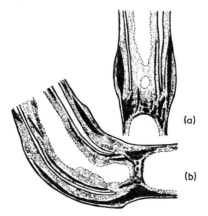

Fig. 1.19 Longitudinal sections through the internodes of a barley stem (a) growing vertically, and (b) laid horizontally, and showing differential growth (after a photograph of Esau [21]).

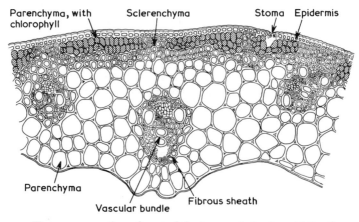

Fig. 1.20 Transverse section of part of the internode in the middle of a young stem (after Lermer and Holzner [59]).

leaf bases and the tattered remains of the leaf blades which remain after cutting and threshing, constitute straw. Its microscopic anatomy has been studied from the point of view of using it as a raw material in paper-making [61], (Chapter 13). The internode is bounded on the outer side by an epidermis, which is silicified [2,59]. Vertical bands of photosynthetic tissue, in communication with the exterior through rows of stomata, alternate with 'girders' of thickened, lignified sclerenchyma fibres (bast), which run between the epidermis and an inner ring of bast (Fig. 1.20). This arrangement gives the stem a pattern of vertical light and dark green stripes. Vascular bundles, each supported by sclerenchymal fibres, occur adjacent to the bast ring, embedded in the ground

The morphology of barley; the vegetative phase

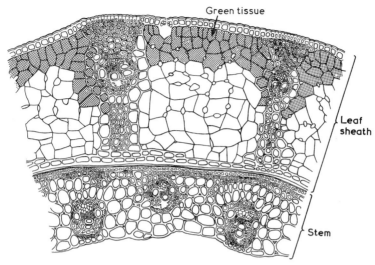

Fig. 1.21 Transverse section through the base of the internode of a barley culm, and the surrounding leaf-sheath (after Lermer and Holzner [59]). Note that in this region the stem is relatively thinner and less well sclerified than higher up (See Fig. 1.20).

parenchyma of the pith. In the inner part of the stem is a pith cavity. Near the base of the internode the sclerenchyma is markedly less well developed (Fig. 1.21). The nodal diaphragms contain a complex anastomosing plexus of vascular bundles from leaves higher up the stem [39,59,78]. In the basal region of the stem where the internodes are close together and the adventitious roots arise the situation is complex and no scheme for the vascular anatomy of this region has been presented. In some varieties the internodes grow in a series of curves instead of growing straight. Types with up to thirteen nodes are not uncommon and among the many dwarf sorts are some which have 30–50 nodes to each stem, and some which have branching stems rather than tillers arising from a crown. Of the dwarf forms some are fully fertile, others are not; some have reduced ears, and shortened awns, others do not (Table 1.1); other, *erectoides* mutations with differing types of stem structure, have value in breeding for lodging resistance (see below and Chapter 12).

The leaves of barley arise as a semicircular ridge around the apical growing point. They grow upward, rolled up within a tube of other leaf sheaths. At first the leaf grows from a basal meristematic zone, but in time this divides so that an additional meristem is carried upwards. The region below the second meristem remains rolled, and is the leaf sheath. The region above grows out as the blade. Young leaves looked at from above spiral in a clockwise direction when an edge is followed from the base up

22 Barley

Table 1.1 A comparison of some dimensions of the six-rowed, rough-awned, naked variety Himalaya and a spontaneous brachytic recessive mutant (from Swenson [88]).

	Himalaya	Brachytic
Height to tip of upper leaf at heading (cm)	65.0	42.7
Height to base of spike at heading (cm)	50.7	34.6
Distance between ligules (cm)	5.5	4.1
Length of leaves (cm)	22.4	15.0
Length of awns (cm)	10.9	5.7
Grain number/plant	85.3	92.4
Weight of grain/25 plants (g)	3.07	2.92
Weight/grain (mg)	36.8	31.4

towards the apex [26]. At the junction between the sheath and the blade two lateral projections are formed, called the auricles or claws. These lack chlorophyll or stomata and are colourless or, in some varieties, purple-red. From the upper surface of the leaf, at the junction, a little papery projection, the ligule, is formed and closely encircles the stem (Fig. 1.22). This little 'tongue' is usually colourless, being made of epidermal tissue only. It forms a close sliding seal around the stem and limits or prevents rain, agricultural sprays and insects getting between the leaf sheath and the stem and reaching the vulnerable meristematic tissue. The leaf sheath is divided to the base. It gives mechanical support to the stem. The basal meristematic zone of the leaf sheath, inserted just above the junction of the leaf base with the node, becomes the swollen, turgid, and shiny pulvinus, which lacks chlorophyll and stomata and contains little sclerenchyma [78]. The changes in epidermal cell types over the surface of the plant are characteristic [78]. The angle of the leaf blade with the stem is typical for a variety. Varieties with differing leaf lengths, widths

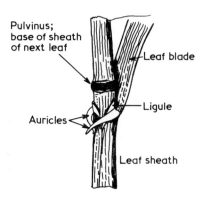

Fig. 1.22 Schematic view of the manner in whch the auricles and ligule are placed at the junction between the leaf sheath and the leaf blade.

The morphology of barley; the vegetative phase

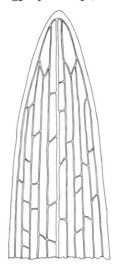

Fig. 1.23 View of the tip of a leaf-blade, showing the venation (after Lermer and Holzner [59]).

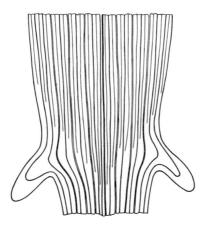

Fig. 1.24 The venation of the leaf in the region of the auricles (after Lermer and Holzner [59]).

and shapes are known. The shape of the leaves varies along the culm [3,88]. Often the flag-leaf is the smallest. Leaves arise on opposite sides of the stem at alternate nodes, giving a two-rowed, 'distichous' arrangement. The elongated leaves have mainly parallel veins cross-linked with minor veins; these fuse near the tip (Fig. 1.23). The veins are carried sideways into the auricles as these projections are formed, and must function while growing from the two meristems (Fig. 1.24). Leaves usually have a conspicuous central rib, from which the blade diverges as a shallow V. Some 16–24 lesser 'nerves' carrying other vascular bundles also run along

24 Barley

each blade [3,26]. In a survey of the leaves of 210 barley cultivars vein frequencies of 2.4–4.5 per mm were encountered, and ranges of 10.6–20 major and 34–46 minor veins per leaf were found [31]. A transverse section of a leaf sheath is shown in Fig. 1.21. Viewed from the inside the sheath looks pale, greeny-white and shiny-brilliant; this is due to the numerous air-spaces beneath the epidermis [78]. The leaf blade is smooth below, but carries grooves on the upper surface, at the base of which the epidermal cells are enlarged, and are clearly analogous to the motor cells which cause rolling in the leaves of many xerophytic grasses [3] (Fig. 1.25). The leaf is mainly composed of green photosynthetic spongy mesophyll bounded by upper and lower epithelia, which are perforated by rows of stomata leading to the intercellular spaces. Here and there vascular bundles occur, running parallel to, or joined with bands of cells with thickened walls, and sclerenchyma (Figs. 1.25 and 1.26). The vascular bundles lie within two encircling 'sheaths' of cells, an outer, thin-walled parenchyma sheath, which contains chloroplasts, and an inner, thick-walled 'mestome-sheath' which is suberized and has the typical structure of an endodermis [23].

The developmental stages of stomata, the two thick-walled guard-cells, and the thin-walled companion cells, have been intensively studied [101,102]. The process is disorganized by mercapto-ethanol, or high temperatures, e.g. 35°C (95°F) [87]. Stomatal frequency varies greatly. In 469 varieties stomatal frequencies varied between 36 and 98 per mm²

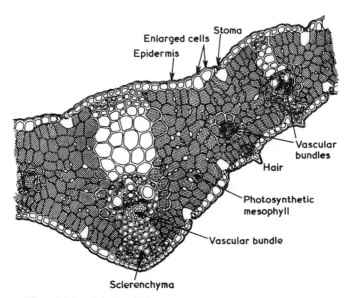

Fig. 1.25 The midrib of the leaf-blade, seen in transverse section (after Lermer and Holzner [59]).

The morphology of barley; the vegetative phase 25

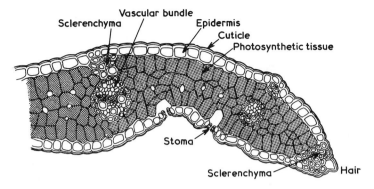

Fig. 1.26 The edge of the leaf blade, seen in transverse section (after Lermer and Holzner) [59].

of flag leaf surface, and varied between 41 and 56 μm in length [65].

In some varieties the leaf sheath is more or less hairy [3,17,79]. In many places it is silicified, as revealed by electron probe micro-analysis or the ash (spodograms) remaining after careful incineration [33,34,51]. The surface of the plant is usually covered with wax to different extents, great variations in wax plate morphology occurring between varieties, parts of one plant, and the same part of a plant under different conditions (Chapter 4). The complex forms of the wax micro-platelets are not 'moulded' by extrusion through the cuticle, but are formed spontaneously when solutions of waxes are allowed to evaporate through membranes [44].

In the phloem elements in barley leaf veins the pores in the sieve plate elements are lined with various amounts of 'callose' and plasmalemma, and are filled with endoplasmic reticulum, which is continuous from cell to cell [22]. The development of chloroplasts from proplastids, has been studied in detail in normal and greening plants, normal leaves fed δ-aminolevulinic acid or in the leaves of a range of pigment-deficient mutants [7,68,75,80,97,99].

1.5. The root system [3,9,40,53,59,93]

When the seed germinates the seminal rootlets (primary or seedling roots, to maltsters the culms, coombes, cummins, or sprouts), usually around 5–7, emerge from the coleorhiza. In the field they grow into the soil where they extend and branch freely, forming a fibrous, branched mass of roots some of which extend deeply downwards. Later the adventitious (coronal, nodal, or crown) root system arises as a series of irregular whorls from the base of the crown, derived from the closely-packed basal nodes. At first many of these roots extend horizontally in the soil. They tend to be thicker, and are less branched than the seminal roots. Some young, waxy-white adventitious roots do not branch for

26 Barley

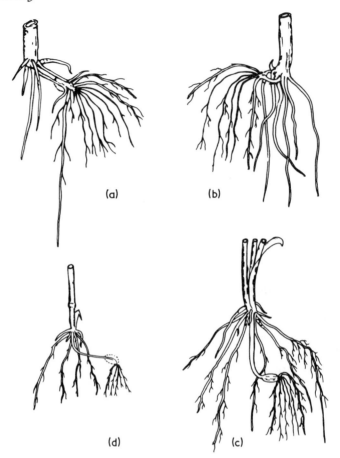

Fig. 1.27 The rooting patterns of (a,b) seedlings of Vega barley 1 month after sowing, and (c,d) of plants of Olli barley, 2 months after sowing (after Kokkonen [53]).

some inches (Fig. 1.27). These root systems receive other names based on ill-conceived suppositions about their functions. Tillers physically separated from the plant can grow supported by adventitious roots only. Sometimes, in drought or other conditions, the adventitious roots do not develop in which case plants may reach maturity growing with only their seminal roots. In other cases the seminal roots cease functioning during the life of the plant. The final extent of the root system depends on many factors (Chapter 6). In very deep soils roots may descend to 1.8–2.1 m (6–7 ft) [49,56,96]. The deepest roots are usually of seminal origin and the upper layers of the soil tend to be packed with adventitious roots (Fig. 1.28). The maximal extent of the root system occurs at about the time of anthesis and it declines thereafter. There are significant differences

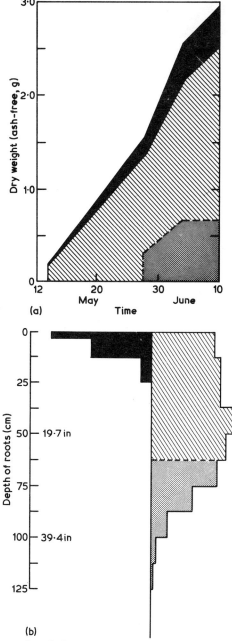

Fig. 1.28 The increase in ash-free dry weight of the roots, until anthesis (a), and the distribution of the root-mass at different depths, at anthesis (b) of 'Breust. Granat' barley (after Gliemeroth) [27]. Black-crown roots; diagonal shading, seminal roots to 62.5 cm (24.6 in.); stippled – seminal roots below 62.5 cm (24.6 in).

28 *Barley*

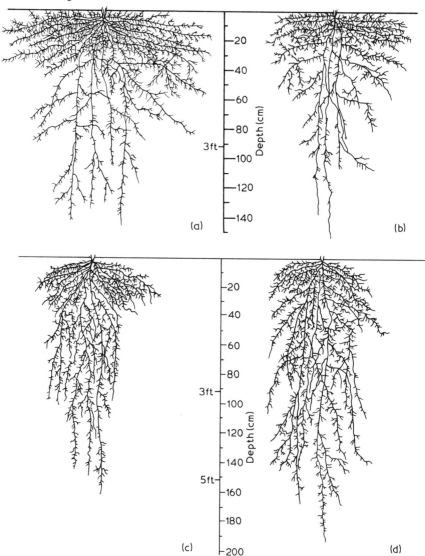

Fig. 1.29 The extents of four types of rooting systems encountered in Indian barleys (after Bose and Dixit [6]). a, Mesophytic; b, semi-mesophytic; c, semi-xerophytic; d, xerophytic.

in the rooting systems of different barleys, which have marked effects on their competitive abilities (Chapter 12). In direct comparisons between barleys from the Indian sub-continent grown at one site, at Pusa, it was shown that barleys from areas with adequate soil moisture tended to have comparatively shallow, 'mesophytic', root systems, while barleys from dry regions produced deep, 'xerophytic' rooting patterns (Fig. 1.29).

The morphology of barley; the vegetative phase

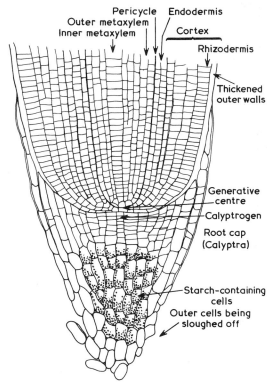

Fig. 1.30 Schematic longitudinal section through the apex of a barley root.

Roots extend by cell divisions, which take place at the apex, and the subsequent growth and differentiation of the new cells (Fig. 1.30). The apex is continually being pushed forward into the soil. At the apex two sets of meristematic cells occur. The outer set, or calyptrogen, divide to produce the root-cap (calyptra). This slimy, loosely-knit, structure protects the apex as it is forced through the soil. It also contains cells rich in starch grains which may act as statoliths. The cells at the generative centre of the root divide and subsequently extend to create the root cylinder, in which the cells are arranged in a series of concentric layers (Fig. 1.30). Damaged root-tips will regenerate; sometimes they become forked, with two root caps, during regeneration [50]. The cells behind the tip expand and vacuolate, and differentiate with increasing age (Fig. 1.31). The cell division rates have been studied in diploid and tetraploid forms, and in different layers of the roots [30,36,84]. The processes of cell differentiation differ a little between roots, and the structures produced differ slightly as seen in transverse sections (Figs. 1.31, 1.32 and 1.33). Apparent discrepancies in the literature may be due to varietal differences, problems of nomenclature, or failure to recognize

30 *Barley*

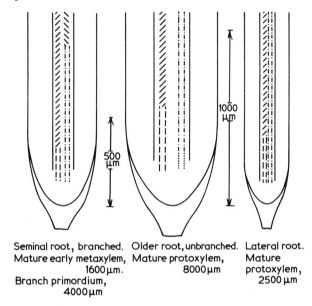

Seminal root, branched. Older root, unbranched. Lateral root.
Mature early metaxylem, Mature protoxylem, Mature
1600 μm. 8000 μm protoxylem,
Branch primordium, 2500 μm
4000 μm

Fig. 1.31 Scheme of the differentiation and maturation of the tissues in three classes of barley roots. The endodermis is represented by solid lines at the periphery of the stele. Casparian strips were detected at 750 μm in the seminal roots (after Heimsch [36]).

the origins of particular roots, or to take account of alterations in structure with advancing age. In transverse section the young, newly differentiated seminal root is bounded by a regular epidermal piliferous layer, carrying many root hairs. Within this is an annulus of thin-walled, loosely-packed parenchymatous cells – the cortex. Bounding the inner face of the cortex is a single layer of cells, the endodermis, the walls of which thicken with increasing age. The pericycle is the next clearly differentiated layer. The walls of the cells of the endodermis and pericycle thicken most rapidly opposite the groups of phloem vessels. The stelar cylinder consists of a small-celled parenchyma in which are alternating xylem and phloem elements, a 'polyarch' arrangement. In young seminal roots there is a large central vessel (Fig. 1.32). Young crown roots have an essentially similar structure except that several scattered large inner metaxylem vessels occur (Fig. 1.33). As the roots age a corky layer develops beneath the epidermis. Sclerenchyma develops in the outer cortex, and the walls of the cells of the stele become thickened. Thus the mechanical strength of the root is increased. In older roots still the cortex is said to wither. In malting barley a bacterial disease can rot the cortex, but is apparently checked by the endodermis [82]. In the soil, roots are associated with a large microflora. They may even be permeated by endotropic mycorrhiza,

The morphology of barley; the vegetative phase 31

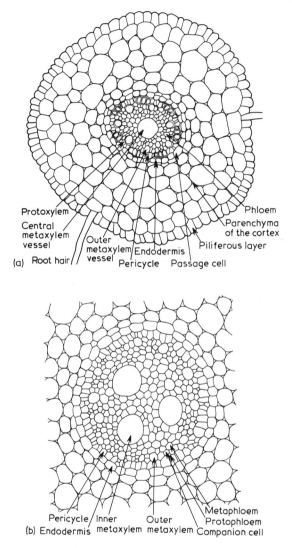

Fig. 1.32 (a) Transverse sections of a young seminal root. (b) Transverse section of the stele of a young adventitious root (after Hagemann [30]).

but whether this association is favourable is unknown. The symbiont is said to gain entry to the roots via the root hairs [93].

The detailed structure of the cells bounding the stele and adjacent to the xylem vessels are of interest in connection with salt uptake and the development of root pressure (Chapter 6). Characteristic living parenchyma cells surrounding the metaxylem vessels could be responsible for 'pumping' ions into the xylem [57]. As the root ages the radial and

32 *Barley*

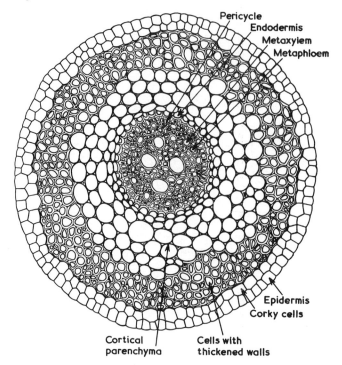

Fig. 1.33 Transverse section of a mature adventitious root (after Lermer and Holzner [59]).

inner tangential walls of the endodermis become thickened, and altered (cutinized, suberized). One study failed to find a modified 'Casparian strip' in the walls of this tissue but others report its presence. The inner walls contain many pits. In the mature roots passage cells are infrequent [18,38].

The parenchyma of the cortex usually contains small air-spaces. Under conditions of poor aeration, in water culture, the roots become shorter and more numerous, and have greater diameters. In these roots the cortex develops large air-passages separated by narrow strands of parenchyma [15]. These enable the roots to receive more oxygen from the upper parts of the plant (Chapter 6).

Patches of potentially meristematic cells occur in the cortex, adjacent to the stele. These may divide and differentiate to give rise to lateral roots, which appear to erode, rather than force, their way to the exterior, through the cortex (Fig. 1.34). The details of the branching of the root vascular systems, and the way(s) in which they merge with the grain, with the 'rhizomatous stem', and with the vascular systems of the culms do not seem to be known.

The morphology of barley; the vegetative phase 33

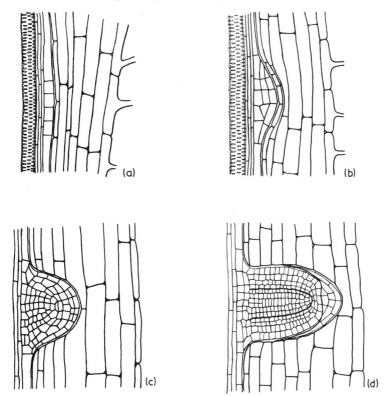

Fig. 1.34 Stages in the development of a lateral root in barley seen in longitudinal section (after Lermer and Holzner [59]). (a) The establishment of a meristematic region. (b-d) Successive stages in the formation of the rootlet, within the cortex.

1.6. Plant morphology and lodging

'Lodged' is the term used to describe the state of a plant when its culms are flattened to the ground, bent over, or laid. When a crop lodges its yield and quality are often reduced, and it becomes more difficult to harvest. Unfortunately the term covers a variety of faults from juvenile lodging – when the pseudostems are prostrated [25], through the roots being inadequate to hold the intact stem upright – to various types of stem failure and breakage involving 'kinking' and bending in the middle or at the soft base of an internode or at the node. The mutant form in which the stem snaps cleanly in two seems to be entirely novel, and not related to the lodging problem. In view of the complexity of the problem it is not surprising that no universal relationship has been found between lodging and straw strength, straw thickness, lignin or cellulose or silica

34 Barley

contents, or height [8,24,73,86]. However, in any particular case lodging is due to some failure in a component part(s) of the plant. In lodging caused by the inadequate development of coronal roots, the cause is apparent. In other cases the causes of failure are less obvious. The above ground, fully grown barley culm is an interesting physical system with a series of hollow internodes, weak in the meristematic region at the base and separated by reinforcing diaphragms at the nodes, supported by ensheathing leaf-bases, and carrying extended leaf blades and a heavy terminal ear, the whole system being anchored and tethered at the base by the roots. The taller the stem the greater the leverage on the basal region. The leverage will be enhanced by increased grain weight or water in the ear, or wind, hail or rain beating and pulling at the structure. Shorter-stemmed plants have a mechanical advantage as leverage at the culm base is smaller. The stem should be tough and elastic. Excess nitrogenous fertilizer favours lodging; it encourages lush, slow maturing, tall 'soft' growth. The strength of the stem is dependent on its outer diameter, wall thickness, and elasticity (Young's modulus) – higher Young's moduli indicate stiffer stems and a greater tendency to break, rather than flex. The physical attributes of the stem change along its length, along the lengths of each internode, and with the stage of maturity. The mechanical principles of stem strength are increasingly being investigated [32,67,71,83].

In 15 varieties of barley the dimensions noted (at the fourth internode) ranged from culm diameter, 3.5–4.8 mm; thickness of the culm wall, 0.38–0.49 mm; number of vascular bundles per stem, 30.1–43.7; total area of bundles, 0.31–0.58 mm^2; total area of sclerenchyma, 0.60–1.03

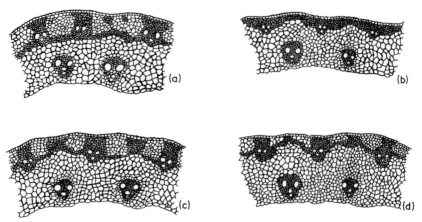

Fig. 1.35 Sections of the internodes of four Indian barleys, showing the disposition of the "mechanical tissues", the sclerenchyma ring and vascular bundles (after Bose *et al.*) [5]. (a) a lodging resistant type; (b) a readily lodged type. (c, d) intermediate types.

Table 1.2 Some details of the stem morphology of the two-rowed, husked Scandinavian barley Bonus, and its stiff-strawed mutant *Erectoides*-23 (*ert*-23) (from von Wettstein [98]).

	Internode number				
	1	2	3	4	5
Internode length (cm)					
Bonus	22.1	17.9	12.5	12.5	9.0
Ert-23	24.2	18.5	11.8	10.5	5.3
Total width (mm, F)					
Bonus	3.62	8.23	8.17	8.45	6.23
Ert-23	4.12	9.10	8.85	8.42	5.58
Width of pith cavity, (mm, f)					
Bonus	2.25	5.64	5.50	5.28	3.05
Ert-23	2.61	6.23	5.87	4.75	2.03
Wall thickness (mm, F-f)					
Bonus	1.37	2.60	2.68	3.18	3.18
Ert-23	1.51	2.87	2.98	3.67	3.55
Sclerenchyma ring thickness (μm)					
Bonus	55.8	34.7	33.0	—	69.8
Ert-23	68.1	54.3	46.9	—	75.1
Tangential width of vascular bundles (μm)					
Bonus	96.2	71.4	97.1	—	93.2
Ert-23	115.1	98.7	111.6	—	101.3

mm^2, yet none of these dimensions clearly correlated with a tendency to resist lodging [24]. However in a comparison of the stems of a limited range of various lodging and non-lodging Indian barleys certain distributions of mechanical tissue were clearly better than others (Fig. 1.35) [5]. Various short-stemmed, stiff-strawed *erectoides* mutant forms are known. Some of these have found use either directly in agriculture or as parents of lodging-resistant varieties (Table 1.2) [98]. As well as a preferable disposition of mechanical tissue in the upper internodes of the mutant it is noticeable that the top internode is longer but the basal internodes are shorter than those of the parental form.

References

1 Abdul-Baki, A.A. and Baker, J.E. (1973). *Seed Sci. Technol.*, **1**, 89–125.
2 Arber, A. (1934). *The Gramineae; a study of cereal, bamboo and grass.* Cambridge: University Press.
3 Bergal, P. and Clemencet, P. (1962). In *Barley and Malt*, (ed. Cook, A.H.) pp. 1–23. London, Academic Press.
4 Bergal, P. and Friedberg, L. (1940). *Ann. Epiphyt.*, **6**, 157–306.
5 Bose, R.D., Aziz, M. and Bhatnagar, M.P. (1937). *Indian J. agric. Sci.*, **7**, 48–88.

6 Bose, R.D. and Dixit, P.D. (1931). *Indian J. agric. Sci.*, **1**, 90–108.
7 Bourdu, R., Mathieu, Y., Miginiac-Maslow, M., Remy, R. and Moyse, A. (1968). *Planta*, **80**, 191–210.
8 Brady, J. (1934). *J. agric. Sci., Camb.* **24**, 209–232.
9 Brenchley, W.E. and Jackson, V.G. (1921). *Ann. Bot.*, **35**, 533–556.
10 Briggs, D.E. (1972). *Planta*, **108**, 351–358.
11 Briggs, D.E. (1973). In *Biosynthesis and its Control in Plants*, (ed. Milborrow, B.V.) pp. 219–277. London: Academic Press.
12 Brown, H.T. (1903). *Trans. Guinness Res. Lab.*, **1**, 96–141.
13 Brown, W.V. (1965). *Phytomorphology*, **15**, 274–284.
14 Brown, H.T. and Morris, G.H. (1890). *J. chem. Soc.*, **57**, 458–528.
15 Bryant, A.E. (1934). *Pl. Physiol., Lancaster*, **9**, 389–391.
16 Buttrose, M.S. (1971). *Planta*, **96**, 13–26.
17 Carson, G.P. and Horne, F.R. (1962). In *Barley and Malt*, (ed. Cook, A.H.) pp. 101–159. London: Academic Press.
18 Clarkson, D.T., Robards, A.W. and Sanderson, J. (1971). *Planta*, **96**, 292–305.
19 Collins, E.J. (1918). *Ann. Bot.*, **32**, 381–414.
20 van der Eb, A.A. and Nieuwdorp, P.J. (1967). *Acta bot. neerl.*, **15**, 690–699.
21 Esau, K. (1965). *Plant Anatomy*, (2nd edition). New York: John Wiley.
22 Evert, R.F., Eschrich, W. and Eichhorn, S.E. (1971). *Planta*, **100**, 262–267.
23 van Fleet, D.S. (1950). *Bull. Torrey bot. Club.*, **77**, 340–353.
24 Garber, R.J. and Olson, P.J. (1919). *J. Am. Soc. Agron.*, **11**, 173–186.
25 Gardener, C.J. and Rathjen, A.J. (1975). *Aust. J. agric. Res.* **26**, 231–242.
26 Gill, N.T. and Vear, K.C. (1969). *Agricultural Botany*. London: Duckworth.
27 Gliemeroth, G. (1957). *Z. Acker-u. PflBau*, **103**, 1–21.
28 Godineau, M.J.-C. (1962). *Revue gén. Bot.*, **69**, 577–622.
29 Hänsel, H. (1952). *Z. PflZücht.* **31**, 359–380.
30 Hagemann, R. (1957). *Kulturpfl.*, **5**, 75–107.
31 Hanson, J.C. and Rasmusson, D.C. (1975). *Crop Sci.*, **15**, 248–251.
32 Hashimoto, T. (1963). *Bull. Hiroshima agric. Coll.*, **2**, 146–191.
33 Hayward, D.M. and Parry, D.W. (1973). *Ann. Bot.*, **37**, 579–591.
34 Hayward, D.M. and Parry, D.W. (1975). *Ann. Bot.*, **39**, 1003–1009.
35 Hector, J.M. (1936). *Introduction to the Botany of Field Crops*, **1**, pp. 234–271. Johannesburg: *Cereals*. Central News Agency.
36 Heimsch, C. (1951). *Am. J. Bot.*, **38**, 523–537.
37 Heinisch, O. (1938). *Z. PflZücht.*, (Reihe A), **22**, 209–232.
38 Helder, R.J. and Boerma, J. (1969). *Acta bot. neerl.*, **18**, 99–107.
39 Hitch, P.A. and Sharman, B.C. (1971). *Bot. Gaz.*, **132**, 38–56.
40 Jackson, V.G. (1922). *Ann. Bot.*, **36**, 21–39.
41 Jacobsen, P. (1966). *Radiat. Bot.*, **6**, 313–328.
42 Jacobsen, J.V., Knox, R.B. and Pyliotis, N.A. (1971). *Planta*, **101**, 189–209.
43 Jakovlev, M.S. (1937). *C. r. (Doklady) Acad. Sci. USSR*, **17**, 69–72.
44 Jeffree, C.E., Baker, E.A. and Holloway, P.J. (1975). *New Phytol.*, **75**, 539–549.
45 Jones, R.L. (1969). *Planta*, **85**, 359–375.
46 Jones, R.L. (1969). *Planta*, **87**, 119–133.
47 Jones, R.L. (1969). *Planta*, **88**, 73–86.
48 Jones, R.L. (1972). *Planta*, **103**, 95–109.
49 Jonker, J.J. (1959). *Jaarb. NaCoBrouw*, **23**, 33–38.
50 Kadej, F. (1956). *Acta Soc. Bot. Pol.*, **25**, 681–712.
51 Kato, H. (1932). *Bull. Miyazaki Coll. Agric. Forestry*, **4**, 87–108. (Engl. Summ.)

52 Konig, F. (1928). *Angew. Bot.*, **10**, 483–576.
53 Kokkonen, P. (1931). *Acta Forestalia Fennica*, **37**, 1–144 (+ Figs.).
54 Krauss, L. (1933). *Jb. wiss. Bot.*, **77**, 733–808.
55 Kudelka, F. (1875). *Landw. Jbr.*, **4**, 461–478. (+ Figs.).
56 Kutschera, L. (1960). *Wurzelatlas-mitteleuropäischer Acker und Kraüter und Kulturpflanzen*. Frankfurt-am-main: DLG-Verlag.
57 Läuchli, A., Kramer, D., Pitman, M.G. and Lüttge, U. (1974). *Planta*, **119**, 85–99.
58 Lehmann, E. and Aichele, F. (1931). *Keimungsphysiologie der Gräser (Gramineen)*. Stuttgart: Ferdinand Enke.
59 Lermer, Dr. and Holzner, G. (1888). *Beiträge zur Kenntnis der Gerste*. München: R. Oldenbourg.
60 Liu, D.J. and Pomeranz, Y. (1975). *Cereal Chem.*, **52**, 620–629.
61 Lloyd, F.E. (1921). *Pulp Paper Mag.*, **19**, 953–954; 1071–1075.
62 MacLeod, A.M. and Palmer, G.H. (1966). *J. Inst. Brew.*, **72**, 580–589.
63 Mann, A. and Harlan, H.V. (1915). *U.S. Dept. Agric. Bull.*, **183**, (via (1916) *J. Inst. Brew.*, **22**, 73–108).
64 Mead, H.W. (1942). *Can. J. Res.*, **20 C**, 501–523.
65 Miskin, K.E. and Rasmusson, D.C. (1970). *Crop Sci.*, **10**, 575–578.
66 Mullenax, R.H. and Osborne, T.S. (1967). *Radiat. Bot.*, **7**, 273–282.
67 Neenan, M. and Spencer-Smith, J.L. (1975). *J. agric. Sci., Camb.*, **85**, 495–507.
68 Nielsen, O.F. (1974). *Hereditas*, **76**, 269–304.
69 Nieuwdorp, P.J. and Buys, M.C. (1964). *Acta bot. neerl*, **13**, 559–565.
70 Noda, A. and Hayashi, J. (1958–9). *Proc. Crop Sci. Soc. Japan*, **27**, 229.
71 Oda, K., Suzuki, M. and Udagawa, T. (1966). *Bull. natn. Inst. agric. Sci. Ser D, Tokyo*, **15**, 55–91.
72 Olsen, O.A. (1974). *Hereditas*, **77**, 287–302.
73 Patterson, F.L., Schafer, J.F. Caldwell, R.M. and Compton, L.E. (1957). *Agron. J.*, **49**, 518–519.
74 Percival, J. (1902). *Agricultural Botany*, (2nd edition). London: Duckworth.
75 Phung nhu Hung, S., Lacourly, A. and Sarda, C. (1970). *Z. Pflphysiol.*, **62**, 1–16.
76 Pomeranz, Y. (1974). *Cereal Chem.*, **51**, 545–552.
77 Pope, M.N. (1945). *J. Am. Soc. Agron.*, **37**, 771–778.
78 Prat, H. (1932). *Ann. Sci. nat. (Bot.)* (10^e Ser.), **14**, 118–324.
79 Reid, D.A. and Wiebe, G.A. (1968). *U.S. Dept. Agric. Tech. Handbook No. 338*; Barley. 61–84.
80 Robertson, D. and Laetsch, W.M. (1974). *Pl. Physiol., Lancaster*, **54**, 148–159.
81 Sawicki, J. (1955). *Acta Agrobotan*, **3**, 129–166.
82 Schnegg, H. (1907). *Z. ges. Brauw.*, **30**, 576–579; 588–591; 600–602; 608–612; 623–625; 630–634.
83 Schwendener, S. (1874). *Das mechanische Princip in anatomischen Bau der Monokotylen mit vergleichenden Ausblicken auf die übringen Pflanzenklassen*. Leipzig: Wilhelm Engelmann.
84 Skult, H. (1969). *Acta Acad. Aboen.* (Ser. B), **29**, 1–15.
85 Smith, N.J.G. (1929). *Ann. appl. Biol.*, **16**, 236–260.
86 Spahr, K. (1960). *Z. Acker-u PflBau.*, **110**, 299–331.
87 Stebbins, G.L., Shah, S.S., Jamin, D. and Jura, P. (1967). *Am. J. Bot.* **54**, 71–80.
88 Swenson, S.P. (1940). *J. agric. Res.*, **60**, 687–713.
89 Symko, S. (1966). *Can. J. Pl. Sci.*, **46**, 206.

38 Barley

90 Taiz, L. and Jones, R.L. (1973). *Am. J. Bot.*, **60**, 67–75.
91 Tallberg, A. (1973). *Hereditas*, **75**, 195–200.
92 Tharp, W.H. (1935). *Bot. Gaz.*, **97**, 240–271.
93 Troughton, A. (1962). *The Roots of Temperate Cereals (Wheat, Barley, Oats and Rye)*. Hurley, Berkshire: Commonwealth Bureau of Pastures and Field Crops.
94 Tschirch, A. and Oesterle, O. (1900). *Anatomischer Atlas der Pharmakognosie und Nahrungsmittelkunde*. pp. 175–180. Leipzig: C.H. Tauchnitz.
95 Vigil, E.L. and Ruddat, M. (1973). *Pl. Physiol., Lancaster*, **51**, 549–558.
96 Weaver, J.E., Jean, F.C. and Crist, J.W. (1922). *Carnegie Inst. Publ. No. 316 The development and activities of the roots of crop plants. A study in crop ecology.*
97 Weier, T.E., Stocking, C.R., Bracker, C.E. and Risley, E.B. (1965). *Am. J. Bot.*, **52**, 339–352.
98 von Wettstein, D. (1952). *Hereditas*, **38**, 345–366.
99 von Wettstein, D., Henningsen, K.W., Boynton, J.E., Kannangara, G.C. and Nielsen, O.F. (1971). In *Anatomy and Biogenesis of Mitochondria and Chloroplasts*. pp. 205–223. Amsterdam: North-Holland.
100 Winton, A.L. and Winton, K.B. (1932). *The Structure and Composition of Foods I. Cereals, Starch, Oil-seeds, Nuts, Oils, Forage Plants*. pp. 50–62; 269–293. London: Chapman and Hall.
101 Zeiger, E. (1972). *Planta*, **108**, 359–362.
102 Zeiger, E. and Stebbins, G.L. (1972). *Am. J. Bot.*, **59**, 143–148.

Chapter 2

The morphology of the reproductive parts in barley

2.1. Introduction

After a number of leaves have been initiated the stem apex gives rise to spikelet initials which, taken together, form the inflorescence of the plant (spike, head, ear). A consideration of various unusual barley types, such as 'many noded dwarf' [38] in which a gradation of forms between leaves and spikelets sometimes occurs, and various other morphological variants emphasizes that a distinction between 'vegetative' and 'reproductive' structures is more or less artificial.

Particular interest is attached to the formation of grains, and the structure of the ear and its parts by those attempting to increase grain yields (Chapters 6 and 12), to classify barley (Chapter 3), and those investigating the morphogenesis of plant parts. Stages of inflorescence development are also used to decide the times of application of agricultural sprays (Chapter 9) [5,12–14,50,54,81].

2.2. The development of the ear

Initially the apex generates single ridges which grow up into leaves (Chapter 1). Normally, by the time two or three leaves are visible all the leaf primordia are formed, and the apex begins to generate the spike (Fig. 2.1). At first the apex lengthens, then double ridges develop. In each case the upper ridge grows fast and gives rise to the floral primordium. The lower ridge appears to give rise to the rachis internode. The rachis is the extension of the stem that is the supporting axis of the ear. At this stage ears of two-rowed and six-rowed cultivars appear similar, but subsequently the development of the lateral florets slows down in two-rowed varieties [14]. By processes of differential growth and folding, series of ridges and papillae form which differentiate into glume initials,

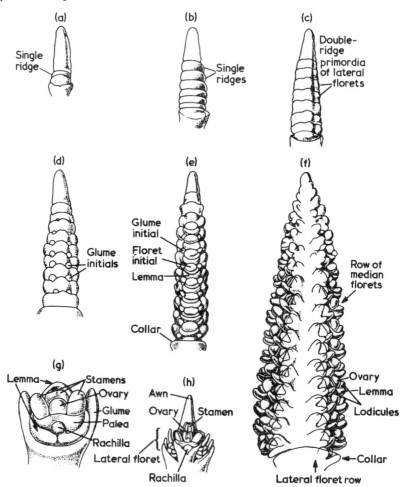

Fig. 2.1 Stages in the differentiation of the ear of a two-rowed barley (after Schuster [81]). (a) Single ridges present; these will form leaf primordia. However the apex is lengthening. (b) The apex elongated further, but still with single ridges, (c) Double ridges forming; initials of the lateral florets are detectable. (d) The initials of the sterile glumes associated with the median floret are visible. (e) A further stage; the collar is visible. (f) A more advanced ear, viewed from the side. (g) A developing median floret, viewed from the rachis side. (h) A more advanced triad of spikelets from a two-rowed barley.

followed by the lemma. The palea develops later, behind the other primordia next to the axis, which is destined to become the rachis. Then three papillae appear and grow into stamens. The pistil develops as a dome between the stamen initials; the styles with their stigmatic hairs form later. In time the rachilla appears on the axis side of the palea, and lodicules form at the floret base. The palea and lemma grow up and eventually enclose the floral parts, although for some time the stamens protrude. The awns begin to grow rapidly after the formation of the anthers, but before the pistil. The oldest, first-differentiated spikelets are at the base of the ear. Eventually the ear terminates with the formation of one or more sterile florets. At the base a ridge, the young collar, encircles the developing ear and represents the first node.

The development of an alternative 'lemma appendage' that sometimes replaces the awn, and is known as a hood is of interest [13,14,35,40,83,96]. Initially lemmae destined to bear hoods develop like the awned forms, but the apex begins to grow more slowly, lateral 'wings' grow out, and the apex is formed into a hollow structure, the hood, which may or may not carry a small awn. At the juncture of the lemma with these extensions a rudimentary 'supernumerary' floret develops, and sometimes a second forms also. These occasionally develop functional anthers and ovaries. Surprisingly the 'teeth' on the epidermis, other structures on the hood, and the course of its development indicate that its structure is 'inverted' relative to that of an awn. Sometimes small grains form in the supernumerary florets. The morphology of hoods is very variable (see below). The expression of the gene for 'hooded' varies not only with the 'genetic background' of the plant, but also with the growth conditions. Thus, subjecting 'hooded' plants to cold shocks, short days, some nutrient stresses, or applications of certain chemicals can cause the partial or complete development of awns in place of hoods.

The stamens differentiate into a four-lobed anther (head) (Fig. 2.2.) and a more delicate filament (stalk). The chambers of the anther lobes are separate at first. Within each lobe a multicellular archesporium develops. These cells divide into an outer layer of parietal cells and an inner region of sporogenous cells. The outer part of the anther consists of an epidermis, the endothecium, a middle layer, and the inner cells of the tapetum. The sporogenous cells divide, forming the pollen (microspore) mother cells. These divide further, by meiosis, giving rise to tetrads of

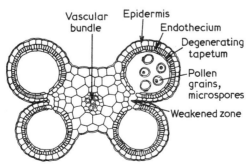

Fig. 2.2 The anther of a barley stamen, near maturity, in transverse section (after Lermer and Holzner [54]).

42 Barley

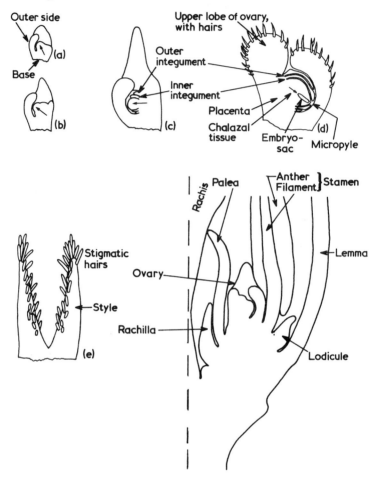

Fig. 2.3 Successive stages in the development of the barley ovary (after Lermer and Holzner [54]). (a-d) Successive stages in median, vertical section. The arrow indicates the approximate position of the embryo-sac mother-cell as the tissue rotates. (e) A view of the ovary apex, (at 90° to sections a-d), showing the developing stigmatic hairs on the under-developed styles. (f) A longitudinal section of half of the ear, indicating the relative positions of the various parts of the spikelet.

microspores. In time these are released, and mature to form pollen. The microspore nucleus divides to form a vegetative nucleus and a generative nucleus. The generative nucleus divides again to produce the two male gametes, so at maturity each pollen-grain contains three nuclei [21]. As the pollen forms so changes occur in the anther walls [74]. The epidermal and endothecial cells elongate, and the latter develop fibrous thickenings. The middle cell layer is crushed. The tapetal cells

The morphology of the reproductive parts in barley

grow, become binucleate and degenerate, only traces remaining when the microspores are mature. Before dehiscence the walls between the spaces in adjacent lobes, the 'thecae', give way. Anthesis is through longitudinal slits that appear in the lateral walls. As maturity approaches the sucrose content of the anther increased markedly, and of the 17 amino acids detected proline, glutamate, aspartate and glutamine increase relative to the others [58].

The development of the ovary is complex (Figs. 2.3 and 2.4). Two lateral extensions grow up and eventually extend to form the styles. At the same time the ovule forms as a bulge of tissue and rotates downwards so that eventually it points nearly towards the base of the ovary with the chalaza on one side next to the funiculus (funicle or stalk), i.e. it is anatropous. The rotation carries the embryo sac mother cell downwards,

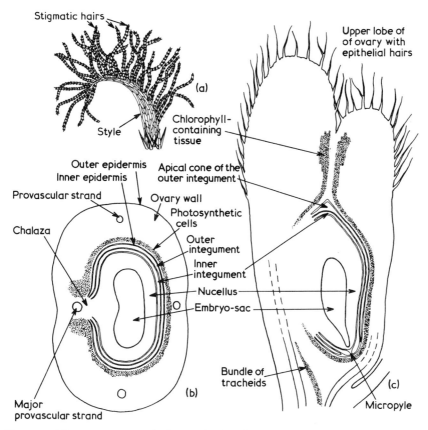

Fig. 2.4 The mature ovary, before anthesis (after Lermer and Holzner [54]). (a) A style with the apical stigmatic hairs, each composed of four columns of cells around a central lumen, and the simple basal hairs. (b) The ovary in transverse section. (c) The ovary in longitudinal section.

44 Barley

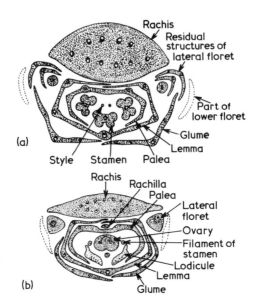

Fig. 2.5 Semi-schematic transverse sections through the young ear of a two-rowed barley (after Schuster [81]). (a) Section at a level near to the base of the anthers. (b) Section lower down, cutting through the ovary and lodicules.

in the nucellar tissue. At the same time the inner and outer integuments grow over the nucellus. The inner nearly closes, leaving only a small pore; the micropyle. The outer integument does not close (Figs. 2.3 and 2.4) [53,54,87]. Initially the integuments are two cells thick. The inner integument, destined to become the testa, is cuticularized on both sides. The epidermis of the nucellus apparently is not. The top of the ovary becomes covered with short hairs. Two styles extend upwards, and are covered with delicate stigmatic hairs, each consisting of four columns of cells, of which the apices turn outwards, arranged around a central lumen (Fig. 2.4). Before anthesis the styles are packed together, with the anthers of the stamens fitting around them. Each ovary has four provascular bundles, the largest of which is situated opposite the chalaza. The lateral bundles extend into the styles. Just within the inner epidermis of the ovary wall is a layer of green cells, containing chlorophyll, which terminates in two thickened bands each side of the chalaza (Fig. 2.4). Immediately before anthesis the embryo-sac contains an egg cell, two synergid cells, two polar nuclei and a variable number of antipodal cells clustered to one side adjacent to the chalaza. These have arisen from the three antipodal cells originally formed in the development of the embryo sac (Chapter 11). The egg cell and synergids are near to the micropyle. Thus each spikelet consists of one ovary and its stigmas, three stamens and two lodicules packed between the palea (palea superior, ventral

The morphology of the reproductive parts in barley 45

palea, pale) and the overlapping lemma (palea inferior, dorsal palea). The rachilla is between the rachis and the palea. Two sterile glumes are situated next to the lemma (Fig. 2.5) [10]. Each lemma is extended as an awn, or more rarely a hood, or is terminated in some other way. Three spikelets occur at each node, which alternate on each side of the rachis. In 'six-rowed' barleys all the spikelets are fertile. In 'two-rowed' barleys only the median spikelet of each triad is fertile. Some of the numerous variations of this arrangement are considered later. The vascular structures of the ear and spikelets have been investigated [2,48,49]. They facilitate the passage of photosynthetic assimilate from the awn and lemma to the subtended grain (Chapter 6).

The pollen and ovules in each floret mature together. The course of flowering differs slightly between open and closed flowering varieties (Fig. 2.6). It usually begins in the florets around the middle of the ear,

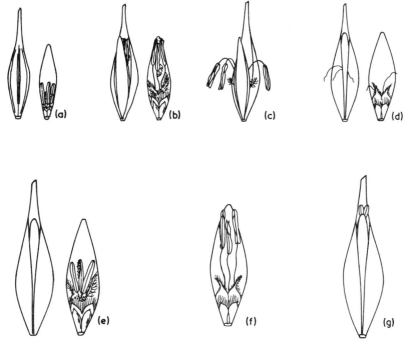

Fig. 2.6 Stages in the flowering of the open and closed-flowering types of barley florets (after Pedersen [63]). (a-d) Open-flowering type. (a) Before anthesis. (b) Separation of the stigmata; the stamen filaments are extending and pollen-release is occurring. The swelling lodicules are separating the palea and lemma. (c) Anthers and stigmata are excerted. (d) Anthers have broken away and the stigmata have withered. The lodicules have collapsed, and the ovary has started to grow. (e,f) Closed-flowering type. Pollen is released and subsequently the filaments carry the anthers upwards, and they protrude between the tip of the palea and the lemma (g).

and spreads upwards and downwards, possibly taking 1, 2 or even 4 days to complete. The median row of florets matures first. However, in some *deficiens* forms blooming begins near the base of the spike. Ears on different tillers may mature at widely differing times. Generally, depending on the weather, blooming is most active from about 5.30 to 8 a.m. and again, to a lesser extent, at 3–5 p.m. [1,29,64].

At anthesis the lodicules become more or less turgid and, in open-flowering types, they push the palea and lemma apart. They deflate after 20–30 min. and afterwards wither. In emasculated plants gaping extends for a longer period [69]. At the same time the styles diverge, and the stamen filaments elongate rapidly, (up to 15.2 cm (6 in) per h) to a length of up to 1 cm (0.39 in) [2,10,25,93]. Most pollen is shed within the spikelet and, as it is shed, the anthers lose their bright yellow colour, and afterwards collapse. Thus self-fertilization is usual in barley, but not inevitable. In closed flowering types the anther tips may be pushed to the level of the palea tip, or later they may be pushed out by the developing grain. In open flowering types the anthers may be totally excerted (Fig. 2.6). Sometimes pollination occurs while the head is in the boot but usually it occurs 3–4 days after it has emerged.

Pollen grains are variously described as 35–40 μm or 40 × 45 μm in diameter, of spheroidal-ovoid shape, with a single circular pore about 4 μm in diameter which has an elevated margin about 3 μm wide, and an exine 1–5 μm thick [28,69]. No feasible way of keeping pollen viable *in vitro* has been found (Chapter 12). At maturity a pollen-grain contains a vegetative nucleus embedded in starch-rich cytoplasm and two male gametes, small cells with conspicuous nuclei, compact cytoplasm containing a variety of subcellular organelles, and some carbohydrate (PAS-positive) component [19,20].

Within about five minutes of adhering to the stigma pollen grains take up moisture and germinate. At the same time the adjacent stigma cells change their staining characteristics and subsequently die and disintegrate [45]. The growth of the pollen tube, and subsequent fertilization, has been described [20,55,65,69,88]. The pollen tube, (5–6 μm in diameter) grows from the pore and meanders over the surface of the stigmatic hair. Normally, guided by the protruding tips of the hair-cells, it penetrates into the central lumen and grows downwards. Rarely it grows upwards away from the style, and then aborts. Although several pollen grains may germinate only one reaches the embryo-sac. The pollen tube grows down the lumen of the stigmatic hair, and through the specialized 'conducting tissue' of the style to the cone-like tip of the outer integument. It then grows between the outer integument and the inner epidermis of the ovary wall, until it reaches the micropyle (Fig. 2.7). All this takes place while the nuclei and starch-filled cytoplasm move along within the growing tube. The tube enters the micropyle, passes

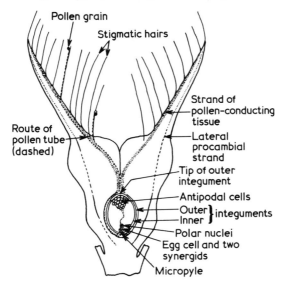

Fig. 2.7 The route of the pollen tube from the stigmatic hair to the egg cell (after Pope [69]: by permission of the American Society of Agronomy).

through the nucellar tissue, and penetrates the embryo-sac where it deposits the vegetative nucleus, the two sperm nuclei, and some starch and cytoplasm into a degenerating synergid cell.

The mature embryo-sac contains one egg cell and two synergids at the micropylar end. Two polar nuclei occur near to the synergids, and a mass of antipodal cells are situated adjacent to the chalaza. The number of antipodal cells may be variable; values recorded are: more than 3; about 15; about 35; 20 to 84; generally 30–60; and up to 100 [20,43,55,65,68,88]. At fertilization one of the synergids begins to degenerate, and receives the contents of the pollen tube. Later the second degenerates. Just before fertilization the polar nuclei begin to migrate away from the egg, towards the antipodal cells [27]. Usually the pollen tube takes about 40–45 min to grow to and enter the micropyle. Despite their migration away from the micropyle the two polar nuclei fuse with one of the male gametes in about 6 h. The result of this triple fusion is the cell which gives rise to the triploid endosperm tissues. The egg cell is fertilized later, and the diploid zygote first divides after about 15–28 h, when the endosperm cells have already divided 3–6 times.

The DNA of the haploid pollen nuclei has doubled, (i.e. is in the 2C state) at fertilization [16,23]. The rates of pollen tube growth, cell division, and other aspects of grain development are strongly temperature dependent. Thus the pollen tube takes 140 min to reach the embryo-sac at $5°$ C ($41°$ F), 30 min at $25°$ C ($77°$ F), and 20 min at $30-35°$ C ($86-95°$ F).

48 Barley

At 40–45°C (104–113°F) development is abnormal. Similarly endosperm cells divide 2–3 times in 6 days at 5°C (41°F), about 8 times in 1 day at 30–35°C (86–95°F), and not at all at 40–45°C (104°F) [68]. At first the antipodal cells divide rapidly, and seem to cause the disorganization and resorption of the adjacent parenchymatous nucellar cells, while the embryo-sac rapidly elongates [33,43,61]. The nucellar epidermis is only resorbed later in grain development. The endosperm cells divide rapidly and at first lie freely in the cytoplasm of the embryo sac. Some come to line the wall, and begin to divide. The future aleurone layer is cut off to the exterior, while the cells of the future starchy endosperm are proliferated inwards. The endosperm cells develop most thickly over the ventral furrow, in contact with the antipodal cells which are resorbed. By the time the embryo-sac has finished growing in length, displacing the nucellus in the process, it is nearly full of endosperm cells (Fig. 2.8). At the same

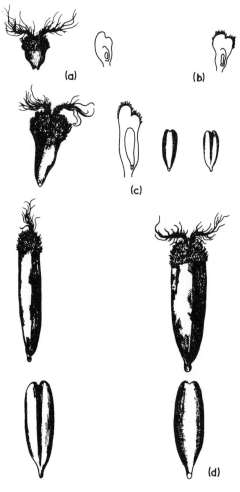

The morphology of the reproductive parts in barley

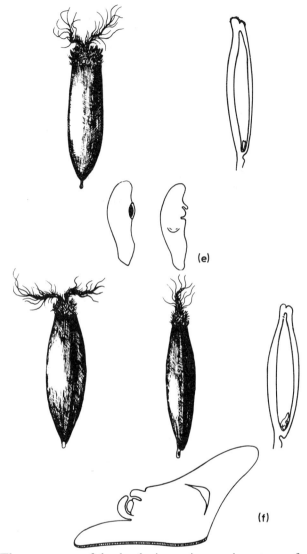

Fig. 2.8 The appearance of the developing grain at various stages after fertilization (after Lermer and Holzner [54]). The diagrams span the first ten days development. Beyond this time growth in width and depth continues, but first the palea then the lemma adhere to the pericarp (ovary wall). (a) The ovary in surface view and median logitudinal section (L.S.) one day after fertilization. (b) Median L.S. of ovary two days after fertilization; the embryo-sac is lengthening. (c) Four days after fertilization; surface view, median L.S., and dorsal and ventral views of the separated embryo-sac. Note the tiny size of the embryo. (d) Six days after fertilization – four views. (e) The 8 day old grain in surface view and median L.S. and a surface view and section of the embryo. (f) The 10 day old grain in plan and side view, and in median L.S. and the embryo (L.S., greatly enlarged). Note the decreasing size of the ovary tip.

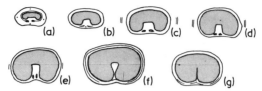

Fig. 2.9 The central region, in transverse section, of the developing barley grain (after Johannsen [43]). The stippled region is starchy endosperm tissue. (b) Grain about 8.5 mm long. (e) Grain about 9 mm long, 52% water; only a trace of green remains each side of the ventral furrow. (f) Grain with about 40% water. The flanks of the starchy endosperm have grown round and are coming together. The gap between the flanks is filled with nucellar cells at the bottom; apparently it is filled with fluid above. (g) The mature grain; air-dry, about 9.5 mm long.

time the outer integument is being resorbed, and the nucellar epidermis is being crushed and bonded to the stretched inner integument, which is becoming the testa. The ovary wall is crushed and stretched – the residues of the inner epidermis become the tube cells, the chlorophyll-containing cells become the cross cells and the outer layers become the outer parts of the pericarp (Chapter 1). As the endosperm cells multiply they meet at a line between the flattened dorsal and ventral sides. Afterwards this join disappears. Continued growth causes the flanks of the starchy endosperm to bulge round until they become pressed together in the ventral mid-line. At one stage liquid, possibly nucellar lysate, occurs at the top of the crease so formed while the nucellar residues form a line of sheaf-cells at the base, in the chalazal region (Fig. 2.9) [43]. At about this stage the cells of the chalazal region, situated between the edges of the developing testa, begin to change into the pigment strand. Brown substances are deposited in the cell lumena and in the cell walls, and the tissue becomes increasingly resistant to dissolution [22]. Starch deposition occurs at different times in the endosperm, in cells which have ceased dividing. It begins at the apex of the grain and around the suture across the central region. Deposition occurs last in the youngest cells immediately below the aleurone layer. The aleurone cells themselves never contain starch, nor do the crushed and depleted endosperm cells which abut the scutellum. Initially the ovary walls contain starch, which first appears near the main vascular bundle, but this disappears during grain development. As the aleurone layer develops the nucellar epidermis is crushed until in the mature grain it appears as part of the outer aleurone cell wall, adjacent to the testa [22,43,53,70,87]. The grain rapidly reaches its maximal length and then most subsequent growth is in girth. As this occurs the starch grains crush and disorganize the nuclei in the cells of the starchy endosperm. In husked varieties first the palea then later the lemma adhere to the pericarp (Fig. 2.8, Chapter 6).

Cell division in the zygote occurs more slowly then in the starchy

The morphology of the reproductive parts in barley 51

endosperm, so initially the embryo is small, compared to its size in the mature grain (Fig. 2.8). At first it grows in a 'lake' of cytoplasm, but when it encounters the endosperm cells its growth slows. The first few planes of cell division are apparently fixed but later divisions seem to occur at random [56,60,86]. At first the embryo is anchored by a small basal protuberance, the suspensor, but after several days this appears to lose its hold, and the embryo is free. The suspensor may give rise to the embryonic appendage of the mature grain [22]. As growth continues so the embryonic organs are differentiated (Fig. 2.8). Plasmodesmata occur elsewhere, but not in the outer cell walls of embryos [60]. Eventually drying, and probably the blocking of the passage of nutrients through the chalaza by the continued deposition of the brown materials that make the 'pigment strand' impermeable in the mature grain, slow and stop grain development.

Grain formation may be abnormal in several ways. Faults in the pericarp and testa may allow the embryo to start growing before the grain is ripe; so-called pregermination (Chapter 5). By keeping the surface layers over the embryo wet viviparous growth is induced (Chapter 12) [66,70]. Sometimes two embryos ('twins') are formed, pressed together in the space normally occupied by one [67]. When one or other of the fertilizations fail, then abnormal grains are formed. If fertilization of the egg cell fails, but that of the polar nuclei is successful, grains are formed that lack embryos. Harlan and Pope [39] found five such grains. Sometimes embryo-less grains are seen in decorticated samples of British barleys. Such grains have a depression where the embryo should be, encircled by a single layer of aleurone cells. When the zygote is successfully formed, but the triple fusion fails, grains contain embryos, but there are no starchy endosperm or aleurone cells, the space being filled with a sweet and watery liquid. Other abnormalities are encountered. Some varieties are, to a variable extent, female-sterile. Sometimes the ovaries are reduced to mere rudiments. Abnormal growing conditions may induce male-sterility. Sometimes this leads to carpelloidy – the more or less complete conversion of stamens to carpels (Fig. 2.10) [31]. A large number of mutations leading to male-sterility are known; these have various morphological effects from producing small, non-functional anthers, through to producing pollen which degenerates [74,75]. Another mutant, 'multiovary' bears four carpels, three taking the place of stamens [44,89]. Another mutant form carries ovaries elevated on stalks [82]. Yet another has two carpels, and so frequently sets twin seeds, as well as having three normal stamens [73]. Several mutants give rise to grains with shrunken endosperms; these may find use in plant-breeding programmes [42].

To summarize the origins of the grain parts; the rachilla, lodicules, palea, lemma, pericarp, testa and the crushed remains of the nucellus are of maternal origin, and are presumably diploid. The embryo (both the axis and the scutellum) is diploid, with equal nuclear contributions from

52 *Barley*

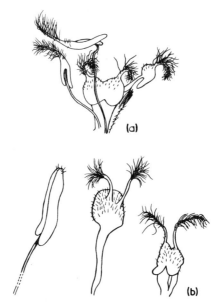

Fig. 2.10 An abnormal flower with three carpelloid stamens (a), and other 'stamens' from near-normal to fully carpelloid types (b) induced, apparently by the weather, in English barley (after Gregory and Purvis [31]).

the egg cell and the male gamete. In contrast the endosperm (both the starchy endosperm and the aleurone layer) is triploid, being derived from a second, triple fusion between one male gamete from the pollen and two polar nuclei derived from the embryo sac. In each case cytoplasmic inheritance, through the egg, would favour the influence of the maternal tissue further. These points are relevant to the genetics of grain formation in connection with breeding programmes (Chapters 11 and 12). The adherence of the pericarp to the testa means that the grain or kernel (naked or husked) is a single-seeded indehiscent fruit, a caryopsis.

2.3. Variations in the form of grains

Barley kernels occur in a wide range of shapes and sizes. In addition to various striking characters, e.g. the possession of awns or hoods (Figs. 2.11 and 2.12), there are many minor characters that are important in recognizing grains of different varieties [7,11,18,91]. Threshed grain, in which the base and apex of the husked forms are roughly broken, is particularly hard to characterize.

The most obvious characteristic is whether grain is husked or naked, i.e. the palea and lemma fail to adhere to the pericarp. The firmness of the attachment of the husk varies between varieties and with the conditions

The morphology of the reproductive parts in barley 53

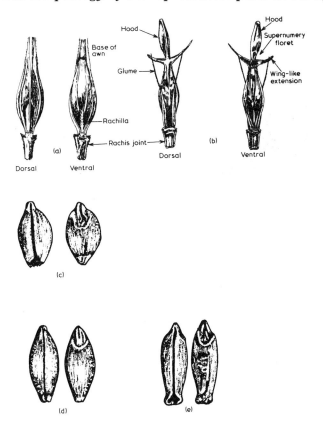

Fig. 2.11 Dorsal and ventral views of various grains (a, b—after Orlov [62] c, d and e, after Bose [15]). (a) Dorsal and ventral views of a husked, awned median grain and (b) a husked hooded median grain, in each case from a two-rowed variety, and attached to a segment of rachis. Note that the sterile glumes are omitted in (a). (c, d and e) Dorsal and ventral views of three types of naked grains. Notice the persistence of the ovary tip.

of growth (Fig. 2.11) [52]. In some forms the attachment of the husk is sufficiently tenuous for them to be called half-naked. Mutations for naked grains are often obtained. Naked barleys are commonly grown in the Himalayan region, China and Japan, where they are used for human food. Similarly naked barleys were grown in the oldest farming communities (Chapter 3).

Commonly grown barleys are two-rowed and six-rowed forms (Figs. 2.11, 2.12 and 2.13). In two-rowed barleys, with only the central spikelets being fertile, the grains are uniformly symmetrical. The lateral florets are reduced, and are often raised on a short pedicel, or stalk. In common forms they are staminate and may produce pollen, but in *deficiens* forms they are reduced even further, to single small 'lemmae'. Their

54 Barley

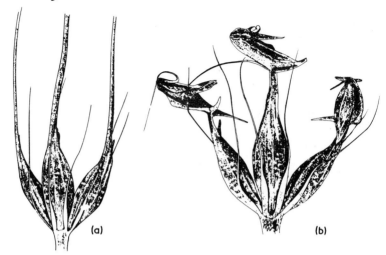

Fig. 2.12 Groups of three kernels (triads) arising at a node of (a) a six-rowed awned, and (b) a six-rowed hooded barley (after Wiebe and Reid [91]).

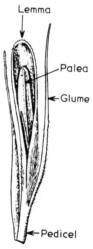

Fig. 2.13 A reduced, male lateral floret from a two-rowed barley (after Bergal and Friedberg [11]).

morphology varies. In six-rowed varieties all three spikelets at each node are fertile. The median grains, one third of the total number, are symmetrical but the remainder, the lateral grains, are unsymmetrical to a greater or lesser extent, each with a right-handed or left-handed bias. Grain shapes vary widely, from the long and slender to short and plump. More subtle differences can also be detected e.g. whether the widest part of the grain occurs midway along the length or towards the base. The shape

The morphology of the reproductive parts in barley 55

of the crease, and the attitudes of grains placed on a horizontal flat surface are also useful recognition characters [18]. Grain sizes vary widely. Dwarf varieties with 5 mg grains are known, while some Abyssinian varieties have 80 mg grains. Typical averages and ranges of grain dimensions for some German barleys are, thousand corn weight, 34 (21–45) g dry weight; length, 8 (6–12) mm; width, 4.1 (2.7–5.0) mm; thickness, 3.5 (1.8–4.5) mm [76]. In any sample of grain, even from one ear, a range of sizes occurs.

Grain colours are characteristic. Naked grains may appear blue, or red-purple, from anthocyanin pigments in the aleurone or pericarp respectively, or they may appear pale yellow though various shades of grey to black, due to an alleged 'melanine-like' pigment in the pericarp. Husked grains may appear greenish if the aleurone is blue and the outer layers are yellow. The husk itself may be several shades of yellow, or even partly orange, brown, various shades of grey or black. The 'nerves' may be purple, with anthocyanins, or the whole husk may be coloured purple. The awns or hoods, lemmae or both may be albino. A little black pigment in an otherwise albino hood produces a greyish awn. Genes for high levels of anthocyanins combined with the 'albino awns' (i.e. absence of other pigments) give forms with blood red awns [4,36,94].

In the mature grain the lodicules are small, withered structures trapped between the husk and the embryo. The manner of their insertion, their size (0.3–1.2 mm^2 in area), their exact shape and hairiness, combined with the fact that their morphology is often characteristic and is stable within one variety and their differing abilities to take up stains, are useful in characterizing grains (Figs. 2.14 and 2.15) [8,9,11,18].

The sterile, outer glumes, two at each spikelet, are normally small, 'lanceolate' structures that may carry fine awns of varying lengths (Figs.

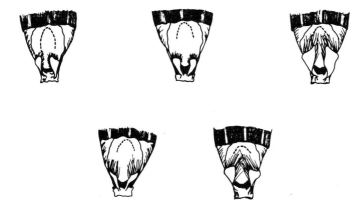

Fig. 2.14 The positions and shapes of five types of lodicules in grains from which the bases of the lemmae have been removed (after Gill and Vera [30]).

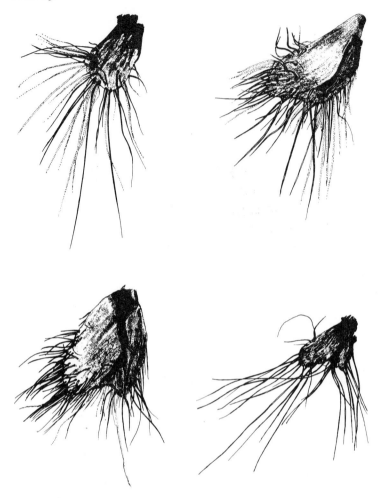

Fig. 2.15 The detailed morphology of four separated lodicules (after Bergal and Friedberg [11]).

2.11, 2.12 and 2.13). In some hooded varieties the glumes can carry small hoods (Fig. 2.16). Yet other forms are known. L'Orge du Prophète, grown in Morocco, is a lax-eared, six-row barley in which the outer glumes are large and lemma-like and carry large awns, so that there are nine striking large awns per ear node, and the ear seems 'bushy' (Fig. 2.17) [57]. A variant of a two-rowed variety also has lemma-like glumes, carrying awns, at each side of the central spikelet. This form had a higher thousand grain weight, (8.7% higher; 49.5g) than comparable normal forms, suggesting that the extra photosynthetic tissue was of value in supplying the grain [77]. Some mutant two-rowed varieties have extra tough bracts

The morphology of the reproductive parts in barley 57

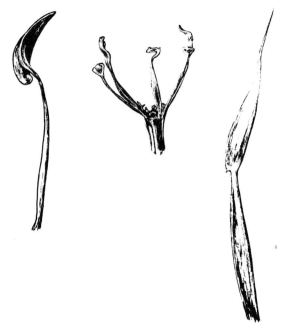

Fig. 2.16 Outer, sterile glumes bearing rudimentary hoods (after Harlan [35]).

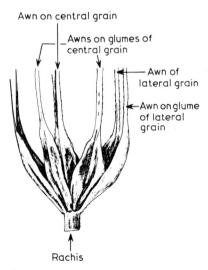

Fig. 2.17 L'Orge du Prophète. A triad of spikelets from this lax-eared, six-rowed Moroccan barley in which the outer, sterile glumes are lemma-like and fully awned (after Miège [59]).

58 Barley

Fig. 2.18 The ear of a variety (*bracteatum*; a mutant), with a large extra basal glume-like bract (1), and progressively smaller bracts (2, 3, etc.) enveloping the median spikelets of the two-rowed ear (after Gustafsson [34]).

below the inflorescence and of progressively lesser size around the upper median spikelets (Fig. 2.18).

Many other mutants forms are known (Chapter 12) [80,85]. Fertile florets may occur in the axils of the glumes each side of the median florets of a six-rowed ear giving rise to grains and a 'ten-rowed ear' (Fig. 2.19) [36]. The occurrence of supernumerary spikelets, and grains, is a common departure from familiar forms.

Awns are apical extensions of the lemmae, and carry three vascular bundles, of which the largest is centrally placed. They are roughly triangular in cross-section and contain two tracts of parenchymatous photosynthetic tissue (Fig. 2.20). The walls of the ground tissue cells, the sclerenchyma, and the epidermis, are thickened. Two rows of stomata connect the photosynthetic tissue with the exterior. As they mature awns become increasingly rigid and brittle and contain heavy deposits of ash (Chapter 6). In some mutants the awns may be curly, but normally they are straight. They vary in robustness, and in length; short awns are termed awnlets. Awnless varieties are known. Under conditions of strong inter-plant competition awn growth may be reduced or prevented [72]. Awns are usually barbed or hairy; the size and disposition

The morphology of the reproductive parts in barley

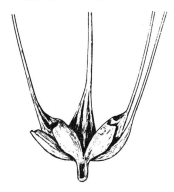

Fig. 2.19 The median spikelet of a six-rowed barley carrying grains in the outer glumes. This is a 'ten-rowed' form as the lateral spikelets do not carry the extra grains (after Harlan [36]).

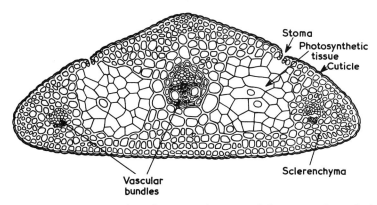

Fig. 2.20 A transverse section of an awn, about 0.3 of the way up from the base (after Lermer and Holzner [54]).

of the teeth, or their absence, on the median or lateral ridges or at the apex or base, are varietal characters. Normally the teeth are angled towards the apex [7,11,18]. Semi-smooth and smooth awned varieties tend to lack sufficient numbers of stigmatic hairs, and show a tendency to sterility in consequence. The toothing on the median and lateral nerves of the lemma, and the hairiness of the inside tips of the lemma, are also useful characters. Awns may be tough, or fragile, or dehiscent. In a Chinese variety the whole lemma is dehiscent, and the naked grains are readily shed in consequence [36]. Varieties are known in which the hood or awn occurs in triplicate giving, e.g., a triple-awned lemma (Fig. 2.21). The lemma is partially divided at the apex in some mutants, and carries two equal or unequal awns (Fig. 2.22). An astonishing variety of other lemma appendages is known, ranging from awns carrying slight lumps or

60 Barley

Fig. 2.21 Triple awns at the apex of a lemma (after Harlan [36]).

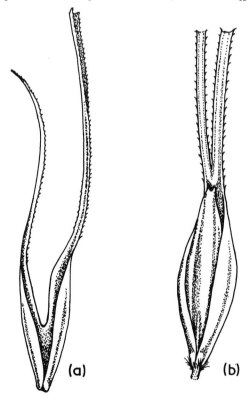

Fig. 2.22 Partially divided lemmae from mutant varieties with two equal or unequal awns (after Scholz and Lehmann [79,80]). (a) A partially divided lemma. (b) A grain carrying two awns, seen from the ventral side.

The morphology of the reproductive parts in barley

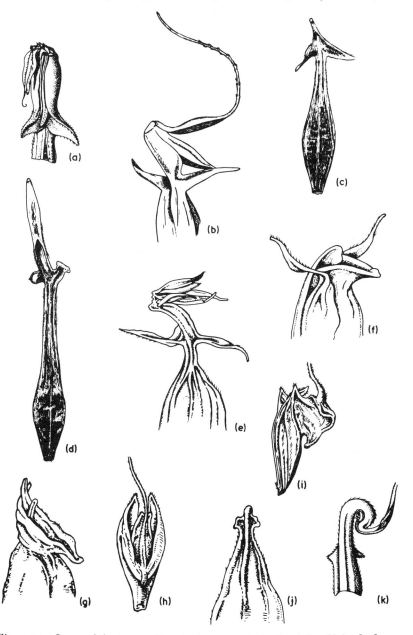

Fig. 2.23 Some of the types of hoods that occur in barley (after Helm [40], Harlan [35], Bandlow [3]). (a, b) *H. vulgare trifurcatum*. (c, d) Hooded grains from various crosses. (e) A modified hood with extra 'wings' at the apex. Sometimes a second rudimentary floret occurs here. (f-k) 'Modified hoods' of various types bent back to various extents exposing the grain tip: (h and i) – *calcaroides* forms. (j) Vestigial hood. (k) Apex of a modified hood bearing two vestigial supernumerary florets (*H, vulgare* va. *urgaicum* (Vav. *et* Orl.))

62 Barley

extensions to highly elaborate structures (Fig. 2.23). Hoods may also be born elevated to varying extents above the lemma tip on an awn, or sessile, or so sub-sessile that the inflated lemma tip is folded back and exposes the grain tip (*calcaroides* and *sub-calcaroides* forms; Fig. 2.23) [2,3,32,34–36,40]. The original hooded types came from the Himalayas (Nepal) but hoods have arisen several times as mutations. Generally the lemma is the hooded structure but sometimes the glumes and rarely even the paleae may carry hoods [95]. In some varieties the lemmae of the median florets have hoods, while those of the lateral florets are awnletted.

The base of the lemma varies in shape in different grains. Where the base is bevelled it is called a *falsum* type; where nicked, *verum* type and neither, a *spurium* type. The shape of the grain base is not a very consistent character; different forms may be found on grains from the same ear. Furthermore intergrades between these types are known (Fig. 2.24) [17,18,26].

The palea is not much used in varietal characterization. Mutants with paleae bearing small awns, or in which the paleae are divided longitudinally into two, nearly to the base, are known. Usually only slight differences in

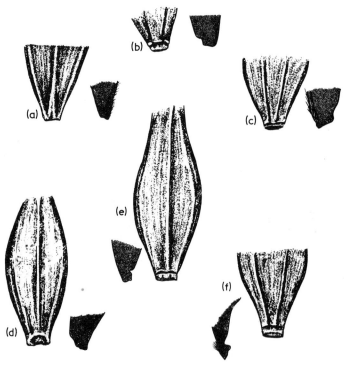

Fig. 2.24 Various forms of the lemma base, seen in surface view and median longitudinal section (after Broili [17]).

The morphology of the reproductive parts in barley 63

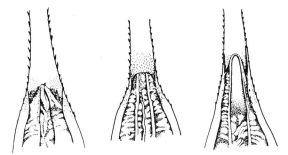

Fig. 2.25 Three types of palea apex (after Bergal and Friedberg [11]). Note the lemma overlapping the palea at the sides, the awn projecting from behind it, and the hairiness within the lemma, at the awn base.

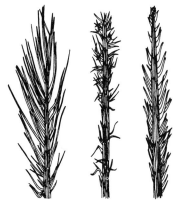

Fig. 2.26 Various common forms of the rachilla (after Ziegler [97]).

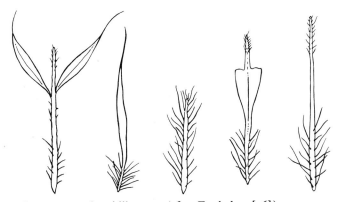

Fig. 2.27 Some unusual rachilla types (after Engledow [26]).

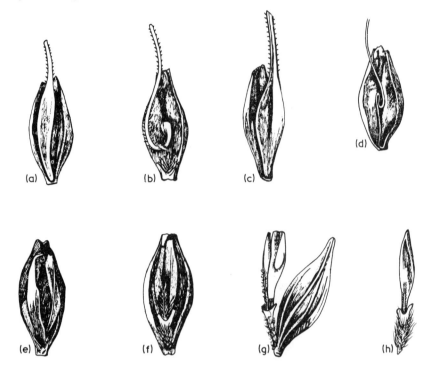

Fig. 2.28 Some unusual rachilla types, and structures that replace them (after Ziegler [97]). Some are replaced with awned, lemma-like structures (a-c) or awns (d), divided lemma-like organs (e) or extra florets, sometimes carrying extra rachillae or even grains (f, g), or rachillae with secondary rachillae (h).

Fig. 2.29 Floret of an induced mutant in which the rachilla has been replaced by a secondary floret which has formed a grain (after Scholz and Lehmann [80]).

The morphology of the reproductive parts in barley

the shape of the tips of the palea are encountered (Fig. 2.25). If the rachilla is removed from the furrow some varieties are found to carry two small tracts of hairs on the sides of the palea, while others do not.

Rachillae arise at the basal end of the grain and lie in the ventral furrow (Fig. 2.11). They may be long or short, or they may abort. They vary in degrees of hairiness, and the hairs themselves may be long or short (Fig. 2.26). There is some correlation between the relative rachilla length and the length of rachis internodes. The lateral grains of six-rowed varieties tend to have larger rachillae than the median grains. The varying types that occur help in identifying grain samples. They may or may not survive threshing. Many aberrant forms of rachilla have been encountered in breeding and mutation programmes, including some which carry grains (Figs. 2.27, 2.28 and 2.29).

2.4. The ear

The ear or head is a terminal spike. Awned ears are often described as 'bearded'. The axis of the ear, the rachis, is usually bilaterally symmetrical, with alternating nodes and internodes (Fig. 2.30). Normally three spikelets

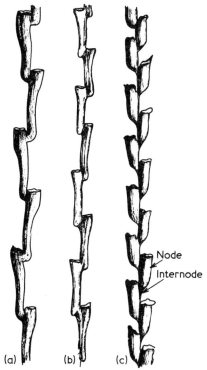

Fig. 2.30 Lengths of rachis with varying internode lengths from (a) lax, (b) dense, and (c) very dense ears (after Wiggans *et al.* [92]). The rachises are viewed from the side, so that in each case the median spikelets would be directed alternately to the right and the left, in the plane of the paper.

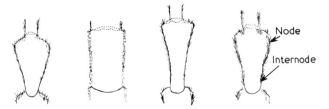

Fig. 2.31 Four types of rachis 'joint', viewed from the front, with differing shapes of node and internode and differing degrees of hairiness (after Bergal and Friedberg [11]).

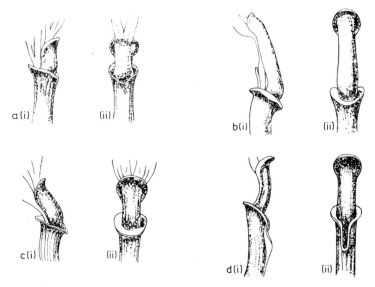

Fig. 2.32 Four of the several types of collar which may be found at the base of the rachis, viewed from the side (i) and the front (ii) (after Bergal and Friedberg [11]).

arise at each node. The shorter the internodes, the more compact the ear, and the more dense it is. Impressions of density are also influenced by grain size and shape. Rachises are normally 2.5–12.7 cm (1–5 in) long and may carry 25–60 kernels in six-rowed varieties or 15–30 kernels in two-rowed varieties. The form of the central part of the rachis, whether viewed from the side (Fig. 2.30) or from the front (Fig. 2.31), and its degree of hairiness, are useful varietal characteristics. The basal node of the rachis is a 'collar'. This is very variable in form and may be flat or cuspid, V-shaped or open. Also the basal internode may be long or short, straight, curved or curled (Fig. 2.32) [11,18]. In other cases the lengths of the central internodes may be variable, giving the spike a tweaked appearance. Rarely two nodes may apparently be opposite, so that in a six-rowed

The morphology of the reproductive parts in barley 67

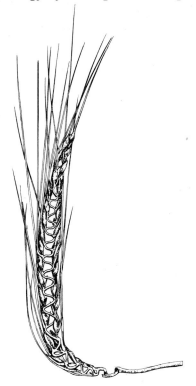

Fig. 2.33 An ear with an 'accordion rachis' (after Harlan [36]).

variety the grains arise in a whorl. In some 'monstrous' forms the rachis and the awns are contorted and twisted. In the mutant variety called 'accordion-rachis' the rachis has a zig-zag shape (Fig. 2.33). Phenocopies of this type have been produced by treating 'normal' growing plants with gibberellic acid [46,84]. In some dwarfs a length of naked rachis, devoid of spikelets, projects above the ear. The rachises of cultivated barleys are tough to varying extents although fragile forms arise during breeding and mutation programmes. A tough rachis is necessary to prevent the ear breaking up before and during harvesting. The rachis of the wild form, *Hordeum spontaneum* readily disarticulates. The dispersal unit is arrow-shaped, consisting of a median awned grain pointed by a rachis segment and barbed by two sterile lateral florets (Chapter 3). Diagrams of *H. spontaneum* are given by Schiemann [78] and Nevski [59].

The fertility of the florets, their degree of development and the closeness of their packing has provided the basis for many barley classification systems (Fig. 2.33–2.37; Chapter 3) [6,10,18,51,91]. Rachis nodes are of various lengths. Grains are loosely packed in lax ears, through closer packing to being very tightly wedged together in dense ears.

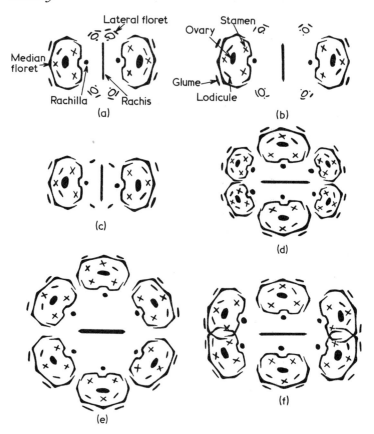

Fig. 2.34 Diagrams of six common types of barley ear. In each case the spikelets from two adjacent nodes are shown from above, as though arranged in one plane around the rachis (after Wiggans [92], Beaven [6]). (a) Two-rowed, lax-eared *distichon (distichum)* type; often nodding *(nutans)*. (b) Dense-eared, two-rowed *zeocrithon* ear type; often erect *(erectum)*. (c) Two-rowed ear with greatly reduced lateral florets, *deficiens (decipiens)* type. (d) A six-rowed ear with lateral florets appreciably smaller than the median ones; *intermedium* type. (e) A dense, six-rowed ear; *hexastichum* type. (f) A lax-eared, six-rowed barley; *vulgare* type. Where the lateral florets overlap such types may be misleadingly called 'four-rowed' or *tetrastichum* forms.

In some six-rowed forms the lateral grains are roughly equal in size to the median grains, while in *intermedium* forms the lateral grains are appreciably smaller. In two-rowed varieties the lateral florets may be present, or they may be greatly reduced and lacking many components. The forms mentioned include the barleys of commerce, and may be summarized by diagrams (Fig. 2.34). The terms six-rowed and two-rowed refer to the number of rows of grain seen when the ears are viewed from above. However in lax-eared, six-rowed forms the lateral grains overlap,

Fig. 2.35 Various two-rowed ears (after Lermer and Holzner [54], Orlove [62], Harlan [36], Gustafsson [34]). (a) A lax ear of the *distichon* type in which the lower awns have been shed. (b) A dense ear of awned barley, a *zeocrithon* form; a type once called Spratt, Fan, Battledore, or Peacock in England. (c) A two-rowed barley of the *deficiens* type, in which the lateral florets are scarcely noticeable. (d) A dense eared, two-rowed hooded barley with large lateral florets. (e) The ear of a mutant (*calcaroides* type) two-rowed barley in which the lemmae of the upper median florets are inflated and turned back – as 'sub-sessile hoods' leaving the grain apices exposed. The lower florets appear more normal, except that the awns are weakly developed.

Fig. 2.36 Some six-rowed barley ears (after Wiebe and Reid [91], Orlov [62], Lermer and Holzner [54]). (a) The awned ear of Trebi barley. (b) The moderately dense ear of a Japanese barley with short, broad awns on the median florets, but awnless lateral florets. (c) A very dense, short-awned ear of a Chinese barley. (d) A small, short-awned, moderately dense ear. (e) An awnless, dense, six-rowed ear. (f) An ear with long median and short lateral awns (var. *nipponicum*). (g) A six-rowed, hooded barley (var. *trifurcatum*).

The morphology of the reproductive parts in barley 71

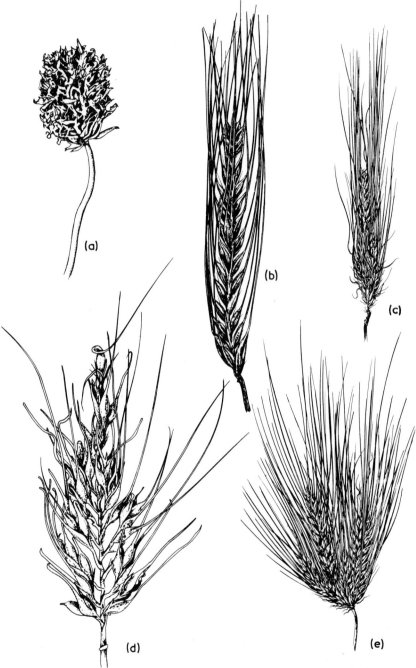

Fig. 2.37 Less usual ear types (after Harlan [36], Wiebe and Reid [91], Lermer and Holzner [54]). (a) A very dense eared, *multiflorus* type, with hooded spikelets. (b) An ear of the *irregulare* type, with two-rowed and six-rowed sections. (c) An ear of the *compositum* type; two-rowed at the tip and apparently *multiflorus* at the base. (d) A branched ear in which secondary rachises, carrying several florets, arise at the nodes of the primary rachis. (e) Another form of branched ear.

giving rise to the misleading name four-rowed (*tetrastichum* forms). Quite clearly this term should be reserved for those rare, genuine four-rowed types in which the central spikelets are sterile and the laterals are fertile [25,71]. 'Two-rowed types' with double (twin) central spikelets might be regarded as other candidates for the title 'four-rowed'. Where the outer glumes of the median florets of six-rowed varieties are fertile the ears are ten-rowed, while fertility in the glumes of all the florets would give rise to eighteen rows! Perhaps the *multiflorus*, pom-pom headed mutants with large, irregular numbers of grains at each node approach this state (Fig. 2.37). Extra grains can also arise on modified rachillae and in hoods.

In irregular barleys (*irregulare* forms) the ear may be partially two-rowed and partially six-rowed at different sites along its length. In *compositum* forms spikes may be two-rowed near the apex, and carry multiple spikelets near the base. In addition some barley ears are branched. Two sorts of branching appear to occur. In one type short extra rachises carrying varying numbers of grains arise alternately from the nodes of the primary rachis (Fig. 2.37). In the other type a group of apparently normal ears arise from a collar on one pedicel (Fig. 2.37).

2.5. Some implications of the wide variety of forms of barley

The wide range of barley types poses problems for the systematist (Chapter 3). These are not made more simple by the plasticity of the plants, as revealed when they are grown under different conditions. Thus under some circumstances, and especially if the central spikelets are removed, seed is occasionally set in the lateral florets of 'two-rowed' barleys thereby moving them to the '*irregulare*' group [37,90]. Wide ranges of genetically determined 'natural' sports and 'artificial' mutations are known. However spraying growing plants with herbicides (Chapter 9), or solutions of gibberellic acid [41,46,84], or exposing them to abnormal day lengths or temperatures or light of abnormal composition [24,46,47], induces a range of alterations in plant phenotypes (appearance) that are strongly reminiscent of mutant types, e.g. carpelloid stamens, accordion rachis, branched spikes, tweaked spikes, multiple spikelet groups and whorls of spikelets, variously situated supernumerary spikes, twinned kernels, blasted (sterile) spikelets, deformed or extra collars or rachis nodes, branching stems, stems carrying adventitious roots from the upper nodes, kernels of excessive size, one-ranked ears (with all spikelets held on one side), and so on. Thus barley types 'almost' grade into each other and are not clear cut, distinct forms; under 'abnormal' conditions the developmental paths may change. Genetic or environmental irregularities may be sufficient e.g. to partially or wholly alter a hooded into an awned

The morphology of the reproductive parts in barley

barley. Evidently the systems which regulate morphological development are readily upset.

The wide range of material available appears to be ideal for comparative or other experimental studies on morphogenesis. A barrier to such work is that unfamiliar types are regarded as 'abnormal' and are referred to as 'freaks', 'monstrosities' or 'oddities'. All that is revealed by these terms is a prejudice against types that do not neatly fall into a classification of familiar, agricultural types. The barleys with branched ears and supernumerary spikelets 'offend' most, as they do not even fit the usual 'definitions' of the genus *Hordeum*. Any weaknesses are in the classification systems and in the terminologies, and not in the plants (Chapter 3).

References

1 Anthony, S. and Harlan, H.V. (1920). *J. agric. Res.*, **18**, 525–536.
2 Arber, A. (1934). *The Gramineae – A Study of Cereal, Bamboo & Grass.* Cambridge University Press.
3 Bandlow, G. (1954). *Züchter*, **24**, 20–27.
4 Bandlow, G. (1959). *Züchter*, **29**, 123–132.
5 Banerjee, S. and Wienhues, F. (1965). *Z. Pflzücht*, **54**, 130–142.
6 Beaven, E.S. (1947). *Barley, Fifty Years of Observation and Experiment.* London: Duckworth.
7 Beil, G.D.H. (1937–8). *Z. Zücht. (Reihe A)*, **22**, 81–146.
8 Bergal, P. (1939). *Ann. Epiphyt.*, **5**, 555–563.
9 Bergal, P. (1948). *Annls Sci. nat. (Bot.) (Sér. 11)*, **9**, 189–269.
10 Bergal, P. and Clemencet, M. (1962). In *Barley and Malt*, (ed. Cook, A.H.) pp. 1–23. London: Academic Press.
11 Bergal, P. and Friedberg, L. (1940). *Ann. Epiphyt.*, **6**, 157–306.
12 Bonnett, O.T. (1935). *J. agric. Res.*, **51**, 451–457 + plates.
13 Bonnett, O.T. (1938). *J. agric. Res.*, **57**, 371–377.
14 Bonnett, O.T. (1966). *Illinois agric. exp. Stn. Bull.*, **721**, 59–91.
15 Bose, R.D. (1931). *Indian J. agric. Sci.*, **1**, 58–89.
16 Brewbaker, J.L. and Emery, G.C. (1962). *Radiat. Bot.* **1**, 101–154.
17 Broili, J. (1908). *Das Gerstenkorn in Bilde.* Verlag Eugen Ulmer, Stuttgart.
18 Carson, G.P. and Horne, F.R. (1962). In *Barley and Malt*, (ed. Cook, A.H.), pp. 101–159. London: Academic Press.
19 Cass, D.D. (1973). *Can J. Bot.*, **51**, 601–605.
20 Cass, D.D. and Jensen, W.A. (1970). *Am. J. Bot.*, **57**, 62–70.
21 Cass, D.D. and Karas, I. (1975). *Can. J. Bot.*, **53**, 1051–1062.
22 Collins, E.J. (1918). *Ann. Bot.*, **32**, 381–414.
23 D'Amato, F., Devreux, M., and Scarascia—Mugnozza, G.T. (1965). *Caryologia*, **18**, 377–382.
24 Dormling, I., Gustafsson, Å. and Ekman, G. (1975). *Hereditas*, **79**, 255–271.
25 Drabble, E. (1906). *New Phytol.*, **5**, 17–21.
26 Engledow, F.L. (1920). *J. Genet.*, **10**, 93–108.
27 Erdelská, (1967). *Ann. Bot.*, **31**, 367–369.
28 Erdtman, G., Berglund, B. and Praglowski, J. (1961). *An Introduction to a Scandinavian Pollen Flora.* Stockholm: Almqvist and Wiksell.
29 Fruwirth, C. (1906). *Fühling's Landswirtschaftliche Zeitung*, **55**, 544–553.
30 Gill, N.T. and Vear, K.C. (1969). *Agricultural Botany.* London: Duckworth.

31 Gregory, F.G. and Purvis, O.N. (1947). *Nature*, **160**, 221–222.
32 Grossin, f. (1961). *Bull. Soc. bot. Fr.*, **108**, 120–125.
33 Guenther, O. (1927). *Bot. Arch.*, **18**, 299–319.
34 Gustafsson, Å. (1947). *Hereditas*, **33**, 1–99.
35 Harlan, H.V. (1931). *J. Hered.*, **22**, 264–272.
36 Harlan, H.V. (1957). *One Man's Life with Barley. The Memories and Observations of Harry V. Harlan*, (ed. Harlan, J.R.) New York: Exposition Press.
37 Harlan, H.V. and Martini, M.L. (1935) *J. Hered.*, **26**, 109–113.
38 Harlan, H.V. and Pope, M.N. (1922). *J. Hered.*, **13**, 269–273.
39 Harlan, H.V. and Pope, M.N. (1925). *Am. J. Bot.*, **12**, 50–53.
40 Helm, J. (1952). *Flora, Jena*, **139**, 96–147.
41 James, N.I. and Lund, S. (1965). *Am. J. Bot.*, **52**, 877–882.
42 Jarvi, A.J. and Eslick, R.F. (1975). *Crop Sci.*, **15**, 363–366.
43 Johannsen, W. (1884). *C. r. Trav. Lab. Carlsberg*, **2**, 60–77 (+ Figs).
44 Kamra, O.P. and Nilan, R.A. (1959). *J. Hered.*, **50**, 159–165.
45 Kato, K. and Watanabe, K. (1957). *Botan. Mag., Tokyo*, **70**, 96–101.
46 Kirby, E.J.M. (1971). *J. exp. Bot.*, **22**, 411–419.
47 Kirby, E.J.M. (1973). *J. exp. Bot.*, **24**, 935–947.
48 Kirby, E.J.M. and Rymer, J.A. (1974). *Ann. Bot.*, **38**, 565–573.
49 Kirby, E.J.M. and Rymer, J.L. (1975). *Ann. Bot.*, **39**, 205–211.
50 Klaus, H. (1966). *Bot. Jb.*, **85**, 45–79.
51 Körnicke, F. and Werner, H. (1885). *Handbuch des Getreidebaues I.* pp. 129–191; 469. Verlag von Paul Parey: Berlin.
52 Kozlov, V. (1927). *Bull. appl. Bot. Plant Breeding*, **17**, 169–181.
53 Krauss, L. (1933). *Jb. wiss. Bot.*, **77**, 733–808.
54 Lermer, Dr., and Holzner, G. (1888). *Beiträge zur Kentniss der Gerste*. München: R. Oldenbourg.
55 Luxová, M. (1968). *Biol. Pl. (Praha)* **10**, 10–14.
56 Merry, J. (1941). *Bull. Torrey bot. Club*, **68**, 585–598.
57 Miège, E. (1928). *Bull. Soc. Sci. natn. Maroc.* **8**, 144–148.
58 Nátrová, Z. (1968). *Biol. Pl. (Praha)*, **10**, 118–126.
59 Nevski, A. (1941). *Act. Inst. Bot. Nom. Komarov; Acad. Sci. USSR. Flora et systematica plantae vasculares, Ser. 1. Fasc.* **5**, 64–255 pp.
60 Norstog, K. (1972). *Am. J. Bot.*, **59**, 123–132.
61 Norstog, K. (1974). *Bot. Gaz.*, **135**, 97–103.
62 Orlov, A.A. (1931). *Bull. appl. Bot. Genet., Plant Breeding*, **27**, 329–381.
63 Pedersen, A. (1960). *Landbrugets Plantekultur.*, (3rd edition) pp. 44, 45 København: Kandrup og Wunsch, Bogtrykkeri.
64 Pope, M.N. (1916). *J. Am. Soc. Agron.*, **8**, 209–227.
65 Pope, M.N. (1937). *J. agric. Res.*, **54**, 525–529 (+ Figs).
66 Pope, M.N. (1942). *J. Am. Soc. Agron.*, **34**, 200–202.
67 Pope, M.N. (1943). *J. Hered.*, **34**, 153–154.
68 Pope, M.N. (1943). *J. agric. Res.*, **66**, 389–402.
69 Pope, M.N. (1946). *J. Am. Soc. Agron.*, **38**, 432–440.
70 Pope, M.N. (1949). *J. agric. Res.*, **78**, 295–309.
71 Prasad, G. and Tyagi, D.V.S. (1975). *Barley Genetics Newsletter*, **5**, 40–42.
72 Qualset, C.O. (1968). *Theoret. appl. Genet.*, **38**, 355–360.
73 Rana, R.S. (1966). *J. Hered.*, **57**, 227–229.
74 Roath, W.W. and Hockett, E.A. (1970). In *Barley Genetics II*, (ed. Nilan, R.A.) pp. 308–315. *Washington State University Press*.
75 Roath, W.W. and Hockett, E.A. (1971). *Crop Sci.*, **11**, 200–203.

76 Rohrlich, M. and Bruckner, G. (1966). *Das Getreide I. Das Getreide und Seine Verarbeitung*, (2nd edition). Berlin: Paul Parey.
77 Sawicki, J. (1953). *Acta Soc. Bot. Pol.*, **22**, 605–615.
78 Schiemann, E. (1948). *Weizen, Roggen, Gerste.* pp. 69–94 Jena: Gustav Fischer.
79 Scholz, F. and Lehmann, C.O. (1961). *Kulturpfl.*, **9**, 230–272.
80 Scholz, F. and Lehmann, C.O. (1965). *Züchter*, **35**, 79–85.
81 Schuster, J. (1910). *Flora, Jena*, **100**, 213–266.
82 Sethi, G.S. and Gill, K.S. (1971). *Curr. Sci.*, **40**, 557.
83 Stebbins, G.L. and Gupta, V.K. (1969). *Proc. natn. Acad. Sci. U.S.A.*, **64**, 50–56.
84 Stoy, V. and Hagberg, A. (1967). *Hereditas*, **58**, 359–384.
85 Stubbe, H. and Bandlow, G. (1946–1947). *Züchter*, **17/18**, 365–374.
86 Suetsugu, J. (1951). *Bull. natn. Inst. agric. Sci. Ser. D Tokyo*, **1**, 49–87.
87 Tharp, W.H. (1935). *Bot. Gaz.*, **97**, 240–271.
88 Thompson, W.P. and Johnston, D. (1945). *Can. J. Res.* **C**, **23**, 1–15.
89 Tsuchiya, T. (1969). In *Induced Mutations in Plants.* pp. 573–590 Vienna: IAEA.
90 Veideman, M. (1927). *Bull. appl. Bot., Leningrad*, **17**, 1–67.
91 Wiebe, G.A. and Reid, D.A. (1961). *U.S. Dept. Agric. Tech. Bull.* **1224**.
92 Wiggans, R.G. (1921). *Cornell agric. exp. Stan. Mem.*, **46**, 365–456.
93 Wilson, A.S. (1874). *Trans. Proc. bot. Soc. Edinb.* **12**, 1876, 84–95.
94 Woodward, R.W. (1941). *J. agric. Res.*, **63**, 21–28.
95 Worsdell, W.C. (1916). *The Principles of Plant Teratology, II.* pp 267–271 London: The Ray Society.
96 Yagil, E. and Stebbins, G.L. (1969). *Genetics*, **62**, 307–319.
97 Ziegler, A. (1911). *Z. ges. Brauw.*, **34**, 513–517; 543–549; 563–568.

Chapter 3
The origin and classification of barleys

3.1. Introduction

Barleys, or other crop plants, are classified for two different purposes. Firstly by the plant taxonomists who wish to decide on the limits – and range(s) – of variability of species to give an indication of their evolutionary relationships. Secondly by 'practical' men (plant breeders, agronomists and technologists), who wish to know whether a particular plant or grain sample is of the desired variety. Thus two types of scheme exist: (a) 'Botanical classifications', hopefully revealing relationships between tribes, subtribes, genera and species, and (b) 'Keys' based on any characters available, designed to allow the identification of barley forms and varieties from the morphology of the plants, or their threshed grains. Unfortunately in the case of barley these objectives have become muddled.

It seems that all cultivated barleys, the wild barley *Hordeum spontaneum* (Kock) and the doubtfully wild *Hordeum agriocrithon* (Åberg) *Hordeum lagunculiforme* (Bakhteev) and *Hordeum paradoxon* (Schiemann) are one species. It has been a habit to classify each minor cultivated variant as a new 'species' and to dignify it with a specific 'Latin' binomial name. Increased knowledge, particularly of experimental genetics, makes these divisions untenable. Furthermore the frequent revision of classification schemes using the same, or very similar, names for slightly dissimilar types makes them confusing. Thus the widely used 'specific' names in fact describe morphological varieties. Many-named agricultural varieties, or cultivars, are bracketed by one 'specific' name – these differ, and must be distinguished. Probably agricultural cultivars are not genetically 'pure' (Chapter 12).

3.2. Classifications of barleys

Cultivated barleys and various wild barleys and barley grasses are classified in the genus *Hordeum*, within the tribe Triticeae (syn. Hordeae, Hordeeae).

The origin and classification of barleys

The fertility (or sterility) of crosses, ploidy levels, and details of morphology show that cultivated barleys are closely related to each other, to *Hordeum spontaneum*, to *H. agriocrithon*, to *H. paradoxon* and to *H. lagunculiforme*, but not to other wild barley grasses. The cultivated barleys have been grouped in the section Hordeum, or Crithe Döll, or Cerealia, Ands. The cultivation, or not, of a species must surely be irrelevant to its botanical classification. Karyotype analysis reveals no differences between these 'species' (large chromosomes, diploids $2n = 14$). They are normally fully interfertile annuals or winter annuals which may be divided and so caused to perennate. Typically they have one floret in each spikelet; three spikelets occur at each ear node of which the central one – or all three – may be fertile (two-row and six-row) [20,62]. Exceptions to this pattern do occur (Chapter 2). Many of the more dramatic differences in plant morphology are due to differences in one, or a few genes, and many have been induced by mutagenic treatments; yet these characters have been used to delineate 'species'. Haploid ($n = 7$) and tetraploid ($4n = 28$) barleys are almost incapable of crossing with the parent forms, yet by adjusting the chromosome complement of their progeny to the usual diploid number ($2n = 14$) full interfertility is restored; are the haploids, diploids and tetraploids to be regarded as different species? Crosses between six-rowed, lax-eared *pallidum* and two-rowed *distichum* varieties readily give rise to true-breeding *intermedium* forms having awnless lateral florets of variable fertility [21,46]. Regularly producing new 'species' by crossing other 'species' seems absurd. The unsatisfactory logic of traditional classifications is further apparent when it is realized that numerous other single-gene differences of a more subtle type, e.g. for disease resistance, or for different isozymes may be present. In 1957, considering relatively few factors, it was noted that from a crossing programme 19 440 000 possible combinations could be obtained. Some classifications would designate all these and even near-isogenic lines, as species! Clearly the classification of a highly variable species really sets very difficult problems. In principle a variety can only be defined by a detailed knowledge of its entire genetic complement. This is unattainable in practice. The classification keys available do group cultivated forms in useful, if 'unnatural' ways. However they rarely take note of *grades* in a character, e.g. the degree of hood elevation; nor do most take account of unusual characters, such as the presence of three awns on a lemma, accordion rachis, or branched ears, or genuine four-rowed ears (Chapter 2). Such omissions are 'justified' on the basis that such forms are 'freaks' or 'mutants' and do not occur in agriculture. Some of these barleys fall outside the accepted morphological limits not merely of the species, but of the genus *Hordeum* [43].

Thus the various schemes of classification separate morphological and physiological types of a single species, whether they are cultivated or not. For the specific name *Hordeum vulgare* (Linn.) emend. or *Hordeum sativum* (Jessen) is variously preferred. Numerous keys for identifying barleys, or classification schemes exist [4,8,9,13,16,17,29,39,42, 54–56, 61, 67, 68, 76, 96]. Many of these publications contain illustrations and list synonyms.

In the absence of a fossil record grasses are classified according to easily detected characteristics that, hopefully, do not alter much with changing conditioning of cultivation or soil type. The type of ear is given prominence in most classification schemes, and especially in keys to cultivated barley. Hundreds of 'Latin' names have been used to characterize varieties ('species'). Thus forms termed *hexastichon* are six-rowed, with dense ears, all spikelets bearing awns or hoods; mean internode lengths on the rachis 1.7–2.1 mm. Six-rowed, medium-dense *parallelum* ears have mean rachis internode lengths of 2.1–2.8 mm, while lax and pendulous-eared, six-rowed *vulgare* forms have internodes of 2.7–4.0 mm on average. For six-rowed lax ears in which the side grains overlap, giving a pseudo four-rowed state, the term '*tetrastichum*' has been used. In *intermedium* forms the lateral florets of the six-rowed ear lack awns and hoods. In *irregulare (mutabile)* forms the ear is variably six-rowed and two-rowed along its length. Very dense two-rowed barleys are termed *zeocriton (zeocrithon, zeocrithum)*, and have mean rachis internode lengths of 1.7–2.1 mm. Medium dense two-rowed ears (rachis internodes 2.1–2.3 mm) are called *erectum* forms while lax, two-rowed barleys are termed *distichon (distichum)* forms, or *nutans* when they are nodding, pendulous and lax (internodes 2.7–4.0 mm).

Ear characteristics may not be sufficient to distinguish between varieties; to achieve this it may be necessary to inspect the morphology of the plants at all stages of their life cycle, determine their disease resistance and so on. For many purposes it is necessary to decide whether a sample of threshed grain is a particular variety, or whether it is a mixture of two or more varieties. Considerable skill is needed for this purpose. It is usually possible to separate two-rowed, regular grain from six-rowed grains in which the lateral grains are asymmetric. Grain and aleurone colour, colour in the 'nerves' of the lemma, and the presence or absence of husk are obvious characters. However the separation of similar varieties requires attention to minute details. The shape of the grain base, even if not removed in threshing, is an unreliable character. Better characters include the presence or absence of teeth on the lemma 'nerves', the occurrence, or not, of hairs inside the ventral furrow, the length and hairiness of the rachilla and the 'attitude' of grain laid on a level surface [29]. The lodicules are useful in separating grain types. After soaking and staining inspection allows the recognition of *parvisquamose* (large lodiculed) and *latisquamose* (small lodiculed) forms, and many subdivisions of these groups [16]. Other tests may include the grains appearance under ultraviolet light, and whether or not they stain with a solution of phenol. The statistical evaluation of pairs of characters may separate similar forms [63], as may differences between isozyme patterns found in resting or germinating grains or their parts. Useful isozyme patterns are given, for example, by the α and β-amylases, peroxidases and esterases [6,7,38,59,71]. Unfortunately isozyme patterns can alter with plant age, and in response to changing conditions. Differences in the leaf flavonoids can distinguish between groups of barleys [37]. Differences in chemical composition can be found between near-isogenic lines selected for morphological differences, for example those with hoods or awns [75].

3.3. The position of barley within the Gramineae

The position of cultivated barleys within the Gramineae, the grass family, is of interest from the evolutionary viewpoint and because, in the absence of direct evidence, the morphological, physiological, biochemical and agricultural characteristics found in closely related grasses can indicate what to expect in barley. However this approach has its limitations. Vavilov [89–91], impressed with the similar ranges of attributes found in different cereals, formulated a 'law' of homologous variation which inferred that any variant found in one species was likely to be matched by an analogous variant in others. This is true of grain colour, hulledness (or nakedness), degrees of plant hairiness, waxiness, size, the presence or absence of hollow stems, and ligules and possibly chemical constitution, e.g. high-lysine grains, the occurrence of ranges of isozymes, and grains with different starch types also. However, the substance avenacin has only been found in oats; fungistatic hydroxamates occur in rye and maize, but have not been detected in barley.

Rye *(Secale cereale)* and the wheats *(Triticum* spp) are the cultivated cereals most closely related to barley and are grouped in the same tribe, the Triticeae (Hordeae). Oats, *(Avena sativa* and related forms) are in a different family, but the same subtribe of 'Festucoid' types, while other cereals are more remote relatives, belonging to different subtribes. The Triticeae is a reasonably homogeneous group in which, for example, seedlings lack a hypocotyl, and the starch type in the caryopsis is unusual. The crown is formed on or near the surface by a node carried up from the grain by the elongation of one or more internodes *above* the coleoptile [41]. The α-amylases of barley, wheat and rye are immunologically related, but are distinct from the rice enzyme [32]. Similarly the β-amylases of wheat, rye and barley are immunologically related to each other but not to the enzymes from oats or sweet potatoes [65]. Hybridization studies with DNA suggest that wheat and rye are more closely related to each other than either are to barley [15]. This degree of relationship is confirmed by the ready occurrence of viable hybrids (after chromosome doubling, *Triticale* forms) between wheats and rye. Fertile hybrids between barley (*Hordeum sativum* complex) and barley grasses have rarely or never been attained. Even obtaining crosses giving rise to infertile hybrids has proved difficult (Chapters 11 and 12). The cause of incompatibility between rye and barley appears to be connected with a failure in endosperm formation a few days after fertilization [85]. Using embryo culture techniques it has proved possible to get infertile hybrids plants from crosses between barley and wheat, or rye, or an *Agropyron* species [58].

The main divisions of the Gramineae are firmly based on a wide range of morphological, biochemical and serological criteria, and statistical analysis of the ways in which characters are associated [51,60,69,70,78,80]. The 'Festucoid' sub-family is made up of species having many characters in common [19,33]. These species are normally herbaceous and come from moist, temperate areas, the fertile florets of which contain two stigmas, three anthers and two lodicules. Many are annuals, but may be induced to perennate by cloning [62]. Generally they are long-day or day-neutral plants. Small numbers of relatively large chromosomes are present – typically the 'basic' chromosome number is seven. The nucleoli do not persist through mitosis [10,23,69]. Culm pulvini are absent, but specialized leaf sheath pulvini are present, and the stems are solid at the nodes [24,26]. Caryopes are small, and the embryo is only a small part of the whole fruit. The embryos have a characteristic structure; leaf margins do not overlap and the pattern of provascular strands between the embryonic axis and scutellum is typical, as is the manner of attachment between these two structures [72]. The embryos typically have ligules, although these are absent from the embryos of barley grains. They sometimes develop on barley embryos grown *in vitro*. Grain germination is inhibited by low concentrations of oxygen, and by isopropyl-*N*-phenyl-carbamate [5]. The rooting pattern of Festucoid types is characteristic [50]. Typically, the roots have files of epidermal cells alternating in size, root hairs grow obliquely towards the apex, arising from the ends of alternate cells [74]. The apex of the plant is covered by a tunica two cell-layers thick [25]. Leaf epidermal cells are a few simple types with no complex hair cells [69]. Vascular bundles in the leaves have a mestome sheath with thickened endodermis-like cell-walls containing proplastids and little or no starch, and a poorly-defined parenchyma sheath containing small chloroplasts with a few starch grains. The mesophyll does *not* radiate from the vascular bundles [22,69]. Festucoid leaves contain a high proportion of air-spaces, approximately 30% in *H. vulgare*, compared to 3–10% in other groups [27]. Festucoid plants contain fructosans, as well as starch, among their storage compounds and their grains appear to be relatively poor in leucine and alanine and rich in lysine and glycine

[31,93]. The prolamine fractions of cereal grains are antigenically related [53]. In the grain the Triticeae contain simple, unusual 'type 1' starch granules in contrast to other members of the Festucoideae which contain compound, 'type 4' starch grains [84]. Allergens, glycoproteins and carbohydrates from pollen of members of the Festucoid group differ from those of grasses from other groups, which explains the taxonomically patterned responses sometimes seen in allergic patients [94]. The members of the Festucoid group differ from some or all of the other groups in the characters mentioned.

The photosynthetic and photorespiratory metabolism of the Festucoid families, and some other groups, differs sharply from that of the Panicoid and Chloridoid families. The Festucoid types have leaves containing many microbodies, having a high rate of photorespiration, and high carbon dioxide compensation points [18,36,49,57]. Increased levels of oxygen stimulate photorespiration and possibly inhibit carbon dioxide fixation [52]. The primary carbon dioxide fixation product is C-3 (contains three carbon atoms) in the Festucoid types; apparently it is a C-4 substance in the Panicoid–Chloridoid group. The temperature optimum for net photosynthesis is lower in the C-3 group, which also discriminates differently to the C-4 group between the heavy and light isotopes of carbon (^{12}C and ^{13}C) and of hydrogen (^{1}H and ^{2}H, D) [14,87,97]. The Festucoid, (C-3) families seem richer in carbonic anhydrase than the C-4 families and the relative levels of the various carbon dioxide fixing enzymes differ [35,40,88]. The division between C-3 and C-4 plants is evidently important and of long standing in evolutionary terms.

3.4. The origin of cultivated barley

Over the years there has been a great deal of disagreement over the origin(s) of cultivated barleys [44,79]. Did six-row and two-row barleys arise separately, or from a common two-row stock, or from a common six-row stock, or in some other way? While the question is not settled the evidence currently available points towards a single origin, from the wild ancestors of the current two-rowed, wild *spontaneum* group. De Candolle [28] argued in 1886 that evidence for the origins of cultivated plants should be sought in tradition, religion, language, archaeology, and the present distribution of related forms. While imprints on coins, and recorded information indicate crop types known at the time of minting or writing it must be remembered that urban civilization came after the beginnings of agriculture. Many ancient religions taught that barley was the first crop, and that it held a special place in the Eleusinian mysteries. It was the cereal gift of the goddess Ceres or, to the Egyptians, the gift of Isis. To the Egyptians the germination of barley symbolized the resurrection of Osiris. Barley was sown and germinated in Nile mud

82 Barley

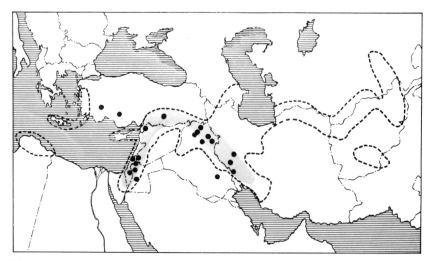

Fig. 3.1 The distribution of *Hordeum spontaneum* Koch (after Harlan and Zohary [47]). The shaded area indicates where the plant occurs in large quantities in primary habitats, and may be truly wild. The dotted line indicates the known limits of the plant's distribution. Outside the shaded area it occurs as a weed in disturbed segetal habitats. The black dots indicate the sites of the earliest known farming settlements. Copyright 1966 by the American Association for the Advancement for Science.

held in pottery trays shaped in the form of the god. The germinating grain symbolized his return to life.

The best evidence available now is the present distribution of wild barleys, the archaeological data and such other information as can explain the spread of the crop. The one undoubtedly wild form of barley (*Hordeum spontaneum* Koch) is concentrated in an arc in the Middle East, with scattered stands over a much wider area from Tunisia to Afghanistan, with doubtful occurrence in Morocco and Abyssinia (Fig. 3.1) [30,47]. It is believed not to extend into Tibet and W. China. In the 'arc' area it occurs wild, in undisturbed habitats as well as in disturbed, 'segetal' habitats such as road verges and the edges of cultivated land. It also occurs as a common weed in the corn crops of both wheat and barley. Outside this main area it occurs only in segetal habitats and as a weed in crops. [47]. Probably the high-density area is its 'natural' distribution zone, and that it has been spread elsewhere by man, possibly by seeds contaminating the seed corn. Unfortunately there seem to be no reports of the frequency of occurrence of *spontaneum* seeds in the seed grain from different agricultural areas. It seems likely that this 'wild' barley occupied at least the 'high-density' zone at the time of incipient agriculture, approximately 10 000 B.C. The wild form has been inadequately investigated, but it is certainly made up of a wide range of races of varying degrees of hardiness, size, leaf-width, grain size, grain colour

The origin and classification of barleys 83

(including brown and black), disease resistance, dormancy and rachis fragility; spring and winter forms are known, and even hooded samples have been found [2,3,20,43–45,81,89]. The 'species' *spontaneum* has unusually large grains for a wild grass; the ear is two-rowed. When the rachis breaks up each segment carries a single grain flanked by two sterile lateral florets, and this 'arrowhead' structure is an efficient dispersal unit which works its way into the soil [98].

The finding of a few grains of a six-rowed, brittle-eared barley mixed among other grains brought from Tibet was held to have revealed the 'wild' ancestor of six-rowed barleys [1,30,76,77]. However this six-rowed form, designated *Hordeum agriocrithon*, (Åberg) has not been seen growing wild in Tibet, or Sinkiang (where other seed originated), but it has been seen growing mixed with six-rowed, naked barleys – apparently as a weed [64]. Another, *'intermedium'*, brittle rachis type from the Far East was also dignified as a wild species *Hordeum paradoxon*, (Schiemann). However *agriocrithon*, *paradoxon*, *lagunculiforme* (bottle-shaped side-grains) and other types have been observed in 'hybrid swarms' formed by accidental crossing between wild two-rowed *spontaneum* and adjacent, cultivated six-rowed, lax-eared *vulgare* forms in Israel, Transcaucasia (USSR) and Turkmenia (USSR). If these 'semi-wild' swarms are left the six-rowed forms disappear during several generations, apparently because the ear-type precludes adequate seed dispersal. These hybrids are, as expected, genetically mixed [3,11,12,83,86,98,99]. The possibility of genetic introgression from the cultivated forms into wild forms, and *vice versa* is obvious emphasizing that modern 'wild' forms, for instance those with unusually large grains, are not necessarily similar to the ancestral forms. Outcrossing can occur quite readily under some circumstances (see Chapter 12). Whether or not *agriocrithon* and *paradoxon* could be formed by repeated hybridization in Tibet–Sinkiang is unclear. It is probable that they could have arisen originally from such a cross, and then been unwittingly selected and propagated as a weed. Alternatively the brittle rachis could have been 'recovered' from a cultivated barley with a tough rachis by back-mutation.

Archaeological evidence on the origin(s) of cereals depends on the recent recognition of the importance of plant remains, the development of the skills needed to recognize grain types by minute inspection of the details of carbonized grains or grain impressions in hardened clay, and on radiocarbon dating which gives a more reliable (but not infallible) chronology than was possible before [30]. Much early work is of doubtful value [34,48,73]. Archaeological evidence indicates that 'western' agriculture began in the 'fertile crescent' where wild barley, and wild emmer and einkorn wheats grow. In the period 10 000–8 000 B.C. semi-nomadic people who were hunters also gathered the wild cereals as has been done at intervals to the present. About 9000 B.C. the 'Natufians', who had a mesolithic culture, apparently gathered and used wild grains, but the argument – based on the finding of pestles and mortars capable of grinding grain, and flint-edged 'reaping sticks' – that they were 'incipient agriculturalists' seems to overweight the evidence as the 'reaping knives' are not very suitable for collecting ripe grain from brittle-eared plants, and could have had other uses. Harvesting is best carried out by sweeping into baskets. However it is possible to harvest unripe *spontaneum* with sickles [44]. Agriculture, the deliberate planting and harvesting of grain, was probably beginning about 8000 B.C. since by 7000 B.C. small farming villages were in existence throughout the 'fertile crescent' from S.W. Iran, through the Taurus and Zagros mountains of Iraq, Turkey, Anatolia and Palestine [47]. At the earliest sites, e.g. Jarmo, only two-rowed barley was cultivated and this was distinguishable from wild *spontaneum* only by the possession of a tough rachis, indicating that it had to be harvested and resown by man [48]. Incidentally, while the presence of a tough rachis proves that agriculture was present the absence of a tough rachis does *not* show the absence of agriculture – it is possible to harvest the fragile-eared form. At later sites occasional samples of naked and/or six-rowed barleys are found. The farmers no doubt selected and propagated these new forms. Gradually mainly six-rowed barleys began to be grown in some areas, but not others.

Finds from Egypt of about 5000 B.C. show the presence of a mixture of two-rowed *distichon* and *deficiens* forms, *irregulare*, and six-rowed lax and dense-eared, husked and naked forms. From there they were presumably carried south to Abyssinia and Eritrea. Subsequently nearly all the barley grown in Egypt was of a six-rowed, lax-eared, husked type. In Mesopotamia, by 5000 B.C., six-rowed, lax-eared, husked barley was being grown and by 2000 B.C. this comprised some 90% of the cereal crop [45]. Possibly the preponderance of six-rowed barleys in the civilizations of the irrigated river valleys, the Tigris–Euphrates and the Nile, was associated with a naturally high state of adaptation and, in the case of Mesopotamia, increasing soil salination. Six-rowed barley, which was being grown in the Indus valley, at Mohenjo-daro and Harappa, by about 2500 B.C., was almost certainly introduced from Mesopotamia. The crop reached China about 2000–1300 B.C. From China it spread to Japan. No two-rowed barleys were grown in India until 1874, or Japan until 1868 [77]. Alexander the Great found barley being grown from Greece to Hindustan [95].

By 3000 B.C. the Swiss lake dwellers were growing husked and naked six-rowed barleys, and these had spread to N. Europe by about 2500 B.C. Two-rowed barley was not introduced to N. Europe until Mediaeval times – perhaps from some middle-Eastern site discovered by the Crusaders. The last, well documented spread of barley to be mentioned is of N. European two-rowed types with the British and N. European settlers to the Eastern Coast of N. America and the six-rowed Spanish and

The origin and classification of barleys

N. African types into California, Mexico and S. America (Chapter 12) [95]. Since that time different genotypes have been collected from most areas, and have been dispersed and tested at sites all round the world.

Before the generalized onset of exchanging barleys it was possible to detect groups of types ('species') characteristic of particular areas [43,66, 67,81,82,89,90,92,95]. Vavilov suggested that the areas where most variants occurred were the 'centres of origin' for a species. This idea seems wrong, but the existence of 'gene centres' or 'centres of diversity' raises the question by what evolutionary mechanism they arise, and provides sources of gene variants for breeding programmes (Chapter 12). Two barley 'centres of diversity' have attracted particular interest: (1) the area including Abyssinia and Eritrea; and (2) the far East, from Nepal, through China to Japan. The highly mixed nature of Ethiopian barleys is staggering; two-rowed, *deficiens* and *irregulare* forms, six-rowed husked and naked types, grains having a range of colours including black, and sometimes very large grains, broad leaves, smoothed awns, open flowering habits and high levels of disease resistance are present. The 'mystery' of their origin has disappeared with the finding that they could have come via Egypt where many of these forms existed about 5000 B.C. Their survival may be explained by the wide range of climatic habitats they occupy, their development in an isolated area under primitive agricultural conditions and tribal preferences – e.g. the Galla tribesmen cultivate *deficiens* forms while the Amharic tribes favour *irregulare* types [43].

Again the Far Eastern centre, with its six-rowed varieties, hooded forms, with naked grains and *intermedium* forms (having smaller awns or other appendages and grains in the lateral florets) is readily understood. An early introduction of six-rowed material from e.g. Mesopotamia, subsequent genetic isolation, and a gradual spread into a range of environments from the Indus valley into the high Himalayas, the Tibetan plateau and into China and Japan with stringent selection by the climate and cultivators at each location, combined with the use of much of the grain for human food could have given rise to the present situation. It has been argued that the genes, e.g. for tough rachis, that are found in the far Eastern region are in some way unique. The frequency of the two recognized genes for tough rachis varies with location [82]. This may merely reflect the genetic nature of the stock originally introduced into a region, and not necessarily that these barleys developed from a separate wild ancestor to other forms.

References

1 Åberg, E. (1940). *Symb. bot. upsal.*, **4**, 1–150.
2 Åberg, E. (1948). *Ann. R. agric. Coll. Sweden*, **15**, 235–250.
3 Åberg, E. (1957). *Ann. R. agric. Coll. Sweden*, **23**, 315–322.
4 Åberg, E. and Wiebe, G.A. (1945). *J. Wash. Acad. Sci.*, **35**, 161–164.

5 Al-Aish, M. and Brown, W.V. (1958). *Am. J. Bot.*, **45**, 16–23.
6 Allard, R.W., Kahler, A.L. and Weir, B.S. (1970). In *Barley Genetics II*, (ed. Nilan, R.A.) pp. 1–13. Washington State University Press.
7 Allison, M.J. and Swanston, J.S. (1974). *J. Inst. Brew.*, **80**, 285–291.
8 Atterberg, A. (1899). *J. Landw.*, **47**, 1–44.
9 Aufhammer, G., Bergal, P., Hagberg, A., Horne, F.R. and Van Veldhuizen, H. (1968). *Barley Varieties EBC.* London: Elsevier.
10 Avdulov, N.P. (1931). *Bull. appl. Bot. (Suppl. 44)* 1–428.
11 Bakhteyev, F. Kh. (1964). In *Barley Genetics I, (Proc. 1st Internatn. Barley Genetics Symp.)*, Wageningen, 1963, pp. 1–18.
12 Bakhteyev, F. Kh. (1970). In *Barley Genetics II*, (ed. Nilan, R.A.) pp. 36–44. Washington State University Press.
13 Bell, G.D.H. (1938). *Z. Zücht., (Reihe, A.)*, **22**, 81–146.
14 Bender, M.M. (1971). *Phytochemistry*, **10**, 1239–1244.
15 Bendich, A.J. and McCarthy, B.J. (1970). *Genetics, Princeton*, **65**, 545–565.
16 Bergal, P. (1948). *Annls. Sci. nat. (Bot.) (11th Sér.)*, **9**, 189–269.
17 Bergal, P. and Friedberg, L. (1940). *Ann. Epiphyt.*, **6**, 157–306.
18 Black, C.C. and Mollenhauer, H.H. (1971). *Pl. Physiol. Lancaster*, **47**, 15–23.
19 Bowden, B.N. (1965). *Outlook Agric.*, **4**, 243–253.
20 Bowden, W.M. (1959). *Can. J. Bot.*, **37**, 657–684.
21 Breitenfeld, C. (1957). *Z. PflZücht.*, **38**, 275–312.
22 Brown, W.V. (1958). *Bot. Gaz.*, **119**, 170–178.
23 Brown, W.V. and Emery, W.H.P. (1957). *Am. J. Bot.*, **44**, 585–590.
24 Brown, W.V., Harris, W.F. and Graham, J.D. (1959). *Southwest. Nat.*, **4**, 115–125.
25 Brown, W.V., Heimsch, C. and Emery, W.H.P. (1957). *Am. J. Bot.*, **44**, 590–595.
26 Brown, W.V., Pratt, G.A. and Mobley, H.M. (1959). *Southwest. Nat.* **4**, 126–130.
27 Byott, G.S. (1976). *New Phytol.*, **76**, 295–299.
28 De Candolle, (1886). *Origin of Cultivated Plants* (2nd edition) reprinted 1967, London: Hafner.
29 Carson, G.P. and Horne, F.R. (1962). In *Barley and Malt*, (ed. Cook, A.H.) pp. 101–159. London: Academic Press.
30 Clarke, H.H. (1967). *Agric. Hist. Rev.*, **15**, 1–18.
31 De Cugnac, A. (1931). *Ann. Sci. nat. (Bot.) (10th Sér.)*, **13**, 1–129.
32 Daussant, J. and Grabar, P. (1966). *Annls. Inst. Pasteur (Suppl.)*, **110**, 79–83.
33 Decker, H.F. (1964). *Am. J. Bot.*, **51**, 453–463.
34 Dimbleby, G.W. (1967). *Plants and Archaeology.* London: John Baker.
35 Everson, R.G. and Slack, C.R. (1968). *Phytochemistry*, **7**, 581–584.
36 Frederick, S.E. and Newcomb, E.H. (1971). *Planta*, **96**, 152–174.
37 Fröst, S., Holm, G. and Asker, S. (1975). *Hereditas*, **79**, 133–142.
38 Frydenberg, O. and Nielsen, G. (1965). *Hereditas*, **54**, 123–139.
39 Grillot, G. (1959). *Annls. amélior. Plantes.*, **9**, 445–552.
40 Gutierrez, M., Gracen, V.E. and Edwards, G.E. (1974). *Planta*, **119**, 279–300.
41 Harberd, D.J. (1972). *Ann. Bot.*, **36**, 599–603.
42 Harlan, H.V. (1918). *U.S. Dept. Agric. Tech. Bull. No. 622.The Identification of the Varieties of Barley.*
43 Harlan, H.V. (1957). *One Man's Life with Barley – The Memories and Observations of Harry V. Harlan*, ed. Harlan, J.R., New York: Exposition Press.

44 Harlan, J.R. (1968). *U.S. Dept. agric., agric. Handbook No. 338, Barley*, pp. 9–31.
45 Harlan, J.R. (1970). In *Barley Genetics II*, (ed. Nilan, R.A.) pp. 45–50. Washington State University Press.
46 Harlan, H.V. and Hayes, H.K. (1920). *J. agric. Res.*, **19**, 575–591.
47 Harlan, J.R. and Zohary, D. (1966). *Science*, **153**, 1074–1080.
48 Helbæk, H. (1966). *Econ. Bot.*, **20**, 350–360.
49 Hilliard, J.H., Gracen, V.E. and West, S.H. (1971). *Planta*, **97**, 93–105.
50 Hoshikawa, K. (1969). *Bot. Gaz.*, **130**, 192–203.
51 Hubbard, C.E. (1948). In *British flowering Plants*, (ed. Hutchinson, J.) London: P.R. Gawthorn.
52 Huber, S. and Edwards, G. (1975). *Biochem. biophys. Res. Commun.*, **67**, 28–34.
53 Kling, H. (1971). *Hoppe-Seyler's Z. physiol. Chem.*, **352**, 1037–1038.
54 Körnicke, F. (1882). *Z. ges. Brauw.*, **5**, 113–129; 305–311; 393–413.
55 Körnicke, F. (1908). *Arch. Biontol. (Berlin)*, **2**, 393–437.
56 Körnicke, F. and Werner, H. (1885). *Handbuch des Getreidebaues. 2 vols.* Bonn: Verlag Emil Strauss.
57 Krenzer, E.G., Moss, D.N. and Crookston, R.K. (1975). *Pl. Physiol., Lancaster*, **56**, 194–206.
58 Kruse, A. (1974). *Hereditas*, **77**, 219–224; 291–294.
59 La Berge, D.E. (1975). *Can. J. Pl. Sci.*, **55**, 661–666.
60 Lowe, J. (1961). *New Phytol.*, **60**, 355–387.
61 Mansfeld, R. (1950). *Züchter*, **20**, 8–24.
62 Morrison, J.W. (1959). *Can. J. Bot.*, **37**, 527–538.
63 Münzer, W. and Fischbeck, G. (1973). *Brauwissenschaft*, **26**, 14–19.
64 Nakao, S. (1958). *Plant Breeding Abst.*, **28**, 3596.
65 Nummi, M., Daussant, J., Niku-Paavola, M.L., Kalsta, H. and Enari, T-M. (1970). *J. Sci. Fd. Agric.*, **21**, 258–260.
66 Orlov, A.A. (1929). *Bull. appl. Bot., Genet. Pl. Breeding (Leningrad)*, **20**, 238–342 (343–345, Engl. Summary).
67 Orlov, A.A. (1931). *Bull. appl. Bot., Genet. Pl. Breeding (Leningrad)*, **27** 329–381 (371–381, Engl. Summary).
68 Orlov, A.A. (1936). *Flora of Cultivated Plants, II, Cereals*. Moskau.-Leningrad, pp. 98–332.
69 Prat, H. (1936). *Annls. Sci. nat. (Bot.) (10e sér.)*, **18**, 165–258.
70 Prat, H. (1960). *Bull. Soc. bot. Fr.*, **107**, 32–79.
71 Przybylska, J., Zimniak-Przybylska, Z. and Dabrowska, T. (1973). *Genet. Pol.*, **14**, 61–69.
72 Reeder, J.R. (1957). *Am. J. Bot.*, **44**, 756–768.
73 Renfrew, J.M. (1969). In *The Domestication and Exploitation of Plants and Animals* (ed. Ucko, P.J. and Dimbleby, C.W.) pp. 149–172. London: Duckworth.
74 Row, H.C. and Reeder, J.R. (1957). *Am. J. Bot.*, **44**, 596–601.
75 Sarkissian, I.V. and Shah, S.S. (1961). *Genetics*, **46**, 895–896.
76 Schiemann, E. (1948). *Weizen, Roggen, Gerste*. Fischer, Jena, 71–94.
77 Schiemann, E. (1951). *Heredity*, **5**, 305–320.
78 Smith, P. (1969). *Ann. Bot.*, **33**, 591–613.
79 Staudt, G. (1961). *Econ. Bot.*, **15**, 205–212.
80 Stebbins, G.L. (1956). *Am. J. Bot.*, **43**, 890–905.
81 Takahashi, R. (1955). *Adv. Genet.*, **7**, 227–266.
82 Takahashi, R. (1964). In *Barley Genetics I. (Proc. 1st. Internat. Barley Genetics Symp.)*, Wageningen, 1963, 19–26.

83 Takahashi, R. and Tomihisa, Y. (1970). In *Barley Genetics II*, (ed. Nilan, R.A.) pp. 51–62. Washington State University Press.
84 Tateoka, K. (1962). *Bot. Mag., Tokyo*, **75**, 377–383.
85 Thompson, W.P. and Johnston, D. (1945). *Can. J. Res.*, **23C**, 1–15.
86 Tovia, T. and Zohary, D. (1962). *Bull. Res. Council Israel*, **11D**, 43–45.
87 Treharne, K.J. and Cooper, J.P. (1969). *J. exp. Bot.* **20**, 170–175.
88 Triolo, L., Bagnara, D., Anselmi, L. and Bassanelli, C. (1974). *Physiologia Pl.*, **31**, 86–89.
89 Vavilov, N.I. (1926). *Studies on the Origin of Cultivated Plants.* pp. 139–245. Leningrad: Inst. of appl. Bot and Improvement of Plants. (English Transl.).
90 Vavilov, N.I. (1951). *Origin, Variation, Immunity and Breeding of Cultivated Plants. Chronica Botanica*, **13**, (No. 1–6).
91 Vavilov, N.I. (1957). *World resources of Cereals, Legumes, Seed crops and Flax, and their utility in Plant Breeding* (Translated 1960), Moscow: Academy of Sciences.
92 Ward, D.J. (1962). *U.S. Dept. Agric. Tech. Bull. No. 1276.*
93 Watson, L. and Creaser, E.H. (1975). *Phytochemistry*, **14**, 1211–1217.
94 Watson, LL. and Knox, R.B. (1976). *Ann. Bot.*, **40**, 399–408.
95 Weaver, J.C. (1950). *American Barley Production.* Minneapolis: Burgess.
96 Wiebe, G.A. and Reid, D.A. (1961). *U.S. Dept. Agric. Tech. Bull. No. 1224.*
97 Ziegler, H., Osmond, C.B. Stickler, W. and Trimborn, P. (1976). *Planta*, **128**, 85–92.
98 Zohary, D. (1960). *Bull. Res. Council Israel*, **9D**, 21–42.
99 Zohary, D. (1970). In *Barley Genetics II*, (ed. Nilan, R.A.) pp. 63–64. Washington State University Press.

Chapter 4
The biochemistry of barley

4.1. Introduction

In this chapter some of the biochemical interconversions that take place in barley are outlined; other aspects are treated elsewhere (e.g. Chapters 5,6, and 14–16). Where data are lacking it has been assumed that 'common' metabolic pathways are operating. There is a deficiency of direct evidence relating to compartmentation of functions between subcellular organelles and the nature of the control mechanisms operating on metabolic sequences in barley. It should be noted that not all biochemical pathways that can operate in barley do so in each tissue, or all of the time. The metabolism of many tissues alters with age.

4.2. Carbohydrates

The monosaccharides detected in barley, free or in combination, are: (1) arabinose, (2) xylose, (3) fucose, (4) ribose, (5) deoxyribose, (6) glucose, (7) fructose, (8) galactose and (9) mannose. The closely related compounds glucuronic acid (10), and galacturonic acid (11) and glucosamine (12) also occur, as do ascorbic acid (13), and its oxidation product dehydroascorbic acid (14), *myo*-inositol (15) (mainly as the hexaphosphate, phytic acid, 16), and a host of intermediary metabolites. Glucose and fructose occur free and in combination. The other monosaccharides occur mainly in various

(1) Arabinose;
α-L-arabinofuranose

(2) Xylose; β-D-xylopyranose

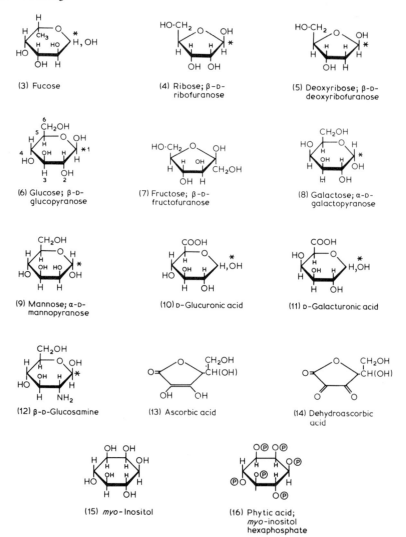

(3) Fucose

(4) Ribose; β-D-ribofuranose

(5) Deoxyribose; β-D-deoxyribofuranose

(6) Glucose; β-D-glucopyranose

(7) Fructose; β-D-fructofuranose

(8) Galactose; α-D-galactopyranose

(9) Mannose; α-D-mannopyranose

(10) D-Glucuronic acid

(11) D-Galacturonic acid

(12) β-D-Glucosamine

(13) Ascorbic acid

(14) Dehydroascorbic acid

(15) *myo*-Inositol

(16) Phytic acid; *myo*-inositol hexaphosphate

combinations in oligosaccharides, polysaccharides, glycosides, glycolipids and glycoproteins. Some of the polysaccharides are linked to phenolic acids, others are probably linked to lignin. The unfamiliar sugar sulphoquinovose occurs combined in a chloroplast lipid. Ribose and deoxyribose occur in nucleosides, nucleotides and nucleic acids. The routes by which glucose and fructose are interconverted are well known (Fig. 4.1). In barley seedlings interconversions and the randomization of the carbon atoms in the hexose molecules largely involve breakdown to, and resynthesis from, the triose-phosphates by way of some steps in the glycolytic sequence. The hexose monophosphate shunt probably operates,

The biochemistry of barley 91

but to a more limited extent [308]. In the case of many other sugars the methods of interconversion are not well characterized, and several routes may operate. Thus various pentoses and other sugars could arise through transketolase and transaldolase acting on other phosphorylated sugars provided by the glycolytic and hexose monophosphate pathways (Fig. 4.2). Alternatively, nucleotide sugar derivatives or related compounds such as UDP-glucuronic acid may be the metabolic intermediates. Glucose and glucuronolactone are the best precursors of xylan in wheat. Xylose and arabinose are used less well. Xylose is converted to hexose phosphates, and is not used directly for xylan synthesis. Arabinose is used more efficiently than xylose. Barley shoots converted glucose, fructose, mannose, galactose, maltose (27) and lactose to sucrose, but did not convert arabinose, xylose, mannitol, sorbitol, gluconic acid or pyruvic acid to the disaccharide [216].

Sucrose (17) is the major sugar of the living tissues in barley. It is an early major product of photosynthesis; in the grain both the aleurone layer and the scutellum convert other sugars to it, and in response to gibberellins glycerides are converted to it in the aleurone layer [57,81, 131,261]. Sucrose is hydrolysed by the enzyme(s) 'invertase' which occurs in solution and in an insoluble form. The sugar is synthesized by uridine diphosphate glucose (18; UDPG) donating glucose to either fructose or fructose-6-phosphate to give either uridine diphosphate (UDP) and sucrose, (17) or sucrose phosphate respectively. The latter is hydrolysed to sucrose and inorganic phosphate [81]. The trisaccharide raffinose (19) occurs in the quiescent embryo, but disappears in the early stages of germination. Its metabolic fate is unknown, as is its mode of synthesis in barley [212].

(17) Sucrose; α-D-Glucopyranosyl-(1→2)
-β-D-fructofuranoside

(18) Uridine diphosphate glucose; UDPG

(19) Raffinose; α-D-Galactopyranosyl-(1→6)-α-D-glucopyranosyl-(1→2)-β-D-fructofuranoside

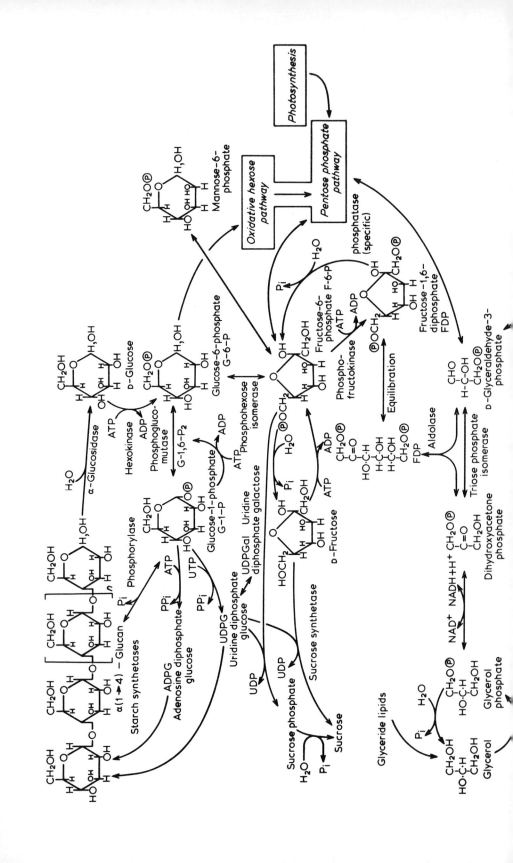

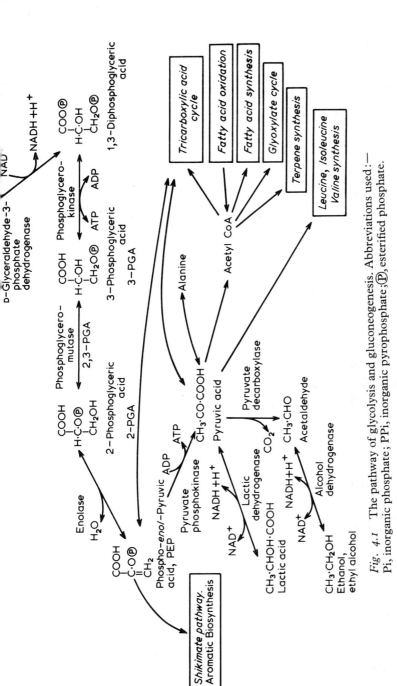

Fig. 4.1 The pathway of glycolysis and gluconeogenesis. Abbreviations used:—
Pi, inorganic phosphate; PPi, inorganic pyrophosphate; ⓟ, esterified phosphate.

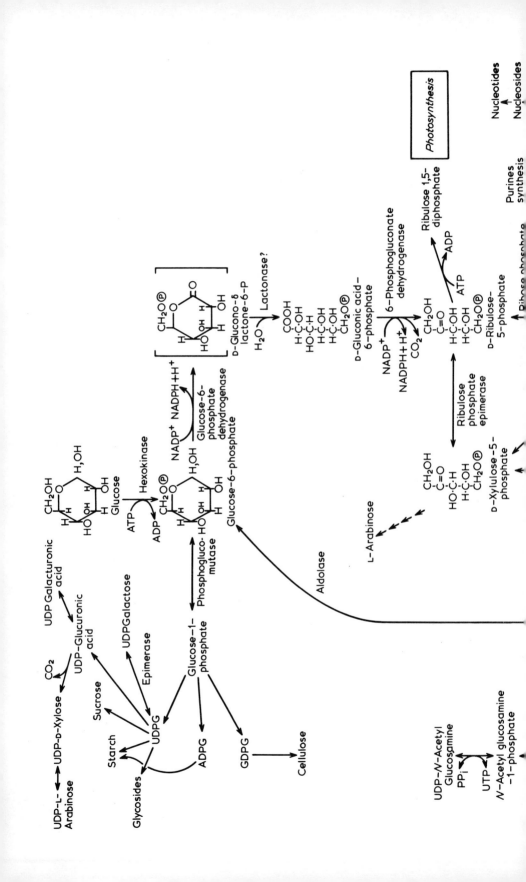

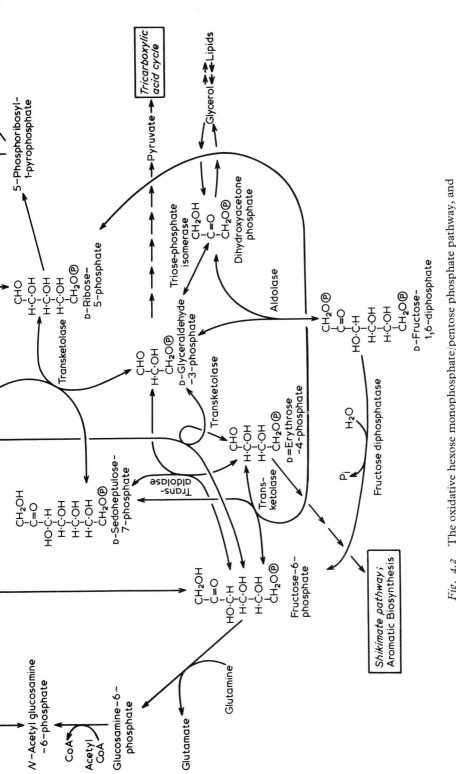

Fig. 4.2 The oxidative hexose monophosphate/pentose phosphate pathway, and some related metabolic interconversions.

96 Barley

Ascorbic acid (13) appears in germinating grains where it occurs in the oxidized and reduced states [334]. Neither its mode of synthesis nor its metabolic fate is known in barley. It has attracted interest both as a vitamin, and, because it may be reversibly oxidized to dehydroascorbic acid (14), it may act as a hydrogen carrier and participate in respiration. Ascorbic acid oxidase is a copper-containing enzyme which is largely associated with the cell walls, at least in roots and is about 50% available to the liquid around the roots [129].

Myo-inositol (15) is synthesized directly from glucose in several plants, via glucose-1-phosphate and *myo*-inositol-1-phosphate. Further phosphorylation can give rise to phytic acid (16). Inositol is degraded by an oxygenase, with the production of glucuronic acid. One metabolic sequence proposed for barley is *myo*-inositol ⟶ glucuronic acid ⟶ glucuronic acid-1-phosphate ⟶ uridine diphosphate glucuronic acid ⟶ uridine diphosphate galacturonic acid [293].

Fructosans are major storage oligosaccharides in the stems and leaves. They also occur in grain. Each fructosan chain is terminated with a glucose residue – i.e. the terminal disaccharide is sucrose. The simplest fructosan trisaccharides are kestose (20) and iso-kestose (21) in which the inter-fructose links are β(2 ⟶ 6) and β(2 ⟶ 1) respectively. The tetra-saccharide bifurcose (22), containing both links, is also present as are two series of higher oligosaccharides in which the linkages are predomi-

(20) Kestose (6-kestose); α-D-Glucopyranosyl-(1→2)-β-D-fructofuranosyl-(6→2)-β-D-fructofuranoside

(21) *Iso*-Kestose, (1-kestose); α-D-Glucopyranosyl-(1→2)-β-D-fructofuranosyl-(1→2)-β-D-fructofuranoside

(22) Bifurcose, the basic unit of two series of fructosans. Extended in direction A, with β-(2→1) linked fructofuranose units kritesin (inulin type) fructosans; extended in direction B with β-(2→6) linked fructofuranose units – hordeacin (phlein type)

The biochemistry of barley

nantly β(2⟶6) (hordeacin, phlein type) and β(2⟶1) (kritesin, inulin type). The series extend to members with molecular weights greater than 8000 [302]. These fructosans are hydrolysed by invertase. It is postulated that they are synthesized from sucrose by transfructosylation [131,301].

Pectin, pure methylated polygalacturonic acid, has not been demonstrated in barley, nor has cellulose, pure poly-β(1 ⟶ 6)-D-glucopyranose. 'Pectin' fractions, defined by the techniques used to prepare them, contain a range of hemicellulosic materials. Where these are readily soluble in hot water they are often called gums; the residue that can only be dissolved in alkali is termed hemicellulose. There is no chemical distinction between these classes. 'Holocellulose' remaining after delignification and the removal of gums and hemicelluloses contains monosaccharide residues other than glucose.

Gums from the endosperm contain glucose, arabinose, xylose, and traces of mannose and galactose. They, and the endosperm cell walls, probably contain esterified ferulic acid [184]. The husk materials contain, in addition, glucuronic and galacturonic acids [8]. Fractionation of endosperm gums with ammonium sulphate can yield a β-glucan and an arabinoxylan. β-Glucan has also been extracted from barley stems [46].

The β-glucan consists of β-D-glucopyranose residues linked (1 ⟶ 3) or (1 ⟶ 4) forming an unbranched chain. There are a few adjacent (1 ⟶ 3) linkages, but the molecule consists predominantly of cellotriose and cellotetraose units, (in which the glucan residues are joined by β(1 ⟶ 4) links) joined by (1 ⟶ 3) links. Another minor glucan component may be slightly branched [8,149,225]. The molecular weight of the β-glucans varies with the method of preparation, and can exceed 200 000 [75]. Still larger β-glucans form part of the hemicellulose fraction. Germinated barley contains enzymes which catalyse the hydrolysis of β-glucans with a rapid reduction of viscosity, indicating that the chains are attacked internally. The products include lower molecular weight β-glucan fragments, cellobiose (23), laminaribiose

(23) Cellobiose
β-D-Glucopyranosyl-(1⟶4)-D-glucopyranose

(24) Laminaribiose
β-D-Glucopyranosyl-(1⟶3)-D-glucopyranose

(25) Xylobiose,
β-D-Xylopyranosyl-(1⟶4)-D-xylopyranose

(24) and glucose (6). β-Glucosidases are present that can hydrolyse the disaccharides to glucose. The mixture of enzymes attacking the glucan contains at least two β-glucosidases, two *endo*-β-glucanases and a laminarinase [206,220,222,245]. The presence of the laminarinase is surprising since laminarin (poly-β(1 $\longrightarrow$ 3)-D-glucopyranose) is not known to occur in barley; possibly it attacks β-glucan to a limited extent [11]. Extracts of barley readily catalyse some transglucosylations. Thus cellobiose, as well as being hydrolysed to glucose, gives rise to oligosaccharides, including gentiobiose, laminaribiose (24), and cellotriose [5,207].

Two pentosan gums have been purified from barley grain. The simpler, isolated from the endosperm, consists of chains of β(1 $\longrightarrow$ 4)-D-xylopyranose units in which about two out of three residues are substituted by L-arabinofuranose residues attached either to the 2' or the 3' position on a xylose residue. The arabinose/xylose ratio varies between different fractions. A barley husk hemicellulose is similar, but also contains non-terminal arabinofuranose residues substituted with xylose, giving 2-O-β-D-xylopyranosyl-L-arabinofuranosyl branches; in addition there are glucuronic acid residues attached to the xylan 'backbone' chain [8,9,131]. A pentosan from barley leaves has a main chain of β(1 $\longrightarrow$ 4)-D-xylopyranosyl residues carrying L-arabinofuranosyl, 4-O-methyl-D-glucopyranosyl, and galactopyranosyl (1 $\longrightarrow$ 4)-xylopyranosyl(1 $\longrightarrow$ 2)-L-arabinofuranosyl substituents [45]. *In vitro* enzymes from germinated grain degrade the backbone chain to xylose (2), xylobiose (25) and higher oligosaccharides. Besides an arabinosidase, inhibitor studies suggest that an *endo*-xylanase, a xylobiase, and possibly an *exo*-xylanase are also present [131].

Germinated barley grains contain hydrolytic enzymes splitting many di- and oligo-saccharides and synthetic substrates such as nitrophenyl glycosides, carboxymethylcellulose and carboxymethylpachyman. A mannosidase has been highly purified from this source [131,147].

The routes by which the gums, hemicelluloses and holocellulose fractions are synthesized are unknown. Many components of these fractions, and of the root slime, remain to be fully characterized. Perhaps nucleotide diphosphate sugar derivatives act as 'high energy' sugar donors acting in sequence with lipid–phosphate–carbohydrate derivatives which serve to transport 'activated' sugars across the bounding cell membranes to the wall. Various nucleotides have been found in barley including UDPG (18) and UDP-*N*-acetylglucosamine [23].

Starch is the major component of the grains. However in green tissues starch only occurs, deposited in the chloroplasts, when conditions are exceptionally favourable for photosynthesis or when sugars are supplied from without. The starch granules in the endosperm are laid down within amyloplasts [80]. Starch grains separated from mature kernels are not pure polysaccharide but contain traces of lipids, minerals, proteins, and nucleotides. The granules fall into two size groups, with diameters mainly in the ranges 1.7–2.5 μm and 22.5–47.5 μm (Chapter 6) [105,106]. The starch polysaccharide is entirely α-glucan and may be divided into the predominantly straight-chain amylose containing D-glucopyranose units linked α(1 ⟶ 4), which may contain as many as 2000 glucose residues, and the amylopectin fraction in which the α(1 ⟶ 4) linked D-glucopyranose chains are branched through α(1 ⟶ 6) linkages [8,131]. The average branch chain lengths are often 24–26 glucose units. Molecular weights may be as high as 10^6–10^8. Both amylose and amylopectin occur in a range of sizes. Amylose contains a proportion of 'anomalous' links, probably α(1 ⟶ 6), which prevent total degradation by β-amylase. Leaching amylose from starch grains at progressively higher temperatures extracts material with differing properties containing more 'anomalous' links [12]. A third minor fraction, distinct from amylose or amylopectin, has also been detected [13]. Starch from most British barleys contains 22–26% amylose, the rest being amylopectin [8,131]. However samples of 11–26% amylose are known and starch from 'waxy' barley contains only 0–3% amylose, while high-amylose starches contain up to 45% [13,105,106]. Amylose and amylopectin are not synthesized simultaneously (Chapter 6). Small starch grains have a higher gelatinization temperature and amylose content (41.3%) than larger grains (26.9%) [14,193]. High-amylose barleys have starch grains of intermediate diameter, 10–15 μm, weighing, for example, 6.3×10^{-10} g each, compared to normal small grains which weigh 2.1–2.3×10^{-11} g, or large grains which weigh 1.3–1.9×10^{-9} g [105,106,109]. It now seems feasible to prepare a wide range of starch types from barley on an industrial scale [71,108]. An unusual form retains sufficient α-amylase which is normally absent from the grain, to be self-liquefying [104]. Fatty acids and other 'impurities' associated with the starch granules markedly alter their cooking properties [107].

Starch is synthesized in plastids, but the route is uncertain. Phosphorylase occurs in barley and catalyses the reversible reaction:

Glucose-1-phosphate + starch chain ⇌ glucose-α(1 ⟶ 4) starch + inorganic phosphate (Pi).

(26) Adenosine diphosphate glucose, ADPG

The equilibrium is unfavourable for starch synthesis. At least two isozymes occur [17]. More probably glucose is added to the starch chains by irreversible reactions using as donors uridine diphosphate glucose (18; UDPG) or adenine diphosphate glucose (26; ADPG). Enzymes catalysing glucose transfer from both of these donors occur in developing grains [16,18]. The sugar and uronic acid nucleotides are formed according to the general scheme:

$$\text{sugar-1-P} + \text{UTP (ATP)} \rightleftharpoons \text{UDP sugar (ADP sugar)} + P P i$$
(nucleotide triphosphates) \hspace{2cm} (pyrophosphate)

The activity of ADP glucose pyrophosphorylase from barley is altered by 'effector' substances, especially 3-phosphoglyceric acid [300]. Indications of how the $\alpha(1 \longrightarrow 6)$ linkages are made or how the amylose/amylopectin ratio is controlled are lacking.

Microscopic evidence indicates that starch grains in the embryo are degraded differently to those in the endosperm (Chapter 1). The method of attack in the embryo is not understood, but the participation of phosphorylase would give rise to G-1-P that entered directly into the glycolytic sequence of reactions. In the endosperm degradation is certainly mainly by hydrolysis of the starch chains, although the participation of phosphorylase is not excluded. Analyses of starch from barley and malted barley indicate that during germination amylose and amylopectin are both partly degraded – probably by β-amylase attacking the chain ends exposed by limited α-amylolysis [122,131]. Kirchoff in 1815 and 1816 [175,176], recorded that in warm water gluten from germinated cereals converted starch to sugars. Despite continuous studies since then the component enzymes of 'diastase' are not adequately characterized. The enzymes believed important are phosphorylase, α-glucosidase ('maltase'), α-amylase, β-amylase, debranching enzyme(s) (possibly two-limit dextrinase and R-enzyme) and transglucosylase(s) [42].

Phosphorylase has been detected in mature grains, but reliable quantification is difficult [284]. β-Amylase occurs as several isozymes [99,186]. It is a thiol-dependent enzyme that associates with other protein molecules via disulphide linkages. It is converted to its monomer, and activated, by agents that reduce these links. During germination the β-amylase monomer appears to be reduced in size [115]. The different forms are

(27) Maltose,
α-D-Glucopyranosyl-(1→4)-D-glucopyranose

The biochemistry of barley 101

immunologically identical [68,270]. The disulphide-bound enzyme occurs in a range of forms of varying solubility, that have been extensively studied [131,264,271,309]. β-Amylase splits alternate α(1 ⟶ 4) inter-glucose linkages with the liberation of maltose (27) from the non-reducing chain ends. Since it cannot bypass α(1 ⟶ 6) linkages amylopectin is converted by the pure enzyme to maltose and a β-limit dextrin in which the outer starch chains are 'pruned' to within two to three glucose residues of the branch-points. Some crystalline preparations of barley β-amylase have been obtained.

α-Amylase, in contrast to β-amylase, is comparatively stable; the activity of one form at least depends upon the presence of calcium ions. The enzyme is able to degrade starch granules. It attacks dispersed starch, in solution, by breaking the α(1 ⟶ 4) linkages at random along the chain. However, links nearer the ends of chains or branch points are less readily attacked [131,305]. Products of starch degradation by α-amylase include glucose (6), maltose (27), maltotriose (28), maltotetraose

(28) Maltotriose, ($n=1$); Maltotetraose, ($n=2$); Maltopentaose, ($n=3$)

(29) *Iso*-Maltose, α-D-Glucopyranosyl-(1⟶6)-D-glucopyranose

(28) and higher straight-chain and branched oligosaccharides. The enzyme occurs as a range of isozymes which vary in their properties, their stability towards acid, their calcium ion requirement, and so on. Indeed two immunologically distinct groups of isozymes appear to exist, and some may be glycoproteins [35,99,156,210,241,339]. Crystals of this enzyme (or one of them) have been prepared. Recently a 'new' amylase was reported [263]; this may be the enzyme that occurs in immature barley [209]. α-Glucosidase (maltase) is obtained in solution with difficulty [162] and more than one form may exist [226]. The enzyme attacks many α-glucosides, including *iso*-maltose (29), and is able to degrade starch at least partially. Debranching enzymes break the α(1 ⟶ 6)-interchain links in amylopectin and smaller dextrins. R-enzyme and limit-dextrinase activity were apparently separated by various techniques

but the enzymes have not been obtained pure and the existence of two enzymes differing in specificity is now in doubt [213,221].

Barley extracts contain enzymes that catalyse transglucosylations, e.g. between maltose (27) molecules, to yield glucose (6) and maltotriose (29). However these may be 'side-reactions' catalysed by hydrolytic enzymes under particular test conditions.

4.3. The glycolytic sequence, the pentose phosphate shunt and the tricarboxylic acid cycle.

Radiochemical and other studies show that these metabolic pathways operate in barley. However, it is not certain whether they operate in all tissues, or what their relative metabolic contributions may be. Under anaerobic conditions in grains hexose is degraded mainly to ethanol and carbon dioxide by the glycolytic sequence of enzyme-catalysed reactions (Fig. 4.1). Most of the enzymes and intermediates of this sequence have been detected in barley [20,81,180,283]. Phosphorylated hexoses are interconverted and fructose-1,6-diphosphate is cleaved to two triose phosphates, dihydroxyacetone phosphate and glyceraldehyde-3-phosphate, which may themselves be interconverted by an isomerase. The glyceraldehyde-3-phosphate is oxidized and NADH is formed. The NADH must be reoxidized to allow the reaction sequence to operate-this is achieved mainly by the reduction of acetaldehyde, the decarboxylation product of pyruvic acid, to ethanol, and to a small extent by the reduction of pyruvate to lactic acid. The enzyme that decarboxylates pyruvate requires the pyrophosphate of thiamine, a vitamin, as a cofactor. Under aerobic conditions NADH is oxidised less wastefully, and the pyruvate is metabolized by a different route. The glycolytic sequence provides a net synthesis of ATP by substrate-level phosphorylation; it provides metabolic intermediates; e.g. dihydroxyacetone phosphate is converted to various glycerol derivatives. Glycerol derived from lipid degradation is phosphorylated and metabolized by oxidation to dihydroxyacetone phosphate. Pyruvate can give rise to acetyl-CoA, alanine and other amino acids. When the sequence is reversed, the irreversible steps being by-passed, it provides the means for carbohydrate synthesis from simpler compounds (gluconeogenesis). It is believed that phosphofructokinase and hexokinase are 'pacemaker' enzymes for glycolysis and that fructose-1, 6-diphosphatase limits gluconeogenesis in barley [101,191]. The replacement of a nitrogen atmosphere by oxygen stops ethanol production in germinating grains, the Pasteur effect. The effect of agents that uncouple oxidative phosphorylation, such as 2,4-dinitrophenol, in allowing aerobic ethanol production shows that this 'Pasteur effect' is normal [230,268].

The biochemistry of barley 103

Glucose-6-phosphate dehydrogenase and 6-phosphogluconic acid dehydrogenase increase in amount in germinating grains [180]. The rest of the 'pentose phosphate pathway' is operating, as indicated by radiochemical studies (Fig. 4.2) [283,310]. This pathway provides a route by which pentoses may be formed and phosphorylated hexoses can be interconverted. Some of the reactions also participate in glycolysis. The reduced nucleotide formed by the initial oxidative steps is NADPH; this may be specifically utilized in a number of biosynthetic reactions.

The oxidative Kreb's tricarboxylic acid cycle is operative in barley as shown, for example, by the redistribution of radioactivity from labelled intermediates (Fig. 4.3) [20,180,283]. The 'glyoxylate by-pass' or cycle also functions, under some conditions at least, in the aleurone layer and possibly in the scutellum and elsewhere (Fig. 4.3) [116,261]. The tricarboxylic acid cycle is confined to mitochondria. The glyoxylate cycle probably occurs in glyoxysomes. Mitochondria capable of oxidative phosphorylation have been separated from barley tissues [233,307]. The interconversion of intermediates in the cycle results in the oxidation of one acetyl equivalent per turn, with the production of water, carbon dioxide, and reduced cofactor, NADH. The acetyl CoA entering the cycle may originate from pyruvate, supplied by the glycolytic sequence or by transamination of alanine, or from the β-oxidation of fatty acids. The cycle also provides carbon skeletons which lead to other biosynthetic sequences, and also receives carbon skeletons from other degradative systems. For example transamination allows the exchange of carbon skeletons between oxaloacetate and aspartate, or glutamic acid and α-oxoglutarate. The glyoxylate cycle assembles acetyl units (C-2) into dicarboxylic acids such as malate (C-4). These can be converted to phospho-*enol*-pyruvate, which may enter the gluconeogenic sequence. Thus acetate from lipids may be converted into carbohydrates.

Many organic acids occur in barley, including acetic, glycollic, fumaric, succinic, oxalic, malonic, malic, citric and isocitric, α-oxoglutaric, lactic, pyrrolidone carboxylic, ferulic, vanillic, *p*-hydroxybenzoic, and *p*-coumaric acids [58,84,146,158,354]. Under some conditions carbon dioxide is fixed non-photosynthetically, especially in roots, and malic acid accumulates (Chapter 6); apparently phospho-*enol*-pyruvate carboxykinase, phospho-*enol*-pyruvate carboxylase and malic enzyme are operative [116].

The reduced pyridine nucleotides produced in metabolism must be reoxidized. NADPH is probably used in reductive biosynthetic mechanisms or it may reduce NAD^+ in a controlled fashion. NAD^+ may also be reduced by a range of other metabolic reactions. NADH may be reoxidized in various reactions, or it may donate its hydrogen, or an electron with an accompanying proton, to the electron transport chain. Electrons from the oxidation of succinate join the chain later (Fig. 4.4).

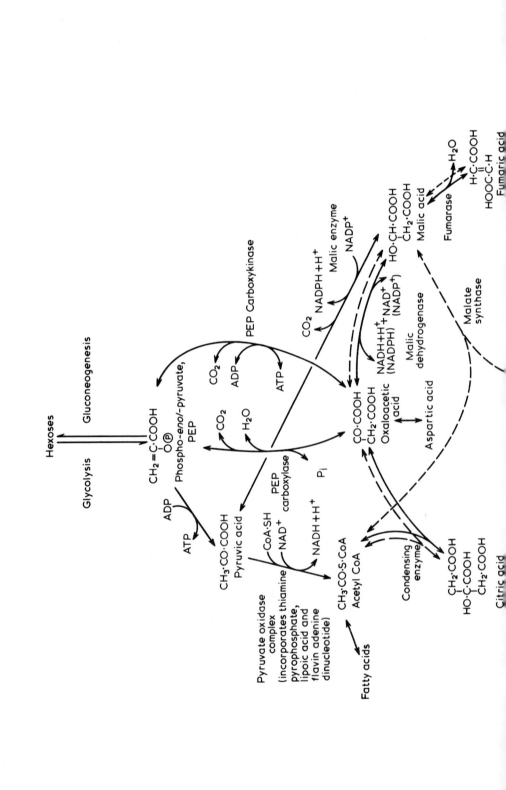

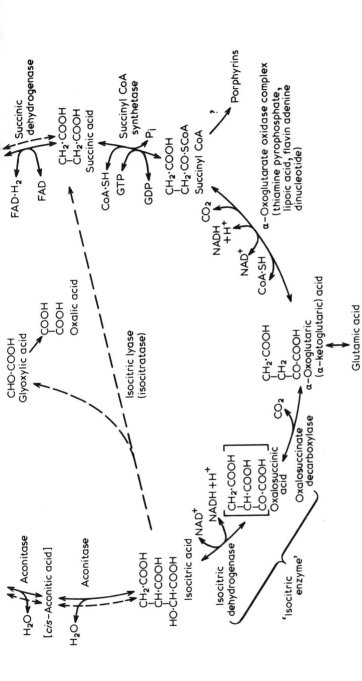

Fig. 4.3 The Kreb's tricarboxylic acid cycle ———, and the glyoxylate cycle, — — —

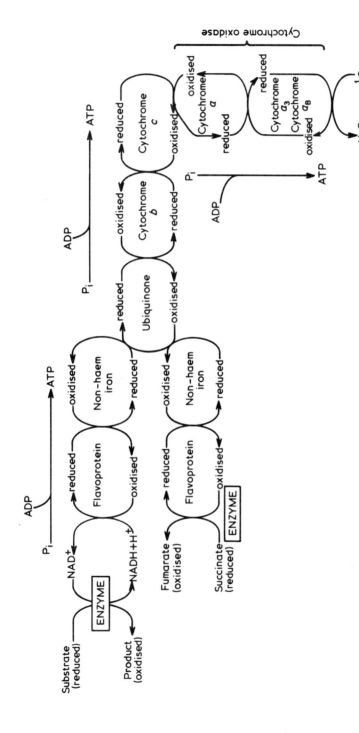

Fig. 4.4 A proposed form of the mitochondrial electron transport chain in plants, indicating which sections probably drive oxidative phosphorylation.

The biochemistry of barley 107

There are uncertainties about the ordering of the components of the electron transport chain, although the presence of the components shown seems certain [25,87,98,140]. As the electrons pass along the chain their potential falls; ultimately they link with oxygen to form water. The energy that becomes available is used to drive the synthesis of 'energy rich' ATP from ADP and Pi. The mechanism of this oxidative phosphorylation, whether it is carried out through the agency of 'energy rich' chemical intermediates or chemical–osmotic potential gradients across membranes, is unclear. However the 'energy rich' phosphate – whether created by oxidative, photosynthetic or substrate-level phosphorylation, as ATP or other nucleotide triphosphates synthesized at the expense of ATP – together with reduced pyridine nucleotides, drive the synthetic and other energy-requiring metabolic processes. Respiration linked to the glycolytic sequence is much more efficient at producing ATP than glycolysis functioning alone, and so is energetically favourable to the plant.

The electron transport chain terminating in cytochrome oxidase is not the only means by which the cells reduce oxygen and form water. Other oxidases occur, such as polyphenol oxidase, ascorbic acid oxidase, and glycollic acid oxidase. The occurrence of cyanide-resistant respiration in roots and other tissues occurs in parallel with oxidation via cytochrome oxidase [20,211]. Succinoxidase, cytochrome oxidase, and cytochrome spectra have been demonstrated in older barley roots with much cyanide-resistant respiration [98,140,159]. Cytochrome oxidase is present in respiring leaves [65]. NADPH-glutathione reductase occurs in barley. The reduced glutathione, or other H-donor, can reduce dehydroascorbic acid to ascorbate. Ascorbate is oxidized back to dehydroascorbate by oxygen in the presence of ascorbate oxidase. It has been proposed that ascorbate + ascorbate oxidase may be the terminal hydrogen donor involved in cyanide-resistant respiration. There are difficulties with this scheme, not least that while some of the enzyme is soluble, much of the ascorbate oxidase is located in the cell walls [20,129,141]. There is no other suggested purpose for ascorbate in barley.

4.4. Barley lipids

Lipids constitute only a small part of the dry matter in most barley tissues yet they do comprise significant reserves in the embryo and aleurone layer of the grain, they are essential for the functional integrity of the cells, and are important as cutin (testa) and cuticular and testa waxes. Consequent upon improved methods of analysis much new information has become available [22,89,131,138,203,204,278,355,369]. While many lipid classes, e.g. cerebrosides, have been detected in barley, individual components such as the poly-hydroxy nitrogenous bases

108 Barley

have often not been characterized, in contrast to the materials from wheat. Many 'unknown' compounds have been noted.

Barley cuticle cutin has not been studied, but the material from the testa contains an estolide that apparently contains fatty acids, dicarboxylic acids, mono-, di-, and tri-hydroxy monocarboxylic acids, hydroxy-dicarboxylic acids, and possibly epoxy-carboxylic acids [43]. The testa wax contains a homologous series of n-alkanes, extending from at least C-11 to C-36, in which the C-29 and C-31 components predominate. Two other homologous series of branched alkanes are present in minor amounts. Other components include esters of alkanols and sterols, free sterols, triglycerides, and traces of other components. The other main constituent is part of a homologous series of 5-n-alkyl resorcinols (30) in which the compounds with 25, 27, 29 and 31 carbon atoms together make up 98% of the total although members with even numbers of carbon atoms are also present [43]. The cuticular wax differs from the testa wax, and contains alkanes, primary alkanols, aliphatic aldehydes, fatty acids, aliphatic esters of primary and secondary alkanols, flavonols, ketones, β-diketones and hydroxy-β-diketones. There are differences in the composition of wax from different varieties, from the same variety grown under different conditions, and from different parts of the same plant. Thus in Bonus plants the C-31 hydrocarbon predominates on the spike, the C-25 hydrocarbon predominates on the internodes [346,359–363]. The C-25, 27, 29, 31 and 33 hydrocarbons usually predominate in the alkane series, 1-hexocosanol is the major aliphatic alcohol, and the major β-diketone is hentriacontan(C-31)-14, 16-dione (31) [154]. The hydroxy-β-diketones are variously reported to be 8- or 9- or 25-hydroxy-hentriacontan(C-31)-14,16-dione (32) [73,154,362]. Probably β-diketones are hydroxylated to form the hydroxy-β-diketones. The hydrocarbons may be formed by the decarboxylation of fatty acids. Other components may arise by the successive reduction of fatty acids to aldehydes and alcohols, and the esterification of the alcohols by fatty acids to form the esters [360–362].

$CH_3 \cdot (CH_2)_n \cdot CH_2$—[resorcinol ring with two OH groups]

(30) 5-n-Alkyl resorcinol
(mainly n = 17, 19, 21 and 23)

$CH_3 \cdot (CH_2)_{12} \cdot \underset{\underset{O}{\|}}{C} \cdot CH_2 \cdot \underset{\underset{O}{\|}}{C} \cdot (CH_2)_{14} \cdot CH_3$

(31) Hentriacontan−14, 16-dione,
(a β-diketone)

$CH_3 \cdot (CH_2)_{12} \cdot \underset{\underset{O}{\|}}{C} \cdot CH_2 \cdot \underset{\underset{O}{\|}}{C} \cdot (CH_2)_8 \cdot \underset{\underset{OH}{|}}{CH} \cdot (CH_2)_5 \cdot CH_3$

(32) 25-Hydroxy-hentriacontan−14,16-dione

The biochemistry of barley 109

Table 4.1 Total fatty acids (%, w/w) in grain of Ingrid barley. Total fatty acids, as methyl esters 1.72–2.49 g/100g dry grain (from Lindberg et al. [203,204]).

Fatty acid	Variety, Ingrid
14:0	0.3–0.5
16:0	21.4–24.0
16:1	0.1–0.3
18:0	0.4–0.9
18:1	10.0–13.7
18:2	56.0–58.3
18:3	5.3–7.1

The whole grain yields a mixture of fatty acids in which linoleic acid predominates (33–39; Table 4.1). On saponification different lipid fractions yield different mixtures of fatty acids. While free fatty acids do occur, in grain mono-, di-, and especially triglycerides predominate (40–43; Chapter 6). Both 1,2- and 1,3-diglycerides occur [278]. The glycerol of lipids is believed to be derived from dihydroxyacetone phosphate, an intermediate of the glycolytic sequence (Fig. 4.3). Similarly it is degraded after phosphorylation and oxidation to the same compound. Lipase, an enzyme catalysing triglyceride hydrolysis, occurs in barley tissues as do various esterases that degrade synthetic chromogenic substrates and may act on lipids [131].

Common fatty acids.

(33) $CH_3 \cdot (CH_2)_{12} \cdot COOH$ Myristic acid (14:0)
(34) $CH_3 \cdot (CH_2)_{14} \cdot COOH$ Palmitic acid (16:0)
(35) $CH_3 \cdot (CH_2)_5 \cdot CH = CH \cdot (CH_2)_7 \cdot COOH$ Palmitoleic acid (16:1)
(36) $CH_3 \cdot (CH_2)_{16} \cdot COOH$ Stearic acid (18:0)
(37) $CH_3 \cdot (CH_2)_7 \cdot CH = CH \cdot (CH_2)_7 \cdot COOH$ Oleic acid (18:1)
(38) $CH_3 \cdot (CH_2)_4 \cdot CH = CH \cdot CH_2 \cdot CH = CH \cdot (CH_2)_7 \cdot COOH$ Linoleic acid (18:2)
(39) $CH_3 \cdot CH_2 \cdot CH = CH \cdot CH_2 \cdot CH = CH \cdot CH_2 \cdot CH = CH \cdot (CH_2)_7 \cdot COOH$ Linolenic acid (18:3)

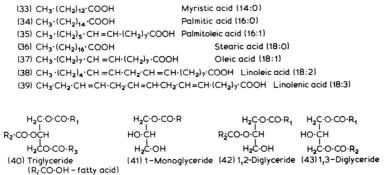

(40) Triglyceride ($R_i \cdot CO \cdot OH$ – fatty acid)
(41) 1-Monoglyceride
(42) 1,2-Diglyceride
(43) 1,3-Diglyceride

Fatty acid synthesis has been studied in barley chloroplasts (Fig. 4.5) [166–168]. The process follows the usual path; two carbon units are added sequentially, from malonyl CoA (C-3) which is decarboxylated, to fatty acids bound through thioester linkages to acyl carrier protein (ACP) molecules. The ACP contains bound phosphopantetheine and

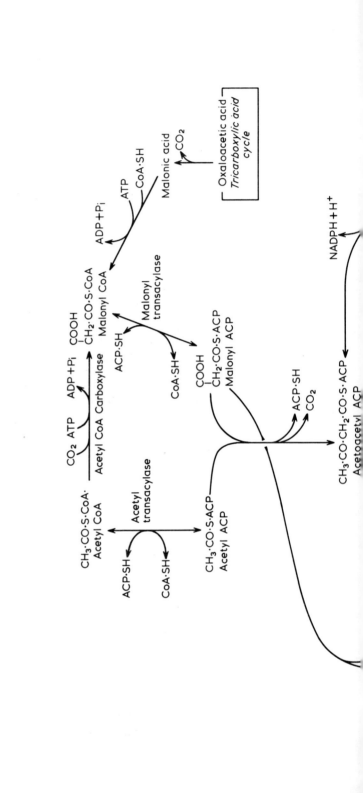

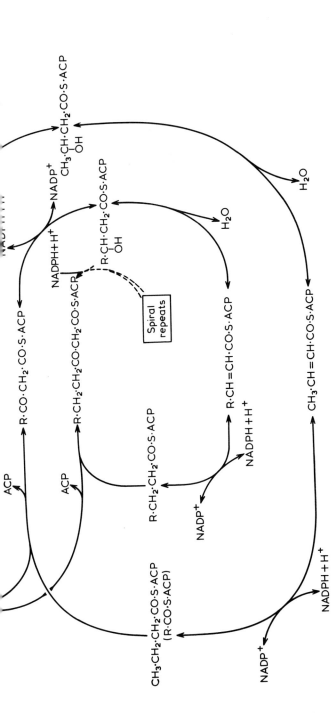

Fig. 4.5 The probable pathway of fatty acid biosynthesis. ACP, acyl carrier protein. CoA, Coenzyme A.

112 Barley

biotin. A series of reductions and dehydrations allows a homologous series of fatty acids to be synthesized. Many problems remain to be resolved, for example how the unsaturated C-18 fatty acids are synthesized, whether by the desaturation of stearic acid or by the addition of three C-2 units to dodecatrienoic acid. Chloroplasts, isolated from barley during a transient period during greening, synthesize 6-methyl-salicylic acid [166].

Degradation of fatty acids is probably by means of the β-oxidation spiral of reactions, which normally occurs in mitochondria and glyoxysomes (Fig. 4.6). Apparently this pathway has not been formally demonstrated in barley. As well as the production of NADH and reduced flavin, β-oxidation produces acetyl-CoA, the acetate moiety of which may be oxidized via the tricarboxylic acid cycle, or may undergo a wide series of other reactions. If it enters the glyoxylate cycle it may be utilized for gluconeogenesis.

In plant leaves fatty acids may undergo α-oxidation. The details may vary, but apparently a hydroperoxy acid is formed and decarboxylated,

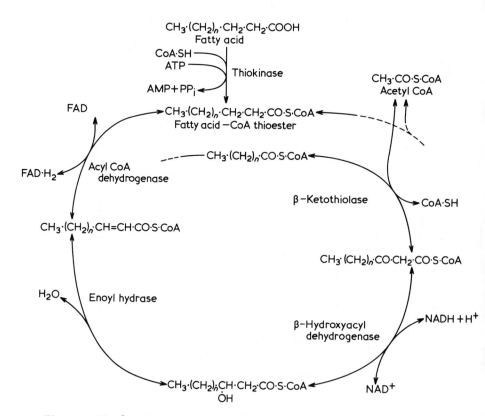

Fig. 4.6 The β-oxidation sequence of fatty acid oxidation. FAD – flavin adenine dinucleotide.

The biochemistry of barley 113

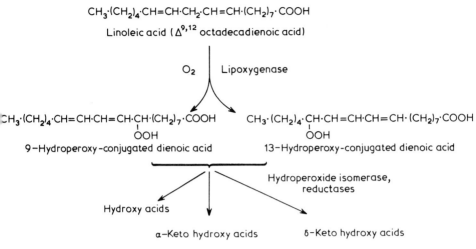

Fig. 4.7 The degradation of linoleic acid by lipoxygenase and hydroperoxide isomerase.

the product being converted via an aldehyde to the next lower fatty acid in the homologous series. Apparently α-oxidation has not been demonstrated in barley.

The enzyme lipoxygenase occurs in barley grains (embryo), in young seedlings, and especially in etiolated leaves [119,366]. The enzyme catalyses the oxidation of linoleic acid to 9- or 13-hydroperoxy unsaturated acids having a conjugated diene system (Fig. 4.7). Another enzyme, hydroperoxide isomerase, converts the hydroperoxy acids to mono-enoic-ketohydroxy fatty acids [368]. Further reactions lead to hydroxy acids, keto-dihydroxy acids and trihydroxy acids. Cutin acids may be formed by this system.

Numerous classes of phospholipids, galactolipids and sulpholipids (presumably sulphoquinovosyl diglyceride) have been detected in barley (44–53) together with lysolecithin, lysocephalin and lysophosphatidyl-inositol. Phosphatidyl serine seems to be only a minor component [10,22,89,120,278]. Chloroplasts are particularly rich in phospholipids, sulpholipids and galactolipids. Enzymes occur that are able to degrade phospholipids [1]. Serine is decarboxylated to ethanolamine which by successive methylations gives rise to choline. Phosphoryl serine, phosphoryl ethanolamine and phosphoryl choline are formed by the transfer of phosphate from ATP to the hydroxyl of the free substances. Reaction of these phosphates with cytidine triphosphate (CTP) gives rise to inorganic pyrophosphate and the cytidine diphosphate derivatives CDP-serine, CDP-choline and CDP-ethanolamine. These react with 1,2-diglycerides to form cytidine monophosphate and phosphatidyl serine, phosphatidyl ethanolamine and phosphatidyl choline. The

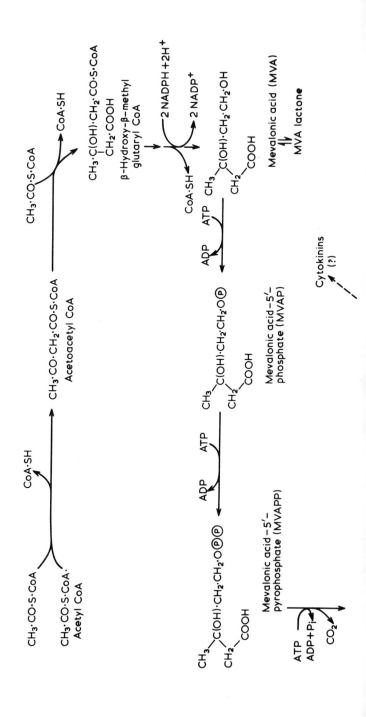

The biochemistry of barley 115

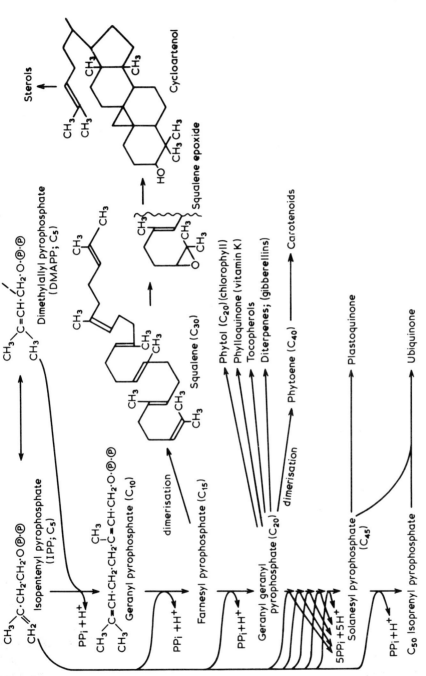

Fig. 4.8 The biosynthesis of terpenes in barley.

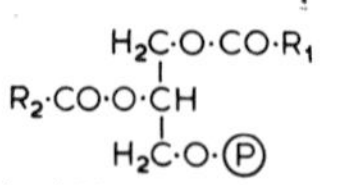

(44) Phosphatidic acid

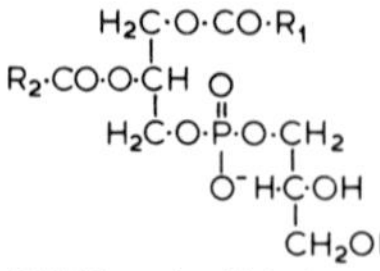

(45) Phosphatidyl glycerol

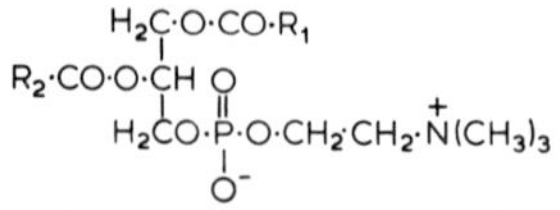

(46) Phosphatidyl choline (Lecithin)

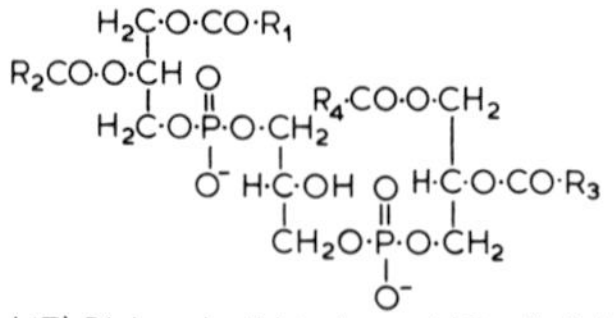

(47) Diphosphatidyl glycerol (Cardiolipin)

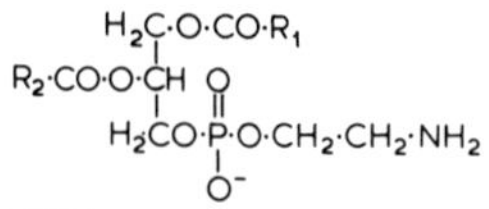

(48) Phosphatidyl ethanolamine (Cephalin)

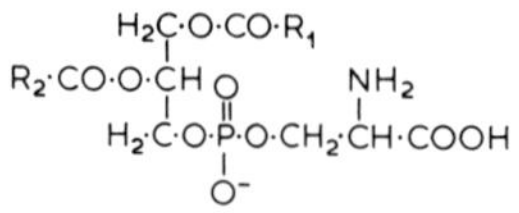

(49) Phosphatidyl serine (Cephalin)

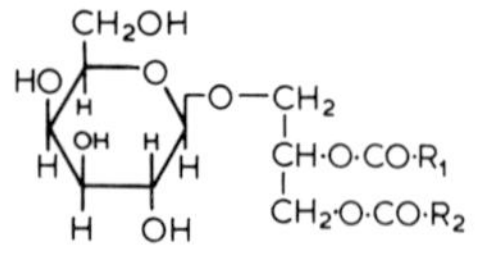

(50) Mono-β-D galactosyl diglyceride

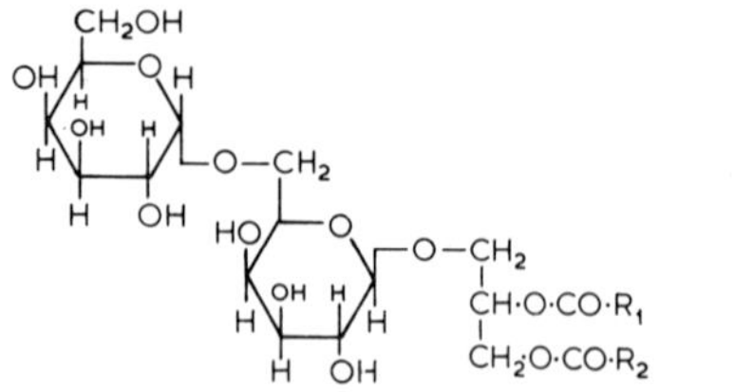

(51) Digalactosyl diglyceride

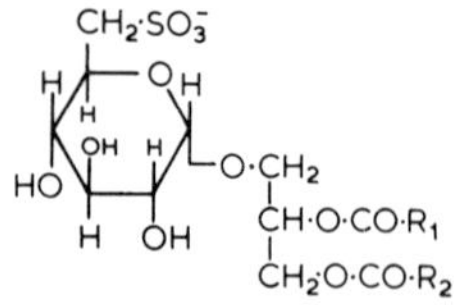

(52) Sulphoquinovosyl diglyceride

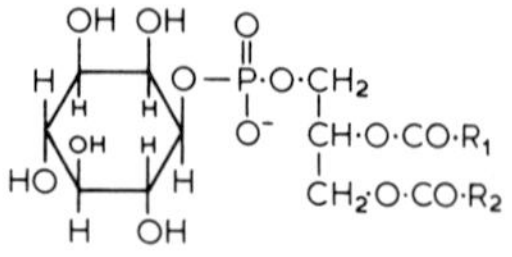

(53) Phosphatidyl inositol

presence of choline kinase, phosphoryl choline-cytidyl transferase and phosphoryl choline glyceride transferase has been demonstrated [161].

Terpenes, or molecules containing terpenoid moieties, are of universal importance. In barley the main path of terpene synthesis – the formation of the C-5 molecules isopentenyl-pyrophosphate (IPP) and dimethylallyl-pyrophosphate (DMAPP) from acetate, via mevalonic acid, and the successive addition of C-5 units, from IPP, to lengthen the chain has been clearly demonstrated by radiochemical studies as has the occurrence of the usual terpenoid intermediates. (Fig. 4.8) [70,124,199,200,252,253]. Farnesol pyrophosphate (C-15) dimerizes to give squalene (C-30) which has been demonstrated in germinating grains. Squalene is oxidized to the 2,3-epoxide which, by cyclization, gives rise to sterols. The first intermediate in plants, in contrast to animals, is cycloartenol. Barley sterols include stigmasterol (54), β-sitosterol (55), campesterol (56), and cholesterol (57). These may occur free, as glycosides (β-glucosides?), esterified with fatty acids, or as acylated glycosides [43,51].

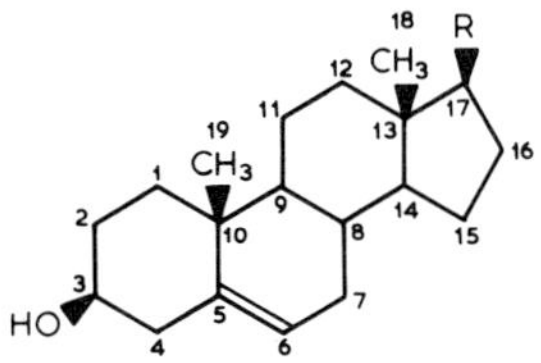

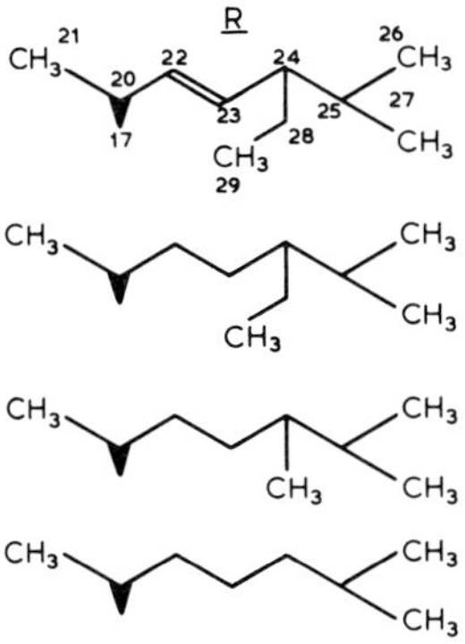

(54) Stigmasterol

(55) β-Sitosterol

(56) Campesterol

(57) Cholesterol

Dimerization of geranyl pyrophosphate (C-20) gives rise to a series of carotenoids and xanthophylls (oxidized carotenoids) of which β-carotene (58), lutein (59), violaxanthin (60), zeaxanthin, neoxanthin (61), isolutein, lutein epoxide and antheraxanthin have been recorded [198,256,356]. The photo-oxidation or oxidation by lipoxygenase of violaxanthin gives rise, among other products, to the growth inhibitory material xanthoxin (62) [47,93]. Xanthoxin has been found in wheat. There is controversy as to whether xanthoxin could give rise to the inhibitory abscisic acid (63) *in vivo*. Abscisic acid is synthesized more directly from mevalonic acid in various species. Apparently neither it nor xanthoxin have been demonstrated in barley.

Various materials occur with cytokinin-like hormonal activity, including zeatin, and zeatin riboside (64) [326,350]. Probably other cytokinins, such as Δ^2-isopentenyl adenosine and perhaps the methyl-thio-derivatives known from other species, will also be identified. The isopentenyl and hydroxymethyl methylallyl side chains are evidently terpenoid in origin.

In other species geranyl geranyl pyrophosphate is cyclized to yield

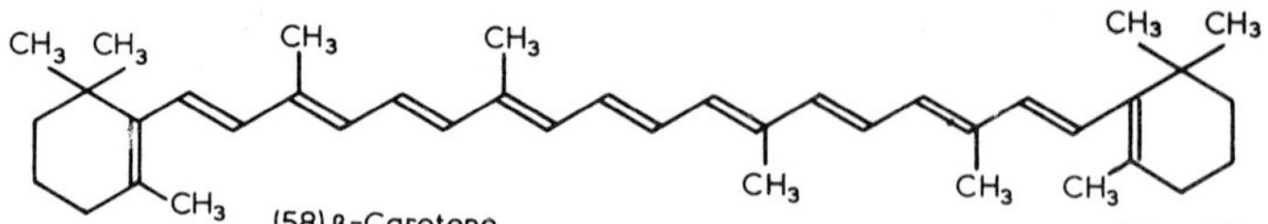

(58) β-Carotene

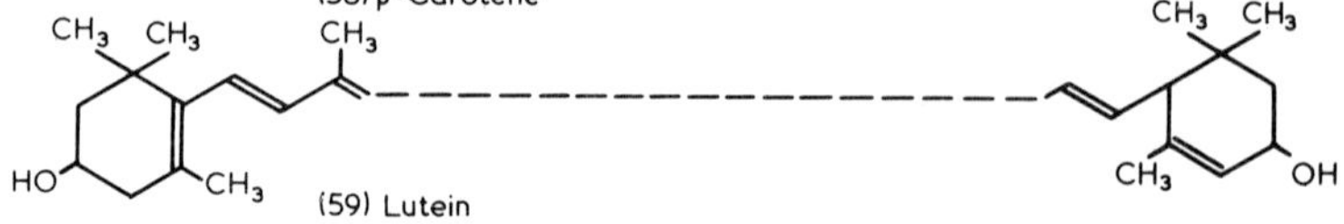

(59) Lutein

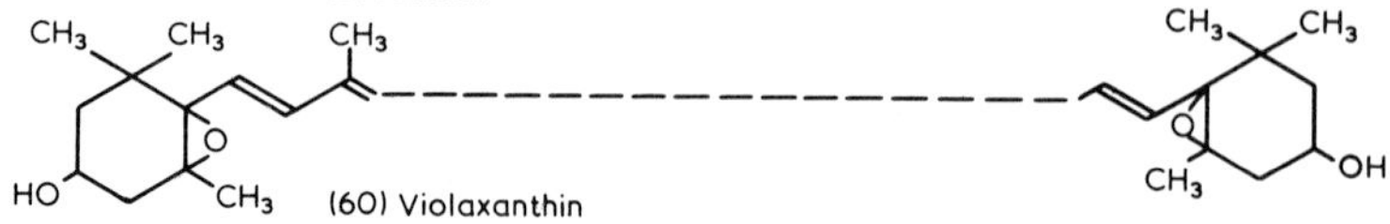

(60) Violaxanthin

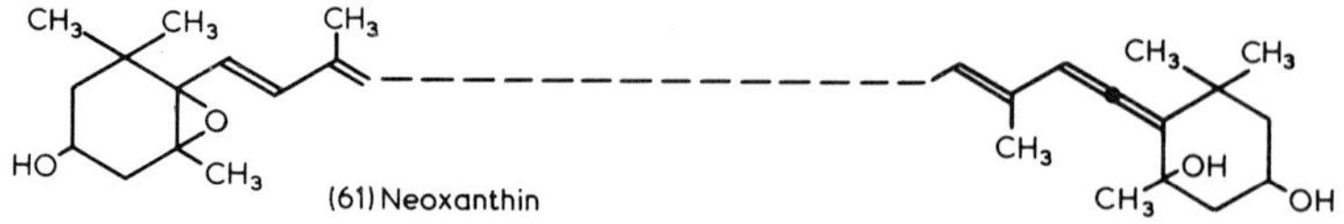

(61) Neoxanthin

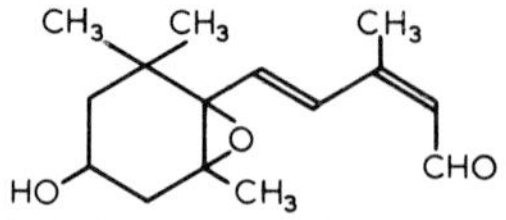

(62) cis, trans –Xanthoxin

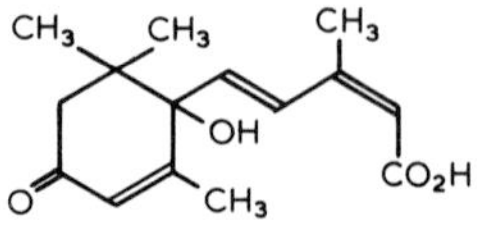

(63) Abscisic acid
(ABA; abscisin II; dormin)

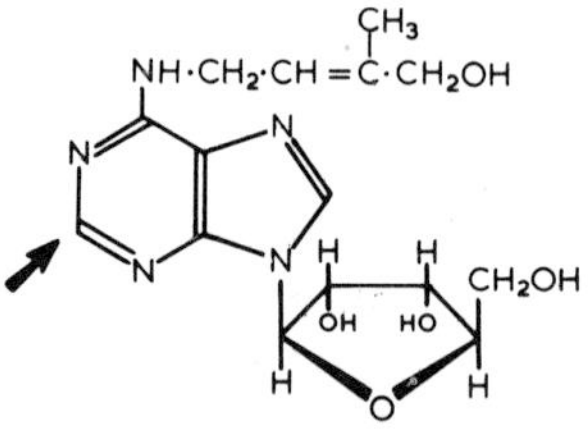

(64) Zeatin riboside.
In other cytokinin bases found in other species the side chain is not hydroxylated and, or may be thiomethylated (CH$_3$·S–) at the site arrowed

ent-kaurene. By a series of steps, not all characterized, kaurene is converted into gibberellins (Fig. 4.9). ent-Kaurene occurs in barley. When kaurene is supplied to embryo homogenates it is metabolized by a side-route to the 16,17-epoxide, the 17-ol, and the 16,17-diol. It is not metabolized by whole grains. However ent-kaurenol has been shown to undergo a series of reactions that are part of the biosynthetic pathway to gibberellins [252,253]. Gibberellic acid (GA$_3$) has been characterized in immature grain, and germinating grain contains at least GA$_1$ (65), GA$_3$ (66), GA$_4$ (67) and GA$_7$ (68) [163,252,253,370]. In germinating grain gibberellin hormonal activity rises and later falls, indicating that binding or destructive mechanisms are operating [125]. When GA$_1$ is supplied to aleurone tissue it is converted to polar metabolites, including GA$_1$-glucoside, GA$_8$ (69) and GA$_8$ glucoside; perhaps 2-O-β-D-glucopyranosyl GA$_8$

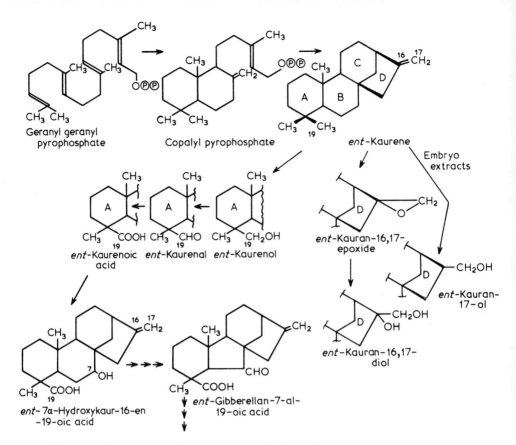

Fig. 4.9 Some steps in the biosynthesis of gibberellins.

that occurs in other species [254,255]. GA_8 is without most forms of biological activity. Abscisic acid accelerates the uptake and metabolism of GA_1 [333].

Terpene units occur in other important classes of compounds. The phytol side chain of chlorophyll is derived from geranyl geranyl pyrophosphate, as are the side chains and parts of the tocopherols (T; 70) and tocotrienols (T-3; 71), which together make up vitamin E, and phylloquinone (Vitamin K; 73).

The side chain of plastoquinone is derived from solanesyl pyrophosphate (C-45) and the side chain of the various ubiquinone isomers

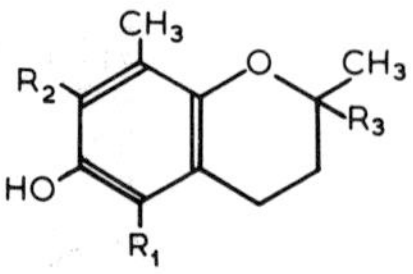

(70) Tocopherols, T; R_3 =

(71) Tocotrienols, T-3; R_3 =

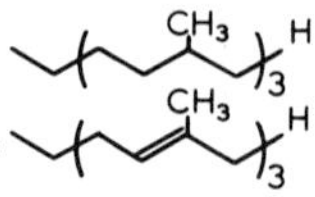

	R_1	R_2
α – Tocopherol	CH_3-	CH_3-
β – Tocopherol	CH_3-	$H-$
γ – Tocopherol	$H-$	CH_3-
δ – Tocopherol	$H-$	$H-$

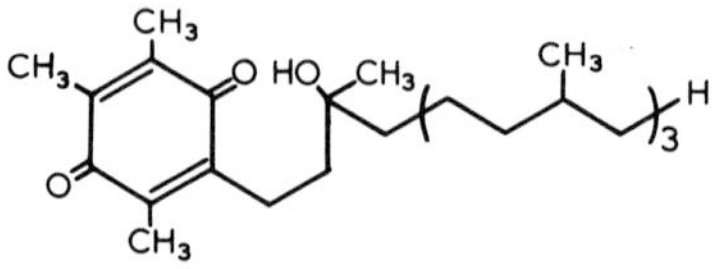

(72) α-Tocopherol quinone

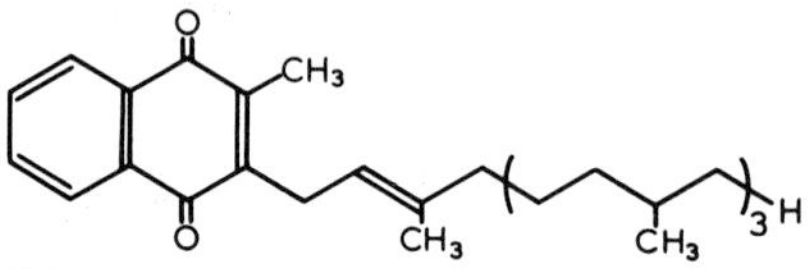

(73) Phylloquinone (vitamin K_1)

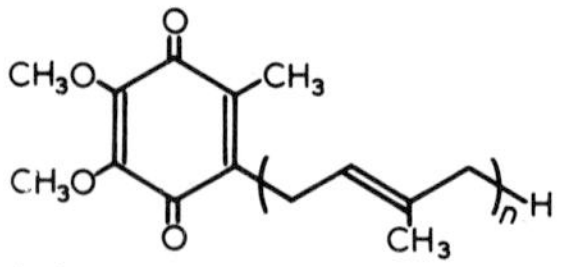

(74) Ubiquinone (Coenzyme Q; CoQ; UQ)
$n=8$ and 9 in barley)

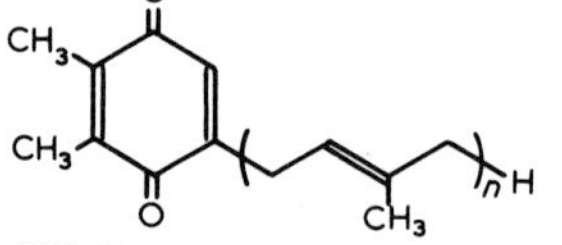

(75) Plastoquinone (PQ; $n=9$. Forms with variously hydroxylated, acylated and partially saturated side chains found in some species)

Table 4.2 Pigments and plastid quinones in ten day old green and etiolated barley seedlings (from Lichtenthaler [198]). The values are µg pigment per 100 plants.

Compound	Etiolated	Green	Green/etiolated
Chlorophyll a	—	16000	—
Chlorophyll b	—	5900	—
β-Carotene	17	900	53
Lutein	149	1400	9.4
Violaxanthin	13	480	37
Neoxanthin	2.7	90	33.3
Antheraxanthin + lutein epoxide	11	6	
Carotenoids	192.7	2876	15
Plastoquinone 45	50	300	6
Plastoquinol 45	54	400	7.4
Total plastoquinone	104	700	6.7
α-Tocopherol	60	450	7.5
α-Tocoquinone	3.5	20	5.7
Total α-tocoquinone	63.5	470	7.4
Benzoquinones	167.5	1170	7
Vitamin K_1	3.2	100	31.5
Total plastid quinones	170.7	1270	7.5

(74) are also derived from long-chain terpenyl pyrophosphates. The cyclic nuclei of the quinones are indirectly derived from shikimic acid. Of the terpenoid quinones, and related compounds, phylloquinone (73), plastoquinone (75) and tocopherylquinone (72) are concentrated in chloroplasts, and perhaps other plastids, whereas mitochondria are rich in tocopherol and ubiquinone [124]. The quinones are thought to act as carriers in electron transfer reactions. Barley grains contain mainly α-T-3 followed in amount by α-T, but β-T, β-T-3, γ-T, γ-T-3 and δ-T are also present [121,313]. In barley ubiquinone has predominantly C-45 side chains [79]. Plastoquinone, and reduced plastoquinol, also occur in barley (Table 4.2). The mixture of quinones may really be more complex, since hydroxylated and acylated forms, and forms with partially saturated side chains occur in other species. The changes in terpenoid and related compounds that occur when leaves undergo greening is striking (Table 4.2).

4.5. Photosynthesis and photorespiration

As usually presented photosynthesis is the process by which plants use the energy from trapped light to synthesize carbohydrate from carbon dioxide and water. While much label from $^{14}CO_2$ is incorporated into sucrose it is also incorporated into many other compounds. In addition

122 Barley

in light the distinct process of photorespiration becomes operative. While most of the reactions of photosynthesis occur in chloroplasts the metabolism of these organelles is linked to that of the rest of the cell and the energy derived from light may be used to synthesize lipids, reduce nitrate, synthesize amino acids and proteins, replicate some DNA and RNA, pump ions, and so on. Probably chloroplasts can carry out these and other metabolic interconversions to some extent.

The energy from light absorbed by chlorophylls a and b, and the accessory pigments is thought to be channelled to sites where the photolysis of water occurs and the energy content of the electrons released from the water is raised (photosystem II; PS II; Fig. 4.10). The electrons (hydrogen equivalents) flow down an energy gradient being transferred between carrier molecules which are alternately reduced and oxidized. In this process ADP is phosphorylated to ATP (photophosphorylation). Energy

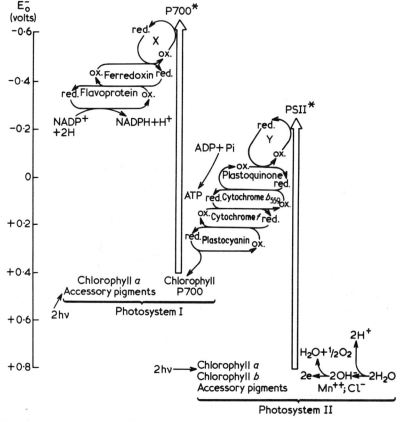

Fig. 4.10 A tentative outline, the 'Z-scheme', of the way in which electrons derived from the photolysis of water are energized by two successive photochemical events, and by passing through a chain of electron carriers drive the phosphorylation of ATP and eventually reduce $NADP^+$.

from more absorbed light again energizes the electrons (photosystem I; PS I) which, after flowing down another carrier chain, reduce $NADP^+$ to NADPH. Thus the system produces NADPH and ATP, both being required for many biosynthetic reactions, and keeps some molecules, such as the strongly reducing ferredoxin, in the reduced state. A tentative formulation of these findings is the 'Z-scheme' (Fig. 4.10). 'Good quality' chloroplasts have been isolated from barley, and photophosphorylation and photoreductive reactions have been shown to occur in them [37,228,281]. Chlorophyll a, chlorophyll b, plastoquinone, cytochromes f, $b559$ and $b6$, the copper-containing protein plastocyanin, and ferredoxin have been demonstrated in barley. Cytochromes are also found in mitochondria and the cytoplasm (P-445-like cytochromes) besides the chloroplasts [32,69,134,246,282]. Photosystem I develops before photosystem II in greening leaves, and has slightly different light absorption characteristics [134,142]. Chlorophyll b is thought to be associated mainly with PS II. However mutants totally lacking chlorophyll b grow well, although their light requirements differ from those of normal

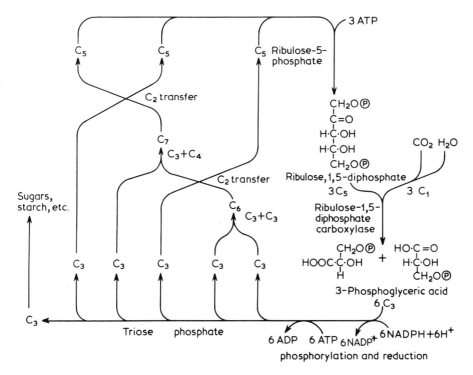

Fig. 4.11 The proposed pathways of carbon, the Benson–Calvin cycle, during photosynthesis in barley and other C-3 plants (after Walker [353]). The 'network' of rearrangements of the carbon chains are catalysed by the enzymes transketolase and transaldolase (compare Fig. 4.2).

plants [33]. Barley is a typical 'C-3' plant. When radioactive carbon dioxide is fixed, the first detectable product is a triose-phosphate (C-3) and the substrate carboxylated, ribulose diphosphate, is renewed by the Benson–Calvin cycle (Fig. 4.11). Etiolated leaves contain other carboxylative enzymes; during greening these decline in amount and ribulose diphosphate carboxylase and phosphoriboisomerase increase. Plants contain carbonic anhydrase, so the interconversion between dissolved carbon dioxide and bicarbonate should be rapid, supplying whichever reactant is needed. Phosphoribulokinase is also present [67,128,165, 173,272]. Ribulose diphosphate carboxylase (RuDP carboxylase) has been greatly purified from barley [178]. As 70–80% of the leaf protein is in chloroplasts [292], and RuDP carboxylase (fraction I protein) is the major protein of the chloroplast, it is undoubtedly the major protein of the plant and, taking plant species together, of the world. It is therefore of nutritional importance also. As expected in the shortest fixation times, 0.1 sec., $^{14}CO_2$ labels only 3-phosphoglyceric acid in barley leaves [182]. In longer term studies $^{14}CO_2$ is incorporated into sucrose, alanine, hexose phosphates, aspartic acid, serine, glycine, glyceric acid, malic acid, glycollic acid and, more slowly, into isocitrate and glutamine [21,269]. The triose phosphate residues produced by the Benson–Calvin cycle could give rise to hexose phosphates and sucrose by the reversed glycolytic sequence. By conversion to pyruvate and entry into the tricarboxylic acid cycle and transamination of the keto acids, alanine (from pyruvate), and aspartate, (from oxaloacetate) and glutamate (from α-oxoglutarate) would be produced. Transaminases occur in barley chloroplasts [170]. The rapid formation of glycollic acid in photosynthesis is not understood. However chloroplasts seem to release this material which is then utilized in photorespiration and perhaps another pathway [338].

During photosynthesis in 'C-3' plants carbon dioxide fixation proceeds simultaneously with carbon dioxide generation, formed wholly or in part by photorespiration, a process distinct from mitochondrial respiration. Lowering the partial pressure of oxygen in the atmosphere slows photorespiration very greatly. Increasing the carbon dioxide concentration increases net photosynthesis, within limits. The carbon dioxide produced in photorespiration is derived from recently fixed carbon dioxide and appears to originate from glycollate, since, when glycollate oxidase is blocked by exposing leaves to sulphur dioxide or α-hydroxy-sulphonic acids, photorespiratory CO_2 output is reduced [325,343]. Glycollic acid is readily oxidized to carbon dioxide by leaves. The glycollate oxidase is located in the peroxisome microbodies of the leaves; it is substrate activated and is dependent on flavin mononucleotide (FMN) for its activity [72,343]. Microbodies isolated from etiolated barley leaves held their glycollate oxidase only loosely when isolated from plants lacking in nitrate, but tightly when nitrate was supplied; however catalase was held

The biochemistry of barley   125

firmly by microbodies from both types of seedlings [297]. The products of glycollate oxidation are glyoxylic acid and hydrogen peroxide; the later is decomposed by the catalase of the microbody. Elsewhere in the cell glyoxylate reductase occurs which reduces glyoxylate to glycollate. Feeding glycollate to barley leaves gives rise to glycine, serine, and an unknown compound [344]. It has been suggested that glycollate is oxidized to glyoxylate, which is converted by transamination to glycine (Fig. 4.18). Glycine is converted to serine by the addition of a C-1 unit, possibly derived from glyoxylic acid and carried on a folate cofactor. Serine is converted via hydroxypyruvate and glycerate to 3-phosphoglycerate, and is then in the glycolytic and gluconeogenic sequence. The purpose of this, apparently wasteful, photorespiratory process, which is not detectable in plants using the 'C-4' pathway in photosynthesis, is unknown. Studies with the inhibitor isonicotinyl hydrazide appear to show that this compound blocks the conversion of serine to pyruvate, and that in chloroplasts the acetyl-CoA used in terpenoid synthesis is derived from the sequence $CO_2 \longrightarrow$ glycollate $\longrightarrow$ glyoxylate $\longrightarrow$ glycine $\longrightarrow$ serine $\longrightarrow$ pyruvate $\longrightarrow$ acetyl-CoA [111].

Various trials have been made to determine the fate of radioactive C-1 compounds, besides carbon dioxide, in barley [77,183,342]. Formic acid is incorporated into serine, especially in the dark, and into glyceric acid in the light and the label is then distributed among other metabolites. Formaldehyde, methanol and carbon monoxide are used to a limited extent by leaves, particularly in the light. As an early metabolic product of these substances is serine they are probably fixed as C-1 derivatives of folic acid.

The reason why the insecticide DDT is toxic to some barley varieties and not others is still unknown. In susceptible varieties DDT harms photosynthesis, plants become chlorotic, the Hill reaction (photochemical reduction of artificial electron acceptors) and photophosphorylation are reduced and the sugar pattern in the leaf is altered [277].

4.6. The formation of porphyrins

The major porphyrin derivatives in barley are chlorophylls *a* and *b*, followed probably by assorted cytochromes, which are iron porphyrin (haem)-proteins and phytochrome, which is probably an open chain tetrapyrrole-protein. Evidence relating to the biosynthesis of cytochromes and phytochrome in barley is lacking.

The formation of chlorophylls has often been studied in greening barley leaves; the process is highly complex (Chapter 6). However the general nature of the pathway of chlorophyll synthesis is known (Fig. 4.12). Nothing seems to be known of its degradation. The first unambiguous precursor of porphyrins in barley is δ-aminolaevulinic acid, ALA. Feeding

ALA to barley leaves in the dark causes protoporphyrin to accumulate. In animals ALA is formed from the condensation of succinate, as the CoA ester, with glycine. In greening barley succinyl-CoA synthetase increases in amount, suggesting its involvement in this process [330]. However, radiotracer experiments show that glycine is a poor precursor of ALA, but glutamine is good – possibly an unrecognized, more important biosynthetic route to ALA exists [55]. In shoots poisoned with caesium salts protochlorophyllide and porphyrins (uroporphyrin, coproporphyrin and protoporphyrin) accumulate [224]. When ALA is fed to a range of barley chlorophyll mutants uroporphyrin, protoporphyrin, magnesium protoporphyrin and its monomethyl ester, and protochlorophyllide were shown to accumulate [113]. Thus the expected biosynthetic intermediates of chlorophyll formation occur in barley. However numerous uncertainties remain; how and when is the magnesium atom inserted into the tetrapyrrole? How and when is the phytyl side chain added, and how and when is the tetrapyrrole linked to the protein component? A metalloporphyrin chelatase, that catalyses the insertion of Zn or Fe into porphyrins, has been found in barley seedlings; this is more likely to be involved in the formation of haem compounds than chlorophylls [110]. Ferrochelatase occurs in etioplasts as well as mitochondria [205]. A geranyl geranyl

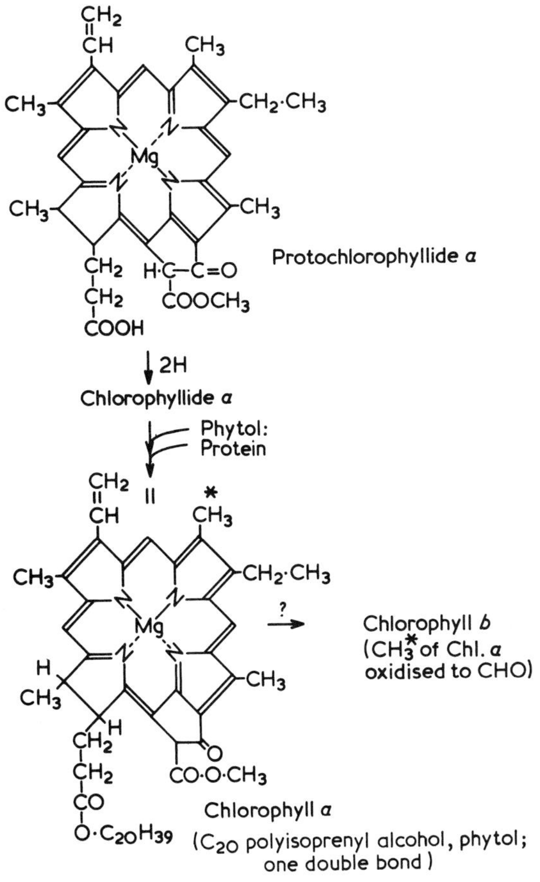

Fig. 4.12 An outline of the biosynthesis of chlorophylls a and b.

protochlorophyllide ester has been detected in etiolated barley leaves [202]. Possibly geranyl geraniol is transferred from the pyrophosphate, and is reduced to phytol while attached to the porphyrin nucleus. The light induced reduction of protochlorophyll(ide) to chlorophyll(ide) has been demonstrated in barley etioplast membranes [123]. Chlorophyll a is probably the precursor of chlorophyll b [3]. The changes that occur in chlorophyll spectra in leaves during greening may reflect not only the assembly of the component magnesium tetrapyrrole-protein, but also the joining of this molecule to the chloroplast membranes, and its orientation relative to other chlorophyll molecules and the accessory pigment molecules [135,265–267]. There are a great deal of data on the temporal relationships of the synthesis and assembly of the parts of chloroplasts [358].

Phytochrome has not been purified from barley. The action spectra of light inhibition of growth of albino barley leaves led to the suggestion that the phytochrome photoreceptor is an open-chain tetrapyrrole [341]. The level of phenylalanine ammonia lyase is also controlled by phytochrome-perceived light [214]. Immunocytochemical studies have demonstrated the subcellular distribution of phytochrome in etiolated coleoptiles [61].

4.7. Phenolic and aromatic substances

A wide range of phenolic materials occurs in barley, from free and combined tyrosine, tyramine and its derivatives, phenolic acids, esters and glycosides, and numerous other types of phenols through lignans and related substances to lignin. The number of soluble phenolic materials recognized is very large and only a fraction of those detected have been characterized [97,117,118,131]. *p*-Hydroxybenzoic acid, vanillic acid, syringic acid, *o*- and *p*-coumaric acids, ferulic, sinapic and chlorogenic acids have been observed free (Figs. 4.13 and 4.14). Water soluble esters of *p*-hydroxybenzoic, protocatechuic, vanillic, syringic, caffeic, ferulic, sinapic and isoferulic acids have been detected as have glycosides of several of these as well as of gentisic, chlorogenic and 2,5-dihydroxybenzoic acids. Chlorogenic acid is itself an ester of caffeic acid with quinic acid. Some other cereals contain esters of caffeic and ferulic acid linked to long-chain *n*-alkanols, glycerol and hydroxy-*n*-carboxylic acids, but these have not been found in barley. Many phenolic acids also occur as coumarins, (lactones), such as coumarin, umbelliferone, herniarin, aesculetin and scopoletin (Fig. 4.14). Phenolic acids occur linked to polymeric and insoluble materials. Acetone powders of barley plants and preparations of lignin, gums, hemicelluloses and cell walls are variously esterified with cinnamic, *p*-coumaric, caffeic and ferulic acids [82,132, 137,184]. *N*-Feruloyl-glycyl-L-phenylalanine appears as a unit in barley proteins [336].

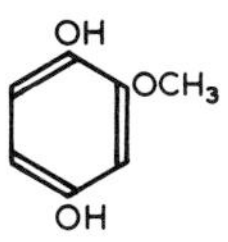

The biochemistry of barley 129

(76) 2-Methoxy-quinol
(77) 2-Methoxy-1,4-benzoquinone 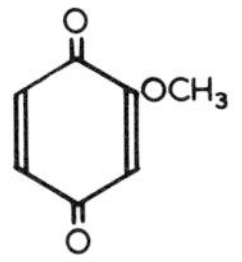
(78) 2,6-Dimethoxy-1,4-benzoquinone 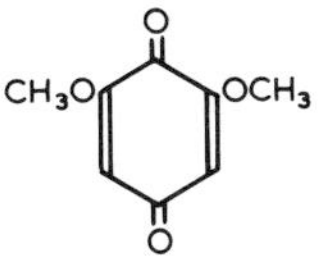

Other types of phenolic materials occur in barley. The presence of 2-methoxyquinol (76), 2-methoxy-1,4-benzoquinone (77) and 2,6-dimethoxy-1,4-benzoquinone (78) in the embryo of the barley grain is inferred from similarities to wheat [208]. 5-n-Alkyl resorcinols occur in the testa wax [43].

Brownish or even black pigments that occur in barleys are probably of phenolic origin, but they seem not to be true melanoidins, despite statements to the contrary [59]. The red and blue colours that are encountered in barley tissues are due to phenolic anthocyanin pigments. At least eight coloured compounds have been found in extracts of coloured grain including derivatives of cyanidin (80), delphinidin (81) and

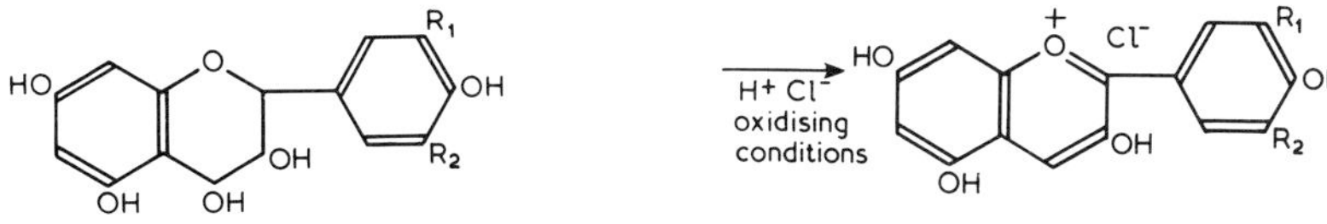

	R_1	R_2	
(88)	H	H	Leucopelargonidin
(89)	H	OH	Leucocyanidin
(90)	OH	OH	Leucodelphinidin

	R_1	R_2	
(79)	H	H	Pelargonidin
(80)	H	OH	Cyanidin
(81)	OH	OH	Delphinidin

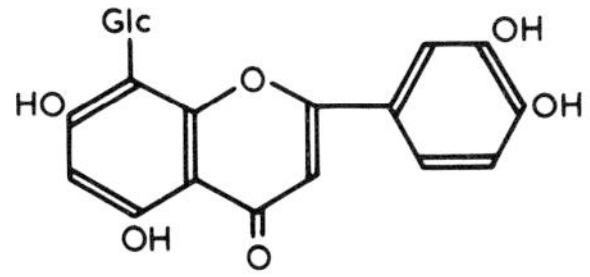

Glc. C-β-D-Glucopyranosyl substituent

(82) Orientin

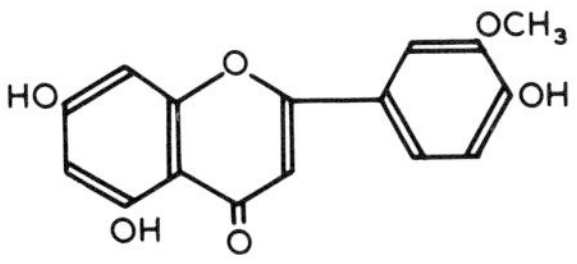

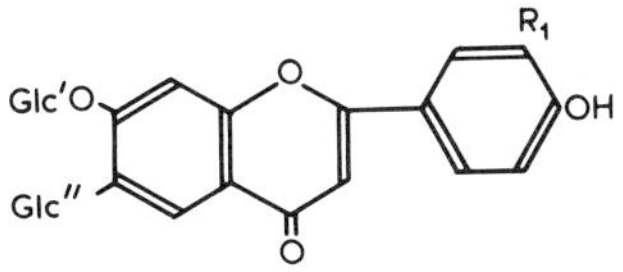

Glc' O-Glucosyl substituent
Glc" C-Glucosyl substituent
(84) R_1=H;   Saponarin
(85) R_1=OH;   Lutonarin
(86) R_1=OCH$_3$; Lutonarin-3-methyl ether

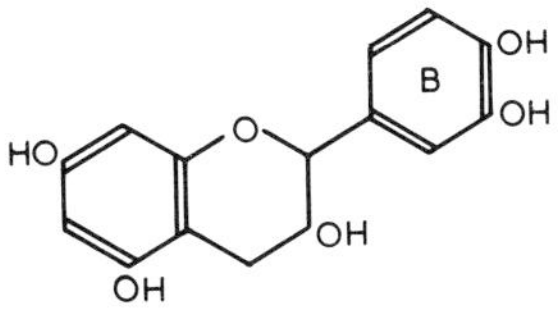

(83) Chrysoeriol

(87) Catechin. Two isomers occur; in D(+)-catechin ring B is below the plane of the molecule, the alcoholic hydroxyl is *trans*. In *epi*-catechin the B ring is above the plane of the molecule, the alcoholic hydroxyl is *cis*.

130 Barley

pelargonidin (79), cyanidin-3-arabinoside and possibly cyanidin-3-glucoside [232,251,349]. Analysis of peeled grain layers shows that the different tissues differ in their complements of anthocyanins [250]. Barley plants contain a range of other flavonoids including orientin (82), chrysoeriol (83) and the C-glycosylflavones saponarin (84), lutonarin (85), and lutonarin-3'-methyl ether (86) [26,215,306]. The formation of the C-glycosylflavones is regulated by phytochrome [54]. The slight oestrogenic activity of barley may indicate the presence of isoflavones [28]. D-(+)-Catechin (87) and *epi*-catechin occur in barley grains [131]. In addition highly reactive leuco-anthocyanins (pro-anthocyanins, anthocyanogens) also occur. These colourless substances, which give rise to the strongly coloured anthocyanin pigments cyanidin, delphinidin and pelargonidin under mildly oxidizing, acidic conditions, have excited interest as tannin precursors which give rise to haze problems in beer and vinegar [117,118,131]. These leuco-anthocyanins (88, 89, 90) may be monomers, dimers, trimers or tetramers which on hydrolysis give rise to catechin as well as anthocyanins (91, 92, 93) [112,117,118,131,311a].

(91) Dimer of leucocyanidin and catechin

(92) Dimer of leucocyanidin and catechin

(93) Triflavan

Chromatographic evidence and the occurrence of methoxyl residues in testinic acid, the crude mixture extracted by alkali from barley grains, suggests that lignans occur in barley [131,329]. The only lignans characterized are apparently hordatines A and B (94, 95) and their glucosides (96), the fungicidal substances formed by the oxidative coupling of coumaroyl agmatine (97), which also occurs in barley (Chapter 9) [332].

(97) Coumaroyl agmatine

oxidative coupling, etc.

(94) Hordatine A $R_1 = R_2 = H$
(95) Hordatine B $R_1 = H$; $R_2 = CH_3O$
(96) Hordatine M Mixture of glucosides of hordatines A and B; $R_1 = $ D-glucopyranosyl

Lignin, as revealed by the red colour given by the phloroglucinol-concentrated hydrochloric acid test, is widespread in barley occurring especially in the vascular bundles and sclerenchyma. Analytical methods for this material are unsatisfactory, since the extractions involved are so vicious that the structure is probably altered [2,280]. Empirical formulae for two caustic-extractable lignin fractions have been given as $C_{36}H_{31}O_6(OCH_3)_4(OH)_5$ and $C_{36}H_{26}O_8(OCH_3)_4(OH)_4$ [280]. Phenolic hydroxyl and methoxyl groups are present. As expected nitrobenzene oxidation gives rise to vanillin, syringaldehyde and p-hydroxybenzaldehyde (Fig. 4.14). Hydrogenolysis and ethanolysis give rise to substituted phenylpropanoid residues. Thus lignin consists of a polymer of phenolic and methoxylated phenylpropanoid units, a concept supported by biosynthetic studies. The structures proposed for lignins are speculative. Probably in the plant the polymeric material is linked to carbohydrate and possibly protein.

Aromatic rings may be formed by at least two routes: (1) the cyclization of chains formed by the condensation of C-2 units – (malonyl-CoA, *minus* CoA and *minus* CO_2), (2) from carbohydrates via the shikimic acid pathway and, often, the amino acids phenylalanine and tyrosine (Figs. 4.13 and 4.14). There is evidence that both pathways operate in

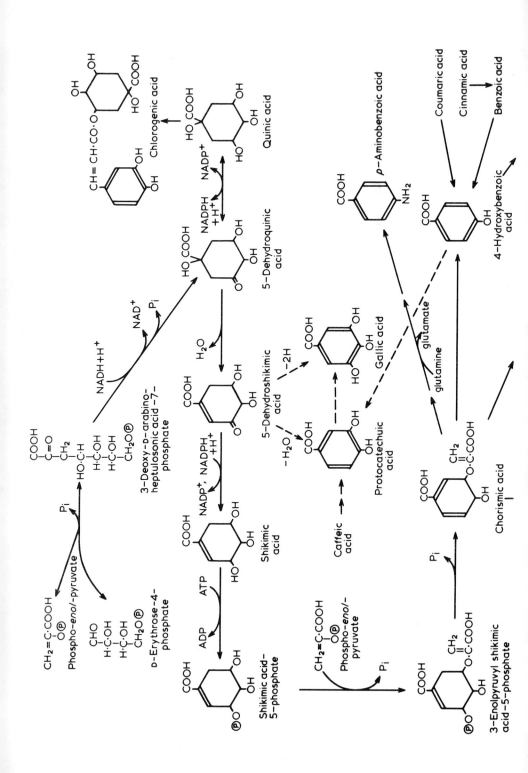

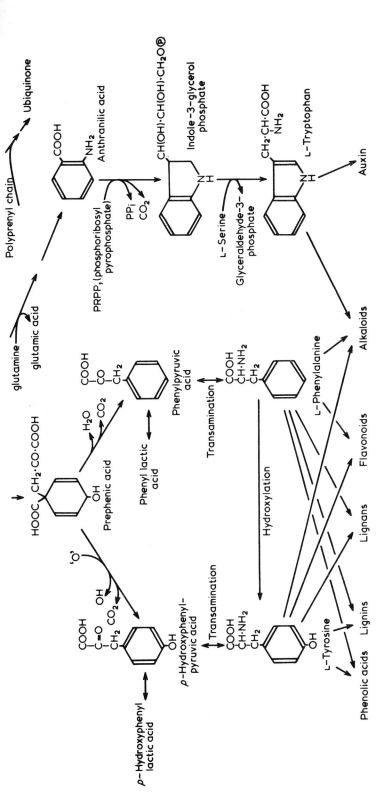

Fig. 4.13 The shikimic acid biosynthetic pathway.

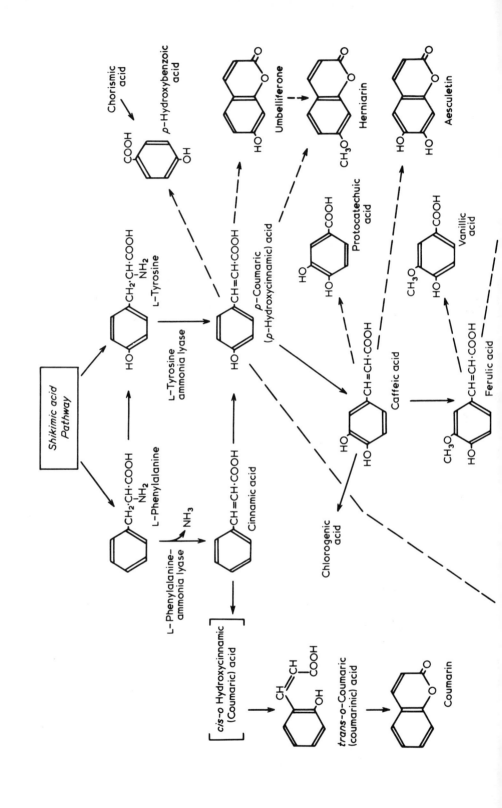

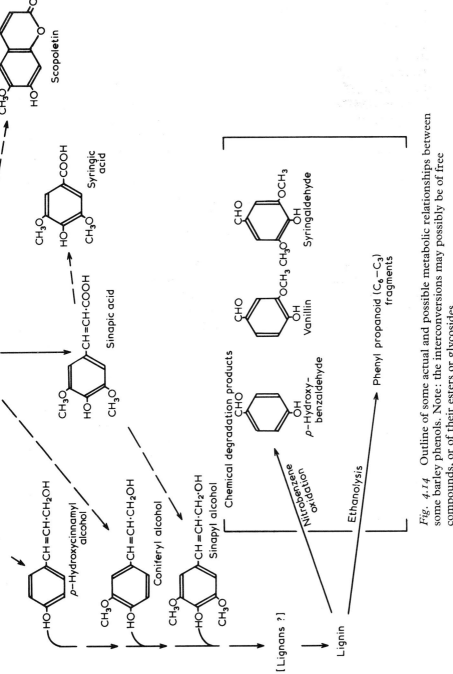

Fig. 4.14 Outline of some actual and possible metabolic relationships between some barley phenols. Note: the interconversions may possibly be of free compounds, or of their esters or glycosides.

barley. Thus the conversion of 'acetate' to methyl salicylate in young chloroplasts has been noted [166]. The 'C-2 equivalents' from malonyl-CoA give rise to the substituted phloroglucinol rings of the flavonoids [27]. There are radiotracer and other pieces of evidence for the presence of the shikimate pathway in barley – such as the occurrence of quinic acid (combined as chlorogenic acid), and the recognition of shikimate-$NADP^+$ oxido-reductase (Figs. 4.13 and 4.14) [348]. The amino acids tyrosine and phenylalanine seem to be necessary intermediates in the formation of most phenolic acids. L-Tyrosine ammonia lyase and L-phenylalanine ammonia lyase convert the amino acids to ammonia and *trans-p*-coumaric acid and *trans*-cinnamic acid respectively [181,260]. Barley extracts are also able to convert L-3,4-dihydroxyphenylalanine (DOPA) to caffeic acid [145]. The level of phenylalanine ammonia lyase is partly under phytochrome control. Tyrosine ammonia lyase exhibits feedback inhibition by various phenolic substances [174]. [^{14}C]-Tyrosine is incorporated into flavonoids, lignin and plastoquinone [27].

The phenylpropanoid acids seem to give rise to the substituted benzoic acids by removal of a C-2 unit e.g. *p*-coumaric acid gives *p*-hydroxybenzoic acid; ferulic acid gives vanillic acid [29,83]. The metabolic pathways are complicated by the existence of glycosides; coumarin is converted to O-coumaroyl-β-D-glucopyranoside in barley [335]. The presence of coumaroyl agmatine has been noted. In other species CoA thioesters of cinnamic acids are reduced to the equivalent alcohols, and this may happen in barley. The CoA esters may also act as 'donors' in forming the numerous other esters that occur. The alcohols and/or their glucosides and perhaps the related aldehydes may be direct precursors of lignin. An $NADP^+$-linked enzyme capable of interconverting cinnamyl alcohol and cinnamaldehyde is known in barley [223]. Numerous oxidase and peroxidase enzyme activities have been noted, which might cause the oxidative polymerization of simple phenolic substances that are supposed to give rise to lignin. Methoxyl groups are formed by the methylation of the free phenolic groups of various substances, some of which are lignin precursors [52]. The most likely methyl donor is *S*-adenosyl methionine. In other grasses demethoxylation, e.g. of ferulic acid to *p*-coumaric acid, has been demonstrated.

Other aromatic substances of unknown provenance, such as benzylamine and benzyl alcohol-β-D-glucopyranoside, also occur. Benzyl alcohol occurs in barleys resistant to greenbugs, and may be responsible for this resistance [164].

4.8. Amino acid metabolism

Tyrosine and phenylalanine are synthesized, by transamination, from the appropriate substituted pyruvic acids formed on the shikimate pathway.

Barley is able to hydroxylate phenylalanine to tyrosine (Fig. 4.13) [196]. Tyrosine may be degraded by way of homogentisic acid, and fumaryl acetoacetate [227]. Tryptophan is synthesized from shikimate via chorismic acid, anthranilic acid, and indolylglycerol phosphate, which reacts with serine to give L-tryptophan and a triose phosphate (Fig. 4.13) [364].

Many transaminations are known, often between glutamic or aspartic acids, as the donors of the amino groups, and other keto acids producing α-oxoglutaric (α-ketoglutaric) or oxaloacetic acids and other amino acids [170,322,365]. Pyridoxal phosphate is probably a general cofactor for transaminations, as well as some other reactions of amino acids. Thus new amino acids may arise by the amination of keto-acids. In the case of some amino acids the presence of the equivalent oxo(keto)-acids has already been noted; namely, glutamic acid, α-oxoglutarate; aspartic acid, α-oxaloacetate; alanine; pyruvic acid; glycine, glyoxylic acid. These keto-acids are components of major metabolic pathways. Barley contains a full range of 'protein' L-amino acids including alanine, arginine, aspartic acid, asparagine, glutamic acid, glutamine, glycine, isoleucine, leucine, lysine, phenylalanine, tyrosine, serine, valine, cysteine, histidine, tryptophan, and threonine as well as the imino acids proline and hydroxyproline and L(+)-pipecolinic acid (piperidine-2-carboxylic acid), β-alanine, α-aminobutyrate, ornithine, citrulline, and probably others that are metabolic intermediates but do not accumulate (Figs. 4.13, 4.15–4.18) [131,171].

Plants utilize nitrate and ammonium ions taken up from the soil. Gaseous nitrogen is neither taken up, nor released, as was shown in about 1890. The respiratory pattern of the roots changes with the uptake and utilization of these ions; the first products of their metabolism appear to be glutamine and glutamic acid (Chapter 6) [24,60,275]. Nitrate is reduced via a complex series of reactions to ammonium ions which may be incorporated into glutamic acid by glutamic dehydrogenase or transamination from glutamine to α-ketoglutarate to produce two glutamate molecules (Fig. 4.15). Glutamine is synthesized from glutamate and ammonium ions. Glutamine also occurs in starving leaves, presumably when toxic ammonium ions, released by the degradation of the amino acids, must be bound [367]. The levels of nitrate and nitrite reductases fluctuate widely under different growth conditions. In photosynthetic tissues the formation of nitrate reductase requires light and the presence of nitrate. In roots, treated correctly, nitrate alone is sufficient to induce the appearance of the enzyme, as it is in aleurone layers [90,296,321,345]. In the leaves nitrite reduction may be closely linked to photochemical events. ATP may be involved in the reduction of nitrite to ammonia [38,53]. Studies with [15N]-labelled materials indicate that the isotopic nitrogen spreads quickly among the amino acids, showing the general

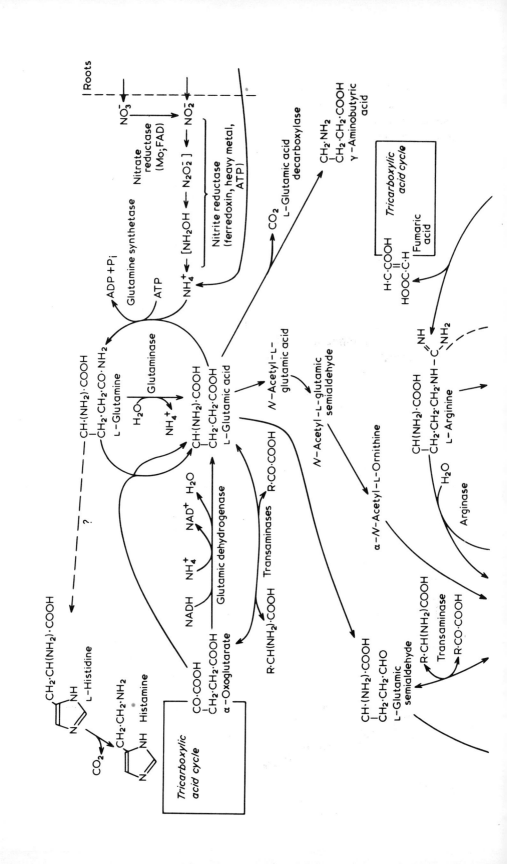

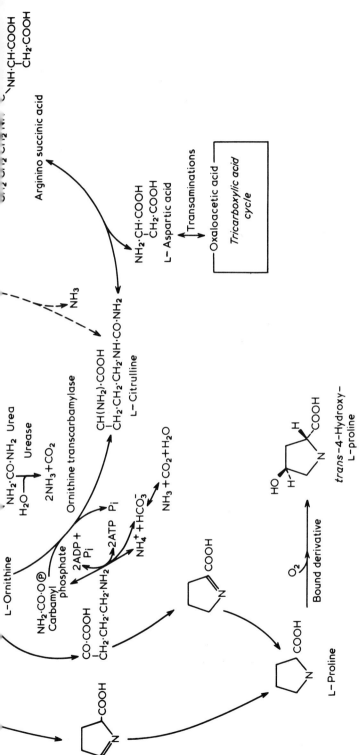

Fig. 4.15 The formation of glutamic acid and some related amino acids in barley.

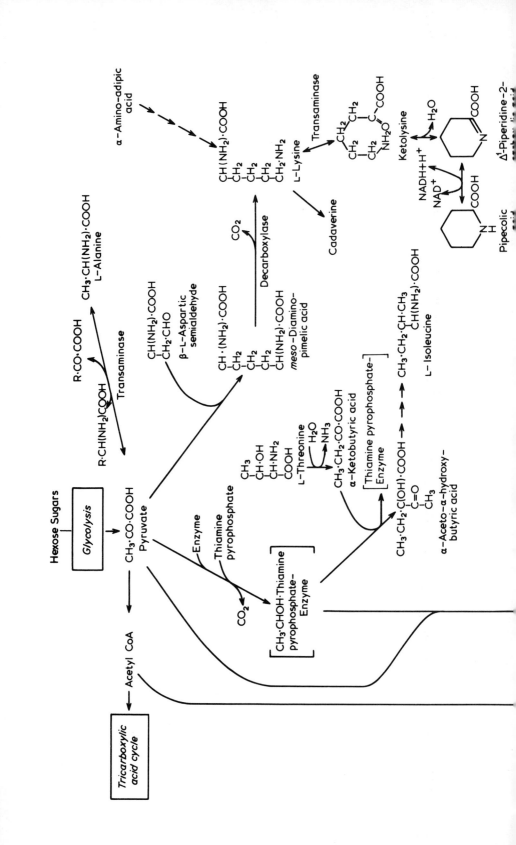

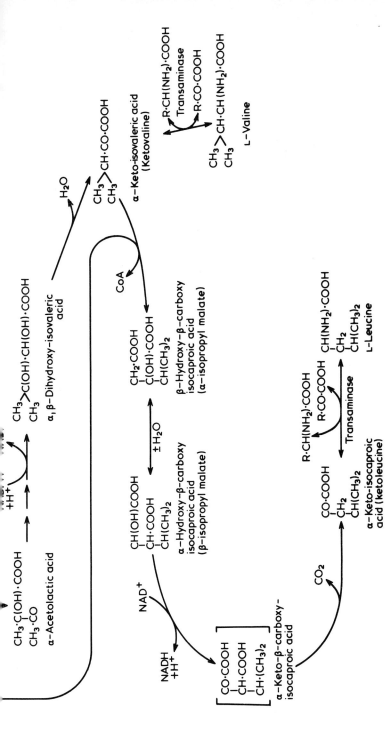

Fig. 4.16 The biosynthesis of alanine, lysine, isoleucine, leucine, valine and pipecolic acid.

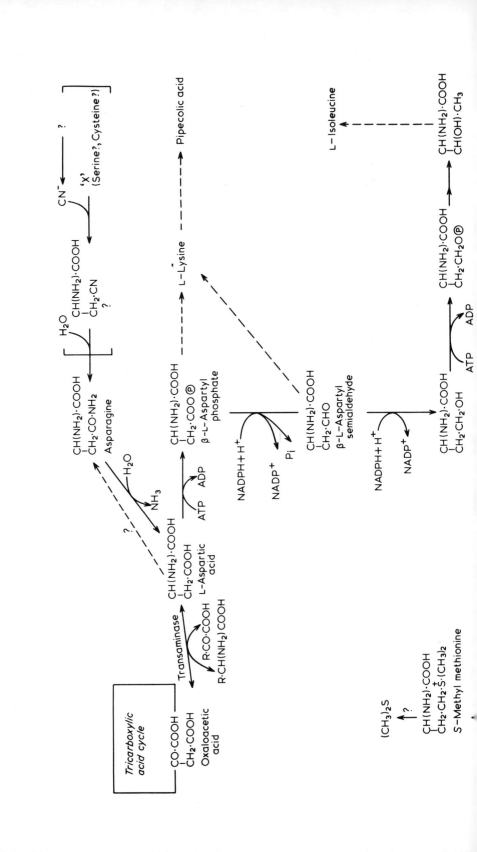

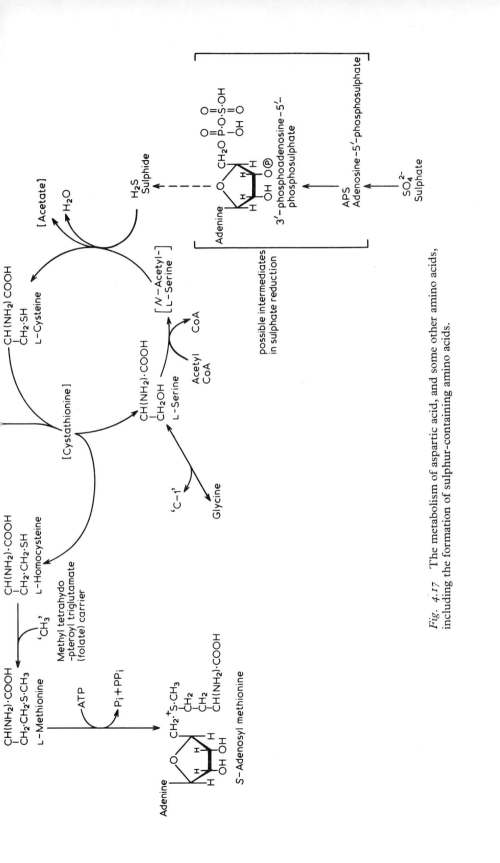

Fig. 4.17 The metabolism of aspartic acid, and some other amino acids, including the formation of sulphur-containing amino acids.

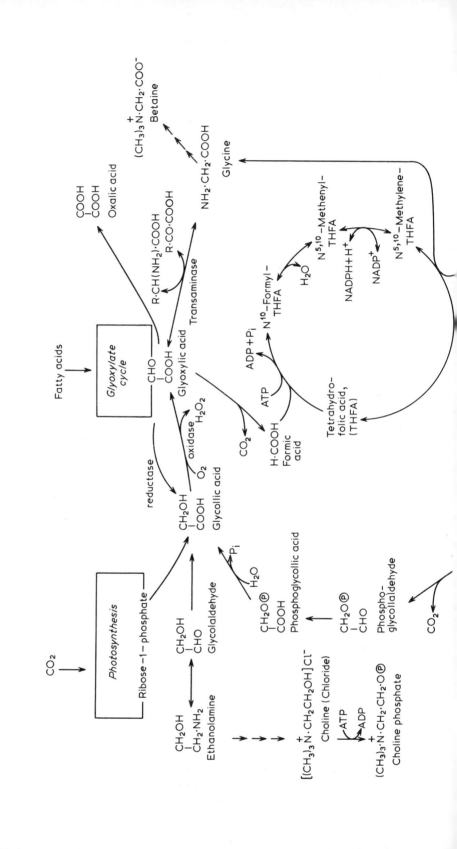

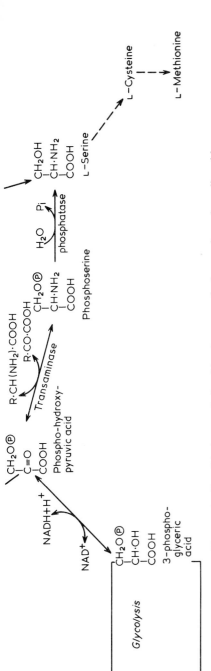

Fig. 4.18 Possible inter-relationships between glycine, serine, glyoxylic acid, glycollic acid, and some other substances.

importance of transaminations and the lability of α-amino groups. Clearly glutamate performs a central function as amino donor and exchanger, α-oxoglutarate being the equivalent receptor molecule [285]. Glutamate is decarboxylated to give γ-aminobutyric acid, which can be metabolized further [19,259]. The decarboxylase responsible has multiple forms and can also decarboxylate γ-methylene glutamic acid [96,150]. Label from [^{14}C]-glutamate enters the acids of the tricarboxylic acid cycle. This amino acid also gives rise to proline, arginine, and other amino acids [259]. Proline is also readily converted to glutamate [299]. Δ^1-Pyrroline-5-carboxylic acid dehydrogenase, which is involved in these interconvertions, occurs in barley [36]. Label from radioactive ornithine enters citrulline, arginine, proline and to a lesser extent glutamate, glutamine and γ-aminobutyrate [62].

The evidence for the routes of formation of some amino acids is poor and so the schemes shown are tentative. In other organisms two biosynthetic routes to lysine are known; labelling studies favour the diaminopimelic pathway in barley (Fig. 4.16) [236,243]. Feeding experiments with barley embryos indicated that the synthesis of some amino acids was subject to feedback control. Thus valine inhibited the synthesis of valine and isoleucine; leucine inhibited the synthesis of leucine, valine and isoleucine (Fig. 4.16) [234,236]. Feedback control evidently occurs at the acetohydroxyacid synthetase step. [235].

Label from aspartic acid readily distributes into other compounds. via oxaloacetate and the tricarboxylic acid cycle. Label is also incorporated into threonine (Fig. 4.17) [258]. Asparagine accumulates in barley under various circumstances but the routes by which it is formed and degraded remain unknown.

The sulphur amino acids are derived eventually from aspartate; the routes are not certain, but O-phospho-homoserine may react with cysteine to give cystathionine, which is cleaved to give L-homocysteine and serine (Figs. 4.17 and 4.18) [103]. The serine is derived from glycine by the addition of a C-1 unit, possibly carried by a folate cofactor (Figs. 4.17 and 4.18). Serine is probably acetylated and the O-acetyl-serine is 'sulphurylated' with sulphide to yield acetate and cysteine [262,320]. Sulphur is taken up by plants mainly as sulphate, which must be reduced before it is incorporated into cysteine. Possibly the sulphate reacts with ATP to form adenosine sulphato-phosphate (APS) which is phosphorylated further to 3'-phosphoadenosine-5'-phosphosulphate (PAPS; Fig. 4.17). This substance, a sulphate donor in various species, may be the intermediate in the reduction of sulphate to sulphide. Various plants are able to reduce sulphur dioxide to hydrogen sulphide. Homocysteine is methylated in barley by N^5-methyl-tetrahydrofolate-triglutamate to yield methionine [50]. Vitamin B_{12}, the cofactor required for this synthesis in animals, does not occur in plants. When methionine reacts with ATP

The biochemistry of barley 147

the very reactive product, S-adenosyl-methionine, is capable of many other reactions, including methylations [248]. It may even methylate methionine in extracts of barley to yield the sulphonium compound S-methyl-methionine [169] a possible precursor of the dimethyl sulphide that occurs in malts [337]. It is curious that while D and L-methionine serve equally well as methyl donors in the synthesis of N-methyl-tyramine and hordenine (in roots), the L-isomer is the better methyl donor for synthesizing gramine in the shoots [197].

The source of asparagine, other than the hydrolysis of some proteins, is a mystery. Feeding [^{14}C]-cyanide to barley gives rise to labelled asparagine. Probably the cyanide condenses with serine, or more probably cysteine, with the formation of cyanoalanine which then undergoes hydrolysis [31,94]. Cyanide is not a recognized metabolic intermediate in barley, although it occurs in some other species.

'Peptides' have been noted in barley extracts, and O-acetyl-serine is a biosynthetic intermediate. When the 'unnatural' D-isomers of several amino acids were fed to barley shoots the N-malonyl derivatives accumulated. In the case of D-phenylalanine, N-acetyl-D-phenylalanine also occurred as a minor product; possibly it was formed by decarboxylation of the malonyl derivative [295]. Traces of malonyl-tryptophan, presumably the L-isomer, occur normally in barley shoots [304].

4.9. The metabolism of some amines

Choline, betaine and ethanolamine occur free and in combination (Fig. 4.18) [131]. Phosphorylcholine is a constituent of barley sap, and roots contain a phosphorylcholine phosphatase, and choline kinase which

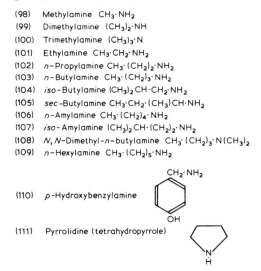

(98) Methylamine $CH_3 \cdot NH_2$
(99) Dimethylamine $(CH_3)_2 \cdot NH$
(100) Trimethylamine $(CH_3)_3 \cdot N$
(101) Ethylamine $CH_3 \cdot CH_2 \cdot NH_2$
(102) n-Propylamine $CH_3 \cdot (CH_2)_2 \cdot NH_2$
(103) n-Butylamine $CH_3 \cdot (CH_2)_3 \cdot NH_2$
(104) iso-Butylamine $(CH_3)_2 CH \cdot CH_2 \cdot NH_2$
(105) sec-Butylamine $CH_3 \cdot CH_2 \cdot (CH_3) CH \cdot NH_2$
(106) n-Amylamine $CH_3 \cdot (CH_2)_4 \cdot NH_2$
(107) iso-Amylamine $(CH_3)_2 CH \cdot (CH_2)_2 \cdot NH_2$
(108) N,N-Dimethyl-n-butylamine $CH_3 \cdot (CH_2)_3 \cdot N(CH_3)_2$
(109) n-Hexylamine $CH_3 \cdot (CH_2)_5 \cdot NH_2$

(110) p-Hydroxybenzylamine

(111) Pyrrolidine (tetrahydropyrrole)

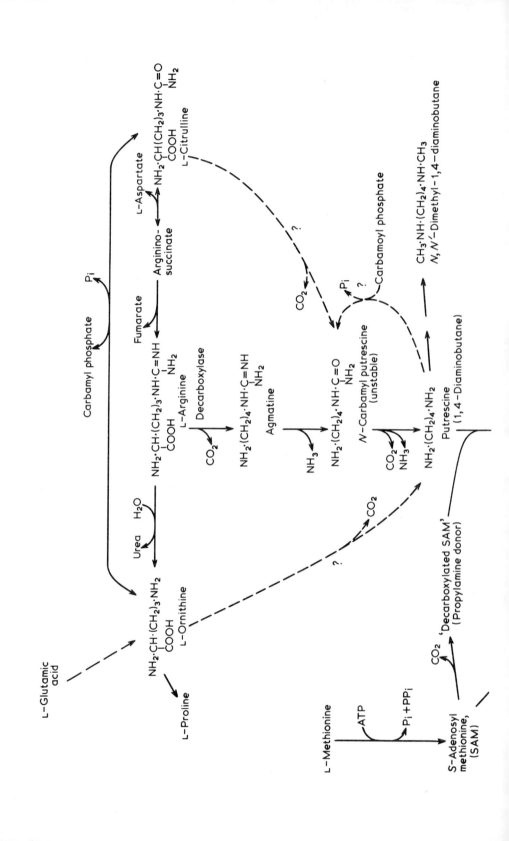

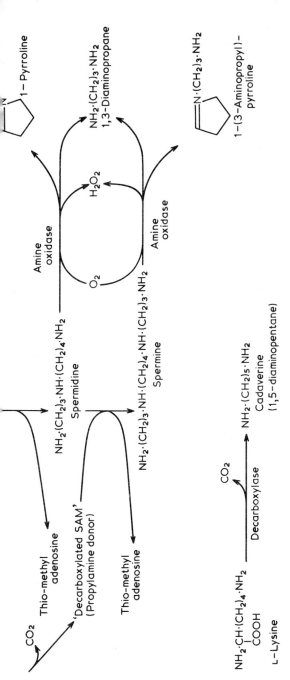

Fig. 4.19 Routes of formation of some polyamines from arginine and lysine.

150 *Barley*

catalyse its hydrolysis and its formation from choline and ATP respectively [217,340]. In various circumstances barley parts contain a wide range of other amines, including ammonia; methylamine (98); dimethylamine (99); trimethylamine (100); ethylamine (101); *n*-propylamine (102); *n*-butylamine (103); *iso*-butylamine (104); *sec*-butylamine (105); *n*-amylamine (106); *iso*-amylamine (107); *N,N*-dimethyl-*n*-butylamine (108); *n*-hexylamine (109); agmatine; 1,3-diaminopropane; putrescine (1,4-diaminobutane); *N,N'*-dimethyl-1,4-diaminobutane; cadaverine (1,5-diaminopentane); spermidine; spermine (Fig. 4.19); histamine (Fig. 4.15); tryptamine; 5-hydroxytryptamine; tyramine; *N*-methyltyramine; hordenine (*N*-dimethyltyramine); candicine (N-trimethyltyramine; Fig. 4.20); 3-aminomethylindole; 3-methylaminomethylindole; gramine (3-methylaminodimethylindole; Fig. 4.21); *p*-hydroxybenzylamine (110); pyrrolidine (tetrahydropyrrole; 111); 1-pyrroline and 1-(3-aminopropyl)-pyrroline (Fig. 4.19) [78,127,131,148,312]. Various of these amines are probably derived by the decarboxylation of the appropriate amino acids (e.g. cadaverine from lysine, Fig. 4.19, or hista-

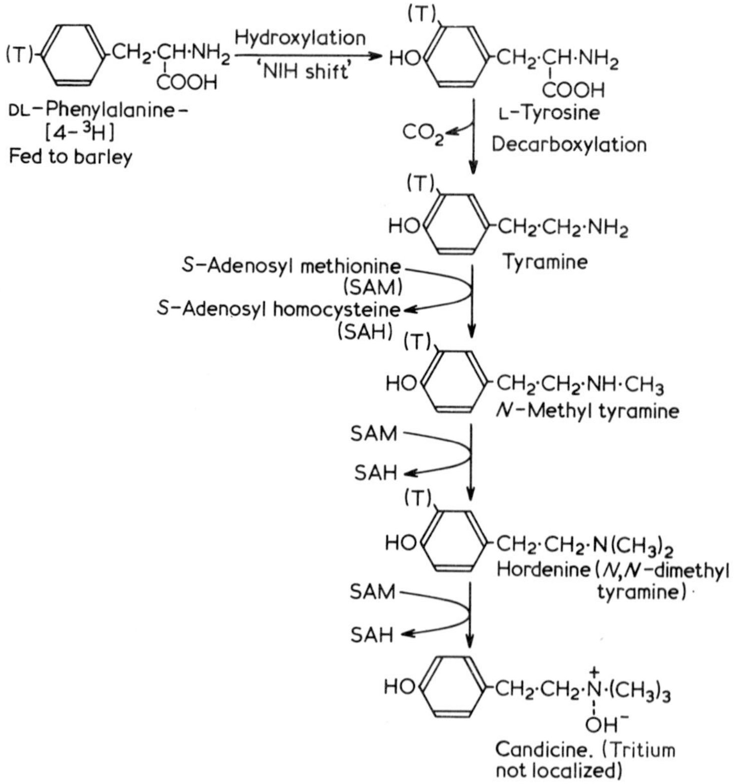

Fig. 4.20   The formation of hordenine, and related substances.

The biochemistry of barley 151

mine from histidine (Fig. 4.15). The origins of some others are discussed below. In still other cases the origins are obscure. Some other plant species are able to transaminate to aldehydes or ketones to produce amines. This type of reaction has not been found in barley. From time to time germinating barley or malt have been reported to contain aldehydes, but others claim that these are formed by the extraction methods used, or during kilning (Chapters 14 and 15). At all events aldehydes, or compounds readily able to yield them, must be present. Compounds recorded are acetaldehyde; acetone; propionaldehyde; butyraldehyde; *iso*-butyraldehyde; valeraldehyde; *iso*-valeraldehyde; furfural; diacetyl; 2,3-pentanedione; caproic aldehyde and methylethylketone [7,66,352]. Keto-acids are also present [179]. Many keto-acids are readily decarboxylated to yield aldehydes.

Agmatine, putrescine, spermidine, spermine, 1,3-diamino-propane, 1-pyrroline and 1-(3-aminopropyl)-pyrroline are normal constituents of barley, some of which increase sharply in amount under conditions of mineral imbalance or deficiency (Chapter 6) [247,323]. Coumaroyl-agmatine occurs in barley [332]. 2-Hydroxyputrescine derivatives have been found in wheat, but not barley. These amines are metabolically related to ornithine and arginine (Fig. 4.19). Ornithine and arginine decarboxylases, and carbamylputrescine amidohydrolase are known in barley; the last two alter in level on feeding acids to the plants [242,314, 319]. Perhaps the direct route from agmatine to putrescine, by the hydrolytic removal of urea, has not been excluded fully [318]. Most probably the aminopropyl groups, added successively to convert putrescine to spermidine and to spermine are donated by 'decarboxylated S-adenosyl-methionine'. The amine oxidase from barley differs significantly from that of some other plant species in attacking only secondary amino-groups. Thus spermine and spermidine are degraded with the production of 1,3-diaminopropane and 1-pyrroline and 1-(3-aminopropyl)-pyrroline respectively (Fig. 4.19) [316,317].

Tyramine, N-methyl-tyramine, hordenine (N-dimethyl-tyramine), and candicine (the quaternary derivative, N-trimethyl-tyramine) have a transient existence in barley roots (Fig. 4.20). The first three compounds have been known for a considerable period but candicine was only located when the traditional methods for dealing with alkaloids were augmented by chromatography [86,131,195,289,290]. Tyramine is formed by the decarboxylation of L-tyrosine. The decarboxylase, which has been partially purified, needs pyridoxal phosphate for its activity, and is inhibited by *p*-coumaric acid [100,144]. Tracer studies showed that the other compounds were synthesized by the successive additions of methyl groups (Fig. 4.20) [196]. The methyl donor is S-adenosyl-methionine. Levels of tyramine-methylpherase, the enzyme that catalyses the formation of N-methyl-tyramine, increase in response to dosing seedlings with kinetin

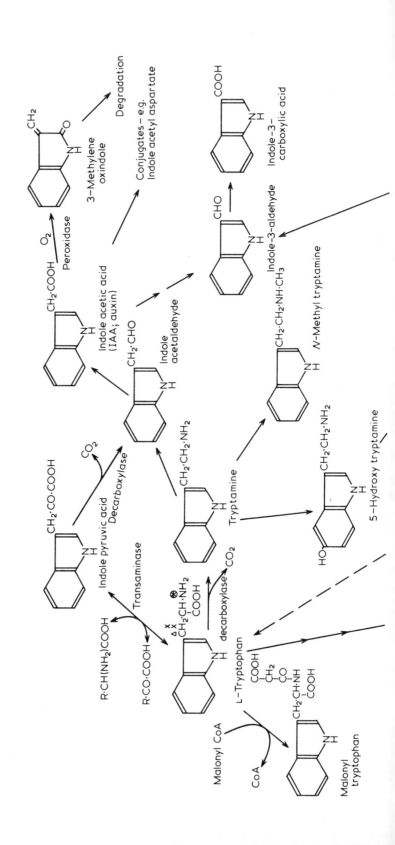

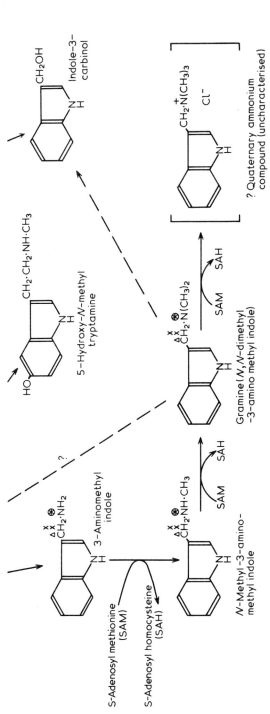

Fig. 4.21 The formation of indoleamines and other indoles in barley.

and benzylaminopurine [218,219,328]. Analysis of the position of tritium label in the tyramine, N-methyl-tyramine and hordenine derived from specifically labelled phenylalanine, showed that during the conversion of phenylalanine to tyrosine an 'NIH-shift' occurred (Fig. 4.20) [196]. In barley grown from grain sterilized with ethylene oxide, N-methyl-tyramine was the main alkaloid accumulated rather than hordenine, which is the usual major component. The occurrence of p-hydroxybenzyl-amine (110), in which the side chain has one less methylene than tyramine, is reminiscent of the relationship between tryptamine and 3-amino-methyl indole. However the route by which p-hydroxybenzylamine is formed is unknown and methylated derivatives of it have not been reported.

Hordenine is the characteristic alkaloid of young roots; gramine is characteristic of the young shoots. It, and a range of other indole derivatives, are formed from L-tryptophan (Fig. 4.21). Barley shoots contain tryptophan, malonyl-tryptophan, 3-aminomethylindole, 3-methyl-aminomethylindole, gramine (3-methyl-amino-dimethyl-indole), tryptamine, N-methyltryptamine, 5-hydroxytryptamine, 5-hydroxy-N-methyl-tryptamine, indol-3-aldehyde, and indol-3-yl-acetic acid (indole acetic acid; IAA; auxin) [249,304]. Indole-3-carboxylic acid has also been reported [229]. Labelling studies show the general correctness of the pathways shown (Fig. 4.21) [114,364]. The reactions leading to gramine formation must be complex since the atoms indicated in the diagram are retained from tryptophan [126,273]. Gramine is degraded in barley, and evidently exerts a feedback control on its own synthesis [126]. In the short term gramine is not readily converted into tryptophan [364]. In long-term experiments the side chain is oxidized to carbon dioxide, probably via indole-3-carbinol, and indole-3-carboxylic acid, and traces of label appear in tryptophan [74]. Barley shoots contain a tryptophan decarboxylase which catalyses the formation of tryptamine; this is converted to hydroxy-tryptamine; the mono-N-methylated derivatives of both substances are also formed [303]. Studies on auxin synthesis have been complicated by the instability of the indoles, and the suspicion that many of the observed metabolic changes were due to bacteria. It seems that in barley auxin (IAA) does occur, and it is synthesized both by way of indole-pyruvic acid and by way of tryptamine [102]. Tracer studies failed to reveal the normal presence of IAA conjugates although in other species conjugates with inositol, arabinose, glucose, lysine and aspartic acid have been reported. However when exogenous IAA is supplied to germinating barley grains indoleacetyl aspartic acid and an uncharacterized product are produced [240].

Barley contains peroxidases that, in the presence of various cofactors, are able to degrade indole acetic acid [190]. Degradation is probably by way of 3-methylene oxindole, a compound lacking auxin hormonal activity.

4.10. Nucleic acids, and some other nitrogenous substances

Nucleic acids from barley appear similar to those from other plants. However there are many gaps in the information available. The extraction and analysis of nucleic acids from grain is not simple, since there are highly active enzymes present that are able to hydrolyse them, and other substances are present which interfere with the usual analyses [131, 151, 192]. The nucleic acids fall into two major chemical groups (Fig. 4.22). DNA, deoxyribonucleic acid, consists of chains of deoxy-

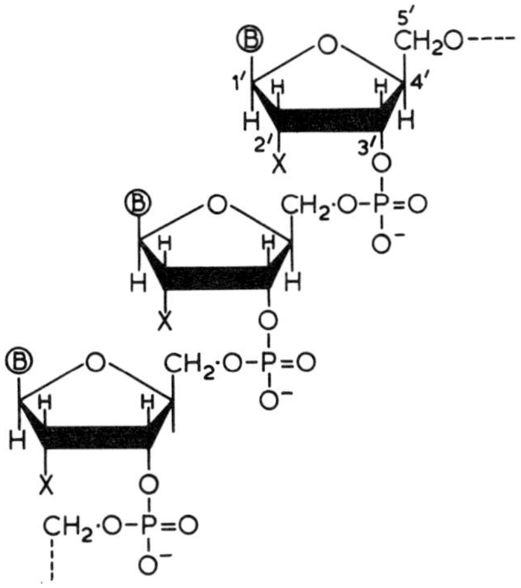

Fig. 4.22 Generalized formula of the nucleic acids. DNA, deoxyribonucleic acid; furanose sugar β-2-deoxy-D-ribose ($x = H$); bases, (B), adenine, guanine, cytosine, thymine and minor bases. RNA, ribonucleic acid; furanose sugar, ($x = OH$) β-D-ribose: bases, (B), adenine, guanine, cytosine, uracil and minor components (possibly including cytokinins and 5-methyl cytosine).

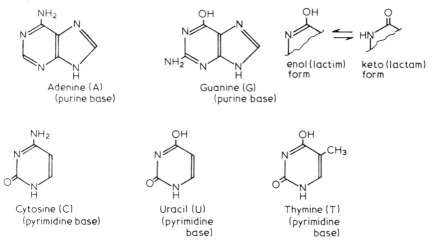

Fig. 4.23 The major purine and pyrimidine bases found in the nucleic acids.

ribofuranose linked by diester phosphate residues between the hydroxyls on positions 3′ and 5′. The 1′-positions on the sugar residues are substituted with organic bases of which adenine, guanine, cytosine, and thymine are the most important, but others may be present in minor amounts (Figs. 4.22 and 4.23). Ribonucleic acids are similar except that the sugar residues are ribofuranose, and the major bases are adenine, guanine, cytosine and uracil. Lesser amounts of other bases such as 5-methyl-cytosine are probably present; cytokinins may be among these minor bases (Figs. 4.22 and 4.23).

In the barley grain about 0.2% of the dry matter is nucleic acid, of which about 30% is DNA and 70% is RNA [131]. The leaf cells of diploid (tetraploid) barley plants are estimated to contain 18.0 (34.2) g × 10^{-12} of DNA-phosphorus, 11.2 (29.3) g × 10^{-11} of RNA-phosphorus; and 27.2 (45.7) g × 10^{-10} of protein [311b].

DNA carries at least most of the genetic information of each individual in its base sequence. Most of the DNA is in the cell nucleus but some is probably in the plastids. There are claims that exogenous bacterial DNA can enter barley roots and become integrated with the plant DNA [194]. RNA occurs at many sites in the cell. Different types of RNA have a range of apparent functions linked with protein synthesis. Evidence exists for the occurrence in barley of messenger (m)-RNA, transfer (t)-RNA, heavy and light ribosomal (r)-RNAs, chloroplastidic RNA, polysome assemblies, and in infected plants viral RNA [41,88,157,185,192,315, 327]. The complexity of the situation is greater than suggested here, as these fractions are families of distinct molecules. For example nine phenylalanine-t-RNA species have been distinguished [136]. Presumably RNA will occur wherever protein synthesis can take place, including the mitochondria. Ribonucleic acid and ribosomes are even associated with cell wall fractions [160,279].

In most species the purine bases are synthesized via orotate, and the pyrimidines are built up by the successive addition of fragments to a structure linked to ribose. Neither these synthetic steps nor the assembly of the nucleotide triphosphates into DNA and RNA have been adequately studied in barley. However wheat has been investigated and where comparative studies have been made barley and wheat are similar [357].

Barley is a rich source of hydrolytic enzymes that degrade DNA, RNA, 3′- and 5′-nucleotides, cylic 2′,3′- and 3′,5′-phosphates and phosphodiesters, and which degrade and deaminate purine and pyrimidine bases. Probably because of the potential industrial use of a cheap supply of such enzymes a number have been studied in detail [40,91,92,131,139, 188,189,257]. Extracts of barley and brewers sweet wort contain free bases, nucleosides, deoxyribonucleosides, and 3′- and 5′-nucleotides testifying to the activities of these enzymes (Chapters 15 and 16) [23]. Purines are probably degraded by way of uric acid to allantoin, then to

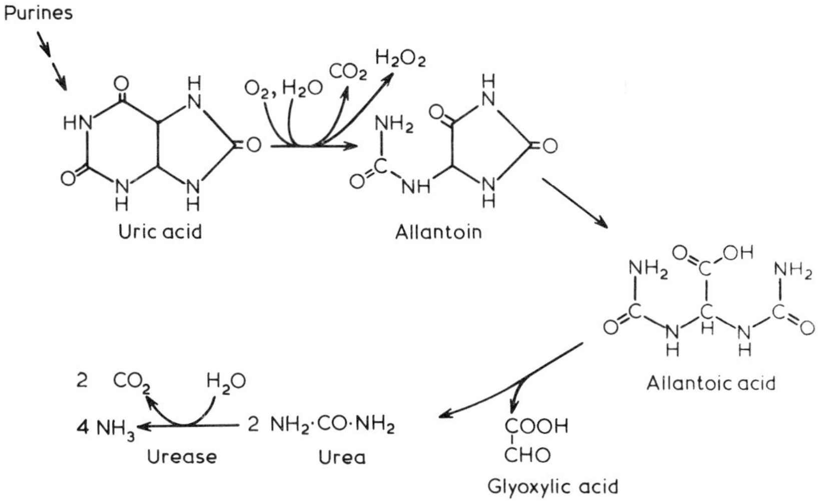

Fig. 4.24 The probable route of purine degradation.

allantoic acid and urea and glyoxylic acid. Urea is hydrolysed to carbon dioxide and ammonia (Fig. 4.24).

ATP is formed by various phosphorylative steps, as already noted. It is hydrolysed by phosphatases and specific ATPases which apparently arise as experimental artefacts from the ATP generating systems when cells are disrupted. It is also of interest that homogenates will convert fluoride to AMP-fluoride in the presence of ATP [4].

Nucleotide cofactors, derivatives of purine or pyrimidine bases with ribose (or less usually deoxyribose) are important in many aspects of metabolism as previously noted (Fig. 4.2). ATP, GTP, CTP, UTP (and the related mono- and di-phosphates), $NADP^+$ (Fig. 4.25) and UDP-N-acetyl glucosamine are present in barley extracts together with various uncharacterized nucleotide derivatives [23]. It seems certain that many more remain to be demonstrated. Phosphate, which is an integral part of nucleic acids and which plays such a vital role in intermediary metabolism, may become esterified, by the phosphorolysis of starch for example, or by substrate level-, oxidative-, or photo-phosphorylation. Inorganic phosphate entering roots is rapidly incorporated into ATP, UDP-glucose, glucose-1-phosphate, glucose-6-phosphate and phosphoglyceric acid [155,347].

Among important cofactors that must be present are coenzyme A (CoA) (112) and the hydrogen carrying cofactors NAD^+, $NADP^+$ (Fig. 4.25), FAD (114) and FMN (115). CoA (112) is a derivative of adenosine containing the vitamin pantothenic acid. NAD^+ and $NADP^+$ both contain the amide of the vitamin nicotinic acid, and adenosine (Fig. 4.25). The probable biosynthetic sequence to NAD^+ is from

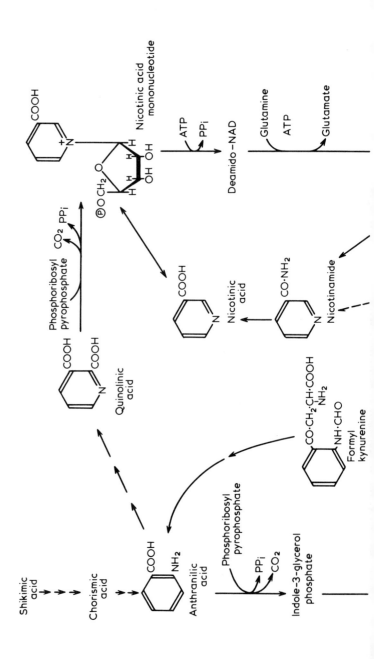

The biochemistry of barley 159

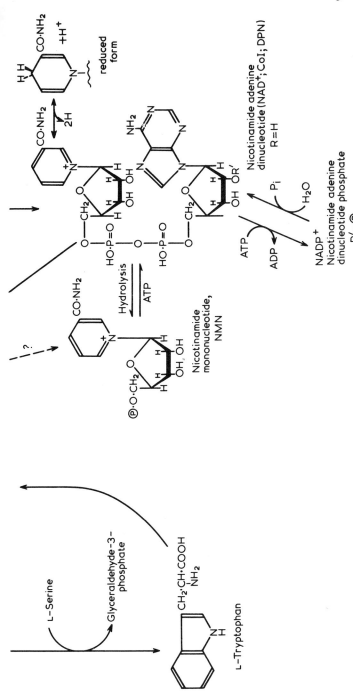

Fig. 4.25 Some biochemical relationships of NAD^+ and $NADP^+$.

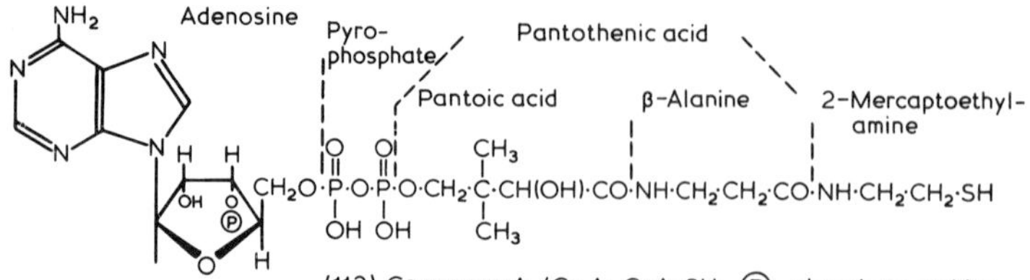

160 Barley

(112) Coenzyme A. (Co A; CoA-SH; P-phosphate residue; pantothenic acid is a vitamin)

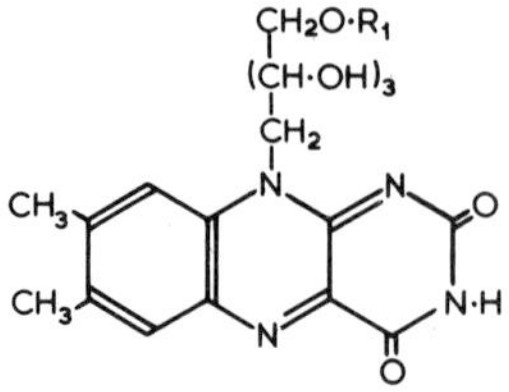

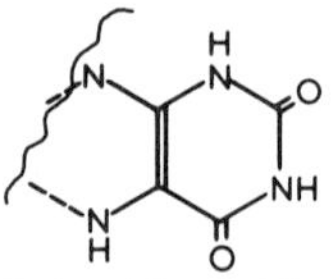

(113) $R_1 = H$ — Riboflavin (a vitamin)
(114) $R_1 =$ Pyrophosphate adenosine – Flavin adenine dinucleotide (FAD)
(115) $R_1 = \text{P}$ – Flavin mononucleotide, (FMN)

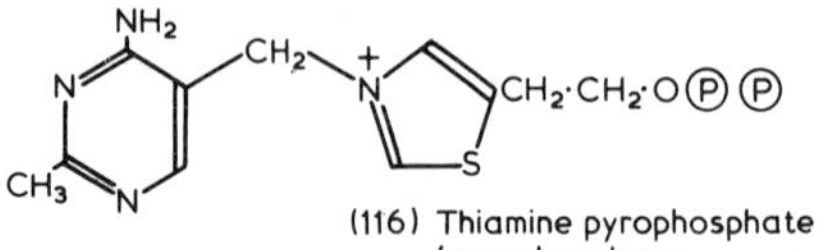

Reduced form of the *iso*-alloxazine ring.

(116) Thiamine pyrophosphate (cocarboxylase; thiamine = aneurine = vitamin B_1)

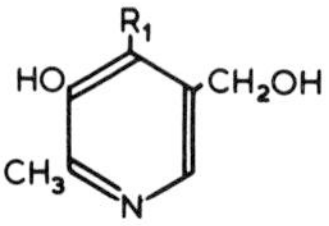

(117) Vitamin B_6
Pyridoxin $R_1 = -CH_2OH$
Pyridoxal $R_1 = -CHO$
Pyridoxamine $R_1 = -CH_2 \cdot NH_2$

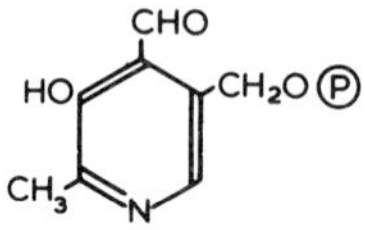

(118) Pyridoxal phosphate, a cofactor in at least some transaminations, and amino-acid decarboxylations

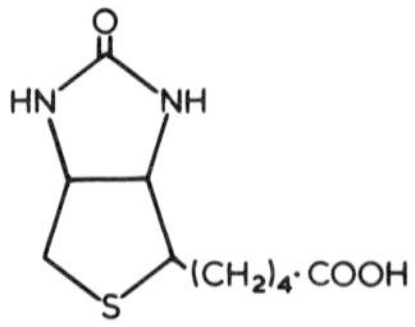

(119) Biotin (a vitamin)

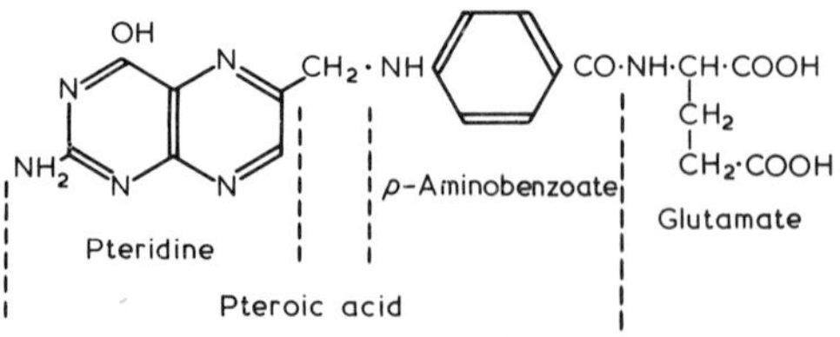

(120) Folic (Pteroyl glutamic) acid

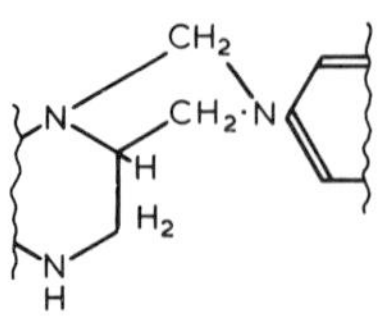

(121) $N^{5,10}$ – Methylene tetrahydrofolic acid

The biochemistry of barley 161

aspartic acid and glycerol, via quinolinic acid, nicotinic acid mononucleotide and deamido – NAD^+ (Fig. 4.25) [298]. $NADP^+$ is formed by the phosphorylation of NAD^+.

Barley contains a range of vitamins (Chapter 14) including riboflavin (113) which probably occurs combined as the cofactors FMN (115) and FAD (114); thiamine which, as the pyrophosphate (116), is a cofactor for some enzymes, as is vitamin B_6 (117; pyridoxin, pyridoxal, pyridoxamine) as the phosphate (118). Biotin (119) and various folate derivatives (120, 121; Fig. 4.18), are also enzyme cofactors.

4.11. Barley proteins

Protein synthesis in barley has been studied by immunological means, and by following the incorporation of amino acids labelled with radioactive or heavy isotopes, and by following changes in the quantities of enzyme proteins. The system most thoroughly investigated is the formation of α-amylase in the aleurone layer, but the synthesis of many other proteins has been demonstrated (Chapter 5) [42,130,288,331]. Some situations are complicated. Thus, as in other species, the large subunits of ribulose diphosphate carboxylase are synthesized in the chloroplast, while the small subunits are synthesized in the cytoplasm and must enter the chloroplasts before the parts can be assembled, and active enzyme formed [63]. The evidence available indicates that protein formation follows the generally accepted pathway, but there are numerous gaps in the evidence. It is believed that DNA, mostly situated in the nucleus, serves as a template for m-RNA, ribosomal RNA and t-RNA. In the cytoplasm, or perhaps within particular organelles, the ribosomal RNA is formed into ribosomes which become attached, at a particular starting-point, to messenger (m)-RNA. The sequence of bases in the m-RNA was determined by the DNA that acted as its template; it, in turn, orders the amino acids as the polypeptide chain is synthesized. In the cytoplasm the amino acids are activated by reaction with ATP, and the activated amino-acyl residues are transferred to transfer (t)-RNA molecules to form a series of amino acid-t-RNAs. These arrange themselves along the m-RNA strand, in an order dictated by the base-sequence. As the ribosome moves along the messenger the amino acid-t-RNA molecules are broken down, the t-RNA is released and the amino acid is linked onto the end of the growing peptide chain. At the end of the sequence the ribosome and the peptide chain are both released. The peptide chain is folded and cross-linked and substitution is carried out to form a protein or protein sub-unit. These may be modified further in the cytoplasm.

Historically the proteins of the grain received the most attention, because of their nutritional and technological significance. This interest continues. In the first studies, by Osborne [276] and others, grain proteins were separated into solubility classes, albumin, globulin, hordein and glutelin (Chapter 6) [30]. This work has been greatly extended subsequently [131,286]. Albumins and globulins are soluble in salt solutions and may be separated by selective precipitation caused by salt removal or other means. The hordeins may be selectively dissolved from the salt insoluble residues in hot, 70% ethanol, the nitrogenous material remaining being the glutelins. Other extractants are now used. Complicating factors are recognized relating to difficulties in extraction, alterations in solubility due to heating while grinding the grain prior to extraction, interference caused by polyphenols and phytate, and the effects of oxidizing or reducing conditions. The solubilities of hordein and glutelin can be increased by the presence of reducing agents, thiols or metabisulphite, which reduce the disulphide bridges which bind the molecules together [286]. Some extractant mixtures, containing detergents, are claimed to extract 94% of grain protein [187]. The crude solubility fractions have their uses, e.g. in preliminary separations in studies on high-lysine barley mutants [39,152,153]. Furthermore they demonstrate the regularity principle of grain composition (Chapter 6). Other solubility fractions have been prepared, e.g. to evaluate the levels of cohesive proteins in different cereals. Barley contains only a little cohesive protein, with little extensibility, which explains why barley flour tends to give flat bread when it is baked [64]. It is clear that the 'classical' Osborne solubility fractions are highly heterogeneous [131]. With more modern techniques even with hordein, an intractable experimental material, at least six components have been demonstrated [231]. The salt-soluble fraction, which was seen to contain about 21 components by immunoelectrophoresis has been shown to contain more than 54 components by crossed immunoelectrophoresis [76,133].

Protein molecules are rarely 'pure'. Thus α-amylase contains bound calcium and possibly carbohydrate residues; plastocyanin contains copper; ferredoxin contains iron; cytochromes contain haem residues; many flavoproteins are present, and nucleoproteins occur [6]. Two hordothionins, separable by electrophoresis, are soluble in petroleum ether and contain protein (45%), carbohydrate (24%), and lipid; they are therefore glycolipo-proteins [143,291].

D-Glucosamine is incorporated into barley glycoproteins as N-acetyl-D-glucosamine [294]. The glycoprotein nature of some esterase and acid phosphatase enzymes has been demonstrated by their retention by concanavalin A on electrophoresis [34]. Some proteins associated with chromatin in the cell nucleus become phosphorylated [56]. Other proteins contain N-feruloyl substituents [336]. A barley protein is known that resembles wheat germ agglutinin (lectin) in being able to coagulate red blood cells, a capability inhibited by N-acetyl glucosamine [95]. Other proteins are present that are able to inhibit proteases.

In addition to proteins and free amino acids barley grains contain various peptides. These may be products of partial proteolysis, or serve some unknown function. Glutathione (γ-glutamyl-cysteinyl-glycine) is present in its oxidized and reduced forms [172]. This may be involved in respiration, or control the levels of, *inter alia*, reduced (free) β-amylase in the germinating grain.

Barley contains enzymes capable of hydrolysing proteins, polypeptides, di- and tri-peptides, amides, and various synthetic substrates. With the exception of autolytic studies the degradation of 'natural' substrates, derived from barley, have rarely been investigated [15,131]. Older work demonstrated the presence of at least two peptidases, soluble thiol-activated and thiol-independent proteases and insoluble proteases. Modern work has revealed the complexity of the mixture of soluble proteinases and peptidases. Unfortunately it is usually not possible to decide when different investigations relate to the same enzyme, or when apparent differences are due to studies being made on essentially different barley varieties, or to different experimental techniques being used. However, up to five endopeptidases have been detected in germinated grain [48,85]. At least three carboxypeptidases, three aminopeptidases and two other peptidases have been distinguished and carboxypeptidases have been purified [49,238,239,244,287,324,351].

Indications that protease inhibitors occur in barley have been found in brewing trials, and cereal protease inhibitors may be significant in reducing the digestibility of foodstuffs (Chapter 16) [201]. Different inhibitors occur in the embryo and endosperm of the grain, and the levels alter during the formation of the grain and during germination. Different inhibitors are active against barley proteases, trypsin, chymotrypsin and *Aspergillus* protease. A protein trypsin inhibitor from barley has been purified [177,237,274]. No less than 22 trypsin inhibitory zones and 11 chymotrypsin inhibitory zones were separated by isoelectric focussing extracts of barley [44].

References

1 Acker, K. (1974). *Ber. Getreidechem.-Tag. Detmold*, 111–127.
2 Adams, G.A. and Castagne, A.E. (1949). *Can. J. Res.*, **B27**, 915–923.
3 Akoyunoglou, G., Argyrondi-Akouyungo, J.H., Michel-Wolwertz, M.R. and Sironval, C. (1967). *Chim. Chrom*, **32A**, 5–8.
4 Alvarez, R., Moore, T.C. and Vandepeute, J. (1974). *Pl. Physiol., Lancaster*, **53**, 144–148.
5 Anderson, F.B. and Manners, D.J. (1959). *Biochem. J.*, **71**, 407–411.
6 Äyräpää, T. (1957). *Acta chem. scand.*, **11**, 1565–1575.
7 Arkima, V. and Ronkainen, P. (1971). *Monats. Brau.*, **24**, 161–163.
8 Aspinall, G.O. and Greenwood, C.T. (1962). *J. Inst. Brew.*, **68**, 167–178.
9 Aspinall, G.O. and Ross, K.M. (1963). *J. chem. Soc.*, 1681–1686.

10 Aylward, F. and Showler, A.J. (1962). *J. Sci. Fd. Agric.*, **13**, 92–95; 492–496.
11 Ballance, G.M. and Meredith, W.O.S. (1976). *J. Inst. Brew.*, **82**, 64–67.
12 Banks, W., Greenwood, C.T. and Thomson, J. (1959). *Makromol. Chem.*, **31**, 197.
13 Banks, W., Greenwood, C.T. and Muir, D.D. (1974). *Stärke*, **26**, 289–300.
14 Bathgate, G.N. and Palmer, G.H. (1972). *Stärke*, **24**, 336–341.
15 Baxter, E.D. (1976). *J. Inst. Brew.*, **82**, 203–208.
16 Baxter, E.D. and Duffus, C.M. (1971). *Phytochemistry*, **10**, 2641–2644.
17 Baxter, E.D. and Duffus, C.M. (1973a). *Phytochemistry*, **12**, 2321–2330.
18 Baxter, E.D. and Duffus, C.M. (1973b). *Planta*, **114**, 195–198.
19 Beevers, H. (1951). *Biochem. J.*, **48**, 132–137.
20 Beevers, H. (1961). *Respiratory Metabolism in Plants*. White Plains, New York: Row, Peterson & Co.
21 Benson, A.A. and Calvin, M. (1950). *J. exp. Bot.*, **1**, 63–68.
22 Benson, A.A. and Maruo, B. (1958). *Biochim. biophys. Acta*, **27**, 189–195.
23 Bergkvist, R. (1957). *Acta chem. Scand.*, **11**, 1457–1464.
24 Berner, E. (1971). *Physiologica Pl.* Suppl. **VI**; 1–56.
25 Bhagvat, K. and Hill, R. (1951). *New Phytol.*, **50**, 112–120.
26 Bhatia, I.S., Kanshal, G.P. and Bajag, K.L. (1972). *Phytochemistry*, **11**, 1867–1868.
27 Bickel, H. and Schultz, G. (1975). *Ber. dt. bot. Ges.*, **87**, 281–290.
28 Bickoff, E.M. (1968). *Am. Perfum. Cosmet.*, **83**, 58–62, 64.
29 Bilbao, J.L.G. (1966). *Cerv. y. Malta*, (Sept.) 7–13 via (1967). *J. Inst. Brew.*, **73**, 318.
30 Bishop, L.R. (1929). *J. Inst. Brew.*, **35**, 316–322.
31 Blumenthal-Goldschmidt, S., Butler, G.W. and Conn, E.E. (1963). *Nature*, **197**, 718–719.
32 Boardman, N.K. and Thorne, S.W. (1968). *Biochim. biophys. Acta.*, **153**, 448–458.
33 Boardman, N.K. and Highkin, H.R. (1966). *Biochim. biophys. Acta.*, **126**, 189–199.
34 Bøg-Hansen, T.C., Brogren, C.-H. and McMurrough, I. (1974). *J. Inst. Brew.*, **80**, 443–446.
35 Bøg-Hansen, T.C. and Daussant, J. (1974). *Analyt. Biochem.*, **61**, 522–527.
36 Boggess, S.F. and Stewart, C.R. (1974 – June). *Pl. Physiol. Abstr.*, p. 52.
37 Bourdu, R., Mathieu, Y., Miginiac-Maslon, M., Rémy, R. and Moyse, A. (1968). *Planta*, **80**, 191–210.
38 Bourne, W.F. and Miflin, B.J. (1973). *Planta*, **111**, 47–56.
39 Brandt, A. (1975). *FEBS Letters*, **52**, 288–291.
40 Brawerman, G. and Chargaff, E. (1954). *J. biol. Chem.*, **210**, 445–454.
41 Breen, M.D., Whitehead, E.I. and Kenefick, D.G. (1971). *Pl. Physiol., Lancaster*, **49**, 733–739.
42 Briggs, D.E. (1973). In *Biosynthesis and its Control in Plants* (Milborrow, B.V. ed.) London: Academic Press, pp. 219–277.
43 Briggs, D.E. (1974). *Phytochemistry*, **13**, 987–996.
44 Bruhn, L.C. and Djurtoft, R. (1976). Personal communication.
45 Buchala, A.J. (1973). *Phytochemistry*, **12**, 1373–1376.
46 Buchala, A.J. and Wilkie, K.C.B. (1970). *Naturwissenchaften*, **57**, 496.
47 Burden, R.S., Firn, R.D., Hiron, R.W.P., Taylor, H.F. and Wright, S.T.C. (1971). *Nature (New Biol.)*, **234**, 95–96.
48 Burger, W.C. (1973). *Pl. Physiol., Lancaster*, **51**, 1015–1021.

49 Burger, W.C., Prentice, N. and Moeller, M. (1971). *J. Inst. Brew.*, **77**, 291–294.
50 Burton, E.G. and Sakami, W. (1969). *Biochem. biophys. Res. Commun.*, **36**, 228–234.
51 Bush, P.B., Grunwald, C. and Davis, D.L. (1971). *Pl. Physiol.*, Lancaster, **47**, 745–749.
52 Byerrum, R.U., Flokstra, J.H., Dewey, L.J. and Ball, C.D. (1954). *J. Biol. Chem.*, **210**, 633–643.
53 Canvin, D.T. and Atkins, C.A. (1974). *Planta*, **116**, 207–224.
54 Carlin, R.M. and McClure, J.W. (1973). *Phytochemistry*, **12**, 1009–1015.
55 Castelfranco, P.A. and Jones, O.T.G. (1975). *Pl. Physiol.*, Lancaster, **55**, 485–490.
56 Chapman, K.S.R., Trewavas, A. and van Loon, L.C. (1975). *Pl. Physiol.*, Lancaster, **55**, 293–296.
57 Chrispeels, M.J., Tenner, A.J. and Johnson, K.D. (1973). *Planta*, **113**, 35–46.
58 Clark, R.B. (1969). *Crop. Sci.*, **9**, 341–343.
59 Claussen, K.A. and Pepper, E.H. (1968). *Cereal Chem.*, **45**, 124–132.
60 Cocking, E.C. and Yemm, E.W. (1961). *New Phytol.*, **60**, 103–116.
61 Coleman, R.A. and Pratt, L.H. (1974). *Planta*, **121**, 119–131.
62 Coleman, R.G. and Hegarty, M.P. (1957). *Nature*, **179**, 376–377.
63 Criddle, R.S., Dau, B., Kleinkopf, G.E. and Huffaker, R.C. (1970). *Biochem. biophys. Res. Commun.*, **41**, 621–627.
64 Cunningham, D.K., Geddes, W.F. and Anderson, J.A. (1955). *Cereal Chem.*, **32**, 91–106.
65 Daly, J.M. and Brown, A.H. (1954). *Archs. Biochem. Biophys.*, **52**, 380–387.
66 Damm, E. and Kringstad, H. (1964). *J. Inst. Brew.*, **70**, 38–42.
67 Dassiou, C. and Akoyunoglou, G. (1969). *Physiologia Pl.*, **23**, 570–574.
68 Daussant, J., Grabar, P. and Nummi, M. (1965). *Proc. eur. Brew. Conv. Stockholm*, pp. 62–69.
69 Davenport, H.E. (1952). *Nature*, **170**, 1112–1114.
70 Davies, B.H., Rees, A.F. and Taylor, R.F. (1975). *Phytochemistry*, **14**, 717–722.
71 De Haas, B.W. and Goering K.J. (1972). *Stärke*, **24**, 145–180.
72 Dézsi L. and Farkas, G.L. (1964). *Acta biol. hung.*, **14**, 325–332.
73 Dierickx, P.J. and Buffel, K. (1972). *Phytochemistry*, **11**, 2654–2655.
74 Digenis, G.A. (1969). *J. pharm. Sci.*, **58**, 39–49.
75 Djurtoft, R. and Rasmussen, K.L. (1955). *Proc. eur. Brew. Conv. Baden-Baden*, p. 17–25.
76 Djurtoft, R. and Hill, R.J. (1965). *Proc. eur. Brew. Conv. Stockholm*, pp. 137–146.
77 Doman, N.G. and Romanova, A.K. (1962). *Pl. Physiol.*, Lancaster, **37**, 833–840.
78 Drews, B., Just, F. and Drews, H. (1957). *Proc. eur. Brew. Conv. Copenhagen*, pp. 167–172.
79 Drews, B., Specht, H. and Hinze, H.J. (1966). *Naturwissenschaften*, **53**, 406–407.
80 Duffus, C.M. (1970). *Phytochemistry*, **9**, 1415–1421.
81 Edelman, J., Shibko, S.I. and Keys, A.J. (1959). *J. exp. Bot.*, **10**, 178–189.
82 El-Basyouni, S.Z. and Neish, A.C. (1966). *Phytochemistry*, **5**, 683–691.
83 El-Basyouni, S.Z., Chen, D., Ibrahim, R.K., Neish, A.C. and Towers, G.H.N. (1964). *Phytochemistry*, **3**, 485–492.
84 Elliott, D.C. (1954). *J. exp. Bot.*, **5**, 353–356.

85 Enari, T.M. (1969). *Valtion Tek. Tutkimuslaitos, Tiedotus*, **103** (4), 1-21.
86 Erspamer, V. and Falconieri, G. (1952). *Naturwissenschaften*, **39**, 431-432.
87 Evans, H.J. (1955). *Pl. Physiol., Lancaster*, **30**, 437-444.
88 Fedorcsák, I., Natarajan, A.T. and Ehrenberg, L. (1969). *Eur. J. Biochem.*, **10**, 450-458.
89 Ferguson, W.S. (1966). *Can. J. Pl. Sci.*, **46**, 639-646.
90 Ferrari, T.E. and Varner, J.E. (1970). *Pl. Physiol., Lancaster*, **46** (Suppl.) 45.
91 Fiers, W. (1962). *J. Chromat.*, **7**, 269-271.
92 Fiers, W. and Vandendriessche, L. (1961). *Archs. int. Physiol. Biochim.*, **69**, 339-363.
93 Firn, R.D. and Friend, J. (1972). *Planta*, **103**, 263-266.
94 Floss, H.G., Hadwiger, L. and Conn, E.E. (1965). *Nature*, **208**, 1207-1208.
95 Foriers, A., De Neve, R. and Kanarek, L. (1975). *Arch. int. Physiol. Biochim.*, **83**, 362.
96 Fowden, L. (1954). *J. exp. Bot.*, **5**, 28-36.
97 Fröst, S. and Holm, G. (1972). *Hereditas*, **70**, 259-264.
98 Fritz, G. and Beevers, H. (1955). *Pl. Physiol., Lancaster*, **30**, 309-317.
99 Frydenberg, O. and Nielsen, G. (1965). *Hereditas*, **54**, 123-139.
100 Gallon, J.R. and Butt, V.S. (1971). *Biochem. J.*, **123**, 5P-6P.
101 Ghadimi, R., Zimmermann, G. and Latzko, E. (1974). *Brauwissenschaft*, **27**, 164-167.
102 Gibson, R.A., Schneider, E.A. and Wightman, F. (1972). *J. exp. Bot.*, **23**, 381-399.
103 Giovanelli, J., Mudd, S.H. and Datko, A.H. (1974). *Pl. Physiol., Lancaster*, **54**, 725-736.
104 Goering, K.J. and Eslick, R. (1976). *Cereal Chem.*, **53**, 174-180.
105 Goering, K.J., Eslick, R. and De Haas, B.W. (1973). *Cereal Chem.*, **50**, 322-328.
106 Goering, K.J., Fritts, D.H. and Eslick, R.F. (1973). *Stärke*, **25**, 297-302.
107 Goering, K.J., Jackson, L.L. and De Haas, B.W. (1975). *Cereal Chem.*, **52**, 493-500.
108 Goering, K.J. and Breisford, D.L. (1965). *Cereal Chem.*, **42**, 15-24.
109 Goering, K.J. and De Haas, B. (1974). *Cereal Chem.*, **51**, 573-578.
110 Goldin, B.R. and Little, H.N. (1969). *Biochim. biophys. Acta*, **171**, 321-332.
111 Gore, M.G., Hill, H.M., Evans, R.B. and Rogers, L.J. (1974). *Phytochemistry*, **13**, 1657-1665.
112 Gorinssen, H., Debeer, L., Vancraenenbroeck, R., Lontie, A. (1970). *Bull. Ass. r. Anciens. Étud. Brass. Univ. Louvain*, **66** (2), 73-89.
113 Gough, S. (1972). *Biochim. biophys, Acta*, **286**, 36-54.
114 Gower, B.G. and Leete, E. (1963). *J. Am. Chem. Soc.*, **85**, 3683-3685.
115 Grabar, P. and Daussant, J. (1964). *Cereal Chem.*, **41**, 523-532.
116 Graham, J.S.D. and Young, L.C.T. (1959). *Pl. Physiol., Lancaster*, **34**, 520-526.
117 Gramshaw, J.W. (1967). *J. Inst. Brew.*, **73**, 258-270; 455-472.
118 Gramshaw, J.W. (1968). *J. Inst. Brew.*, **74**, 20-38.
119 Graveland, A., Pesman, L. and van Eerde, P. (1972). *Master Brewers Ass. Am. Tech. Q.*, **9** (2), 98-104.
120 Gray, I.K., Rumsby, M.G. and Hawke, J.C. (1967). *Phytochemistry*, **6**, 107-113.
121 Green, J., Maicinkiewicz, S. and Watt, P.R. (1955). *J. Sci. Fd. Agric.*, **6**, 274-282.

The biochemistry of barley 167

122 Greenwood, C.T. and Thomson, J. (1961). *J. Inst. Brew.*, **67**, 64–67.
123 Griffiths, W.T. (1975). *Biochem. J.*, **152**, 623–635.
124 Griffiths, W.T., Threlfall, D.R. and Goodwin, T.W. (1967). *Biochem. J.*, **103**, 589–600.
125 Groat, J.I. and Briggs, D.E. (1969). *Phytochemistry*, **8**, 1615–1627.
126 Gross, D., Lehmann, H., Schütte, H.-R. (1970). *Z. PflPhysiol.*, **63**, 1–9.
127 von Haartmann, U., Kahlson, G. and Steinhardt, C. (1966). *Life Sci.*, **5**, 1–9.
128 Hall, D.O., Huffaker, R.C., Shannon, L.M. and Wallace, A. (1959). *Biochim. biophys. Acta*, **35**, 540–542.
129 Hallaway, M., Phethean, P.D. and Taggart, J. (1970). *Phytochemistry*, **9**, 935–944.
130 Hardie, D.G. (1975). *Phytochemistry*, **14**, 1719–1722.
131 Harris, G. (1962). In *Barley and Malt*, (ed. Cook, A.H.) pp. 431–582; 583–694. London: Academic Press.
132 Harris, P.J. and Hartley, R.D. (1976). *Nature*, **259**, 508–510.
133 Hejgaard, J. and Bøg-Hansen, T.C. (1974). *J. Inst. Brew.*, **80**, 436–442.
134 Henningsen, K.W. and Boardman, N.K. (1973). *Pl. Physiol., Lancaster*, **51**, 1117–1126.
135 Henningsen, K.W., Thorne, S.W. and Boardman, N.K. (1974). *Pl. Physiol., Lancaster*, **53**, 419–425.
136 Hiatt, V.S. and Snyder, L.A. (1973). *Biochim. biophys. Acta.*, **324**, 57–68.
137 Higuchi, T., Ito, Y., Shimada, M. and Kawamura, I. (1967). *Phytochemistry*, **6**, 1551–1556.
138 Hølmer, G., Ory, R.L. and Høy, C.E. (1973). *Lipids*, **8**, 277–283.
139 Holbrook, J., Ortanderl, F. and Pfleiderer, G. (1966). *Biochem. Z.*, **345**, 427–439.
140 Honda, S.I. (1955). *Pl. Physiol., Lancaster*, **30**, 174–181; 402–410.
141 Honda, S.I. (1957). *Pl. Physiol., Lancaster*, **32**, 23–31.
142 Horak, A. and Zalik, S. (1973). *Pl. Physiol., Lancaster*, **51** (Suppl.) 66.
143 Hoseney, R.C., Pomeranz, Y., Hubbard, J.D. and Phinney, K.F. (1971). *Cereal Chem.*, **48**, 223–229.
144 Hosoi, K. (1974). *Pl. Cell Physiol.*, **15**, 429–440.
145 Hosoi, K., Yoshida, S., Hasegawa, M. (1969). *Shokubutsugaku Zasshi*, **82**, (972). 239 via (1969). *Chem. Abs.*, **71**, 87866z.
146 Houston, D.F., Hill, B.E., Garrett, V. and Kester, E.B. (1963). *J. Agric. Fd. Chem.*, **11**, 512–517.
147 Houston, C.W., Latimer, S.B. and Mitchell, E.D. (1974). *Biochim. biophys. Acta.*, **370**, 276–282.
148 Hrdlička, J., Dyr, J. and Zimova, L. (1968). *Sci Papers Inst. Chemical Technol., Prague*, **E.20**, 27–32.
149 Igarashi, O., Noguchi, M. and Fujimaki, M. (1968). *Agric. biol. Chem., Tokyo*, **32**, 272–278.
150 Inatomi, K. and Slaughter, J.C. (1975). *Biochem. J.*, **147**, 479–484.
151 Ingle, J. (1963). *Phytochemistry*, **2**, 353–370.
152 Ingversen, J. and Køie, B. (1971). *Hereditas*, **69**, 319–323.
153 Ingversen, J. and Køie, B. (1973). *Phytochemistry*, **12**, 73–78; 1107–1111.
154 Jackson, L.L. (1971). *Phytochemistry*, **10**, 487–490.
155 Jackson, P.C. and Hagen, C.E. (1960). *Pl. Physiol., Lancaster*, **35**, 326–332.
156 Jacobsen, J.V. and Knox, R.B. (1973). *Planta*, **112**, 213–224.
157 Jacobsen, J.V. and Zwar, J.A. (1974). *Aust. J. Pl. Physiol.*, **1**, 343–356.
158 Jacobson, L. (1955). *Pl. Physiol., Lancaster*, **30**, 264–269.
159 James, W.O. and Lundegårdh, H. (1959). *Proc. R. Soc. B.*, **150**, 7–12.

160 Jervis, L. and Hallaway, M. (1970). *Biochem. J.*, **117**, 505–507.
161 Johnson, K.D. and Kende, H. (1971). *Proc. natn. Acad. Sci., U.S.A.*, **68**, 2674–2677.
162 Jørgensen, B.B. and Jørgensen, O.B. (1963). *Acta chem. Scand.*, **17**, 1765–1770.
163 Jones, D.F., Macmillan, J. and Radley, M. (1963). *Phytochemistry*, **2**, 307–314.
164 Juneja, P.S., Pearcy, S.C., Gholson, R.K., Burton, R.L. and Starks, K.J. (1975). *Pl. Physiol., Lancaster*, **56**, 385–389.
165 Kannangara, C.G. (1969). *Pl. Physiol., Lancaster*, **44**, 1533–1537.
166 Kannangara, C.G., Henningsen, K.W., Stumpf, P.K. and von Wettstein, D. (1971). *Eur. J. Biochem.*, **21**, 334–338.
167 Kannangara, C.G., Henningsen, K.W., Stumpf, P.K., Appelqvist, L.A. and von Wettstein, D. (1971). *Pl. Physiol., Lancaster*, **48**, 526–531.
168 Kannangara, C.G. and Jensen, C.J. (1975). *Eur. J. Biochem.*, **54**, 25–30.
169 Karr, D., Tweto, J. and Albersheim, P. (1967). *Archs. Biochem. Biophys.*, **121**, 732–738.
170 Kasperek, M. (1968). *Physiol. Vég.*, **6**, 19–25.
171 Kasting, R. and Delwiche, C.C. (1957). *Pl. Physiol., Lancaster*, **32**, 471–475.
172 Kauppinen, V., Nummi, M. and Enari, T.-M. (1967). *Brauwissenschaft*, **20**, 1–4.
173 Keller, C.J. and Huffaker, R.C. (1967). *Pl. Physiol., Lancaster*, **42**, 1277–1283.
174 Kindl, H. (1970). *Hoppe-Seyler's Z. physiol. Chem.*, **351**, 792–798.
175 Kirchoff, H.C. (1815). *Schweigger's J. Chemie Physik*, **14**, 389–398.
176 Kirchhoff, M. (1816). *J. Pharm., Anvers*, **2**, 250–258.
177 Kirsi, M. (1974). *Physiologia Pl.*, **32**, 89–93.
178 Kleinkopf, G.E., Huffaker, R.C and Matheson, A. (1970). *Pl. Physiol., Lancaster*, **46**, 204–207; 416–418.
179 Knorr, F. (1956). *Brauwissenschaft*, **9**, 286–287.
180 Kotzé, J.P. and Latzko, E. (1965). *Z. PflPhysiol.*, **53**, 320–333.
181 Koukol, J. and Conn, E.E. (1961). *J. biol. Chem.*, **236**, 2692–2698.
182 Krall, A.R. (1955). *Pl. Physiol., Lancaster*, **30**, 269–271.
183 Krall, A.R. and Tolbert, N.E. (1957). *Pl. Physiol., Lancaster*, **32**, 321–326.
184 Kringstad, H. (1972). *Brauwissenschaft*, **25**, 272–277.
185 Kummert, J. and Semal, J. (1972). *J. gen. Virol.*, **16**, 11–20.
186 La Berge, D.E. and Meredith, W.O.S. (1971). *J. Inst. Brew.*, **77**, 436–442.
187 Landry, J., Moureaux, T. and Huet, J.C. (1972). *Bios.*, **3**, 281–292.
188 Lantero, O.J. and Klosterman, H.J. (1973). *Phytochemistry*, **12**, 775–784.
189 Laufer, L. and Gutcho, S. (1968). *Biotechnol. Bioengn.*, **10**, 257–275.
190 Laurema, S. (1974). *Physiologia Pl.*, **30**, 301–306.
191 Latzko, E. and Kotzé, J.P. (1965). *Z. PflPhysiol.*, **53**, 377–387.
192 Lázár, G., Fedorcsák, I. and Solymosy, F. (1969). *Phytochemistry*, **8**, 2353–2365.
193 Le Corvaisier, H. (1939). *Bull. Soc. Sci. Bretagne*, **15**, (extra) 1–74.
194 Ledoux, L. and Huart, R. (1969). *J. mol. Biol.*, **43**, 243–262.
195 Lee, S.R. (1958). *Seoul Univ. J. Nat. Sci.*, **7 B**, 24–33.
196 Leete, E., Bowman, R.M., Manuel, M.F. (1971). *Phytochemistry*, **10**, 3029–3033.
197 Leete, E. and Marion, L. (1954). *Can J. Chem.*, **32**, 646–649.
198 Lichtenthaler, H.K. (1969). *Biochim. biophys. Acta*, **184**, 164–172.
199 Lichtenthaler, H.K. (1973). *Ber. d. bot. Ges.*, **86**, 313–329.

200 Lichtenthaler, H.K. (1975). *Physiologia Pl.*, **33**, 241–244.
201 Liener, I.E. (Ed. 1969). *Toxic constituents of plant foodstuffs.* London: Academic Press.
202 Liljenberg, C. (1974). *Physiologia Pl.*, **32**, 208–213.
203 Lindberg, P., Tanhuanpää, E., Nilsson, G. and Wass, L. (1964). *Acta. Agric. scand.*, **14**, 297–306.
204 Lindberg, P., Bingefors, S., Lannek, N. and Tanhuanpää, E. (1964). *Acta. Agric. scand.*, **14**, 3–11.
205 Little, H.N. and Jones, O.W.G. (1976). *Biochem. J.*, **156**, 309–314.
206 Luchsinger, W.W., Chen, S.C. and Richards, A.W. (1965). *Archs. Biochem. Biophys.*, **112**, 531–536.
207 Luchsinger, W.W. and Richards, A.W. (1968). *Cereal Chem.*, **45**, 115–123.
208 Mace, M.E. and Hebert, T.T. (1963). *Phytopathology*, **53**, 692–700.
209 MacGregor, A.W., Thomson, R.G. and Meredith, W.O.S. (1974). *J. Inst. Brew.*, **80**, 181–187.
210 MacGregor, A.W., La Berge, D.E. and Meredith, W.O.S. (1971). *Cereal Chem.* **48**, 490–498.
211 Machlis, L. (1944). *Am. J. Bot.*, **31**, 183–192; 281–282.
212 MacLeod, A.M. (1957). *New Phytol.*, **56**, 210–220.
213 MacWilliam, I.C. and Harris, G. (1959). *Archs. Biochem. Biophys.*, **84**, 442–454.
214 McClure, J.W. (1974). *Phytochemistry*, **13**, 1065–1069; 1071–1073.
215 McClure, J.W. and Wilson, K.G. (1970). *Phytochemistry*, **9**, 763–773.
216 McReady, R.M. and Hassid, W.Z. (1941). *Pl. Physiol., Lancaster*, **16**, 599–610.
217 Maizel, J.V., Benson, A.A. and Tolbert, N.E. (1956). *Pl. Physiol., Lancaster*, **31**, 407–408.
218 Mann, J.D. and Mudd, S.H. (1963). *J. biol. Chem.*, **238**, 381–385.
219 Mann, J.D., Steinhart, C.E. and Mudd, S.H. (1963). *J. biol. Chem.*, **238**, 676–681.
220 Manners, D.J. and Marshall, J.J. (1969). *J. Inst. Brew.*, **75**, 550–561.
221 Manners, D.J. and Rowe, K.L. (1971). *J. Inst. Brew.*, **77**, 358–365.
222 Manners, D.J. and Wilson, G. (1974). *Carb. Res.*, **37**, 9–22.
223 Mansell, R.L., Gross G.G., Stöckigt, J., Franke, H. and Zenk, M.H. (1974). *Phytochemistry*, **13**, 2427–2435.
224 Marschner, H. (1965). *Flora, Jena*, **155**, 558–572.
225 Marshall, J.J. (1975). *Carb. Res.*, **42**, 203–207.
226 Marshall, J.J. and Taylor, P.M. (1971). *Biochem. biophys. Res. Commun.*, **42**, 173–179.
227 Massicot, J. and Marion, L. (1957). *Can. J. Chem.*, **35**, 1–4.
228 Mathieu, Y. (1967). *Photosynthetica*, **1**, 57–63.
229 Méndez, J. (1967). *Phytochemistry*, **6**, 313–315.
230 Merry, J. and Goddard, D.R. (1941). *Proc. Rochester Acad. Sci.*, **8**, 28–44.
231 Mesrob, B., Petrova, M. and Ivanov, C.P. (1970). *Biochim. biophys. Acta.*, **200**, 459–465.
232 Metche, M. and Urion, E. (1961). *C. r. Acad. Sci. Paris*, **252**, 356–358.
233 Miflin, B.J. (1968). *Biochem. J.*, **108**, 49P–50P.
234 Miflin, B.J. (1969). *J. exp. Bot.*, **20**, 810–819.
235 Miflin, B.J. (1971). *Archs. Biochem. Biophys.*, **146**, 542–550.
236 Miflin, B.J. (1973). In *Biosynthesis and its Control in Plants*, (ed. Milborrow, B.V.) pp. 49–68. London: Academic Press.
237 Mikola, J. and Kirsi, M. (1972). *Acta Chem. scand.*, **26**, 787–795.

238 Mikola, J., Pietilä, K. and Enari, T.M. (1972). *J. Inst. Brew.*, **78**, 384–388.
239 Mikola, J. and Kolehmainen, L. (1972). *Planta*, **104**, 167–177.
240 Minchin, A. and Harmey, M.A. (1975). *Planta*, **122**, 245–254.
241 Mitchell, E.D. (1972). *Phytochemistry*, **11**, 1673–1676.
242 Mizusaki, S., Tanabe, Y., Noguchi, M. and Tamaki, E. (1973). *Pl. Cell Physiol., Tokyo*, **14**, 103–110.
243 Møller, B.L. (1974). *Pl. Physiol., Lancaster*, **54**, 638–643.
244 Moeller, M., Robbins, G.S., Burger, W.C. and Prentice, N. (1970). *J. Agric. Fd. Chem.*, **18**, 886–890.
245 Moffa, D.J. and Luchsinger, W.W. (1970). *Cereal Chem.*, **47**, 54–63.
246 Morgan, N.L. and Griffiths, W.T. (1975). *Biochem. Soc. Trans.*, **3**, 391–392.
247 Moruzzi, G. and Caldarera, C.M. (1964). *Archs. Biochem. Biophys.*, **105**, 209–210.
248 Mudd, S.H. (1960). *Biochim. biophys. Acta.*, **38**, 354–355.
249 Mudd, S.H. (1961). *Nature*, **189**, 489.
250 Mullic, D.B. and Brink, V.C. (1966). *Crop Sci.*, **6**, 204–206.
251 Mullick, D.B., Faris, D.G., Brink, V.C. and Acheson, R.M. (1958). *Can. J. Pl. Sci.*, **38**, 445–456.
252 Murphy, G.J.P. and Briggs, D.E. (1973). *Phytochemistry*, **12**, 1299–1308; 2597–2605.
253 Murphy, G.J.P. and Briggs, D.E. (1975). *Phytochemistry*, **14**, 429–433.
254 Musgrave, A., Kays, S.E. and Kende, H. (1972). *Planta*, **102**, 1–10.
255 Nadeau, R., Rappaport, L. and Stolp, C.F. (1972). *Planta*, **107**, 315–324.
256 Nakayama, T.O.M. (1962). *Proc. ann. mtg. Am. Soc. Brew. Chem.*, 137–139.
257 Nakagiri, Y., Maekawa, Y., Kikara, R., Miwa, M. (1968). *Hakko Kogaku Zasshi*, **46**, 605–609; 610–615; 616–625. via *Chem. Abs.* (1969), **70**, 843 x.
258 Naylor, A.W., Rabson, R. and Tolbert, N.E. (1958). *Physiologia Pl.*, **11**, 537–547.
259 Naylor, A.W. and Tolbert, N.E. (1956). *Physiologia Pl.*, **9**, 220–229.
260 Neish, A.C. (1961). *Phytochemistry*, **1**, 1–24.
261 Newman, J. and Briggs, D.E. (1976). *Phytochemistry*, **15**, 1453–1458.
262 Ngo, T.T. and Shargool, P.D. (1974). *Can. J. Biochem.*, **52**, 435–440.
263 Niku-Paavola, M.-L. and Nummi, M. (1971). *Acta. Chem. scand.*, **25**, 1492–1493.
264 Niku-Paavola, M.-L., Skakoun, A., Nummi, M. and Daussant, J. (1973). *Biochim. biophys. Acta.*, **322**, 181–184.
265 Nielsen, O.F. (1974). *Archs. Biochem. Biophys.*, **160**, 430–439.
266 Nielsen, N.C. (1975). *Eur. J. Biochem.*, **50**, 611–623.
267 Nielsen, O.F. and Gough, S. (1974). *Physiologia Pl.*, **30**, 246–254.
268 Nordheim, W. (1965). *Getreide und Mehl*, **15**, 87–92; 99–110.
269 Norris, L., Norris, R.E. and Calvin, M. (1955). *J. exp. Bot.*, **6**, 64–74.
270 Nummi, M., Vilhunen, R. and Enari, T.M. (1965). *Proc. eur. Brew. Conv. Stockholm*, p. 52–61.
271 Nummi, M., Niku-Paavola, M.-L. and Enari, T.-M. (1972). *Acta. Chem. scand.*, **26**, 1731–1732.
272 Obendorf, R.L. and Huffaker, R.C. (1970). *Pl. Physiol., Lancaster*, **45**, 579–582.
273 O'Donovan, D. and Leete, E. (1963). *J. Am. Chem. Soc.*, **85**, 461–463.
274 Ogiso, T., Noda, T., Sako, Y., Kato, Y. and Aoyama, M. (1975). *J. Biochem., Tokyo*, **78**, 9–17.
275 Oji, Y. and Izawa, G. (1972). *Pl. Cell. Physiol.*, **13**, 249–259.
276 Osborne, T.B. (1895). *J. Am. Chem. Soc.*, **17**, 539–567.

277 Owen, W.J., Delaney, M.E. and Rogers, L.J. (1974). *Biochem. Soc. Trans.*, **2**, 1109–1112.
278 Parsons, J.G. and Price, P.B. (1974). *Lipids*, **9**, 804–808.
279 Phethean, P.D., Jervis, L. and Hallaway, M. (1968). *Biochem. J.*, **108**, 25–31.
280 Philips, M. and Goss, M.J. (1934). *J. Am. Chem. Soc.*, **56**, 2707–2710.
281 Phung nhu Hung, S., Hoarau, A. and Moyse, A. (1970). *Z. PflPhysiol.*, **62**, 245–258.
282 Plesnicar, M. and Bendall, D.S. (1970). *Biochim. biophys. Acta.*, **216**, 192–199.
283 Pollock, J.R.A. (1962). In *Barley and Malt*, (ed. Cook, A.H.) pp. 303–398. London: Academic Press.
284 Porter, H.K. (1950). *Biochem. J.*, **47**, 476–482.
285 Pragnell, M.J., Jones, M. and Pierce, J.S. (1969). *J. Inst. Brew.*, **75**, 511–514.
286 Préaux, G. and Lontie, R. (1975). In *The Chemistry and Biochemistry of Plant Proteins*, (eds. Harborne, J.B. and Van Sumere, C.F.) pp. 89–111. London: Academic Press.
287 Prentice, N., Burger, W.C. and Moeller, M. (1971). *Cereal Chem.*, **48**, 587–594.
288 Quail, P.H. and Varner, J.E. (1971).*Analyt. Biochem.*, **39**, 344–355.
289 Rabitzsch, G. (1959). *Planta Med.*, **7**, 268–297.
290 Raoul, Y. (1937). *Ann. Sci. Ferm.*, **3**, 129–148; 193–218; 385–405.
291 Redman, D.G. and Fisher, N. (1969). *J. Sci. Fd Agric.*, **20**, 427–432.
292 Rhodes, M.J.C. and Yemm, E.W. (1963). *Nature*, **200**, 1077–1080.
293 Roberts, R.M. (1971). *J. biol. Chem.*, **246**, 4995–5002.
294 Roberts, R.M., Connor, A.B. and Cetorelli, J.J. (1971). *Biochem. J.*, **125**, 999–1008.
295 Rosa, N. and Neish, A.C. (1968). *Can. J. Biochem.*, **46**, 797–806.
296 Roth-Bejerano, N. and Lips, S.H. (1973). *New Phytol*, **72**, 253–257.
297 Roth-Bejerano, N. and Lips, S.H. (1975). *Pl. Physiol., Lancaster*, **55**, 270–272.
298 Ryrie, I.J. and Scott, K.J. (1969). *Biochem. J.*, **115**, 679–685.
299 Sane, P.V. and Zalik, S. (1968). *Can. J. Bot.*, **46**, 1331–1334.
300 Sanwal, G.G., Greenberg, E., Hardie, J., Cameron, E.C. and Preiss, J. (1968). *Pl. Physiol., Lancaster*, **43**, 417–427.
301 Schlubach, H.H. (1965). *Fortschr. Chem. Org. Naturstoffe*, **23**, 46–60.
302 Schlubach, H.H. and Grehn, M. (1968). *Hoppe-Seyler's Z. physiol. Chem.*, **349**, 1141–1148.
303 Schneider, E.A. and Wightman, F. (1974). *Can. J. Biochem.*, **52**, 698–705.
304 Schneider, E.A., Gibson, R.A. and Wightman, F. (1972). *J. expt. Bot.*, **23**, 152–170.
305 Schwimmer, S. (1950). *J. biol. Chem.*, **186**, 181–193.
306 Seikel, M.K., Bushnell, A.J. and Birzalis, R. (1962). *Archs. Biochem. Biophys.*, **99**, 451–457.
307 Sen, D. (1975). *Phytochemistry*, **14** 1505–1506.
308 Shibko, S. and Edelman, J. (1957). *Biochim. biophys. Acta.*, **25**, 642–644.
309 Shinke, R. and Mugibayashi, N. (1972). *Agric. biol. Chem.*, **36**, 378–382.
310 Shiomi, N. and Hori, S. (1974). *Pl. Cell. Physiol.*, **15**, 559–564.
311a Silbereisen, K. and Kraffczyk, F. (1967). *Monats. Brau.*, **20**, 217–223.
311b Skult, H. (1968). *Acta Acad. Abo., Math. Phys.* **28** (4), 12 pp.
312 Slaughter, J.C. and Uvgard, A.R.A. (1972). *Phytochemistry*, **11**, 478–479.
313 Slover, H.T. (1971). *Lipids*, **6**, 291–296.
314 Smith, T.A. (1965). *Phytochemistry*, **4**, 599–607.

315 Smith, H. (1970) *Phytochemistry*, **9**, 965–975.
316 Smith, T.A. (1975). *Phytochemistry*, **14**, 865–890.
317 Smith, T.A., Brown, D., Cavender, R.W. and Sarginson, C. (1974). *Biochem. Soc. Trans.*, **2**, 99–101.
318 Smith, T.A. and Garraway, J.L. (1964). *Phytochemistry*, **3**, 23–26.
319 Smith, T.A. Sinclair, C. (1967). *Ann. Bot.*, **31**, 103–111.
320 Smith, I.K. and Thompson, J.F. (1969). *Biochem. biophys. Res. Commun.*, **35**, 939–945.
321 Smith, F.W. and Thompson, J.F. (1971). *Pl. Physiol., Lancaster*, **48**, 219–223.
322 Smith, B.P. and Williams, H.H. (1951). *Archs. Biochem. Biophys.*, **31**, 366–374.
323 Smith, T.A. and Wilshire, G. (1975). *Phytochemistry*, **14**, 2341–2346.
324 Sopanen, T. and Mikola, J. (1975). *Pl. Physiol., Lancaster*, **55**, 809–814.
325 Spedding, D.J. and Thomas, W.J. (1973). *Aust. J. biol. Sci.*, **26**, 281–286.
326 Srivastava, B.I.S. (1963). *Archs. Biochem. Biophys.*, **103**, 200–205.
327 Srivastava, B.I.S. and Arglebe, C. (1967). *Pl. Physiol., Lancaster*, **42**, 1497–1503.
328 Steinhart, C.E., Mann, J.D. and Mudd, S.H. (1964). *Pl. Physiol., Lancaster*, **39**, 1030–1038.
329 Stevens, R. (1958). *J. Inst. Brew.*, **64**, 470–476.
330 Stobart, A.K. and Pinfield, N.J. (1970). *New Phytol.*, **69**, 31–35.
331 Stoddart, J.L., Thomas, H. and Robertson, A. (1973). *Planta*, **112**, 309–321.
332 Stoessl, A. (1967). *Can. J. Chem.*, **45**, 1745–1760.
333 Stolp, C.F., Nadeau, R. and Rappaport, L. (1973). *Pl. Physiol., Lancaster*, **52**, 546–548.
334 Sugawara, T. (1953). *Jap. J. Bot.*, **14**, 125–146.
335 Van Sumere, C.F. and Kint, J. (1964). *Archs. Int. Physiol. Biochim.*, **72**, 706–708.
336 Van Sumere, C.F., de Pooter, H., Haider Ali and Degrauw-van Bussel, M. (1973). *Phytochemistry*, **12**, 407–411.
337 Szlavko, C.M. and Worrall, R.J. (1975). *J. Inst. Brew.*, **81**, 438.
338 Tamàs, I.A. and Bidwell, R.G.S. (1971). *Can. J. Bot.*, **49**, 299–302.
339 Tanaka, Y. and Akazawa, T. (1970). *Pl. Physiol., Lancaster*, **46**, 586–591.
340 Tanaka, K., Tolbert, N.E. and Gohlke, A.F. (1966). *Pl. Physiol., Lancaster*, **41**, 307–312.
341 Todd, G.W. and Galston, A.W. (1954). *Pl. Physiol., Lancaster*, **29**, 311–318.
342 Tolbert, N.E. (1955). *J. biol. Chem.*, **215**, 27–34.
343 Tolbert, N.E. (1971). *Ann. Rev. Pl. Physiol.*, **22**, 45–74.
344 Tolbert, N.E. and Cohan, M.S. (1953). *J. biol. Chem.*, **204**, 639–648.
345 Travis, R.L., Jordan, W.R. and Huffaker, R.C. (1970). *Physiologia Pl.*, **23**, 678–685.
346 Tribe, I.S., Gaunt, J.K., and Wynn Parry, D. (1968). *Biochem. J.*, **109**, 8p–9p.
347 Tyszkiewicz, E. (1959). *Compt. rend.*, **249**, 1926–1928.
348 Udvardy, J. and Farkas, G.L. (1968). *Acta biochem. biophys. hung.*, **3**, 153–164.
349 Urion, E. and Metche, M. (1961). *Brauwissenschaft*, **14**, 227–232.
350 Van Staden, A. and Drews, S.E. (1975). *Pl. Sci. Lett.*, **4**, 391–394.
351 Visuri, K., Mikola, J. and Enari, T.-M. (1969). *Eur. J. Biochem.*, **7**, 193–199.
352 Wagner, B. (1962). *Monats. Brau.*, **24**, 285–287.
353 Walker, D.A. (1970). *Nature*, **226**, 1204–1208.

354 Wall, J.S., Swango, L.C., Tessari, D. and Dimler, R. J. (1961). *Cereal Chem.*, **38**, 407–422.
355 Walsh, D.E., Banasik, O.J. and Gilles, K.A. (1965). *J. Chromatog,.* **17**, 278–287.
356 Warner, H.R., Tuleen, N.A., Bell, W.D. and Snyder, L.A. (1969). *Molec. gen. Genet.*, **104**, 241–252.
357 Wasilewska, L.D. and Reifer, I. (1968). *Acta Soc. Bot. Pol.*, **37**, 647–656.
358 von Wettstein, D. (1974). *Biochem. Soc. Trans.*, **2**, 176–179.
359 von Wettstein-Knowles, P. (1971). In *Barley Genetics II*, (ed. Nilan, R.A.) pp. 146–193. Washington State University Press.
360 von Wettstein-Knowles, P. (1972). *Planta*, **106**, 113–130.
361 von Wettstein-Knowles, P. (1974). *FEBS Letters*, **42**, 187–191.
362 von Wettstein-Knowles, P. (1976). *Molec. gen. Genet.*, **144**, 43–48.
363 von Wettstein-Knowles, P. and Netting, A.G. (1976). *Lipids*, **11**, 478–484.
364 Wightman, F., Chisholm, M.D. and Neish, A.C. (1961). *Phytochemistry*, **1**, 30–37.
365 Wilson, D.G., King, K.W. and Burris, A.H. (1954). *J. biol. Chem.*, **208**, 863–874.
366 Yabuuchi, S. and Amaha, M. (1975). *Phytochemistry*, **14**, 2569–2572.
367 Yemm, E.W. (1949). *New Phytol.*, **48**, 315–331.
368 Zimmerman, D.C. and Vick, B.A. (1970). *Pl. Physiol.*, *Lancaster*, **46**, 445–453.
369 Zürcher, C. (1971). *Monats. Brau.*, **24**, 276–284.
370 Murphy, G.J.P.—(unpublished).

Chapter 5

Grain quality and germination

5.1. Introduction

The importance of different aspects of quality depend upon the grain's intended use. Viability of feed grain is immaterial, unlike its nutritional value and freedom from toxic substances. Seed grain should have a high viability, give vigorous seedlings and be free of grains of other varieties, or species (Chapter 9). However, the rate of seed germination is usually not critical. Seed grain is normally dressed with toxic chemicals to control pathogens. Malting barley has a preferred chemical composition, should be adequately uniform, must germinate rapidly and evenly and should not contain poisonous materials (Chapter 15). For grain which is to be put into store the moisture content, temperature, and freedom from pests are of immediate importance (Chapter 10). In some situations agreed, or legally defined, tests of quality are applied to grain intended for particular purposes. The tests used often differ between countries, and even between different groups of users in one country. Yet others tests are employed as aids by grain users for their own purposes, and not as a basis for commercial transactions.

5.2. Sampling tests with small numbers of grains

Grains in a single lot from one field, planted to one variety, differ in size, shape, physical intactness, viability and chemical composition. Even grains from one ear differ. Mixing grain lots often increases the variability of the sample. As it is impossible to test every grain, samples for evaluation must be taken so that, as far as is possible, they represent the bulk. Samples from different parts of a grain lot may be pooled, or analysed separately to gain an estimate of its inhomogeneity. Moving grain streams may be sampled at intervals, either automatically or by hand. Stationary grain

Grain quality and germination 175

in sacks or heaps, in lorries or wagons may be sampled using 'spears' (triers). These are driven into the grain then, by turning a handle, an outer metal sheath is revolved uncovering pockets along the length of the device. When the pockets have filled they are closed, and the spear is withdrawn. The grain samples from different depths are emptied into one or more container(s). Samples must be taken from all depths and at many locations in each bulk [10,70,121,122]. The uncleaned sample should at once be put in a labelled, airtight container. Each sample, at least 1 kg, is many times larger than is immediately needed for analysis. Without due care in sampling subsequent analyses are valueless. The sample is well mixed, by repeated passage through a sample divider, then sub-samples are taken. Dividers eliminate bias e.g. by selection for intact grains only, or for a particular grain size or colour. Each consists of a container divided by partitions, each compartment going to a different outlet spout. The grain may be poured directly among the partitions or it may be delivered onto a rotating wheel which scatters it into the compartments. Tests based on small numbers of grains are statistically unreliable.

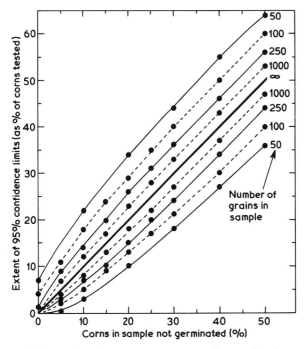

Fig. 5.1 The relationship between the percentage of corns showing a character (e.g. failure to germinate) and the 95% confidence limits of that estimate for samples of differing sizes. The heavy central line, ∞, is when an infinite number of grains, or all the grains in a bulk are evaluated so that there can be no sampling error (plotted from a table of Urion and Chapon [231]).

Weighing individual 1000-grain samples from an ungraded bulk can easily give an estimate of the mean bulk value that is 12% in error. Weighing 5000 grains reduces this uncertainty to about 4% [119]. Triplicate samples of *at least* 50 g, and preferably 100 g, should be weighed out and *then* counted [70]. The result is given as the thousand corn (dry) weight. In germination or other tests where small numbers of grains are used, the statistical uncertainty of the results reaches high levels [231]. If one in ten grains in a sample do not germinate then the 95% confidence limits for the percentage of dead grains in samples of different sizes are: 50 grains, 3–22%; 100 grains, 5–18%; 250 grains, 7–14%; 1000 grains, 8–12% (Fig. 5.1). Where the grain has previously been mixed and sorted, e.g. graded for corn size, the variability between sub-samples is often reduced.

5.3. Grain evaluation (other than viability and germinability)

Inspection of a sample detects dirt, trash, broken grains, grains of other cereals and weed seeds, and sometimes insects, insect fragments, rodent hairs, urine, or faeces (Chapter 10). Good quality, undressed grain should be plump, and free from taints or musty or sour odours. The husk should appear 'fine', 'bright' and clean and not be greenish (unripe grain unless the variety is pigmented) or dull, grey tinged, mottled, discoloured or stained in ways indicating that overheating or microbial attack has occurred. The grains should be of the correct variety (Chapters 2 and 3). The extent of husk damage and the number of broken, split and half-grains may be noted, and whether damage has occurred mainly at the apex or at the base of the grains [224]. The presence of awn stubs shows that a sample has been threshed gently. Husk (or husk and pericarp) contents may be determined by removing the outer layers by pearling, or with chemical reagents. Thus sulphuric acid or sodium hypochlorite solutions may be used or, with less convenience, an alkaline steep followed by grain dissection [119,199,238]. Decortication with sulphuric acid allows the selection and surface sterilization of uniform, undamaged grains for experimental purposes. Abraded or harvest damaged grains may be detected since punctures or faults in the testae allow the penetration of iodine from a strong solution of potassium iodide, causing the starch below each fault to stain blue-black [23,173]. Mealy or steely endosperms can be seen in decorticated grains, as can pregerminated grains. Decorticated grains are particularly prone to microbial attack [18,173].

Samples of 50 grains may be transected in a corn-cutter (farinator) and the cut endosperms scored for mealiness (floury or opaque appearance) and steeliness (flinty or glassy appearance). Steely grains are less desirable for malting. Mealy grains have a lower density (Chapter 1). The weight per unit volume (lb/bu or kg/hl) is sometimes determined. The 'assort-

Grain quality and germination 177

ment' or grading is found by passing the grain over a series of mechanically shaken slotted sieves and noting the proportions by weight that occur in each fraction [70], e.g. < 2.2 mm, 2.2–2.5 mm, 2.5–2.8 mm and ⩾ 2.8 mm.

The genetic purity of a sample, whether grains of more than one cultivar are present, can be difficult to assess. Judgement is made on the general appearance of the grain; the grain size; the occurrence of the ratio of one plump and symmetrical grain to two narrower grains twisted in opposite senses, indicating the presence of a six-rowed barley; the ratio of length to breadth, and so on (Chapter 2). The morphology of the plants grown from the grains may be noted. Isozyme patterns obtained on grain extracts may prove helpful.

Particular micro-organisms, e.g. *Fusaria* and ergot sclerotia may be sought. Grains may be shaken in liquids and, the microbe numbers assessed in the washing fluid. To detect loose smut embryos must be separated from the grains, optically cleared, and selectively stained. Fungal mycelium is sought by inspecting the exposed scutellar face under a microscope [122]. High levels of grain fat acidity are an indication of grain deterioration. Other analyses are occasionally made. However, determinations of water (moisture) content and nitrogen, ('crude protein') content are made frequently, and deserve special comment.

The moisture content is usually expressed as a percentage of the present fresh weight (% fr. wt or 'as is'), i.e. water × 100/(water + dry matter). Several other units (e.g. % dry weight or d.b., dry basis, or 'on dry') are sometimes employed. The moisture is associated, with varying degrees of firmness, with the grain structure (Chapter 10). The last traces of water are difficult to remove. For simplicity grain moistures are determined by arbitrary methods – for example by heating a ground sample for exactly 4 h in a slow stream of chemically dried air at 105°C. At higher temperatures thermal decomposition of the grain may occur. Such methods give slightly low estimates of moisture content compared to 'absolute' determinations made by heating ground grain for some months in a vacuum connected to a water trap. Other methods, which require regular standardization, are based upon measurements of the dielectric constant or electrical resistance of ground samples, or the measurement of the water in a methanol extract of grain by infrared spectrophotometry, gas–liquid chromatography, or the Karl Fischer titration; by measuring the volume of water carried over by distillation with a high boiling-point immiscible liquid; the use of rapid high temperature (130°C), direct or infrared drying methods; the release of acetylene from calcium carbide; the temperature rise that occurs on mixing with concentrated sulphuric acid; or n.m.r. (nuclear magnetic resonance) spectroscopy [119,120].

Standard methods for determining nitrogen in barley are based on

178 Barley

destructive procedures in which the nitrogenous compounds are decomposed to ammonium ions. These are assayed either colorimetrically or by titration after distillation, as in the Kjeldahl method [119]. By convention crude protein = $N \times 6.25$. However, dye-binding methods for estimating protein are also used, as are a range of other techniques. Generally they must all be calibrated against a standard method and may be unreliable if applied indiscriminately, e.g. to varieties with differing lysine contents (Chapter 12).

5.4. The penetration of water, and other substances, into grain

Immersion of grain in water immediately establishes a surface film that can be removed by blotting, or centrifugation. This film represents 2–5% (fr. wt) of the grain, and must be removed for accurate studies of water uptake. Where present, as in malting at the end of steeping, it is a reserve of moisture that is taken up by the grain as germination begins; it retards grain respiration and germination. The surface layers, the husk and pericarp, are hydrated in the first 2 h of immersion. (Table 5.1) [205]. In naked grain the pericarp is hydrated in about 30 sec. It conducts water so well that the grain interior hydrates at the same rate whether the apex, the base, or the whole grain is immersed in water [30,58]. During imbibition the moisture content increases rapidly, then at a progressively declining rate until, if germination does not occur, as in a malting steep, it approximates to a limiting value (Fig. 5.2). However, if germination begins water uptake continues as the seedling grows. The initial hydration process is purely physical; it is accompanied by an output of heat, as when starch is hydrated; is not dependent on the viability of the grain. As the grain hydrates, first the embryo and later the endosperm swell. Mealy, mature grains tend to take up moisture faster than steely grains; freezing or 'frosting', which increases mealiness also increases the rate of

Table 5.1 The moisture content of the parts of barley grains steeped for various times at 13°C. Initially the moisture content of the whole grain was probably about 14% (from Reynolds and MacWillian [205]).

	% Moisture after steeping at 13°C for			
	2 h	4 h	6 h	24 h
Scutellum	18.5	36.8	50.0	57.9
Embryo axis	33.0	43.6	51.3	57.9
Pericarp + testa + aleurone	11.4	15.6	26.8	36.4
Starchy endosperm	10.0	12.5	21.4	35.0
Husk	40.5	43.1	44.9	47.1
Whole grain (by direct estimation)	21.6	25.9	28.6	38.4
Whole grain (by summation)	20.9	24.8	27.1	37.5

Grain quality and germination

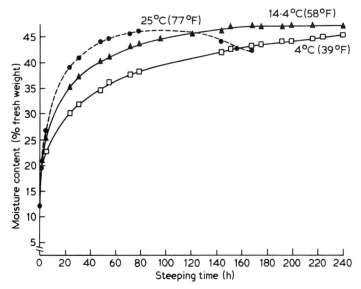

Fig. 5.2 The water content of ungraded samples of Proctor barley, steeped at three different temperatures. In each case the surface moisture (2–5%) was first removed by centrifugation. The apparent decline in moisture content of the sample steeped (soaked by immersion) at 25° C (77° F) was due to the partial decomposition of the grains (after Hough *et al.* [118]).

water uptake (as in wheat), presumably because mealy grains have less physical cohesion to resist swelling [210]. Smaller grains take up moisture faster and to a higher final level than large grains, the effect being most noticeable in grains of less than 2.4 mm width [250]. During prolonged steeps the surface layers of grain tend to decompose and dry matter is lost, with consequent distortions of the 'apparent' water uptake curves. The rate of grain imbibition increases rapidly with increasing temperature (Fig. 5.2) [33]. Probably all barleys have about the same temperature coefficient of imbibition. There is a linear inverse relationship between the temperature and the log of the time for steeping grain to reach any particular moisture content [22,33]. Heavy water (deuterium oxide, D_2O) is taken up more slowly than normal water (protium oxide). Grains imbibed with deuterium oxide subsequently germinate and grow poorly. When barley is steeped 0.5–1.5% of its dry matter is lost. Dust and dirt are removed and considerable quantities of uncharacterized pigments and other substances, minerals, inorganic and organic phosphates, sugars, phenols, phenolic acids and amino acids are leached from the grains [137,199]. Most of this material is leached from the husk and pericarp; less is dissolved from naked grains. Some comes from the interior of damaged grain and, in long steeps, from the decomposition of the grain surface layers. As the malting grain germinates the growing embryo first takes water from the surface film and later withdraws moisture from the endosperm. When this has dried sufficiently, i.e. to about 36% moisture, growth is checked [142]. As spraying water accelerates embryo growth proportionately more than endosperm breakdown possibly little water passes from the embryo back into the dried endosperm.

The surface layers of grains are selectively permeable. Some organic solvents, when dissolved in water, penetrate the grain and accelerate water uptake without, however, altering the temperature coefficient of imbibition [30,33]. Other substances when in solution, such as sugars, sodium chloride and sulphuric acid are excluded. Grains act as osmometers and take up less water from strong solutions of excluded substances. As the water is withdrawn, so the solutions become concentrated. Sulphuric acid causes the husk and pericarp to disintegrate, and so it is used in decorticating barley [30,103,199]. It is excluded from the grain interior by the testa, and the pigment strand. If the testa is punctured the acid rapidly gains entry, as can be seen most easily in grains (*caerulescens* varieties) having blue pigments in the aleurone layer; such pigments turn red in contact with acid. When grain is damaged water is taken up rapidly and so the surface layers must also limit water uptake [219]. Grain heated to 100°C or permeated with mercuric chloride still excludes sulphuric acid, so the selective permeability is not dependent on vitality. The site(s) of water penetration into grains are disputed. Water has free access to all parts outside the testa. During imbibition the embryo swells first in whole or decorticated grains; furthermore the basal, embryo end of the grain is always the wettest [58,209]. The embryo hydrates sooner than the endosperm and reaches a higher final moisture content; also the embryonic axis hydrates before the scutellum [46,205]. After warm-water steeps, used to control *Ustilago nuda*, the grain is redried and it is found that the embryo dehydrates first [151]. Some dyes penetrate the micropyle and stain the embryo locally – this is prevented by capping the micropyle with wax [58]. Thus most water probably enters and, on drying, leaves the grain through the micropylar region, via the suspensor, and the embryonic axis. Thus the micropyle and/or suspensorial remains must exhibit selective permeability. The energies of activation of the hydration of the embryo and endosperm are similar, 3.2–3.4 kcal/g mole/°C [212]. The barley testa itself is probably relatively impermeable to water [58,113]. Possibly relatively slow water uptake also occurs at the pigment strand.

The need for surface disinfection in experimental studies, the necessity to apply fungicidal seed dressings, and the use of additives to break dormancy or control malting has led to many studies of the penetration or non-penetration of substances into grains. Conclusions are tentative since commercial lots always contain grains in which the pericarp and testa are weakened or broken in various ways, and so represent less of a barrier.

Further, the failure to remove all surface residues invalidates results based on subsequent germination tests. Grains matured in a wet environment seem to have thinner testae, and are more permeable than those grown in a dry season [227]. Generally strong mineral acids are excluded, but after a lag nitric acid penetrates the grain. Trichloracetic acid (a strong acid) penetrates readily, as do the weak organic acids formic, acetic, etc. from dilute solutions [30,31]. Aqueous ethanol, acetaldehyde, acetone, and ethyl acetate penetrate grains readily, ethylene glycol and urea penetrate slowly, phenols enter at various rates, while sugars and glycerol do not penetrate at all. Absolute alcohol and anhydrous acetic acid penetrate only slowly [32,37]. Formalin (aqueous formaldehyde) has often been used to disinfect grain surfaces but it often damages them. Imbibed grains are less readily damaged than dry ones by formalin. The ease of penetration of salts from solution into undamaged grains varies. Thus copper sulphate, potassium chromate, silver nitrate, sodium chloride, sodium nitrate, and potassium sulphate are excluded, while potassium nitrate may penetrate slowly. When grains are soaked in a solution of silver nitrate, followed by sodium chloride, and the precipitated silver chloride is blackened by light, the extent of staining shows that the testa and pigment strand limit salt penetration. Copper sulphate was used as a surface fungicide, but it harms mechanically damaged grains. Precipitation of residual copper by a lime water treatment minimizes damage. Phenols, aniline and ammonia enter the grain readily. Sodium, potassium and lithium hydroxides in weak solution are excluded for a time, but in strong solution they quickly enter grains and destroy the selective permeability [103]. Sodium carbonate is toxic to barley, potassium carbonate less so; sodium bicarbonate is not toxic [103]. Nitric acid and strong alkalis first gain entry near the embryo [58]. Strongly ionized mercuric nitrate and mercuric sulphate are excluded from grains; the weakly ionized mercuric cyanide and mercuric chloride are not. Sodium bromate penetrates grain quickly, within an hour, and mainly gains entry at the embryo, although it probably also penetrates slowly over the whole surface [28]. Hydrogen peroxide, which is used to break dormancy, presumably enters the grain but direct evidence is lacking. Gibberellic acid also enters ungerminated grain, since it helps to break dormancy. In grain that is germinating it seems to penetrate at the embryo end since experiments with decorticated grain show that the testa at least restricts penetration [21]. Penetration may be accelerated by the presence of accidental or deliberate abrasions in the grain [34]. Tests for abrasion refer to faults either in the testa or in the pericarp; presumably only lesions in the testa permit the accelerated uptake of gibberellin [194, 218,219].

Iodine, dissolved in water, penetrates the testa as shown by the blue–black colour it gives with the endosperm starch. However, in a solution with an excess of potassium iodide the I_3^- ion predominates, and in this case the iodine penetrates the intact grain slowly at the basal end, and is first easily detected by a blue–black ring in the endosperm, around the periphery of the scutellum. The coloured zone slowly spreads up the grain towards the apex [58]. Thiosulphate ions cannot enter the grain to decolourize the complex. However, the I_2–KI solution quickly penetrates faults in the testa, and the blue–black stains so produced usefully demonstrate such faults in decorticated grains [23]. Other stains, e.g. acid fuchsin, have also been used to locate faults in the testa.

5.5. Testing for grain germinability

Dry, 'resting' barley grains are quiescent, having minimal metabolic activity. When wetted, and held at a suitable temperature, with a sufficient supply of oxygen and in the absence of toxic agents, mature grains germinate. Some grains will not germinate, either because they are dead, or because they are dormant – i.e. they are alive, but will not germinate under the conditions prevailing. Specific conditions for germination have been quoted: for example the minimum temperature for germination, 5°C; the optimum temperature, 29°C; the maximal temperature, 38°C. Such values are misleading, since samples of grain vary widely in their responses to different germination conditions. The percentage of grains germinating in a test may be called the germinative energy (G.E.) while the proportion of living grains (germinated + dormant) is the viability or germinative capacity (G.C.). The tests and the criteria used differ. Maltsters require grain to germinate rapidly (at least 50% in 1–2 days; 95–100% in 3 days), and grains are scored as 'germinated' as soon as the coleorhiza (chit) has appeared. For seed-testing each grain must germinate to form a whole, normal, vigorous seedling with adequate roots, a shoot and a first leaf. To determine viability (germinative capacity) dormant grains must be compelled to germinate or their viability must be detected in some other way. For reliable results replicated sub-samples should be set to germinate under rigidly standardized conditions. Mechanical devices or vacuum heads (with, say, 100 holes to which grains adhere, to be released by releasing the vacuum) may be used to count the grains, and to deposit them uniformly on the germination substratum. Numerous 'standard' test conditions are in use [70,119,121,122]. When tests are carried out under inadequately controlled conditions they give different results when applied to immature, dormant grains. Micromalting trials, using deliberate variations in the conditions, are particularly helpful to maltsters (Chapter 15). In the Schönfeld test 1000 grains are held in a filter-funnel, housed in a cabinet at 18–20°C (64–68°F) in an atmosphere saturated with water vapour. The grain is steeped and drained, mixed and re-wet at specified intervals. After a certain time the percentage of

Grain quality and germination

germinated grains is noted [121]. In the Schönjahn or Coldewe methods grains are held, embryo-ends down, in holes perforating thick porcelain plates. Each plate is housed in a container, covered with coarse sand, which is watered. Surplus water drains into the bottom of the container, keeping the atmosphere moist. Germination is assessed by counting the numbers of bunches of rootlets growing through the holes, beneath the plates. In the Aubry test grains are set between two sheets of absorbent paper, the assembly is held in a moist chamber and is watered 'at discretion'. More satisfactory tests are carried out in flat dishes (often Petri dishes), on filter-papers, graded sand or soil that is as uniform as possible, and to which exactly known proportions of water are added, e.g. 19% water to sand [15], or 4 ml or 8 ml of water to particular sizes and grades of filter paper discs [121,199]. Samples with dormant grains germinate less well with an excess of water e.g. 8 ml of water against 4 ml on filter paper discs the difference being defined as water-sensitivity (Fig. 5.3). The difference between the viability and the germination on 4 ml of water is called 'dormancy'. Good malting barley germinates rapidly e.g. on four successive days, 80%, 94%, 98%, 98%; 2% dead [231]. Generally 96–100% germination in three days is the preferred range. For seed

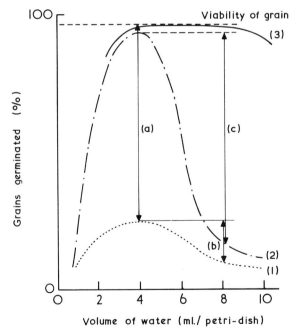

Fig. 5.3 The germination of a sample of barley, on paper discs in Petri dishes at three stages of maturity, in the presence of differing quantities of water.
(1) Freshly harvested, dormant (a), and water-sensitive (b). (2) Partly matured grain, showing water-sensitivity (c). (3) Fully mature grain.

testing the grain is covered with sand or soil and must emerge into light and be normally developed after a period of 7–14 days. The light must be diffuse, from cool white fluorescent lamps rich in red, and poor in far-red wavelengths, of 750–1250 lux switched on for at least 8h in each 24; tungsten lamps are *not* suitable. Soil reduces the phytotoxicity of many seed dressings, and so may be preferred to sand [122,155].

Some tests for viability rely on the mistaken assumption that all living seeds will germinate in tests prolonged to 2–3 weeks. More or less adequate tests involve treatments intended to break dormancy, or indirect criteria. Pre-treatments for reducing dormancy are:

(i) warm-drying grain, e.g. at 38–40°C (100–104°F) for up to 10 days;

(ii) stratifying the grain – keeping it cold, in air on a wet substratum – e.g. at 5°C (41°F) for 5–7 days;

(iii) setting grain to germinate on a substratum wet with a solution of potassium nitrate (0.2%) [122], or rarely dilute hydrogen peroxide or gibberellic acid;

(iv) steeping the grain in dilute hydrogen peroxide (0.75%; the Thunaeus test); after chitting the grains die. Grains not chitted at the end of the test may be peeled and set to grow on a wet substratum [121].

(v) peeling, i.e. the removing the husk, pericarp and testa from above the embryo, by hand, is generally effective in breaking dormancy, but is tedious [15];

(vi) decorticating (removing the husk and pericarp) with sulphuric acid (50%) is generally effective in breaking dormancy [199]; however, damaged grains may be killed by the acid treatment. Particularly deeply dormant, decorticated grain may also need to be stratified before it will germinate [41];

(vii) cutting dormant grains in half, transversely, enhances their germination (the Eckhardt test); the treatment is not completely effective and tests are frequently invalidated by the growth of micro-organisms on the cut surfaces.

Various indirect measures of viability have been proposed. Thus dead grains take up barium chloride, which can be detected by X-rays [155] and they are also able to take up the stains resazurin and indigo carmine. The most successful indirect methods detect living tissues by the ability of their endogenous enzymes and substrates to reduce materials such as *m*-dinitrobenzene, salts of selenium or tellurium and tetrazolium salts to coloured, insoluble substances which, by staining the tissues, give 'topographical' indications of the viability of the grain parts [119,156]. Triphenyl-tetrazolium chloride and the more rapidly reduced 2-(*p*-iodophenyl)-3-(*p*-nitrophenyl)-5-phenyl tetrazolium chloride are mainly used; being reduced to highly insoluble, bright red formazans (TTC tests) [155,156]. Grains are split longitudinally, often with a special tool, and one half of each is immersed in a solution of a tetrazolium salt. After a period of incubation, about 45 min at a fixed temperature, the embryos are inspected [121]. Viable aleurone tissue stains, as do embryos and micro-organisms. Faint or no staining in the coleorhiza or root initials may be traced to earlier physical damage to the base of the grain. Faint staining is given by old, 'weak' seeds. Heat-damaged grains may stain, and so escape detection, in which case the viability of a seedlot is over-estimated. High concentrations of contaminating microbes may give visible staining. Damage by toxic agents, such as seed dressings, can also escape detection. An alternative technique is to stain the separated embryo ends of grains and, after removing the residual endosperm, inspect the outer embryonic axis and the inner, scutellar face of the embryo. Where grain is heat-damaged, with free evaporation (and not by heat applied to grain held in a closed container) there is an unstained area in the centre of the scutellum opposite the embryonic node [224,237].

The detection of damaged or deteriorated grain presents difficulties, not least distinguishing them from dormant grains. Many factors can alter seedling growth – for example stressing the grain in the ear by over-heating. Infections with micro-organisms can cause grains to rot during tests. Seed dressings that are badly formulated or applied, or are applied to damaged grains, can prevent germination or cause seedling abnormalities [147]. A grossly battered sample can be recognized by the presence of skinned, split and broken grains. Slightly bruised grain cannot be detected by inspection, but will deteriorate rapidly in store; such damage may be detected by reduced resistance to 50% sulphuric acid [42,173]. Browning following acid or acid and hypochlorite treatments has been used to detect abrasion of the pericarp at the apex or embryo damage. Staining starch, where a solution of iodine in potassium iodide penetrates faults in the testa, is often used to detect damage. Faulty or old seeds deteriorate rapidly when stressed in rapid 'ageing' tests e.g. storage for several weeks at 40–45°C (104–113°F) in a water saturated atmosphere. 'Normal' seedlings, by seed testing criteria, should have at least two vigorous roots with root hairs, a well-grown coleoptile, with one or more seedling leaf(s) that is half, or more, of the coleoptile in length, and is green [122,236]. Mechanically damaged seeds may produce seedlings with few or no roots, reduced shoot growth, and split coleoptiles [155,237]. Limited heat damage can produce seedlings with few or no roots, but with enhanced coleoptile growth. Similar effects may be caused by toxic agents such as copper sulphate or some seed dressings. These may cause other seedling abnormalities, such as short, thickened rootlets, and shortened and swollen coleoptiles with a glossy appearance. Freezing damage may cause malformed, thickened and curved coleoptiles with a grainy appearance; few, malformed and twisted or no foliar leaves, and spindly seedlings with longitudinal splitting, and a dry and wilted appearance, and few roots [122,147,236]. Some types of

Japanese threshing equipment preferentially damage the basal, embryo end of the grain [224,242]. This damage may be detected (i) by a phenol staining method; (ii) by the rise in pH of the embryo from about 4.2–4.5, when bromocresol green is yellow, to over pH 5.2 when bromocresol green turns blue, (iii) by the use of the rH-indicator neutral violet; (iv) by tetrazolium staining, and (v) by the use of acid fuchsin, which only enters dead grains. A sensitive test for embryo damage is based on the reduced ability of separated embryos, incubated for 8h on water at 25°C (77°F), to synthesize starch from their endogenous reserves [88].

5.6. Vigour

'Vigour' is used to mean different things, e.g. the relative extent of seedling growth, or the intensity of some biochemical characteristic [240]. A major objective is to find tests that will decide whether grain, when sown in the field, will emerge as quickly, grow as strongly and become as well established as conditions permit. In this sense dormant seed tends to lack vigour and, as dormancy declines, vigour increases [72]. An inverse correlation has been detected between the degree of water sensitivity of some grain lots and emergence in the field [165]. In one test grains are germinated between two vertical rolls of wet paper. The length of the coleoptile and the length and number of roots produced by each grain is taken as a measure of vigour. In such a test larger grains from a particular lot will be more vigorous. Large seeds which contain more nutrient reserves are known to give seedlings which establish better under some conditions (Chapter 7). Seeds with a high nitrogen content respire faster than seeds of equal size having a low nitrogen content. At least on soils short of nitrogenous nutrients, N-rich seeds establish seedlings better than N-poor seeds [153]. In many regions the best depth for sowing Barley is about 4cm (1.57in); grain sown deeper emerges less well [185]. Other tests evaluate 'vigour' by the ability of seedlings to emerge through layers of graded brick-dust, gravel, or paper weighted with sand. An indirect measure of 'vigour' is the respiration rate of grain 2–6h after steeping [240]. Deteriorating grain may have an unusually high R.Q. [4], show fat rancidity, or leak sugars or electrolytes (detected by increased conductivity) into steeping water. Leakage is not a reliable indication of low vigour [1]. A useful indication of deterioration would be a change that occurs regularly and early on during grain senescence and can be determined with precision. The reduced incorporation of [^{14}C]-glucose into proteins and polypeptides that occurs in older grain seems suitable [5].

5.7. Dormancy

Dormancy varies between samples according to variety, the weather conditions during growth and harvest, the way in which the grain has

Grain quality and germination

been handled and stored, and the methods used to test germination [15–17,107,159,199,230,231]. Dormancy may be increased by post-harvest treatments. A distinction is often made between 'primary' or 'intrinsic' dormancy and 'secondary' dormancy, imposed by external factors. Sometimes forms of 'secondary' dormancy, such as water-sensitivity, have been regarded as a separate entity. This seems unjustified, although the distinction has operational merit. The causes of dormancy are unknown, but are probably a mixture of factors occurring within the grain, in its surface layers, and imposed by the test conditions. The expression of dormancy as a single number (the percentage of living grains failing to germinate within a set time in a particular test) is unsatisfactory, since dormant grain may germinate more slowly than fully mature grain containing a proportion of dead corns, yet the final percentage germination may be the same, or less, or more (Fig. 5.4). The proportion of the corns germinating alters with the nature of the test. Grain may be 'mature' according to a test, yet the rate of germination can continue to increase during storage e.g. a particular lot with 100% germination achieved 50% germination in 31h immediately after harvest but in 15.5h after 3 weeks storage [95]. Thus a description of dormancy should indicate the rate as well as the final percentage of germination, under a defined set of conditions, as well as the percentage of grains germinating in the chosen standard time. Numerous attempts to provide single descriptive numbers have been made [81,95,97]. The best type of presenta-

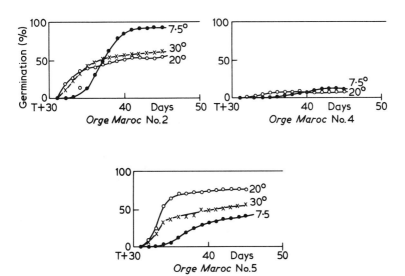

Fig. 5.4 The influence of temperature on the germination of three barleys (from Urion and Chapon [231]). Temperatures shown are °C. Equivalents: 7.5°C, 45.5°F; 20°C, 68°F; 30°C, 86°F.

tion seems to be a graph of cumulative percentage germination against time, for each germination condition. Most results found in the literature are single figure values based on counts taken at an arbitrary time and, in consequence, their meaning is often dubious.

The weight, size, colour and water uptake rate of grains are not connected with their rapidity of germination. However, dormancy differs in intensity between varieties, between different samples of one variety and in response to the agencies and treatments that more or less break it [144,163]. The varieties Old Scotch Common and Domen have little dormancy, while Trebi and especially the wild *Hordeum spontaneum* show intense dormancy. In a survey of about 4000 varieties, the 111 most dormant came from the winter-barley area of the Near East and N. Africa, near to the present distribution of *H. spontaneum*, while the least dormant varieties came from Ethiopia [35]. Extreme dormancy is unacceptable to maltsters but some is necessary to minimize pregermination (sprouting in the ear). Freshly harvested, immature milk-ripe grain is the most dormant, and dormancy declines as the grain ripens, so a variable amount survives in the harvested grain. The degree of recovery partly depends on the weather, which may also induce secondary dormancy. Dormancy is least when barley has ripened in warm, dry sunny weather; it is most pronounced when the weather has been cool and damp. This correlation allows the degree of dormancy to be forecast from the weather of the pre-harvest period [83]. During storage dormancy normally declines, and the germination of the grain improves. The rate of this post-harvest maturation can be altered by the manner in which the grain is handled. Drying grain in warm air – e.g. at 40°C (104°F) to about 10–12% moisture – and keeping the grain dry and warm speeds the decline in dormancy [15,199]. Warm storage also hastens subsequent grain deterioration (Chapter 10). Storing dormant grain cold and damp can delay the disappearance of dormancy for at least three years [35,199]. Warming in a closed container, for example at 40°C (104°F) for some days, reduces dormancy immediately. Drying cool, in a vacuum desiccator, has almost no immediate effect, but the dried grain recovers from dormancy more quickly during subsequent storage [107,199]. Recovery is faster if drying takes place in the presence of air or oxygen rather than nitrogen [199]. Judged by the 4 ml and 8 ml tests 'dormancy' declines before 'water sensitivity' on storage, and desiccation favours the disappearance of water sensitivity. However, these 'scores' probably estimate the same phenomenon under two levels of stress, since the activation energies for their disappearance are virtually identical (17.5 and 18.3 kcal/mole) [94]. Dormancy is usually overcome by warming grain in air at 40°C (104°F) for periods of up to 10 days. However a transient 'secondary' dormancy may be induced if water is allowed to evaporate freely; e.g. one sample of barley germinated 55% before drying, 38% 5 days after, and 98%

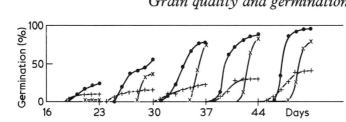

Fig. 5.5 The increasingly ready germination (decline in dormancy) of a barley, Aurore Bretagne, during storage. At each sampling time germination tests were carried out at three temperatures; × —— × 6°C, (43°F); ● —— ●, 13°C (55°F); + —— + 20°C, (68°F); (from Urion and Chapon [231]).

after 16 days [9,176]. The effect is more marked when grains are heated more strongly, e.g. to 50–70°C (122–158°F) [84]. Within limits fully mature grain germinates more rapidly at higher temperatures. Haberlandt indicated that rootlets appeared in 6 days at 4.8°C (about 41°F); 3 days at 10.5°C (51°F); 2 days at 15.6°C (60°F) and 1.75 days at 18.5°C (64°F). Given time dormant grains achieve higher percentages of germination at lower temperatures (Fig. 5.5). Most dormant barleys will germinate at 2–7°C (about 36–45°F). As the grain matures, the temperature range over which it will germinate widens (Table 5.2) [9,98,107,231]. Often grains will only germinate uniformly at 33°C (91.4°F) if the husk is removed [230]. An extremely dormant sample of *H. spontaneum* grain, which was 100% viable, germinated better at 5°C (41°F) than 20°C (68°F) which was better than at 25°C (77°F). At no time did more than 30% of intact grains germinate at 20°C (68°F) and none germinated at 30°C (86°F) [190]. 'Fixed time' germination tests can be very misleading. In 5 days a sample of grain germinated best (67%) at 12°C (53.6°F); however, in 11–14 days 95% germination occurred at 5°C (41°F) whereas at 12°C (53.6°F) the germination was only 86% [107]. Perhaps only fully after-ripened barleys germinate at 38°C (100.4°F) [35]. By this

Table 5.2 The germination of a partially dormant barley at different temperatures, with increasing time (from Atterberg [9]).

Duration (days)	Germination (%)							
	°C 7	10	13	15	17	19	22	25
	°F (44.6)	(50)	(55.4)	(59)	(62.6)	(66.2)	(71.6)	(77)
4	31	75	94	95	94	92	80	41
6	84	97	97	96	96	94	83	48
8	92	98	98	—	—	96	87	51
10	95	—	—	—	—	—	91	58
12	96	—	—	—	—	—	92	67
16	97	—	—	—	—	—	—	74

criterion many malting barleys are used well before they are mature. Freezing is reported to remove 'secondary dormancy'. Stratification, i.e. holding barley on a moist substratum in air at 5°C (41°F) for up to 8 days, will often allow subsequent germination at 20°C (68°F) [98,122]. Some process is set in train at 5°C (41°F) which is not stopped by transfer to the higher temperature.

Excess water depresses the germination of dormant grain (Fig. 5.3). The water sensitivity, determined as the difference between germination (%) on paper at two moisture levels (4 ml and 8 ml tests), is greater at 25°C (77°F) than at 18°C (64.4°F) [199]. Placing grain dorsal (embryo-side) downwards, rather than upwards and crowding the grains together depresses germination further [211]. The film of water over the grains restricts the access of oxygen, and this checks germination. Numerous agents, when added to the water, reduce water sensitivity to a greater or lesser extent. These include salts of heavy metals, formaldehyde, calcium hypochloride and substances that react with thiols [199]. While single antibiotics are usually unhelpful (an antibacterial agent merely allows fungi to overgrow the grain, presumably by reducing bacterial competition), a mixture of up to four antibiotics, chosen to antagonize a wide spectrum of microbial types, can alleviate water sensitivity [90]. Many chemical agents which can break dormancy, and which in excessive doses often kill the grain, also kill micro-organisms. It seems likely that microbes play a large part in enhancing grain dormancy. No known agents can sterilize husked grain without causing some damage. Warm drying reduces the microbial population and enhances germination in the 8 ml test. Storing the grain damp and warm increases the microbial population and increases water sensitivity [95]. Microbes alter the maltability of grain and also hasten its deterioration on storage (Chapter 10) [15,217]. Treating grain with plant extracts and nutrients that encourage microbial growth (e.g. sugars, ammonium lactate and urea) encourages dormancy [15,48, 62,90]. Nevertheless it has been claimed, surprisingly, that micro-organisms play no part in water sensitivity.

Steeping damage is probably related to water sensitivity. Mature grain, steeped anaerobically for extended periods, will not germinate quickly after draining. In contrast mature grain may begin germinating under water in a well aerated steep. The inclusion of hydrogen peroxide in steep liquor can induce most grains to germinate. The infiltration of water into grain under a partial vacuum reduces subsequent germination [16,230]. A delay in germination is caused if wet grain is first held in an atmosphere of nitrogen – the check is probably due to the accumulation of toxic anaerobic metabolites such as ethanol; such grain appears to be water sensitive [230]. Steeping grains in dilute solutions of ethanol, acetic acid or coumarin also induces 'water sensitivity'; unimpeded access of oxygen is needed for the metabolic removal of these toxic substances [129]. Anaerobic, moist grain produces ethanol some of which in steeping leaks into the water; acetic acid occurs in steep water, presumably produced from the ethanol by micro-organisms. Steep water is often inhibitory to grain growth. Microbes produce many substances that are toxic to grain (Chapter 10).

Short preliminary steeps (e.g. 6h) can encourage subsequent rapid germination. More prolonged steeps are generally harmful and may reduce the vigour of plants coming up in the field [140]. Brief hot water steeps can break dormancy [112,199]. Longer cold water steeps or warm water steeps, as used to control loose smut, can induce water-sensitivity whether the water is stagnant or running [48,129,154]. Agents such as silver nitrate or even sodium chloride can reduce or prevent steep damage. Maltsters successfully grow water-sensitive grain by steeping it for limited periods, with various air-rests. This allows the germination of grain that will not grow in running water (Chapter 15) [199]. Below a moisture content of about 30% grain will not germinate; steeped by constant immersion to 45% moisture water-sensitive grain germinates very badly; steeped to about 37% moisture (the value reached in the 4ml test) then air rested for 8–24h, then re-immersed until the moisture is about 45% the grain will subsequently grow well, although no sign of germination is apparent at the end of the air-rest (Fig. 5.6) [215]. Some change connected with the 'air-rest' allows the grain to germinate subsequently, a situation reminiscent of the state at the end of the 'cold-rest' when seeds are stratified. The optimal water supply for germination decreases with increasing grain maturity [84]. Well after-ripened grain of *H. spontaneum* is less damaged than dormant grain by steeping at 30°C (86°F) [190].

Physical treatments to dormant grain can enhance germination. Moderate battering or abrasion can improve germination as can lifting the husk, or (better) removing the husk and pericarp, i.e. decorticating with sulphuric acid, or peeling away the husk, pericarp and testa, or lifting the husk and scarifying the pericarp and testa [15,16,34,107,174, 199,231]. Comparison of results obtained with decorticated and peeled grains suggests that usually the testa has little effect on dormancy. Perforating, pricking or slitting the grain coats, especially near the embryo, or cutting the grain transversely into two enhances germination, supposedly reducing 'dormancy' and not 'water sensitivity' [199]. In experiments with *H. spontaneum* germination increased in the order intact grains < dehusked grains < cut grains < cut and dehusked grains. In grain with intact husks and entire corns 6% germinated; in those with 0.25, 0.5, and 0.75 of the endosperm removed 12%, 46% and 48% germinated respectively; in those with dehusked grains the values were 12%, 58%, 65% and 81%; and in those with isolated embryos 100% germinated [190]. Embryos so dormant that they will not germinate even when

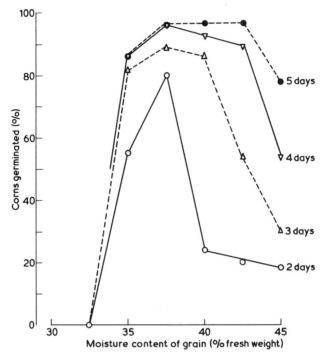

Fig. 5.6 The germination on successive days, under malting conditions, of samples of a barley first steeped to different moisture contents (after Sims [215]).

isolated from grain seem to be rare [15]. Deeply dormant, separated embryos may germinate better after the removal of the scutellum [98]. Some grains, in addition to decortication, need chilling or treatment with gibberellic acid to induce germination, suggesting that either 'embryo dormancy' is more common than generally believed or that the testa can have a slight inhibiting effect [41,199].

Pregermination of grains in the ear before harvest is usually associated with splitting in the pericarp and testa, suggesting that these layers are responsible for inhibiting germination in the maturing grain [96,241]. Dried and after-ripened immature grain will germinate [105]. Vivipary can be induced by removing the lemma over growing grains in the ear and keeping them moist (Chapter 12) [201]. Under these conditions the seed coats may be weakened, as they are known to be thin in grain formed in wet conditions [227]. Alternatively germination may be permitted by a reduction in the internal osmotic pressure. In culture immature embryos only continue embryonic development – rather than growing into small seedlings – when the osmotic pressure is high [249].

Exposure of grains to a.c. and d.c. electric currents, to ultrasonication (23–960 kHz), to radiofrequency electric fields, to shortwave radiations,

Grain quality and germination 193

to high magnetic fields and to ultraviolet radiation (sometimes with protection for the embryo) are all reported to stimulate the germination of immature grains. In nearly every case prolonged treatment or increased intensity of treatment harms the grain.

Despite an early belief that grain was best malted in buildings with blue glass, to filter the daylight, the effect of light on the germination of mature barley is at most marginal. However some dormant barleys are sensitive to light [40,99,216]. Far-red, tungsten, and sometimes blue lights are inhibitory; red light, and some other colours, can stimulate the germination of these grains. As dormancy declines, so does the light sensitivity. Phytochrome may control this response.

Gibberellic acid readily breaks dormancy, although apparently not water-senstivity [199]. Sometimes kinetin, or benzyl aminopurine, enhance germination slightly [198]. In a particular barley kinetin did not break dormancy, while gibberellic acid did; however, kinetin did reverse the inhibitory effect of abscisic acid (ABA). Thus it seems unlikely that an accumulation of ABA causes dormancy [139]. However, an inhibitory factor chromatographing with 'inhibitor-β' declines with the decline of dormancy [204]. Cytokinins do occur in barley (Chapter 4). Other compounds that occur in barley, such as coumarin, umbelliferone and herniarin, can check germination when applied exogenously in sufficient amounts, but their function *in vivo*, or that of their metabolites, is uncertain [232]. At very low concentrations exogenous coumarin stimulates barley germination [143]. Apparently, endogenous auxins are not involved in dormancy [90]. At particular dose levels, or when applied for short periods of time, numerous chemical substances relieve dormancy to a greater or lesser extent; at higher doses they frequently damage, or kill, the grain. In different cases they may be effective used in a grain pretreatment, or in solution in the water used to wet the grain. Among the best known agents for breaking dormancy are hydrogen peroxide and a high partial pressure of oxygen; an atmosphere of 36% oxygen is optimal; at higher pressures the gas is damaging or lethal, especially in prolonged experiments [15,107]. More deeply dormant grain needs higher levels of oxygen for germination. Conversely well matured grain may germinate with as little as 0.3–0.5% of oxygen in the atmosphere [46]. Dormancy-breaking agents include nitrates, nitrous acid, dilute mineral acids to acidify the germination substrate to about pH 3, (hydrochloric acid, sulphuric acid, phosphoric acid, nitric acid for a brief exposure only), citric acid (but *not* acetic acid or other fatty acids which are generally toxic), potassium dihydrogen phosphate and sulphur dioxide. Other effective reagents include sodium metabisulphite, calcium hydroxide, sodium carbonate, sodium hydroxide, hydrogen sulphide, various organic mercaptans, sodium sulphide, thiourea, hydroxylamine, azide ions, cyanide ions, carbon monoxide, sodium chloride, ammonium sulphate, sodium hypochlorite, calcium hypochlorite (bleaching powder), silver nitrate, and many other substances [84, 107, 144, 163, 190, 199, 200, 230]. While *some* of these agents *may* penetrate ungerminated grains, the only action they could have in common with each other and with several of the physical treatments would be to alter the structures of the surface layers and, like mixtures of antibiotics, be more or less harmful to the micro-organisms present there.

Any hypothesis concerning dormancy must recognize that it is not an 'all-or-none' phenomenon since even though particular grains do not germinate in one test they almost certainly would in another. Furthermore, the rate of germination, and presumably 'vigour', increase with increasing post-harvest maturation after 'dormancy' has disappeared. An attractive hypothesis is that the embryos of dormant grains have a higher oxygen requirement than those of non-dormant grains and experiments with separated embryos suggest that this is true. The reason for the reduced oxygen requirement is unknown. During after-ripening the ability of moistened embryos to release more thiol-containing compounds progressively increases [11]. The supply of oxygen to the embryo is limited by the surface structures, by any surface moisture, and by the associated mixed microbial population. Various combinations of these factors and the variable damaging effects on the grain of treatments used to break dormancy can account for most observations, if allowance is also made for the accumulation of toxic substances either from microbial metabolism or anaerobic grain metabolism under unfavourable conditions. The husk, the pericarp and perhaps the testa are barriers to rapid oxygen penetration into the grain, and also a population of micro-organisms can readily consume considerable quantities of oxygen – and will consume more at higher temperatures. A film of surface moisture on grain reduces oxygen uptake – presumably by the time taken for oxygen to dissolve in the water, the time taken for it to diffuse into the grain, and the lower concentration of oxygen in air-saturated water compared with air. The level of dissolved oxygen in water saturated with air falls sharply with rising temperatures. Metabolic interpretations of whole grain respiration must be treated with caution since, in the absence of non-damaging, reliable sterilizing agents or grain produced under aseptic conditions, it is impossible to separate the contributions made by the grain, by the associated microbial population and the restrictions imposed by the grain structures and the surface moisture. It is usually reported that dormant and non-dormant grains respire at about the same rate, or that dormant grains respire slightly faster; mitochondria from water-sensitive grains take up oxygen and phosphate at the same rate as those from mature grains [130,163]. Grain maturation and the microbial population can vary independently and different microbes will be harmed to different extents by different agents – facts that could explain why treatments vary in their ability to

Grain quality and germination

break the dormancy of different seed lots. Low internal oxygen concentrations could explain why water-sensitive grains are relatively less damaged by γ-irradiation [222]. Sliced endosperms from water-sensitive grain respond badly to added gibberellic acid, but respond better if the grain is decorticated. The improved response to gibberellic acid during maturation is not entirely dependent on surface sterility [61]. Endosperms from undried, immature grain respond less well to gibberellic acid than endosperms from grains that have dried on the ear [76]. Reduced oxygen availability may explain why sterile, decorticated grain makes less α-amylase when germinated on wet, rather than drier paper [23]. Enhanced oxygen availability can explain the faster development of malting extract in decorticated compared to entire barley grains [218].

5.8. The gas exchange of germinating grains

Stored grain respires very slowly when it is dry and the rate of respiration increases with increasing grain moisture content and temperature (Chapter 10). An uncertain, variable proportion of this is due to micro-organisms. The oxygen uptake and carbon dioxide output – which may be due to different metabolic changes – are indiscriminately called 'respiration'. The respiratory quotient – the ratio of carbon dioxide released to oxygen taken up (R.Q. = $+CO_2/-O_2$) can sometimes give an indication of the type of metabolism occurring. For example the conversion of hexose to ethanol involves the release of carbon dioxide but no oxygen uptake, so gives an R.Q. of infinity, and a ratio of carbon dioxide produced to ethanol produced of 1. In the total oxidation of hexose, the R.Q. = 1; higher values indicate that aerobic fermentation is occurring together with respiration, when hexose is the substrate; an R.Q. of 0.57 is given by the total oxidation of ethanol; the partial oxidation of ethanol to acetic acid involves oxygen uptake and no carbon dioxide output; the oxidation of triglycerides gives an R.Q. of about 0.7; the conversion of triglycerides to sugars, an R.Q. of about 0.4; numerous other values are known, but those mentioned are of immediate interest. Contamination by micro-organisms distorts these values. The carbon dioxide output of stored grain, and grain during germination, is greater for samples with higher, rather than lower, nitrogen contents and is greater for thin rather than plump grain. Heat output also increases with increasing nitrogen content (Table 5.3) [2,116]. There are minor differences between measured heat outputs and those calculated from the chemical changes observed [141]. As grains hydrate when they are immersed in water (steeped) oxygen uptake and carbon dioxide emission accelerate. If the steep is not aerated it rapidly becomes anaerobic and alcoholic fermentation occurs. More than half of the ethanol formed enters the steep water [46]. In air some ethanol evaporates and more is oxidized; if the

Table 5.3 The respiration (CO_2 output) rate of different classes of grain germinating, under approx. malting conditions, at 12°C (from Abrahamsohn [2]).

Thousand corn weight (g)	Crude protein (%)	Carbon dioxide produced, (CO_2 mg/h/100g grain)
31.7	9.57	31.3
31.7	11.34	39.5
43.1	9.58	24.8
43.1	12.09	31.1

husks are removed the R.Q. is about 0.8 as the ethanol is oxidized. Otherwise, when the grain is transferred from the steep to air the alcohol content rises at an enhanced rate until the grain chits, after which the level falls. In exceptionally well aerated steeps the R.Q. may be reduced to about 1, and the grain will chit under water (Table 5.4) [73]. As wheat and rye grains hydrate their respiration roughly parallels the content of unbound ('freezable') water in the grain. As it germinates malting barley, with its limited supply of 'internal' water, respires faster until about days 3–5, after which the respiration rate declines; enzyme activities (diastase, protease) do not increase in parallel to each other, or to the respiration rate (Fig. 5.7) [53,54]. Within limits respiration is faster at elevated

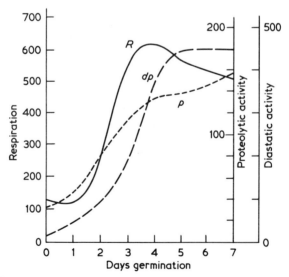

Fig. 5.7 The carbon dioxide output rate, R (mg CO_2/h/kg grain), the diastatic power, dp (Windisch–Kolback), and the proteolytic activity, p (Idoux), of barley during a 7 day malting period (after de Clerck and Cloetens [54]).

Table 5.4 The gas exchange of barley immersed in flowing water, equilibrated with different gas mixtures. A half or more of the oxygen was taken up as the liquid flowed through the bed of grain. Most of the germinated grains occurred near the water inlet (from Ekström et al. [73]).

O_2 (%) in circulating gas	CO_2 output (mmole/kg dry matter/h) at			O_2 uptake (mmole/kg dry matter/h) at				R.Q. $(+CO_2/-O_2)$ at				Grains germinated (%)
	3h	24h	48h	72h	3h	24h	48h	72h	3h	24h	48h	72h	
5.8–6.3	0.65	1.73	2.33	2.35	0.24	0.52	0.84	1.05	2.73	3.37	2.83	2.26	12–14
10.2–11.4	0.57	1.81	2.50	2.55	0.37	0.90	1.22	1.41	1.51	2.01	2.05	1.83	13–14
21.0	0.73	1.80	2.47	2.70	0.80	1.65	2.25	2.55	0.90	1.09	1.09	1.06	89–95
37.0–40.5	0.56	2.15	2.99	3.23	0.69	1.85	2.90	3.32	0.82	1.03	1.03	0.97	94–96

Table 5.5 The compositions of barley samples after germination, under malting conditions, for 10 days at the temperatures indicated. Results are expressed as % dry matter, original barley (from Day [64]).

°C	°F	Fatty matter	Sugars	Soluble carbohydrates other than sugars	Starch	'Cellulose'	Nitrogenous bodies soluble at 40°C and permanently soluble after boiling (N × 6.33)	Nitrogenous bodies soluble at 40°C coagulated on boiling	Solids permanently soluble at 40°C	Starch conversion products formed by action of diastase at 40°C	Carbon oxidized	Carbon oxidized reckoned as $C_6H_{10}O_5$
4.5	40	2.03	2.75	4.84	57.81	7.77	2.59	0.30	15.65	5.47	0.57	1.28
10.0	50	1.78	6.54	7.40	49.70	6.60	3.77	0.39	30.46	12.74	1.46	3.28
12.8	55	1.71	7.37	7.15	47.20	6.72	3.72	0.40	32.33	14.08	2.02	4.54
15.6	60	1.61	5.64	7.26	48.71	6.55	3.35	0.45	26.41	10.16	2.21	4.97
18.3	65	1.50	4.86	6.31	48.76	6.38	3.04	0.44	23.56	9.35	2.47	5.56
21.0	70	1.50	5.10	6.30	49.00	6.23	2.87	0.42	21.74	7.48	2.66	5.98

198 Barley

temperatures or in grains with higher moisture contents (Table 5.5; Chapter 15) [46,47,64,65,184,196]. Temperature has a dramatic effect, e.g. on carbon dioxide output (mg/h/500 plants), 0.3°C, 10.5; 5°C, 15.0; 16°C, 26.6; 26°C, 64; 33.6°C, 94. Analyses of grains, after germination at different temperatures, suggest that the balance of metabolic changes alters with temperature, and does not merely change in rate, a conclusion in agreement with malting experience (Table 5.5; Chapter 15). The energy of activation of oxygen uptake from water (12–14 kcal/°C/mole) is sufficiently like that from air (10–13 kcal/°C/mole) to show that the same mechanism is involved [63]. Cyanide-resistant respiration occurs in grain [167].

When grain hydrates and grows on a continuously wet substratum the time course of carbon dioxide emission is significantly different to that found in malting barley (Fig. 5.8) [4,87,128,167]. In the dry grain respiration is low (e.g. $0.06\mu l\ O_2$/g dry wt/h; 8.3% H_2O), and the R.Q. is slightly below 1. As hydration proceeds, the carbon dioxide output rises sharply, and the R.Q. exceeds 2, as aerobic fermentation occurs, and ethanol accumulates. When the grain chits the oxygen uptake rate rises sharply and the R.Q. falls, settling eventually to about 1, indicating that (i) hexose is the major substrate, and (ii) before chitting the surface layers of the grain restricted gas exchange. This is confirmed by removing the husk, when the oxygen uptake rate increases 2–3 times. Grains separated into embryos and endosperms have higher oxygen uptakes than intact grains. Oxygen is found to suppress glycolysis, so the Pasteur effect is operative. Wetting the surface of the grain, or transferring the grain from air into aerated water, reduces oxygen uptake to about 0.3–0.5 of the original value [63]. The maximum respiration rate coincides with the highest rate of transfer of nitrogenous substances from the endosperm to the embryo [86]. In time the depletion of the endosperm reserves causes a check, then a decline in the respiration rate. The check in the

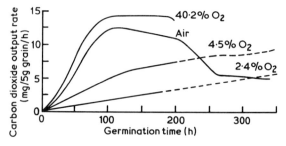

Fig. 5.8 The rates of carbon dioxide output of barley seedlings, growing in the dark on a wet substratum at 22.5°C (about 72°F) in air and gas mixtures containing various proportions of oxygen. Water was added to the substrata at zero time. Some curves are reconstructed (after Forward [87]).

decline of respiration coincides with a change in respiratory substrate; protein breakdown begins. A final 'senescent rise' may be due to microbes growing on the dying seedlings (Fig. 5.8) [86,87,128,243]. The quantity of alcohol produced in the initial stages of germination can be surprisingly high. If the grain is kept in a stream of nitrogen gas, ethanol is found in the condensate, in concentrations of 3.2, 4.9 and 6.6% (wt/wt) on successive days [230]. A 1:1 correspondence between the amounts of ethanol and carbon dioxide produced is *not* found. In some conditions the alcohol yield was 87% of that expected from the carbon dioxide produced. In a fast-flowing nitrogen gas stream more 'extra' carbon dioxide is produced – probably due to the reversible decarboxylation of organic acids [46,47,182].

The respiration of roots is readily altered for example by the uptake of ions (Chapter 6) [244]. The respiration of green shoots in light is difficult to evaluate because of the intervention of photosynthesis and photorespiration (Chapter 4). Increasing oxygen partial pressures increase grain respiration, within limits, and carbohydrate utilization in roots (Fig. 5.8) [87,114,235]. At 23–25°C (73–77°F), as the oxygen (%) in the atmosphere increased, the relative respiration rates for some barley samples were: 0.2%, 15; 1%, 23; 2.7%, 54; 5.2%, 61; 9.5%, 75; 21%, 100. Oxygen levels above atmospheric do slightly increase respiration rates [235]. Ionizing the air, possibly due to the formation of ozone (O_3), increases oxygen uptake to a limited extent [169]. Carbon dioxide in the atmosphere checks grain respiration and, in the presence of 80% carbon dioxide and 20% oxygen, grain chits but will not develop further. In 1799 Rollo (quoted de Saussure) [207] showed that in anaerobic conditions barley gives off carbon dioxide and eventually rots, while in air it takes up oxygen and has a R.Q. of approximately 1. R.Q. values above 1 are taken to indicate aerobic fermentation, although contributions from unrelated decarboxylation almost certainly contribute 'extra' carbon dioxide. The R.Q. of samples of grain germinated at intervals increases during storage at room temperature, and under conditions chosen to accelerate deterioration [5,60]. R.Q. values of 0.6–0.9 in quiescent grain and in the first stages of imbibition, of 0.7–0.8 in 'stripped' grains, of as little as 0.3–0.4 (generally 0.6–0.8) in separated endosperms, and 0.6 (generally 0.8–1.0) in starved isolated embryos, in which the carbohydrate reserves are depleted, indicates that lipids can be respiratory substrates, or may be converted to sugars [4,128,167,231]. In the aleurone layer the glycerides are probably respiratory substrates initially but later, some at least, are converted to sucrose [183].

5.9. The chemical composition of the quiescent barley grain

During the initial stages of germination either in malting, or on a wet substratum, before the onset of photosynthesis or the uptake of nutrients by the roots, the embryonic plant grows at the expense of the reserves in the grain. A knowledge of the biochemical changes which occur is essential for an understanding of seedling establishment and malting. These changes have been intensively studied for about 200 years and consequently the following account embodies only a tiny part of the information available [25,108,199].

Barley grains vary widely in size (5 to 80mg dry weight) and considerably in chemical composition. An 'average' British two-rowed barley grain weighs 35–40mg. Decortication, (with 50% sulphuric acid), shows that the husk and pericarp together contribute some 10% to the dry weight. The embryo (about 1.2–1.7mg) contributes about 3% dry weight, the remaining 87% dry weight being made up of the testa, aleurone layer, starchy endosperm, and the associated remains of the nucellus and the pigment strand. Indirect evidence suggests that the aleurone is 7–13% dry weight of the grain; more direct estimates give values of about 14% dry weight for separated aleurones with the associated testae, pigment strands and some nucellar tissue. Thus the tissues remaining – the starchy endosperm and the nucellar remains in the sheaf cell region – comprise about 73% dry weight of the grain. The actual values vary for grains of different sizes and conformations. An estimate of the proportions of the parts of particular grains, and the contribution of the cell walls and gums to these parts, are given in Table 5.6.

The husk (palea + lemma) is made up of heavily thickened cell walls containing nearly all the lignin of the grain, pentosans (some of which are extractable as 'gum'), mannan, uronic acids, hemicelluloses and crude

Table 5.6 The proportions of lightweight British two-rowed barly grains passed by a 2.8mm slotted sieve but retained by a 2.5mm sieve. The weight of 1000 grains was 30.8mg (freeze dried, about 8% moisture; from Morrall and Briggs [175]).

Parts separated	Proportions by weight (freeze dried; %)	Parts of the fractions recovered as	
		cell-wall (%)	gum (%)
Husk + pericarp	13	– *	– *
Embryo and adhering testa	3.6	15.5	0.9
Aleurone + testa + pigment strand†	11.6	42.0	7.1
Starchy endosperm + sheaf cells	71.8	7	3.0

*Not determined

† The testa + pigment strand are about 1.5% of the grain-weight, so the aleurone layer would be about 10% of the whole.

cellulose (about 30%) [158,234]. This tissue contains no starch and practically no water-soluble sugars. Its outer cell walls (and the awn) are heavily loaded with silica. In mature, ripe grain the pericarp is dead and starch-free; nothing is known of its chemical composition. In immature, growing grain this tissue is photosynthetically active, and contains starch for a period; possibly some enzyme activities may survive in the desiccated tissue of the mature grain. The contaminating microflora contributes its enzyme complement to this tissue.

The testa, with the associated pigment strand, is the major barrier to the penetration of chemical substances and microbes into the grain. It also prevents soluble substances leaking from the grain interior. It appears to contain 'crude cellulose', and uncharacterized material, associated with an estolide of fatty acids, and polyhydroxy-carboxylic acids. It is coated with a wax containing alkanes, 5-n-alkyl resorcinols and various trace components (Chapter 4) [26].

The quiescent embryo rarely, or never, contains starch. Its cell walls contain uronic acids, pectin, hemicelluloses, traces of galactan and araban and crude cellulose (about 7% dry weight). It is rich in petrol-extractable lipid (14–17%), sucrose (14–15%) and raffinose (5–10%) as well as containing higher fructosans; however, it is poor in free hexoses (about 0.2%) [108,158,159]. More complete extraction shows that the whole grain contains about 3.5% (even 4.6% in particular cases) of total lipids divided, in one case, into neutral lipid, 70.7%; neutral glycolipid, 10.2%, and phospholipid, 19.1% [195]. The ash content of the whole grain is 2–3%; in the embryo it is 5–10% (Chapter 14) [56]. The nitrogen content of the embryo is about 5.5% ('crude protein' about 34%) [36].

The composition of the aleurone layer is uncertain. It contains most (about 90%) of the neutral lipid of the degermed, decorticated grain, which is about 67% of the grain's total [108,183]. Much of the lipid is stored in spheroplasts (Chapter 1). The aleurone grains that are present within the cells contain stores of protein, phytic acid, and apparently a glycoprotein [125]. Estimates of the 'crude protein' content of preparations of aleurone, with the adhering testa, are 17–20% [21]. The tissue also contains sucrose, and perhaps some fructosan and raffinose [108]. The cell walls contain arabinoxylan (85%), 'cellulose' (8%) and protein (6%); hydroxy-proline was not detected [166]. It contains more than 75% of the ash of the degermed, decorticated grain, which in turn contains about 83% of that found in the intact, decorticated grain [56].

Starch comprises 63–65% dry weight of plump, two-rowed barleys; in quiescent grain it is confined almost entirely to the starchy endosperm, making it about 85–89% of this tissue. Analyses of pearled barley give values of about 85% (dry weight; Chapter 14), in reasonable agreement since many 'starch' analyses are slightly unreliable and pearl barley is not pure starchy endosperm. A further difficulty in making quantitative estimates is the inhomogeneity of the starchy endosperm as exemplified by the lack of starch in the 'sheaf cells' and the compressed cell layer, and the relatively low starch content, but high protein content, of the sub-aleurone layer (Chapter 1) [149,234]. The sub-aleurone layer is also rich in 'saccharogenic amylase', mainly β-amylase, where the bound form is associated with the hordein-rich protein granules [74,229]. Pearl barley contains about 9% (dry weight) of crude protein. The material unaccounted for as starch and protein is probably mainly the water-soluble gum, (β-glucan, 1.5%; arabinoxylan, about 1% of the whole grain) and cell wall polysaccharides (about 7%; Table 5.6). The cell walls contain a fibrillar phase of crude 'cellulose' rich in mannan, with bound arabinoxylan and an amorphous phase of arabinoxylan (about 23% dry weight) and β-glucan (71% dry weight) [80,108,158]. No pectic material or 'true' cellulose is present, and the trace of protein that occurs is devoid of hydroxyproline. The starchy endosperm is poor in simple sugars, but of these hexoses are prominent [108]. The levels of soluble sugars in quiescent barley vary to a significant extent.

5.10. Biochemical changes in germinating grain

The restoration of growth to the quiescent embryo, the mobilization of the food reserves of the endosperm, the onset of photosynthesis and the beginnings of transpiration and nutrient uptake by the roots involve all the biochemical capabilities of the grain (Chapters 4 and 6). If the grain is hydrated with tritiated water some amino acids, and subsequently some acids of the tricarboxylic acid cycle are labelled in 120–360 min indicating that metabolic changes are occurring [59]. In this section only some few specialized aspects of grain germination are discussed.

Whether barley is malted at 14–18° C (about 57–64° F) or is grown on a wet substratum at a higher temperature, e.g. 25° C (77° F), the initial trends of the changes in carbohydrate levels are similar although the rates are different. (Table 5.7; Chapter 15) [23,108,118,127,157,199]. Raffinose and sucrose decline to low levels in the first day or two of germination. Subsequently, while raffinose continues to decline and finally disappears, the levels of sucrose, glucose, fructose and maltose rise. At the same time the starch content of the grain declines; physical, microscopic and chemical investigations show that the endosperm cell-walls are hydrolysed with falls in the contents of insoluble glucose, arabinose and xylose; the levels of soluble pentosans and glucans (dextrins *plus* β-glucans) rise [175]. The massive degradation of reserve starch in the endosperm may be marginally offset by starch formation in the growing embryo. In growing seedlings synthetic processes are more evident than in malting grain and more cellulose and hemi-cellulose are laid down in the embryo. Eventually the endosperm starch becomes exhausted and, after a maximum

Table 5.7 Changes in the carbohydrates of barley during commercial malting (the roots were retained), results expressed as % dry weight at time of sampling. The loss of dry matter in converting the barley to malt + roots was probably about 7% (from Harris and MacWilliam [109]).

Sugar	Initial barley	Steeped barley	Days on malting floor							Kilned malt
			1	2	3	4	7	9	10	
Sugars soluble in 80% EtOH										
Fructose	0.07	0.16	0.07	0.09	0.16	0.31	0.27	0.47	0.53	0.28
Glucose	0.02	0.08	0.05	0.10	0.22	0.36	0.73	1.05	1.13	1.54
Sucrose	1.07	0.75	0.51	0.72	0.42	0.58	1.09	1.71	1.85	4.24
Maltose	Trace	0.04	0.01	0.01	0.08	0.11	0.43	0.62	0.67	0.68
Maltotriose	Nil	Nil	Nil	Nil	Nil	0.03	0.09	0.42	0.53	0.51
Raffinose	0.29	0.19	0.17	0.02	Nil	?	?	?	?	?
Glucodifructose	~0.1	~0.1	~0.05	~0.01	0.01	?	?	?	?	?
Xylose + Arabinose	Nil	Nil	Nil	Nil	Trace	0.01	0.01	0.05	0.05	?
Polysaccharides soluble in 80% EtOH										
Fructosan	0.43	0.21	0.07	0.11	0.17	0.25	0.61	0.81	0.84	0.30
Glucosan	0.26	0.17	0.08	0.21	0.47	0.73	1.33	1.98	2.29	0.53
Araban	0.04	0.02	0.01	0.02	0.05	0.07	0.10	0.17	0.18	0.19
Xylan	0.04	0.02	0.02	0.03	0.06	0.08	0.10	0.18	0.19	0.21
Polysaccharide soluble in water, after alcoholic pre-extraction										
Fructosan	0.05	0.04	0.04	—	0.03	0.06	0.09	0.05	0.08	Trace
Glucosan	1.01	2.60	1.70	1.26	1.61	1.78	1.96	3.68	4.64	1.14
Araban	0.08	0.14	0.14	0.09	0.14	0.19	0.36	0.34	0.46	0.20
Xylan	0.08	0.12	0.12	0.09	0.13	0.20	0.28	0.44	0.42	0.21
Galactan	Trace	Trace	Trace	Trace	Trace	Trace	Trace	Trace	Trace	Trace
Extraction with chloral, precipitation with acetone										
'Starch'	57.3	—	56.3	—	55.3	—	54.9	—	52.9	53.1

Table 5.8 The lipids of barley and green malt. In other trials the phospholipids altered less (from Holmberg and Sellmann-Persson [117]).

	Composition of lipid extract (%)	
	Barley	Malt
Hydrocarbons	0.3	2.0
Sterol and wax esters	3.3	18.3
Triglycerides	55.1	55.9
Fatty acids	3.6	0.9
Sterols and fatty alcohols	3.3	5.4
Diglycerides	5.2	6.5
Monoglycerides	8.0	7.9
Phospholipids	21.1	4.0

on day 4 at 25°C (77°F), in all tissues except the shoots where the maximum is at 5 days, the levels of soluble sugars decline in all parts of the grain [23]. The sucrose level in the endosperm does not follow the trends of the other sugars, indicating that it is not all derived from the products of the hydrolysis of starch or the other polysaccharides. The sucrose level in the starchy endosperm may be augmented by the hydrolysis of fructosans as it is by the conversion of other sugars to sucrose in the aleurone layer, and by the conversion of neutral glycerides into sucrose also in the aleurone layer by way of the glyoxylate cycle, a conversion which explains the low R.Q. of this tissue [50,101,183]. Glucose and maltose are converted to sucrose in the scuttellum [38,71]. During germination the levels of mono-, di- and tri-glycerides decline in the aleurone layer. The overall lipid content of barley, about 3%, declines by about 0.1–0.2% during malting, or more in experimental germinations, and the proportions of the lipid classes change (Tables 5.8) [108,115,117]. The changing levels of citric, malic and gluconic acids and of glycerol have also been followed during germination [91]. There is no loss of nitrogen from the grain but there is a massive movement of nitrogenous substances from the endosperm to the embryo. For example, while the nitrogen contents of the embryo and the endosperm *plus* integuments were 85 and 543 mg/1000 grains respectively before germination, after 11 days malting the values were – embryo, 228; roots, 76 and endosperm *plus* integuments 323 mg/1000 grains [36]. The proportions of the nitrogenous fractions alter during growth; the substances soluble in water and salt solutions increase in amount at the expense of the less soluble hordein and glutelin. However, the net changes observed in analysis of whole grains are made up of mainly degradative changes in the endosperm and synthetic changes in the embryo (Table 5.9; Fig. 5.9) [14,85,108]. In malting a high level of free proline occurs, apparently formed in the

Table 5.9 Alterations in the nitrogen fractions of an British two-rowed barley, during malting. The grain was steeped for 60 h and kilned after 314 h germination. Roots were included in all analyses. The salt-soluble fraction includes the non-protein, proteose, albumin and globulin groups of substances (from Bishop [14]).

		Nitrogen in fraction (mg/1000 corns)							
Time from start of steep (h)	1000 corn dry weight (g)	Total nitrogen	Salt soluble	Hordein	Glutelin	Albumin	Non-protein	Globulin	Proteose
70	42.9	635	185	206	245	39	62	51	35
141	42.2	634	223	191	220	60	86	44	35
189	42.6	631	286	160	186	67	125	49	49
218	41.7	642	324	138	180	67	167	48	45
241	40.7	626	356	113	159	67	191	58	48
290	40.4	638	350	95	192	51	207	54	44
410 (finished malt)	38.3	622	367	75	182	48	208	48	63

206   Barley

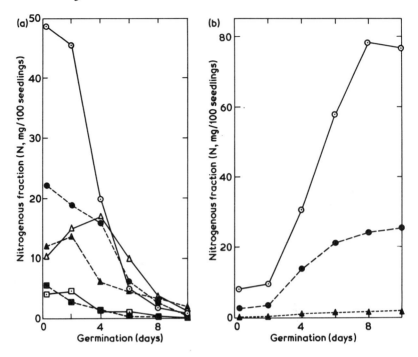

Fig. 5.9 Alterations in some of the nitrogenous fractions of the embryo and endosperm of Spratt-Archer barley, germinated on a wet substratum, in light, at 22.5° C (about 72° F) (from Folkes and Yemm [85]). The barley was rich in protein, having a nitrogen content of 2.5%. (a) Fractions from the endosperm; ⊙——⊙, hordein; ●--●, hordenin; □——□, albumin; ■---■, globulin; △——△, non-protein nitrogenous substances; ▲---▲, residual, alkali-insoluble nitrogenous substances. (b) Fractions from the embryo: ⊙——⊙, protein; ●---●, water-soluble, non protein nitrogenous substances. ▲----▲, chlorophyll.

embryo; its rise is selectively depressed by adding potassium bromate to the grain and is relatively insensitive to exogenous gibberellic acid [135]. The enhanced level of proline in malting grain (in which water availability limits growth) invites comparison with the rise in proline that occurs in water-stressed plants (Chapter 6).

In grain grown on a wet substratum, without an exogenous source of nitrogen, the depletion of the nitrogenous reserves of the endosperm is ultimately nearly complete. When the decline of particular amino acids (free + protein bound) in the endosperm is balanced against the rise in the embryo it is seen that a major overall decline occurs in 'amide', glutamic acid and proline, (which occur in high levels in the reserve proteins of the endosperm), while there is a net conversion of these into other substances [85]. In the endosperm the peak level of soluble nitrogenous compounds, including 'amides' occurred on about day 4, or earlier in grains grown at a higher temperature [23,85]. The aleurone layer converts various amino-acids to glutamine [183]. While the free amino acids of the seedling axis continue to rise, the level in the scutellum 'peaks', then declines in parallel with that occurring in the starchy endosperm [23]. In dry barley practically all the protein potential thiol groups are oxidized and occur as disulphides. In the first few hours of malting they are rapidly reduced to thiols, presumably with the consequent release of bound β-amylase and activation of the thiol-dependent proteases [138]. Increasing levels of 'protease', a total activity of numerous enzymes, have often been noted in germinating grain [108]. In malting carboxypeptidases increase 10–20 times, naphthylamidases double in amount, and aminopeptidases increase 3–6 times [172]. Also during germination the levels of proteolytic inhibitors decline [170]. Numerous other quantitatively minor groups of substances alter in germinating barley, although they have received relatively little attention. For example, more cutin appears covering the coleoptile [128]. The root-slime appears, and there is a transient occurrence of the basic substances tyramine, N-methyltyramine, hordenine (N,N-dimethyltyramine) and candicine (N,N,N-trimethyltyramine) in the roots (Chapter 4). There is a major movement of mineral substances from the endosperm, mainly the aleurone layer, to the embryo. The phytic acid, which is stored in the aleurone grains in the aleurone layer is hydrolysed to release inorganic phosphate [56,152]. Grains germinated at different moisture contents, or at different temperatures, do not merely undergo the same chemical changes at different rates. The chemical compositions and the enzyme complements of the seedlings are appreciably different (Table 5.5; Chapter 15) [64, 148, 213,214].

5.11. Embryo culture *in vitro*

Embryos have been cultured *in vitro* to allow studies of their nutrient requirements, their biochemistry, their ability to produce enzymes and to obtain plants from genetically abnormal grains in which the endosperms are deficient. Since embryos can readily be grown separated from the endosperm, when nutrients are provided, it is evident that the endosperm is a foodstore. Separated embryos from mature grains grow, under aseptic conditions, on simple media containing a sugar and a nitrogen source. Compounds vary in their suitability as nutrients; sucrose is the best carbohydrate nutrient found, followed by glucose and maltose [38]. In the earlier studies asparagine was the best nitrogen source tested, with potassium nitrate and ammonium sulphate being less good and some amino acids, such as tyrosine, phenylalanine and leucine, tested singly, actually being toxic. Later glutamine was shown to be better than asparagine; a complete mixture of amino acids supports enzyme production

208 Barley

better than any single one in isolated embryos [19,206]. Feeding experiments have given indications of the regulatory mechanisms of amino-acid synthesis (Chapter 4). The best pH for embryo culture is about 5 [146]. Gibberellic acid enhances the growth of isolated embryos [208].

Very small embryos taken from immature grains in the ear are comparatively difficult to culture without the formation of morphological abnormalities. Complex media containing salts, a range of amino acids, vitamins, a high concentration of sucrose (e.g. 9%), careful pH adjustment, and light of a particular composition may be needed to achieve normal growth [45,187,188]. Immature embryos, in intact florets on the spike, develop inadequately *in vitro* unless the palea, lemma (and flag leaf) are removed, suggesting that a 'hull factor' is needed for normal development [145].

Larger immature embryos are less fastidious in their nutrient requirements, but on many media they display 'precocious germination', that is they grow into small, spindly plantlets instead of continuing normal embryonic development [249]. This tendency is overcome by media with high osmotic pressures, caused by the addition of casein hydrolysate, phosphate, sodium chloride, mannitol or sucrose (12.5%); the incorporation of *Hordeum* endosperm in the medium adjacent to the embryos is also favourable; reports conflict on whether or not it is advantageous to add coconut milk to the medium [186,249]. Precocious germination is also checked by low partial pressures of oxygen, elevated culture temperatures, and red (but not far-red or blue) light. The parallel between effects that enhance dormancy and depress precocious germination is even greater than suggested by these results since embryos collected from grains in the early spring, when dormancy is favoured, do not pregerminate [188]. The *in vitro* culture of separated embryonic axes, (i.e. in contrast to axis + scutellum) or barley roots has not been very successful [3, 168].

5.12. The mobilization of the endosperm reserves

As the embryo grows, so the degradation of the cell walls, protein and starch grains of the endosperm, together with changes in the aleurone layer, give rise to low molecular weight substances that diffuse in solution to the scutellum and serve as nutrients. In malted grain some structural changes can be detected by the failure of partly degraded starchy endosperm to stain with dyes or, after drying, by the friability of the tissue. In decorticated grain grown on a wet substratum degradation of the starchy endosperm is so complete that parts of the endosperm liquefy completely [23,24]. The exact physical pattern of endosperm degradation, (Chapter 1) [24], is entirely consistent with the mechanisms that are believed to regulate breakdown processes.

In malting grain the level of soluble substances (cold water extract;

mainly soluble carbohydrates), soluble nitrogen, α-amylase and 'diastase', all increase first at the embryo end of the grain, as does malt extract [21,67,239]. α-Amylase is formed in roughly equal amounts on the dorsal and ventral sides of the grain. The 'skew' pattern of modification is due to the scutellum being mounted at an angle to the long axis of the grain, and to the presence of the ventral furrow with its crest of nucellar 'sheaf cells' which are relatively resistant to enzymic degradation (Chapter 1).

When the germ is removed from a dry grain and the hydrated endosperm is held under aseptic conditions any changes that occur are minimal, and are not easily detected. The aleurone layer can be peeled from the starchy endosperm after 3 days, if the incubation has been aerobic. If the degermed grain is held in contact with saturated calcium sulphate solution then in 2–3 weeks the endosperm begins to soften and limited starch degradation occurs. These changes are prevented by anaerobiosis or 'anaesthetics' [37,39,225]. They are probably due to some carbohydrase(s) ('β-glucanase' and 'laminarinase') synthesized *de novo*, at least in part, in the hydrated aleurone layer in the *absence* of hormone from the embryo [12,132,161], and their leakage from the tissue, together with α-glucosidase that is already present. The release of these enzymes may be due to the autolysis of the exhausted tissue. They, and the preformed β-amylase of the sub-aleurone layer, attack their respective substrates and degrade the cell walls and starch to a limited extent, their action being facilitated by the high salt concentration. These processes are dependent on an aleurone layer that is viable, at least initially; they are however slow, and are largely independent of the normal, rapid, controlled changes occurring during modification. No α-amylase is produced under these conditions.

Isolated embryos, cultured *in vitro*, release hydrolytic enzymes from the scutellar epithelium [19,27,38,102]. This mixture of enzymes is complex, and includes α-amylase; it is able to hydrolyse starch, disaccharides, oligosaccharides, barley endosperm cell walls, β-glucans, pentosans, peptides, proteins, nucleic acids, various phosphates, and some artificial substrates. Embryos can successfully be transplanted to other degermed grains – either to barley, or wheat, or rye or oats [37,38,221]. In these cases modification proceeds more or less normally. However, if the aleurone layer of the recipient degermed grain has been killed, then enzyme production in the grain is abnormally low, and modification is much reduced, as is embryo growth [21,37]. Thus the embryo is needed to initiate normal modification but a viable aleurone layer is needed to carry it out. Originally the embryo was thought to synthesize most of the required hydrolytic enzymes, but isolated embryos (however well supplied with nutrients) never make nearly as much α-amylase (the most thoroughly studied enzyme) as is found in whole grain. Endosperms, separated from dry grains, hydrated, and incubated under 'germination' conditions make none. The explanation is that a hormonal substance, a gibberellin, moves from the embryo during germination and triggers a massive production of enzymes in the aleurone layer [245]. This embryo gibberellin can be replaced or augmented by exogenous gibberellic acid [25,191,192,246,247]. Although some other cereal grains, e.g. wheat, behave in a similar way to barley, it is also clear that others do *not*. For example, grains of *some* maize varieties contain enough gibberellin in the endosperm to trigger modification when hydrated in the absence of the embryo [110]. Gibberellic acid added to whole grain not only breaks dormancy; it also accelerates embryo growth and grain respiration, increases the transaminase content, enhances the production of hydrolytic enzymes and consequently the breakdown of the reserve substances of the endosperm and increases the levels of simple soluble substances — such as sugars, amino acids, nucleotides, and inorganic phosphate (Chapter 15) [6,20,23,111].

When degermed grain is incubated under aseptic conditions in water containing gibberellic acid then amylases, other carbohydrases, proteases, peptidases, phosphatases, peroxidases and nucleases appear in solution and the starchy endosperm is so thoroughly degraded that it liquefies and high levels of simple sugars, amino acids, polypeptides and inorganic phosphate appear in solution [20,162]. The starchy endosperm begins to break down immediately below the aleurone layer. In response to gibberellins isolated aleurone layers release many hydrolytic enzymes, minerals, and sucrose [20,55,133,183,220].

α-Amylase is normally absent from sound, ungerminated grain and is synthesized *de novo* in the embryo, and in the aleurone layer in response to gibberellins. Not all enzymes formed in the aleurone layer are similar in this respect. Nitrate reductase is induced by its substrate and not by gibberellin, which may marginally impede its production [78]. α-Amylase is formed only in response to some gibberellins, helminthosporol, helminthosporic acid, some related substances, and possibly some solvent residues [25]. Indole acetic acid (auxin) is not required for enzyme formation in isolated aleurone layers. Auxin is without effect or (in larger doses) is toxic to whole grains [52,68]. Kinetin and benzylaminopurine, two synthetic cytokinins, do not enhance α-amylase production by the aleurone layer but they do alter the pattern of its release [55].

Abscisic acid reduces sugar release from endosperms incubated with gibberellic acid and also blocks α-amylase formation [51,228]. Ethylene in low doses sometimes enhances the production of α-amylase by aleurone layers in response to gibberellic acid; high doses of the gas reduce the hormonal response. The variable nature of the response may be due to irregular concentrations of this gaseous hormone in the tissue. Ethylene overcomes the inhibitory effects of abscisic acid [123]. In the absence of measurements on the levels of abscisic acid, cytokinins, and ethylene in whole grain, the significance of these observations is unknown. 2-Chloroethyl phosphonic acid, which can decompose to release ethylene, depresses the response of aleurone tissue to GA_3 [66].

In intact, malted grain α-amylase originates from the scutellum (7% retained, 8% released into the starchy endosperm) and the aleurone layer (85%) [21,27]. Enzyme accumulates in the aleurone layer before it is released. Enzyme production in the scutellum is strongly dependent on a supply of amino-acids; when these are present its production is enhanced by added gibberellic acid. Enzyme production in the scutellum is repressed by sugars in low concentrations (e.g. 5 mM). Presumably endogenous gibberellins trigger the production of this enzyme in the embryo. In contrast the aleurone layer, having no endogenous source of gibberellins, is entirely dependent on an outside supply to trigger α-amylase production which, as in the embryo, occurs *de novo* in this tissue [25]. The evidence for *de novo* synthesis of α-amylase in aleurone tissue includes the incorporation of radioactive amino-acids, density labelling with ^{18}O or D (^2H) and the effects of various inhibitory agents and amino-acid analogues [25,79,104]. The process of 'induction' can be divided into stages, which may be detected experimentally by adding inhibitors at various times, or by altering the supply of ions to the tissue, the pH, or the temperature [43,92,93]. Enzyme is released from the tissue by a separate, energy-dependent mechanism [233]. Enzyme is synthesized more rapidly in aleurone that has been hydrated at least 24h before gibberellin is supplied, and synthesis is often most rapid if hydration has taken place before the removal of the embryo. Thus hydration of the tissue, to allow the establishment of a basal metabolic rate, and the arrival of a hypothetical 'embryo factor' are pre-requisites for maximal enzyme production [25]. Enzymes are synthesized in the aleurone at the expense of endogenous reserve substances; externally supplied amino-acids are normally without effect on enzyme yield. At low concentrations sugars are without effect; at high concentrations (e.g. 100 mM), they are inhibitory, perhaps due to an osmotic effect [7,27,49,55].

The gibberellin 'trigger substance' moves from the embryo to the endosperm in about the second day of malting, at about 14° C, so that the removal of the embryo, or damage to it (e.g. by local burning, acid or γ-irradiation), from the third day onwards has little effect on subsequent enzyme production [25,100]. Initially gibberellin production is in the scutellum, probably at the expense of stored precursors such as *ent-*kaurene [178,179,203]. Barley embryos, apparently devoid of conjugates of gibberellins, nevertheless release gibberellin for up to 60 h when cultivated *in vitro*, indicating that synthesis is occurring [57,248]. The gibberellin moving to the aleurone is predominantly GA_3 (gibberellic acid) [177]. The peak level of GA_3, in malting barley, is about 1–2 ng/grain [100,177,178]. Damaging the embryo early in germination reduces or prevents gibberellin production, as do a range of toxic chemicals [23]. When agents are added to germinating grains, those that check embryo growth only often enhance the survival of α-amylase, others check embryo growth and gibberellin production, while a third group inhibits embryo growth, gibberellin production *and* the ability of the aleurone layer to synthesize α-amylase in response to an exogenous supply of gibberellin. Compounds such as chlorocholine (CCC), which block gibberellin production, are not specific in their action; they have other toxic effects on grain [23]. Substances which block gibberellin formation do not necessarily block the response of the aleurone layer to added gibberellins [193]. At a later stage the embryonic axis begins to make gibberellins, presumably in connection with its extension growth, and these are mainly GA_1, GA_4 and GA_7 [177]. In malting grain, with limited embryo growth, the correlation between endogenous gibberellin levels and α-amylase production is excellent. The gibberellin content peaks and declines, being followed by the rate of α-amylase synthesis; the decline indicating the destruction or conversion of gibberellin to inactive forms [100]. It is the massive production of gibberellin in connection with elongation growth, rather than endosperm degradation which conceals this correlation in grain grown on a wet substratum at higher temperatures [100,177]. In general, claims that substances other than gibberellins induce α-amylase formation are invalid [55].

The production, or possibly the release of gibberellin from the embryo is prevented by sugars [203]. The onset of gibberellin release in malting coincides with the fall in endogenous sugar levels caused by the embryo utilizing the sugars available preceding the mobilization of the endosperm reserves. The 'cut off' in gibberellin supply coincides with the rise in sugar levels consequent on the degradation of the endosperm, which is triggered mainly through the arrival of gibberellins. Thus gibberellin levels in the endosperm are probably controlled by a 'feed-back loop', as gibberellins regulate α-amylase formation, and this enzyme largely controls sugar levels in the endosperm [25]. Low levels of sugars cut off gibberellin supplies; high levels check the ability of the aleurone to respond to gibberellins [27,49,134]. Externally applied gibberellins can increase the production of α-amylase by up to three times, and so the production of this enzyme in malting grain is limited by the gibberellin supply, and not the ability of the grain tissues to respond to the hormone. However, isolated aleurone layers 'saturated' with gibberellin do not make more than a certain amount of enzyme; when the reserve substances of the tissue are exhausted enzyme production stops, even when gibberellins, calcium ions (an integral part of the enzyme), and amino acids are supplied.

Some aspects of aleurone metabolism have been extensively studied. Gibberellic acid (GA_3) is taken up and converted to polar and other metabolites, such as GA_8 – processes that are accelerated by abscisic acid [180,181,223]. This mechanism can explain the decline in biologically active gibberellins seen in germinating barley [100]. Gibberellins enhance the rate of nucleic acid metabolism in the tissue, although the net quantity may be unaltered [44,126]. Ribonucleic acid methylation is enhanced by GA_3; this process is inhibited by abscisic acid [44]. More ribosomes and polysomes are formed in response to GA_3, processes also blocked by abscisic acid [77]. GA_3 induces the elevation of enzymes of the CDP-choline pathway, and more material is incorporated into the endoplasmic reticulum; this may represent enhanced turnover rather than net synthesis [13,82]. Gibberellin also reduces the transfer of radioactive carbohydrate to the cell walls [131].

Although other enzymes have not been studied as thoroughly as α-amylase it is clear that many behave in different ways. β-Amylase occurs preformed in the quiescent grain, and is mainly situated in the sub-aleurone region; no *de novo* synthesis of this enzyme occurs during germination [74,104]. During germination the 'bound', insoluble enzyme becomes free and soluble, due to the reduction of disulphide bridges between enzyme molecules and insoluble proteins; limited proteolysis apparently also plays a part in enzyme release [25,108]. The reductive and proteolytic changes are apparently indirectly dependent on the arrival of gibberellins at the aleurone layer, which releases proteases. α-Glucosidase occurs in the aleurone of quiescent grain and β-glucanase and laminarinase appear in the hydrated tissue. The levels of α-glucosidase and β-glucanase are further enhanced when gibberellin reaches the tissue. The release of these enzymes, like that of α-amylase and mineral substances, is also gibberellin-dependent [55,132,133,161]. In malting grain 'cytolytic' activity, due to enzymes such as 'laminarinase' and β-glucanase, rises faster when respiration is declining in intensity [118]. Enzyme activity may decline while the malting grain is still germinating, before it is kilned [89]. The release of phosphatase from the aleurone layer is also gibberellin dependent, but the degree of dependence is greatly influenced by the age of the aleurone and the previous history of the grain and it is formed, at least in part, in response to tissue hydration, before gibberellins arrive [29,189]. Gibberellins also induce the formation and release of peroxidase [106]. The glutamic decarboxylase levels of the aleurone alter relatively little during germination [69]. The respiration of the aleurone layer rises to a constant value after hydration and gibberellins increase the respiration rate further [25]. Different enzymes are released at different rates from aleurone layers 'triggered' with gibberellic acid [136,160,197].

Enzymes and metabolic functions are not uniformly distributed throughout the grain. The capacity for respiration is confined to the embryo and aleurone layer [225]. Cytochemical techniques have demonstrated the appearance of α-amylase at particular sites inside the aleurone

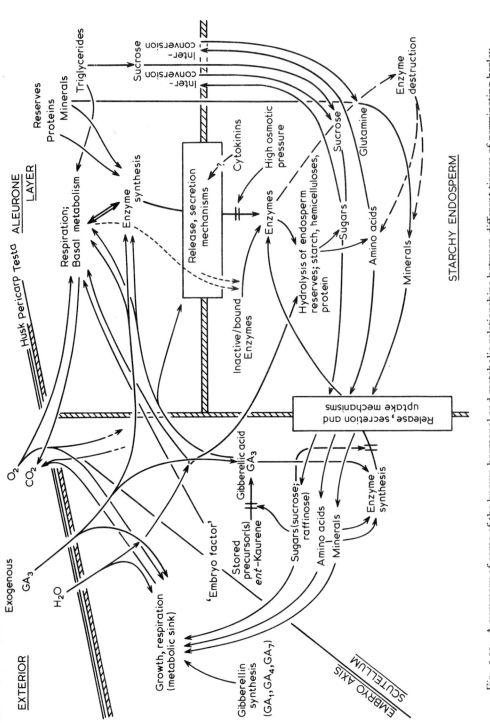

Fig. 5.10 A summary of some of the known hormonal and metabolic relationships between different tissues of germinating barley grains. Other inter-relationships are not excluded. GA_3, gibberellic acid; GA_1, GA_4, GA_7, other gibberellins (Chapter 4); O_2

cells, where phosphatase and other enzymes are concentrated [8,124]. In the resting grain esterases and proteases are confined to the embryo and aleurone layer [74,75]. During germination the aleurone releases several proteases and the distribution of peptidases alters. The activities of peptidases even vary along the length of a rootlet [150,171,226]. The changes and distribution of many other enzymes have been studied [25,108,202]. Probably many more variations on the mechanisms by which enzyme levels are controlled will be discovered. Reports describing the nature of the changes occurring in levels of particular enzymes frequently disagree. These disagreements may often be due to technical difficulties which make many 'quantitative' studies untrustworthy.

Enzymes also undergo destruction and when this occurs in the endosperm the products of their degradation are presumably used as nutrients by the embryo. In the case of α-amylase enzyme destruction is probably mediated by proteolytic enzymes which can act on the amylase because the pH of the endosperm falls to a level where α-amylase is unstable and the levels of calcium ions and the substrate, starch, which both tend to stabilize the enzyme, also decline. These alterations are directly or indirectly caused by the embryo. Stimulating embryo growth by late additions of gibberellic acid accelerates the rate of enzyme breakdown; treatments that check embryo growth reduce the rate of α-amylase degradation [23,25].

A schematic summary of some of the interactions between the parts of the grain is given in Fig. 5.10. The observation that short, broad grains with wide, overlapping scutella are best for malting is explained equally well by large scutella producing more gibberellin and this having a smaller distance to diffuse to induce modification up to the apex of the grain as by the original explanation that the large scutella produced more enzymes of modification, and that these had to travel less far to reach the apex of the grain [164]. The pattern of endosperm breakdown is adequately explained by the initial release of enzymes from the scutellum, and the subsequent production or activation and release of enzymes (whether synthesized *de novo* or not) as the GA_3 diffuses down the grain and encounters the aleurone layer, and the hydrolysis of the endosperm components by these enzymes [24].

References

1 Abdul-Baki, A.A. and Anderson, J.D. (1970). *Crop Sci.*, **10**, 31–34.
2 Abrahamsohn, B. (1911). *Z. ges. Brauw.*, **34**, 179–182.
3 Almestrand, A. (1951). *Physiologia Pl.*, **4**, 224–241.
4 Anderson, J.D. (1970). *Crop. Sci.*, **10**, 36–39.
5 Anderson, J.D. and Abdul-Baki, A.A. (1971). *Pl. Physiol., Lancaster*, **48**, 270–272.
6 Aramaki, K. and Kuroiwa, Y. (1967). *Rep. Res. Lab. Kirin Brewery Co., Japan*, **10**, 29–37.

7 Armstrong, J.E. and Jones, R.L. (1973). *J. Cell Biol.*, **59**, 444–455.
8 Ashford, A.E. and Jacobsen, J.V. (1974). *Planta*, **120**, 81–105.
9 Atterberg, A. (1907). *Landw. Vers. Sta.*, **67**, 129–143.
10 Bauwin, G.R. and Ryan, H.L. (1974). In *Storage of Cereal Grains and their products* (2nd edition). (ed. Christensen, C.M.) pp. 115–157. Am. Ass. Cereal Chemists, St. Paul, Minn.
11 Belderok, B. (1968). *J. Inst. Brew.*, **74**, 333–340.
12 Bennett, P.A. and Chrispeels, M.J. (1972). *Pl. Physiol., Lancaster*, **49**, 445–447.
13 Ben-Tal, Y. and Varner, J.E. (1974). *Pl. Physiol., Lancaster*, **54**, 813–816.
14 Bishop, L.R. (1929). *J. Inst. Brew.*, **35**, 323–338.
15 Bishop, L.R. (1944). *J. Inst. Brew.*, **50**, 166–185.
16 Bishop, L.R. (1945). *J. Inst. Brew.*, **51**, 166–175; 213–224.
17 Bishop, L.R. (1946). *J. Inst. Brew.*, **52**, 273–282.
18 Briggs, F.N. (1927). *J. agric. Res.*, **35**, 907–914.
19 Briggs, D.E. (1962). *J. Inst. Brew.*, **68**, 470–475.
20 Briggs, D.E. (1963). *J. Inst. Brew.*, **69**, 13–19; 244–248.
21 Briggs, D.E. (1964). *J. Inst. Brew.*, **70**, 14–24.
22 Briggs, D.E. (1967). *J. Inst. Brew.*, **73**, 33–34.
23 Briggs, D.E. (1968). *Phytochemistry*, **7**, 513–529; 531–538; 539–554.
24 Briggs, D.E. (1972). *Planta*, **108**, 351–358.
25 Briggs, D.E. (1973). In *Biosynthesis and its Control in Plants*, ed. Milborrow, B.V. pp. 219–277. London: Academic Press.
26 Briggs, D.E. (1974). *Phytochemistry*, **13**, 987–996.
27 Briggs, D.E. and Clutterbuck, V.J. (1973). *Phytochemistry*, **12**, 1047–1050.
28 Brookes, P.A. and Martin, P.A. (1974). *J. Inst. Brew.*, **80**, 294–299.
29 Brookes, P.A. and Martin, P.A. (1975). *J. Inst. Brew.*, **81**, 357–363.
30 Brown, A.J. (1907). *Ann. Bot.* **21**, 79–87; *J. Fed. Inst. Brew.*, **13**, 658–673.
31 Brown, A.J. (1909). *Proc. R. Soc. Ser. B*, **81**, 82–93.
32 Brown, A.J. and Tinker, F. (1916). *Proc. R. Soc. Ser. B*, **89**, 373–379.
33 Brown, A.J. and Worley, F.P. (1912). *Proc. R. Soc. Ser., B*, **85**, 546–553.
34 Brown, C.R. (1974). *J. Inst. Brew.*, **80**, 471–473; 483–486.
35 Brown, E., Stanton, T.R., Wiebe, G.A. and Martin, J.H. (1948). *U.S. Dept. Agric. Tech. Bull.* No. 953.
36 Brown, H.T. (1909). *J. Inst. Brew.*, **15**, 184–286.
37 Brown, H.T. and Escombe, F. (1898). *Proc. R. Soc.*, **63**, 3–25.
38 Brown, H.T. and Morris, G.H. (1890). *J. chem. Soc.*, **57**, 458–528.
39 Bruschi, D. (1908). *Ann. Bot.*, **22**, 449–463.
40 Burger, W.C. (1965). *J. Inst. Brew.*, **71**, 244–250.
41 Caldwell, F. (1957). *J. Inst. Brew.*, **63**, 340–344.
42 Caldwell, F. (1963). *J. Sci. Fd Agric.*, **14**, 765–769.
43 Carr, D.J. and Goodwin, P.B. (1972). In *Plant Growth Substances, 1970*, (ed. Carr, D.J.) pp. 378–387. Berlin: Springer Verlag.
44 Chandra, G.R. (1972). In *Plant Growth Substances, 1970*, (ed. Carr, D.J.) pp. 365–370. Berlin: Springer Verlag.
45 Chang, C.W. (1963). *Bull. Torrey bot. Club*, **90**, 385–391.
46 Chapon, L. (1959). *Brasserie*, **14**, 222–231.
47 Chapon, L. (1961). *Brauwissenschaft*, **14**, 457–465.
48 Chinn, S.H.F. and Russell, R.C. (1958). *Phytopathology*, **48**, 553–556.
49 Chrispeels, M.J. (1973). *Biochim. biophys. Res. Commun.*, **53**, 99–104.
50 Chrispeels, M.J., Tenner, A.J. and Johnson, K.D. (1973). *Planta*, **113**, 35–46.
51 Chrispeels, M.J. and Varner, J.E. (1967). *Pl. Physiol., Lancaster*, **42**, 398–406; 1008–1016.

Grain quality and germination 217

52 Cleland, R. and McCombs, N. (1965). *Science*, **150**, 497–498.
53 Clerck, J. de. (1938). *Bull. Assoc. Anc. Étud. École Sup, Brass., Louvain.*, **38**, 85–100.
54 Clerck, J. de. and Cloetens, J. (1940). *Bull. Assoc. Anc. Étud. École Sup. Brass. Louvain*, **40**, 41–48.
55 Clutterbuck, V.J. and Briggs, D.E. (1973). *Phytochemistry*, **12**, 537–546.
56 Clutterbuck, V.J. and Briggs, D.E. (1974). *Phytochemistry*, **13**, 45–54.
57 Cohen, D. and Paleg, L.G. (1967). *Pl. Physiol., Lancaster*, **42**, 1288–1296.
58 Collins, E.J. (1918). *Ann. Bot.*, **32**, 381–414.
59 Collins, D.M. and Wilson, A.T. (1975). *J. exp. Bot.*, **26**, 737–740.
60 Crabb, D. (1970). *J. Inst. Brew.*, **76**, 469–476.
61 Crabb, D. (1971). *J. Inst. Brew.*, **77**, 522–528.
62 Crabb, D. and Kirsop, B.H. (1970). *J. Inst. Brew.*, **76**, 158–162.
63 Dahlstrom, R.V. (1965). *Cereal Sci. Today*, **10**, 466–472.
64 Day, T.C. (1891). *J. chem. Soc.*, **59**, 664–677.
65 Day, T.C. (1896). *Trans. Proc. bot. Soc. Edinb.*, **20**, 492–501.
66 Devlin, R.M. and Cunningham, R.P. (1970). *Econ. Bot.*, **24**, 369–373.
67 Dickson, A.D. and Burkhart, B.A. (1942). *Cereal Chem.*, **19**, 251–262.
68 Dickson, A.D., Shands, H.L. and Burkhart, B.A. (1949). *Cereal Chem.*, **26**, 13–23.
69 Duffus, C.M., Duffus, J.H. and Slaughter, J.C. (1972). *Experienta*, **28**, 635–636.
70 E.B.C. (1963). *Analysis Committee of the European Brewery Convention Analytica—EBC* (2nd edition). Amsterdam: Elsevier.
71 Edelman, J., Shibko, S.I. and Keys, A.J. (1959). *J. exp. Bot.*, **10**, 178–189.
72 Eifrig, H. (1961). *Proc. int. Seed. Test. Ass.*, **26**, 761–771.
73 Ekström, D., Cederqvist, B. and Sandegren, E. (1959). *Proc. eur. Brew. Conv. Rome*, pp. 11–26.
74 Engel, Chr. (1947). *Biochim. biophys. Acta*, **1**, 42–49; 278.
75 Engel, Chr. and Heins, J. (1947). *Biochim. biophys. Acta*, **1**, 190–196.
76 Evans, M., Black, M. and Chapman, J. (1975). *Nature*, **258**, 144–145.
77 Evins, W.H. and Varner, J.E. (1972). *Pl. Physiol., Lancaster*, **49**, 348–352.
78 Ferrari, T.E., and Varner, J.E. (1970). *Proc. natn. Acad. Sci. U.S.A.*, **65**, 729–736.
79 Filner, P. and Varner, J.E. (1967). *Proc. natn. Acad. Sci. U.S.A.*, **58**, 1520–1526.
80 Fincher, G.B. (1975). *J. Inst. Brew.*, **81**, 116–122.
81 Finlay, K.W. (1960). *J. Inst. Brew.*, **66**, 51–57; 58–64.
82 Firn, R.D. and Kende, H. (1974). *Pl. Physiol., Lancaster*, **54**, 911–915.
83 Fischbeck, G. and Reiner, L. (1967). *Brauwelt*, **107**, 1278–1279.
84 Fischnich, O., Thielbein, M. and Grahl, A. (1962). *Saatgut-Wirtsch.*, **14**, 12–14; 39–42.
85 Folkes, B.F. and Yemm, E.W. (1958). *New Phytol.*, **57**, 106–131.
86 Folkes, B.F., Willis, A.J. and Yemm, E.W. (1952). *New Phytol.*, **51**, 317–341.
87 Forward, D.F. (1951). *New Phytol.*, **50**, 297–324; 325–356.
88 French, R.C. (1959). *Pl. Physiol., Lancaster*, **34**, 500–505.
89 Fujii, T. and Horie, Y. (1972). *Rept. Res. Lab. Kirin Brewery Co., Japan*, **13**, 37–42.
90 Gaber, S.D. and Roberts, E.H. (1969). *J. Inst. Brew.*, **75**, 299–302; 303–314.
91 Gehloff, G., Schlosser, A. and Piendl, A. (1972). *Master Brewery Ass. Am. Tech. Quart.*, **9**, 144–147.

92 Goodwin, P.B. and Carr, D.J. (1972). *Planta*, **106**, 1–12.
93 Goodwin, P.B. and Carr, D.J. (1972). *J. exp. Bot.*, **23**, 1–7; 8–13.
94 Gordon, A.G. (1968). *J. Inst. Brew.*, **74**, 355–360.
95 Gordon, A.G. (1969). *Proc. Ass. Off. Seed Anal. N. Am.*, **59**, 58–72.
96 Gordon, A.G. (1970). *Can. J. Pl. Sci.*, **50**, 191–194.
97 Gordon, A.G. (1973). In *Seed Ecology*, (ed. Heydecker, Proc. 19th Easter School in Agric. Sci., University Nottingham, 1972) pp. 391–410. London: Butterworths.
98 Grahl, A. (1970). *Proc. int. Seed Test Ass.*, **35**, 427–438.
99 Grahl, A. and Thielbein, M. (1959). *Naturwissenschaften*, **46**, 336–337.
100 Groat, J.I. and Briggs, D.E. (1969). *Phytochemistry*, **8**, 1615–1627.
101 Grüss, J. (1898). *Woch. Brau.*, **15**, 81–84; 269–275.
102 Grüss, J. (1928). *Woch. Brau.*, **45**, 497–500; 506–509; 539–542.
103 Grüss, J. (1930). *Woch. Brau.*, **47**, 249–252; 257–261; 269–272.
104 Hardie, D.G. (1975). *Phytochemistry*, **14**, 1719–1722.
105 Harlan, H.V. and Pope, M.N. (1926). *J. agric. Res.*, **32**, 669–678.
106 Harmey, M.A. and Murray, A.M. (1968). *Planta*, **83**, 387–389.
107 Harrington, G.T. (1923). *J. agric. Res.*, **23**, 79–100.
108 Harris, G. (1962). In *Barley and Malt*, (ed. Cook A.H.) pp. 431–582; 583–694, London: Academic Press.
109 Harris, G. and MacWilliam, I.C. (1954). *J. Inst. Brew.*, **60**, 149–157.
110 Harvey, B.M.R. and Oaks, A. (1974). *Planta*, **121**, 67–74.
111 Hayashi, T. (1940). *J. agric. chem. Soc. Japan*, **16**, 531–538.
112 Heinisch, I. (1931). *Woch. Brau.*, **48**, 293–297; 307–311.
113 Hinton, J.J.C. (1955). *Cereal Chem.*, **32**, 296–306.
114 Hoagland, D.R. and Broyer, T.C. (1936). *Pl. Physiol., Lancaster*, **11**, 471–507.
115 Hølmer, G., Ory, R.L. and Høy, C.E. (1973). *Lipids*, **8**, 277–283.
116 Hoffmann, J.F. and Sokolowski, S. (1910). *Woch. Brau.*, **27**, 469–471; 483–486; 498–503.
117 Holmberg, J. and Sellmann-Persson, G. (1967). *Proc. eur. Brew. Conv., Madrid*, 213–217.
118 Hough, J.S., Briggs, D.E. and Stevens, R. (1971). *Malting and Brewing Science*, (Revised –1975). London: Chapman and Hall.
119 Hudson, J.R. (1960). *Development of Brewing Analysis, a Historical Review.* Institute of Brewing, London.
120 Hunt, W.H. and Pixton, S.W. (1974). In *Storage of Cereal Grains and their Products* (2nd edition) ed. Christensen, C.M. pp. 1–55. St. Paul, Minnesotta; Am. Ass. Cereal Chemists.
121 I.o.B. (1971). Analysis Committee of the Institute of Brewing (1971). *J. Inst. Brew.*, **77**, 181–226.
122 I.S.T.A. (1966). *Proc. int. Seed Testing Ass.*, **31** 152 pp.
123 Jacobsen, J.V. (1973). *Pl. Physiol., Lancaster*, **51**, 198–202.
124 Jacobsen, J.V. and Knox, R.B. (1973). *Planta*, **112**, 213–224.
125 Jacobsen, J.V., Knox, R.B. and Pyliotis, N.A. (1971). *Planta*, **101**, 189–209.
126 Jacobsen, J.V. and Zwar, J.A. (1974). *Aust. J. Pl. Physiol.*, **1**, 343–356.
127 James, A.L. (1950). *New Phytol.*, **39**, 133–144.
128 James, W.O. and James, A.L. (1940). *New Phytol.*, **39**, 145–176.
129 Jansson, G. (1961). *Ark. Kemi*, **17**, 281–289.
130 Jansson, G. (1962). *Ark. Kemi*, **19**, 141–148.
131 Johnson, K.D. and Chrispeels, M.J. (1973). *Planta*, **111**, 353–364.
132 Jones, R.L. (1971). *Pl. Physiol., Lancaster*, **47**, 412–416.
133 Jones, R.L. (1973). *Pl. Physiol., Lancaster*, **52**, 303–308.

134 Jones, R.L. and Armstrong, J.E. (1971). *Pl. Physiol., Lancaster*, **48**, 137–142.
135 Jones, M. and Pierce, J.S. (1967). *J. Inst. Brew.*, **73**, 577–583.
136 Jones, R.L. and Price, J.M. (1970). *Planta*, **94**, 191–202.
137 Kamra, Om P., Kamra, S.K., Nilan, R.A. and Konzak, C.F. (1960). *Hereditas*, **46**, 152–170; 261–273.
138 Kauppinen, Y., Nummi, M. and Enari, T.-M. (1967). *Brauwissenschaft*, **20**, 1–4.
139 Khan, A.A. and Waters, E.C. (1969). *Life Sciences*, **8**, 729–736.
140 Kidd, F. and West, C. (1918). *Ann. appl. Biol.*, **5**, 1–10, 112–142; 157–170; 220–251.
141 Kieninger, H. (1969). *Brauwissenschaft*, 6–13; 62–69; 100–104.
142 Kirsop, B.H., Reynolds, T. and Griffiths, C.M. (1967). *J. Inst. Brew.*, **73**, 182–186.
143 Knypl, J.S. (1964). *Naturwissenschaften*, **51**, 117–118.
144 Kudo, S. and Yoshida, T. (1958). *Rep. Res. Lab. Kirin Brewery Co., Japan*, **1**, 39–45.
145 La Croix, L.J., Naylor, J. and Larter, E.N. (1962). *Can. J. Bot.*, **40**, 1515–1523.
146 van Laer, H., and Lomaers, R. (1922). *Annls Brass. Dist.*, **20**, 184–185.
147 Lafferty, H.A. (1953). *Proc. int. Seed Test. Ass.*, **18**, 239–247.
148 Lefebvre, J.M. (1959). *Annls Physiol. Végétale, Paris*, **1**, 93–109.
149 Levy, M. (1936). *C. r. Trav. Lab. Carlsberg, (Sèr. Chim)*, **21**, 101–110.
150 Linderstrøm-Lang, K. and Holter, H. (1931). *C. r. Trav. Lab. Carlsberg*, **19**, 1–39.
151 Linskens, H.F. (1950). *Züchter*, **20**, 168–187.
152 Liu, D.J., Pomeranz, Y. and Robbins, G.S. (1975). *Cereal Chem.*, **52**, 678–686.
153 Lopez, A. and Grabe, D.F. (1971). *Agron. Abs.*, p. 44.
154 Lubert, D.J. and Pool, A.A. (1964). *J. Inst. Brew.*, **70**, 145–155.
155 MacKay, D.B. (1972). In *Viability of Seeds*, (ed. Roberts, E.H.) pp. 172–208. London: Chapman and Hall.
156 MacLeod, A.M. (1952). *Trans. Proc. bot. Soc. Edinb.*, **36**, 18–33.
157 MacLeod, A.M. (1957). *New Phytol.*, **56**, 210–220.
158 MacLeod, A.M. (1960). *Wallerstein Lab. Commun.*, **23**, 87–98.
159 MacLeod, A.M. (1967). *J. Inst. Brew.*, **73**, 146–162.
160 MacLeod, A.M., Duffus, J.H. and Johnston, C.S. (1964). *J. Inst. Brew.*, **70**, 521–528.
161 MacLeod, A.M., Duffus, J.H. and Millar, A.S. (1963). *Proc. eur. Brew. Conv., Brussels*, pp. 85–100.
162 MacLeod, A.M. and Millar, A.S. (1962). *J. Inst. Brew.*, **68**, 322–332.
163 Major, W. and Roberts, E.H. (1968). *J. exp. Bot.*, **19**, 77–89; 90–107.
164 Mann, A. and Harlan, H.V. (1916). *J. Inst. Brew.*, **22**, 73–108 (reprint 1915, *U.S. Dept. Agric. Bull. No. 183*).
165 Matthews, S. and Collins, M.T. (1973). *Proc. 7th British Insecticide and Fungicide Conf.*, **1**, 135–141.
166 McNeil, M., Albersheim, P., Taiz, L. and Jones, R.L. (1975). *Pl. Physiol., Lancaster*, **55**, 64–68.
167 Merry, J. and Goddard, D.R. (1941). *Proc. Rochester Acad. Sci.*, **8**, 28–44.
168 Michejda, J. (1966). *Acta Soc. Bot. Pol.*, **35**, 71–77.
169 Middleton, N.I. (1927). *Ann. Bot.*, **41**, 345–356.
170 Mikola, J. and Enari, T.-M. (1970). *J. Inst. Brew.*, **76**, 182–188.
171 Mikola, J. and Kolehmainen, L. (1972). *Planta*, **104**, 167–177.
172 Mikola, J. and Pietilä, K. and Enari, T.-M. (1972). *J. Inst. Brew.*, **78**, 388–391.

220 *Barley*

173 Mitchell, F.S., Caldwell, F. and Hampson, G. (1958). *Nature*, **181**, 1270–1271.
174 Moormann, B. (1942). *Kühn – Arch.* **56**, 41–79.
175 Morrall, P and Briggs, D.E. (in preparation).
176 Murphy, A.J. (1904). *J. Inst. Brew.*, **10**, 99–148.
177 Murphy, G.J.P. – (unpublished).
178 Murphy, G.J.P. and Briggs, D.E. (1973). *Phytochemistry*, **12**, 1299–1308; 2597–2605.
179 Murphy, G.J.P. and Briggs, D.E. (1975). *Phytochemistry*, **14**, 429–433.
180 Musgrave, A., Kays, S.E. and Kende, H. (1972). *Planta*, **102**, 1–10.
181 Nadeau, R. and Rappaport, L. (1974). *Pl. Physiol.*, Lancaster, **54**, 809–812.
182 Nance, J.P. (1949). *Am. J. Bot.*, **36**, 274–276.
183 Newman, J. and Briggs, D.E. (1976). *Phytochemistry*, **15**, 1453–1458.
184 Nielsen, N. (1937). *C. r. Trav. Lab. Carlsberg*, (Sèr. Physiol.), **22**, 49–60.
185 Nobbe, F. (1876). *Handbuch der Samenkunde*, Verlag Wiegandt, Hempel und Parey, Berlin, 631 pp.
186 Norstog, K. (1961). *Am. J. Bot.*, **48**, 876–884.
187 Norstog, K. (1967). *Bull. Torrey, bot. Club.*, **94**, 223–229.
188 Norstog, K. and Klein, R.M. (1972). *Can. J. Bot.*, **50**, 1887–1894.
189 Obata, T. and Suzuki, H. (1976). *Pl. Cell Physiol.*, **17**, 63–71.
190 Ogawara, K. and Hayashi, J. (1964). *Ber. Ohara Inst.*, Landw. Biol., **12**, 159–188.
191 Paleg, L.G. (1960). *Pl. Physiol.*, Lancaster, **35**, 293–299; 902–906.
192 Paleg, L.G. (1961). *Pl. Physiol.*, Lancaster, **36**, 829–837.
193 Paleg, L.G., Kende, H., Ninnemann, H. and Lang, A. (1965). *Pl. Physiol.*, Lancaster, **40**, 165–169.
194 Palmer, G.H. (1969).*J. Inst. Brew.*, **75**, 536–541.
195 Parsons, J.G. and Price, P.B. (1974). *Lipids*, **9**, 804–808.
196 Pedersen, M.R. (1878).*C. r. Trav. Lab. Carlsberg*, **1**, 44–48 (Résumé).
197 Pollard, C.J. and Nelson, D.C. (1971).*Biochim. biophys. Acta*, **244**, 372–376.
198 Pollock, J.R.A. (1959). *J. Inst. Brew.*, **65**, 334–337.
199 Pollock, J.R.A. (1962). In *Barley and Malt*, (ed. Cook, A.H.) pp. 303–398; 399–430. London: Academic Press.
200 Pollock, J.R.A. and Pool, A.A. (1962). *J. Inst. Brew.*, **68**, 427–431.
201 Pope, M.N. (1949). *J. agric. Res.*, **78**, 295–309.
202 Prentice, N. (1973). *Cereal Chem.*, **50**, 346–353.
203 Radley, M. (1969). *Planta*, **86**, 218–223.
204 Rejowski, A. and Kulka, K. (1967). *Acta Soc. Bot. Pol.*, **36**, 221–234.
205 Reynolds, T. and MacWilliams, I.C. (1966). *J. Inst. Brew.*, **72**, 166–170.
206 Rijven, A.H.G.C. (1956). *Aust. J. biol. Sci.*, **9**, 511–527.
207 de Saussare, T. (1804 Edit; year XII). *Recherches Chimiques sur la Végétation* (Reprod. in fac-simile by Gauthier-Villars, Editeur – Imprimeur – Libraire, Paris; 1957).
208 Schooler, A.B. (1960). *Agron. J.*, **52**, 411.
209 Schulz, K.G. (1934). *Woch. Brau.*, **51**, 201–204.
210 Schulz, K.G. (1935). *Woch. Brau.*, **52**, 118–119.
211 Scriban, R. (1965). *Pet. J. Brasseur*, **73**, 39–43.
212 Sfat, M.R. (1966). *Master Brewers Ass. Am. Tech. Quart.*, **3**, 22–26.
213 Shands, H.L., Dickson, A.D., Dickson, J.G. and Burkhart, B.A. (1941). *Cereal Chem.*, **18**, 370–394.
214 Shands, H.L., Dickson, A.D. and Dickson, J.G. (1942). *Cereal Chem.*, **19**, 471–480.
215 Sims, R.C. (1959). *J. Inst. Brew.*, **65**, 46–50.
216 Singh, O. (1969). *Indian J. Sci. Ind.* (a), **3**, 35–38.

217 Sloey, W. and Prentice, M. (1962). *Proc. Ann. Mtg. Am. Soc. Brew. Chem.*, pp. 24–29.
218 Sparrow, D.H.B. (1964). *J. Inst. Brew.*, **70**, 514–521.
219 Sparrow, D.H.B. (1965). *J. Inst. Brew.*, **71**, 523–529.
220 Srivastava, B.I.S. (1965). *J. Inst. Brew.*, **71**, 21–25.
221 Stingl, G. (1907). *Flora*, **97**, 308–331.
222 Stoilov, M., Jansson, G., Eriksson, G. and Ehrenberg, L. (1966). *Radiat. Bot.*, **6**, 457–467.
223 Stolp, C.F., Nadeau, R. and Rappaport, L. (1973). *Pl. Physiol., Lancaster*, **52**, 546–548.
224 Stow, I. (1967). *Bull. Brew., Sci.*, **13**, 29–43.
225 Stoward, F. (1911). *Ann. Bot.*, **25**, 799–841; 1147–1204.
226 Sundblom, N.O. and Mikola, J. (1972). *Physiologia Pl.*, **27**, 281–284.
227 Tharp, W.H. (1935). *Bot. Gaz.*, **97**, 240–271.
228 Thomas, T.H., Wareing, P.F. and Robinson, P.M. (1965). *Nature*, **205**, 1270–1272.
229 Tronier, B. and Ory, R.L. (1970). *Cereal Chem.*, **47**, 464–471.
230 Urion, E. and Chapon, L. (1955). *Am. Brewer*, **88**, (3) 41–85.
231 Urion, E. and Chapon, L. (1955). *Proc. eur. Brew. Conv. Baden-Baden*, pp. 172–202.
232 Van Sumere, C.F., Cottenie, J., de Greef, J. and Kint, J. (1972). *Recent Adv. Phytochem.*, **4**, 165–221.
233 Varner, J.E. and Mense, R.M. (1972). *Pl. Physiol., Lancaster*, **49**, 187–189.
234 Vine, H.C.A. (1913). *J. Inst. Brew.*, **19**, 413–448.
235 Vlamis, J. and Davis, A.R. (1943). *Pl. Physiol., Lancaster*, **18**, 685–692.
236 Wellington, P.S. (1970). *Proc. int. Seed Test Ass.*, **35**, 449–597.
237 Wellington, P.S. and Bradnock, W.T. (1964). *J. natn. Inst. agric. Bot.*, **10**, 129–143.
238 Whitmore, E.T. (1960). *J. Inst. Brew.*, **66**, 407–408.
239 Windisch, W. and Kolback, P. (1929). *Woch, Brau.*, **46**, 459–463.
240 Woodstock, L.W. (1973). *Seed Sci. Technol.*, **1**, 127–157.
241 Yamamoto, K. (1951). *Rep. inst. agric. Res. Tohoku Univ.*, **3**, 13–27.
242 Yamashita, A. (1967). *Bull. Brew. Sci.*, **13**, 25–28.
243 Yemm, E.W. (1937). *Proc. R. Soc. Ser B*, **123**, 243–273.
244 Yemm, E.W. and Willis, A.J. (1956). *New Phytol.*, **55**, 229–252.
245 Yomo, H. (1958). *Hakkô Kyôkaishi*, **16**, 444–448.
246 Yomo, H. (1960). *Hakkô Kyôkaishi*, **18**, 494–499; 500–502; 600–602; 603–605.
247 Yomo, H. (1961). *Hakkô Kyôkaishi*, **19**, 284–285.
248 Yomo, H. and Iinuma, H. (1966). *Planta*, **71**, 113–118.
249 Ziebur, N.K. and Brink, R.A. (1951). *Am. J. Bot.*, **38**, 253–256.
250 Žila, V.V., Trkan, M. and Škvor, F. (1942). *Woch. Brau.*, **59**, 63–65.

Chapter 6
The growth of the barley plant

6.1. The description of growth

A satisfying numerical description of the growth of barley plants has not yet been given. Each part of the plant is initiated, undergoes a surge of growth, ceases growing, senesces and dies. Its demands upon, and contribution to the whole plant vary continuously with age, weather, soil conditions, and the alternation of day and night. Probably no two plants are ever truly identical. Over limited periods growth has been described by various formulae, but at best these seem to have descriptive value. Difficulties arise from (i) the plasticity of the plant, i.e. the dramatically altered forms achieved under different growing conditions (Table 6.1), and (ii) the difficulties of making adequate quantitative studies, especially on roots in the soil. To circumvent this last, and to provide uniform growing conditions, water culture is often used for experimental purposes. An acre (0.405 ha) of Plumage–Archer barley contained 14.7 miles (23.7 km) of coulter row, with 6.75 in (17.2 cm) between rows and contained $1.5-2 \times 10^6$ plants (Table 6.2) [102]. Of the seeds sown about 87% grew. While the average value was 22.6 grains/ft (30.5 cm) coulter row, the range was from 3 to 35 plants/ft (30.5 cm) row. The more widely spaced plants tillered more, but did not fully compensate for the effects of excessive spacing (however see Chapter 7). As the plant density increased, so the number of ears/plant declined.

The ratio of 1 : 2 : 3-eared plants varied between seasons. In a dry summer, on light soil, the proportions (%) were 84.6 : 14.7 : 0.71 and in a wetter summer on better soil 39.4 : 34.6 : 21.5. In the first instance the 1-eared plants produced 74% of the total grain harvested; in the second, 20.4%. Ears on main tillers usually gave the plumpest grain so that the proportion of small grain found on succeeding ears of 3-eared plants were 1st ear, 18.8%; 2nd ear, 41.8%; 3rd ear, 64.6%

Table 6.1 Comparison of single Hannchen barley plants grown (i) in rows [6 in (15.2 cm) apart, 18–20 plants/foot (30.5 cm)] and (ii) without competition [i.e. 10 ft (3.05 m) apart in all directions] (from Pavlychenko and Harrington [315]; by permission of the Agricultural Institute of Canada).

	Plant spacing	
	6 in (15.2 cm) rows	Single plants
1932 SEASON		
Tops (g. dry weight)	4.7	43
Roots – total length (ft)	770 (234.7 m)	6917 (2108.3 m)
diameter (ft)	—	2.3 (0.701 m)
1933 SEASON		
Top growth—Height (in)	20 (0.508 m)	25 (0.635 m)
Tillers (number)	1	127
Seminal root system		
Number	6	9
Length (no. branches; in)	192 (4.88 m)	378 (9.60 m)
Penetration (in)	30 (0.762 m)	63 (1.60 m)
Branches (1st order, longest; in)	14 (0.356 m)	37 (0.940 m)
Number	2782	6930
Crown root system		
Number	4	83
Length (no. branches; in)	5 (0.127 m)	1909 (48.49 m)
Penetration (in)	0	63 (1.600 m)
Branches (1st order, longest; in)	0	32 (0.813 m)
Number	0	23738
Total root system		
Length (without branches; in)	194	2287

Table 6.2 Some effects of variations in the density of plants in the coulter row (from Engledow [102]).

Number of plants/ft (30.5 cm)		Proportion planted at this density (%)	Yield	
Range	Mean		(cwt/acre)	(kg/ha)
0–15	13.4	27.5	12.3	1544
16–19	17.5	30.0	15.3	1920
20–23	21.5	26.5	17.4	2184
24–35	27.4	16.0	20.8	2611
Average whole field				
3–35	19.0	100.0	15.9	1996

(Table 6.3). Thus the plumpest and thinnest grains came from 3-eared plants. On *average* there was little difference in the quality of the grain from the different plant types. These results emphasize another dilemma; whether to consider a crop collectively, or as individual plants, or tillers.

224 Barley

Table 6.3 Characteristics of grain from successive ears of Plumage–Archer barley (from Engledow [102]).

Type of plant	1 ear	2 ear		3 ear		
Ear number	1	1	2	1	2	3
Grain no./ear	15.3	19.0	13.0	19.9	15.8	12.2
Thousand corn weight (g)	34.6	34.2	29.1	38.8	33.5	28.9
Grain weight/ear (g)	0.54	0.66	0.37	0.77	0.53	0.35

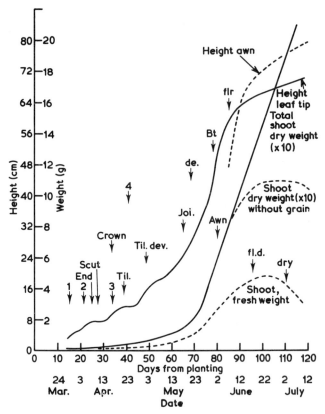

Fig. 6.1 The growth of the main shoots of the two-rowed, spring variety Hannchen, of plants initially selected for uniformity. (Rosslyn, Va., U.S.A.; sown 14 March; (after Pope [328]). Key to times marked by arrows. 1, 2, 3, 4, the 1st, 2nd, 3rd and 4th leaves showing; End, endosperm nearly exhausted; Scut, scutellum shrivelled; crown, crown developed and crown roots appeared; Til, tiller visible; Til. dev., tillers well developed; Joi, jointing (shooting, stem extension); de, basal leaves mainly dead; Bt., boot (flag) leaf showing; awn, awns showing; flr, first flowering, anthesis; fl. d., tip of flag leaf dying; dry, spikes nearly dry.

6.2. Sequential changes in the growth of the plant

The rate of growth depends on the weather, the climate, the water supply, soil fertility, the degree of competition with weeds and other plants, the depredations of pests and diseases, and whether the crop is autumn or spring-sown. Initially growth is slow, while the seedlings become established and the tillers form (Fig. 6.1) [328]. In the 'grand period of growth' the stems lengthen, and dry weight increases rapidly. Not all viable tillers carry ears, and not all the tillers survive to maturity (Fig. 6.2) [405, 406]. The growth curves of successive leaf blades and sheaths are sigmoid [382]. The grains are the last part to cease growing, and sometimes may actually lose weight before maturing. These losses, and weight losses from the rest of the plant, have been attributed to respiration exceeding photosynthesis, breakage and loss of the dried leaves and awns, and leaching of soluble materials by rain. The duration of the growth period also varies, being about 105 days in Britain, and 45–60 days on the Canadian Prairies.

Seedling growth is checked when the endosperm reserves are exhausted.

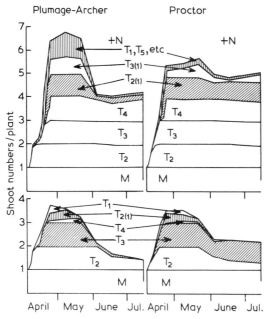

Fig. 6.2 Changes in the number of living shoots per plant with the varieties Plumage–Archer and Proctor grown with (+ N) or without extra nitrogenous fertiliser (after Thorne [405,406]). The main stem is designated M. The tillers, T, are numbered with respect to their origins. Secondary tillers are indicated by the numbers in brackets. Note that both varieties produce more tillers in response to the extra nitrogen, and the relatively greater occurrence of tiller death in the older variety, Plumage–Archer.

226 Barley

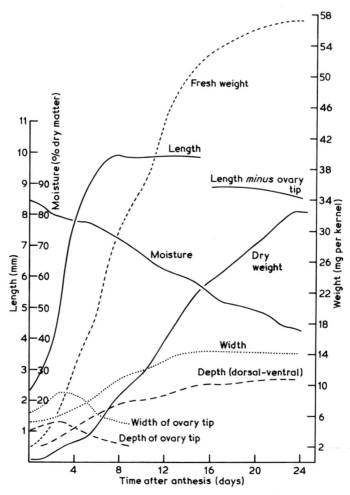

Fig. 6.3 Changes in the grains of Hannchen two-rowed barley grown on irrigated plots in Idaho (after Harlan [158]). Flowering began in the middle of the ear, and had extended to the tip and the base by day 1. The sterile, lateral florets flowered on days 1 and 2. The paleae began to adhere to the kernels, and the kernels (*not* the husk elements) turned yellow on days 9 and 10. On days 14 and 15 the kernel became tough and the ovary tip was resorbed. The lemmae began to lose their colour on days 15 and 16, and started to adhere to the kernels on days 19 and 20. By Days 23 and 24 the awns were yellow, and growth had probably ceased, although traces of chlorophyll were present in the creases.

Leaves appear, grow, and senesce in sequence. The dead leaves may break up. Eventually the flag-leaf dies and then the awns, in all cases from the tip backwards; lastly the grain loses its green colour, and is ripe. The crown (adventitious, coronal) root system is established shortly after the crown becomes well defined. The period from sowing to the onset of extension growth varies very much in length, including the whole

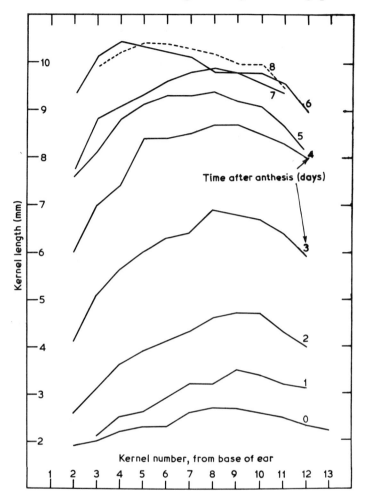

Fig. 6.4 The mean length, on successive days after anthesis, of kernels (with the ovary tip) of Hannchen barley at different positions in the ear (after Harlan [158]).

winter in autumn-sown 'winter barleys'. From anthesis onwards the gain in the dry weight of the ear is mainly due to the growing grains.

The length of the grain ceases to increase about 7 days after pollination, but width increase continues longer, and the increase in dry weight continues for longer still (Fig. 6.3) [158, 162, 264]. Eventually there is a small decline in grain length. The dimensions of the 'ovary tip' increase and decline while the true seed is still lengthening. While the fresh weight continues to rise nearly to maturity the moisture content of the grain declines, and falls sharply during ripening. In some climates there is a fall in mineral content and dry matter if over-ripe grain is allowed to remain in the field [369]. There are diurnal variations in grain moisture content, which is higher in the morning than in the evening. Kernels grow more evenly

Table 6.4 Growth stages in barley (Subdivisions of these are possible; see Fig 6.6) [38,85,107,239].
The numbers refer to the Large–Feekes scale, the letters to the Dommergues system. The Large–Feekes scale was originally developed to describe wheat and has been modified.

Stage

Germination and seedling establishment
0. Initial growth—below soil surface.
1. Coleoptile reaches soil surface. Brairding (L). Leaves appear, and are numbered in sequence. (F-1, F-2, F-3 etc.). Each leaf is 'visible' when the auricles appear. Note when (i) endosperm is exhausted; (ii) the crown is defined; (iii), crown-roots appear; (iv) other.
2. Tillering begins (T). The main stem appears at ground level (T-1). (N.B. in some numbering systems the main stem is M, and subsequent tillers T_1 T_2 etc.).
3. Secondary tillers grow from the crown, (T-2, T-3, etc.). These vary in number; tertiary and even quaternary tillers may arise. In some varieties the plant is still more or less prostrate.
4. The 'pseudo-stem' of rolled leaf sheathes begins to erect, and the leaf sheathes begin to lengthen.
5. Pseudostem is erect.

Stem extension (jointing; shooting)
6. The first stem node is visible at the shoot base.
7. The second stem node is visible.
8. The tip of the last leaf (flag leaf) is visible; the leaf is still rolled.
9. The ligule of the flag leaf is visible (D). Basal leaves are beginning to die. Meiosis usually occurs at about this stage.
10. The boot stage. The sheath of the flag leaf is fully visible and is swollen by the ear; the ear (head) is 'in boot'.

Heading (earing) Some barleys retain the ear in the boot. The more usual sequence of events is given

10.1 The tips of the awns just emerge (B)
10.2 One quarter of the head showing
10.3 Half the ear emerged
10.4 Ear three-quarters emerged
10.5 Ear fully emerged; still erect (E). Kernels are watery ripe.

} Blooming (anthesis) begins in the central florets, often with the ear in the boot. Stamens may or may not be excerted.

Ripening
11.1 Kernels milky – ripe.
11.2 Mealy-ripe – kernels soft and dry.
11.3 Kernel is hard to thumb-nail (J)
11.4 Dead-ripe (R).

} Neck of ear continues to lengthen – angle of head to peduncle may alter (A).

(Over-ripe). Shedding ears, grains; straw breaking.

in the centre of each spike. Larger kernels are usually subtended by larger awns. The growth rates vary with time along the ear (Fig. 6.4) [158,159,425]. Under uniform growing conditions in Idaho, the period from flowering to maturity was 26 days in three successive years, compared

The growth of the barley plant

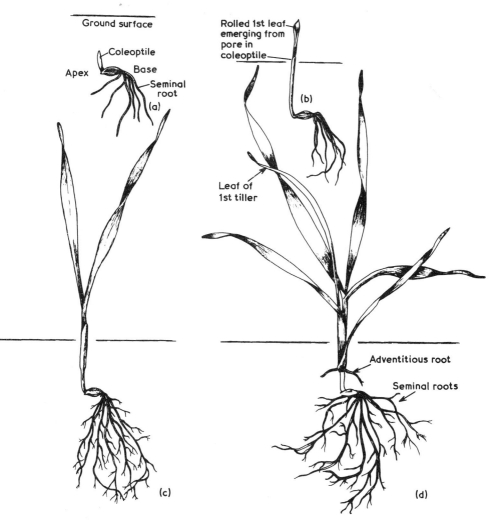

Fig. 6.5 Diagrams of the appearance of barley plants at successive stages of growth (after Pedersen [317]). (a) Germinating grain, with the coleoptile emerging at the apex of the grain, the seminal roots at the base. (b) The first leaf emerging from the coleoptile, which has cleared the soil surface. (c) The emergence of the second leaf. (d) The appearance of the first tiller-leaf. The basal region is swelling to form the crown and the first adventitious roots are appearing. (e) The onset of stem elongation (jointing, shooting); one main shoot and three tillers are present. (f) Ears emerged and erect. Two sterile tillers are present. (g) Plants approaching maturity. The leaves are withering and the ears are beginning to nod.

with periods of 20, 22 and 32 days in Denmark [158,369]. Growth stages vary in duration according to circumstances, and the different aspects of growth also may change their temporal relationships – indeed non-

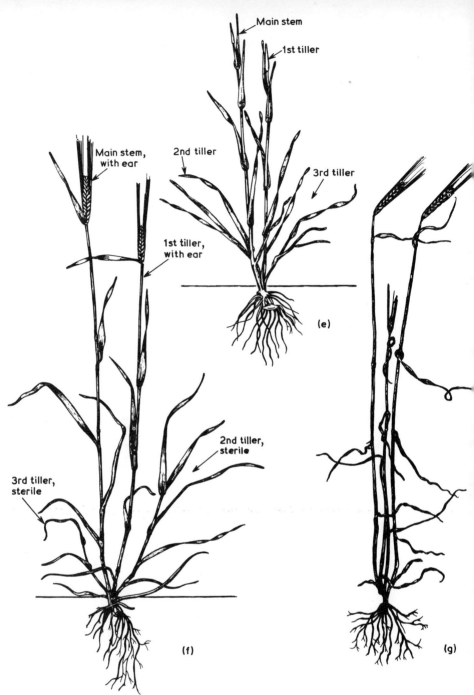

vernalized winter cultivars sown in the spring may never head. 'Scales' of growth, based on plant morphology, are used for experimental purposes and to decide when fertilizers and sprays should be applied. The exact morphology of the dissected stem apex may be used to establish 13–17 developmental stages (Chapter 2; Table 6.4) [6,22]. The number of seedling leaves visible is a useful guide, but is not absolutely linked to the development of the apex. Grain development may be specified by its moisture content. Many terms are used to describe grain development, e.g. the liquid endosperm stage; the soft dough stage. Four other stages are: (1) green ripeness – the upper leaves and awns are green, but the lower leaves are yellow or withered; all nodes are juicy, the grains are pale green, and easily crushed, with contents soft and waxy; (2) yellow ripeness – all leaves are yellow and withered; the lowest nodes are dry, while the upper ones are juicy; the awns are greenish yellow; (3) full ripeness – chlorophyll has left all parts; (4) over-ripeness – the plants are brittle and are beginning to break up and shed grains and ears; they may become discoloured. Several formal 'growth scales', with illustrations, have been presented (Table 6.4; Fig. 6.5). A modified scale, designed to allow the storage in a computer of growth data for small grains grown in all parts of the world, has been suggested [452].

The 1–3 tillers of field-grown plants ripen at about the same time; with isolated greenhouse-grown plants the main stem may be ripening while the last tillers are heading. This type of 'discrepancy' complicates many studies.

6.3. The composition of the growing plant

The chemistry of the growing plant is best known in relation to its value as fodder or silage (Chapter 14). As growth proceeds dry matter accumulates, and the composition of the plant alters. The nitrogen content goes through a maximum in the early stages of growth (6–7% dry matter), then declines to a lower value (0.6–0.8% dry matter). True and crude protein contents decline [88,319]. Similarly the ash content peaks, and later declines when the sulphur and phosphorus levels also decline. The fat content falls from 4.8% to 1.5% [381]. In the whole plant, less the ear, the nitrogen content declines from about 5.6% to 1.6%, the phosphorus from 4% to 1.9% and the sulphur from 2.4% to 1.3%. In the ears the final values were nitrogen, 2.5%; phosphorus 2.9% and sulphur, 1.9% [88].

Amino acid levels, both free and combined, have been followed (Chapter 14) [98]. Nitrogenous fractions, minerals, ash, some vitamins, carotene and reducing sugars have also been determined [95,166]. Succinic, fumaric, malic, isocitric, trans-aconitic, citric and quinic acids have been found in appreciable quantities in leaves [73]. The cellulose and pentosan contents rise while the uronic acid level declines. In the hemicellulose fraction xylose increases, arabinose, glucose and uronic acids decline and galactose alters little with increasing age [63]. Estimates of lignin content disagree but probably the proportion increases steadily from a very low initial value [319]. The carbohydrate content of the plant

232 Barley

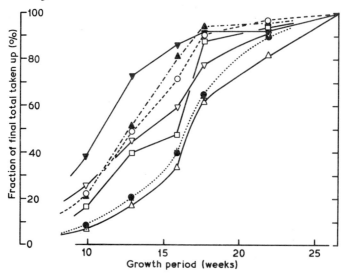

Fig. 6.6 Increases in dry matter, in mineral content and in the total water transpired by barley growing in S. Australia. Results are expressed as percentages of the values reached at maturity (after Richardson and Trumble [351]). ▼—▼, nitrogen, N; ▲–··–▲, potassium, K; ○---○, calcium, Ca; ▽—▽, phosphate, P; +, silicate, SiO_2; □——□, total ash; ●····● dry matter; △—△, water transpired. The SiO_2 points mostly coincide with those for dry matter. Published with the permission of the Waite Agricultural Research Institute.

rises from about 50% to 80–88% [88]. As the grain fills its crude fibre level falls, e.g. from 6.3 to 5.1%, while that of the straw rises from about 30% to about 40% [234].

The uptake of minerals is roughly sigmoid with time, but the uptake of water, nitrogen, phosphorus, and potassium are apparently independent (Fig. 6.6) [351,431]. When the dry weight, starch content and contents of potassium, sodium, nitrogen, and phosphate were followed, potassium was lost during maturation, as were sodium and nitrogen and phosphate to lesser extents.

6.4. The composition of the growing grain

The rather irregular and prolonged increases in ash dry matter found in the grain in England, with a fluctuating climate, contrast with the shorter more regular period of grain filling encountered elsewhere (Fig. 6.7) [53,158,161,260]. While the total ash of the grain rises, then levels out, the percentage ash content falls to a low value of about 2.0–3.8% or less, at maturity. The ash content of the grain varies with the nutrient status of the plant [53]. The ash contents of the palea plus lemma and rachis are higher, with maximum values of about 14% and 17% respectively, while that of the awn is higher still, up to about 35%, with

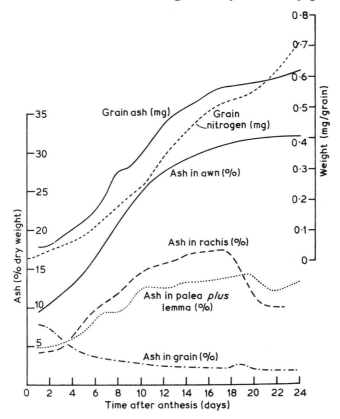

Fig. 6.7 The composition of Hannchen barley grains, and their parts, growing under full irrigation at Idaho, U.S.A. (data of Harlan [158], Harlan and Pope [161]).

a wide range of values occurring (e.g. rachis, 4.8–13.6%, awn, 17.2–37.8% [1,161]. Awns contain less ash at the base than at the tip; e.g. in a sample of Hannchen: base, 33%; middle, 39%; tip, 41%. The amount of ash in the awn depends upon the supply of irrigation water. The rachises of hooded and awnless barley, and heads from which the awns have been clipped, are rich in ash, and tend to be brittle. In general, conditions which favour low ash contents favour tough rachises [1]. The moisture content of the grain remains approximately constant for a short time, then falls until maturity, but the pattern observed varies with geographical location [53,158,162,260].

The total nitrogen content of the grain increases almost continuously and may still be rising or be steady at maturity. On a percentage basis the nitrogen content may fluctuate more. The total contents (wt/grain) of potash, phosphorus, calcium, magnesium, zinc, iron, manganese and

234 Barley

Table 6.5 Mineral contents of some dry, above-ground parts of two different barley samples (from Burd [64]).

Sample			Mineral content (% dry matter)				
			N	P	K	Ca	Mg
Stems and leaves		(a)	0.24	0.13	1.74	0.28	0.24
		(b)	0.50	0.12	1.94	0.57	0.23
Heads		(a)	1.22	0.40	0.51	0.06	0.17
	*	(b)	1.69	0.42	0.50	0.09	0.17
Grain		(a)	1.36	0.49	0.44	0.05	0.18
		(b)	1.92	0.49	0.51	0.05	0.13
Entire above-		(a)	0.71	0.28	1.19	0.18	0.22
ground parts		(b)	1.22	0.31	1.22	0.31	0.18

*Head sample taken earlier than rest of series.

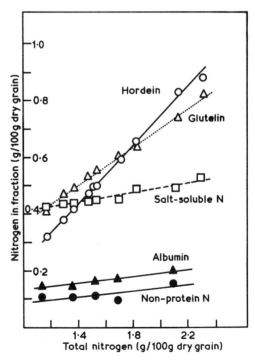

Fig. 6.8 The regular relationships between the quantities of nitrogenous fractions, separated by their differing solubilities, and the total nitrogen contents of a series of samples of mature Plumage–Archer grain (after Bishop [45]).

copper usually increase steadily, but may go through a maximum; on a percentage basis they fall [95,251]. Some values found at maturity were: starch, 62%; K_2O, 0.74%; Na_2O, 0.15%; N, 1.42%; P_2O_5, 0.96% (Chapter 14; Table 6.5) [440]. In arid conditions the phosphate content of grain increases with increasing applications of irrigation water, e.g. from 0.31% to 0.40% [130]. Ranges of micro-elements, per kg dry grain, have been reported as manganese, 28–31 mg; copper, 5.6–7.5 mg; iron, 37–48 mg; zinc, 27–30 mg.

The crude 'protein' (N × 6.25) that accumulates in the grain contains about 80% true protein. Different nitrogenous fractions accumulate at different rates [163,191]. Analysis of different grain lots of particular barley varieties grown under a range of conditions shows that with increasing nitrogen contents the levels of the different crude nitrogenous fractions increase in an ordered manner, hordein and glutelin increasing proportionately more than the other fractions (Fig. 6.8) [45]. Similarly, in samples of grain with increasing total carbohydrate contents the 'extract-yielding carbohydrates', mainly starch, increase proportionately

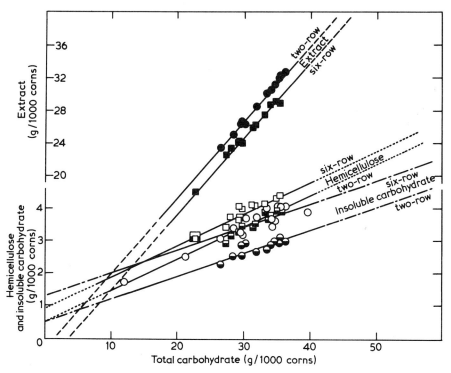

Fig. 6.9 The regular relationship between the total carbohydrate content and the carbohydrate fractions of a range of mature grain samples of two-rowed and six-rowed varieties (after Bishop and Marx [46]).

more than hemicelluloses or insoluble carbohydrates (Fig. 6.9) [46]. The 'regularity principle' of grain composition revealed by these results, and the realization that different varieties varied in their relative compositions, provided the basis for the 'extract prediction' equations used in malting (Chapter 15). The feed value of different barley lots varies in a way not immediately expected from their gross chemical composition; for example when barleys are grown at higher levels of N fertilization the nitrogen ('crude protein') of the grain increases but the proportion of lysine in the protein declines. Thus for the variety Maris Mink, protein 7.5% dry matter, lysine 5.5 g/100 g amino acids, and in another sample, protein 23.1% dry matter, lysine 3.7 g/100 g amino acids; comparable figures from the high-lysine mutant Risø 1508, protein 7.7% dry matter; lysine 7.1 g/100 g amino acids and protein 22.9% dry matter, lysine 6.0/100 g [347]. The nitrogenous solubility fractions usually studied are not of constant composition. The amino acid composition of maturing and mature grain, grown under different conditions, has been studied repeatedly (Chapter 14; Table 6.6) [98,327,347,376]. The late accumulation of glutelin and hordein (prolamin) is reflected in the enhanced levels of glutamate and proline. All the common

Table 6.6 The total amino acid composition in developing grains of Proctor barley, 1969 (from Smith [376]). The samples were hydrolysed before analysis. Results are expressed as g amino acid/1000 corns. By permission of the Cambridge University Press.

Days before harvest	42	35	28	21	14	7	0
Amino acids							
Aspartic acid	0.042	0.045	0.088	0.169	0.242	0.250	0.233
Threonine	0.021	0.023	0.051	0.105	0.139	0.149	0.140
Serine	0.024	0.023	0.054	0.116	0.166	0.191	0.183
Glutamic acid	0.071	0.085	0.234	0.571	0.981	1.068	1.036
Proline	0.034	0.037	0.090	0.226	0.376	0.439	0.410
Glycine	0.023	0.025	0.057	0.126	0.171	0.180	0.177
Alanine	0.036	0.038	0.066	0.148	0.185	0.182	0.179
Valine	0.028	0.029	0.064	0.140	0.208	0.221	0.216
Cystine	0.001	0.002	0.006	0.036	0.066	0.066	0.062
Methionine	0.008	0.010	0.024	0.049	0.063	0.066	0.070
Isoleucine	0.022	0.021	0.052	0.106	0.149	0.154	0.153
Leucine	0.035	0.036	0.088	0.197	0.290	0.304	0.304
Tyrosine	0.016	0.017	0.042	0.094	0.143	0.151	0.150
Phenylalanine	0.021	0.021	0.057	0.132	0.202	0.221	0.214
Ammonia	0.009	0.011	0.026	0.065	0.099	0.097	0.104
Lysine	0.028	0.032	0.062	0.123	0.166	0.165	0.159
Histidine	0.009	0.011	0.026	0.058	0.086	0.090	0.091
Arginine	0.028	0.033	0.077	0.153	0.224	0.238	0.236
% Nitrogen (S.D.)	1.84	1.69	1.63	1.74	1.94	1.90	1.85
	(0.04)	(0.09)	(0.07)	(0.10)	(0.11)	(0.07)	(0.09)
% recovery of N as amino acids	73.2	77.0	77.9	81.2	86.4	92.4	92.1
1000 corn weight	4.04	4.80	11.82	24.20	32.50	35.0	35.4
(S.D.)	(0.06)	(0.11)	(1.64)	(1.97)	(0.66)	(1.80)	(1.71)

The growth of the barley plant   237

amino acids have been found free in the developing grain [163]. The nitrogen contents of individual grains formed on one ear may vary substantially (e.g. 10–50% in one trial) and the differences between grains from different ears are even greater. While values are irregular there is a trend for the grains at the tips and bases of each ear, which are usually the smallest, to have the highest nitrogen percentage although the largest grains generally contain the largest absolute amounts [112,299].

Grain carbohydrates alter in the period from anthesis to maturity [71,163,231,260]. On a percentage basis the total sugars and reducing sugars (including glucose and fructose) decline, while non-reducing sugars are roughly constant in amount. While glucose and fructose decline from an early peak, sucrose 'peaks' later, followed by raffinose which peaks a little before maturity (Fig. 6.10). Fructosans, 'glucodi-

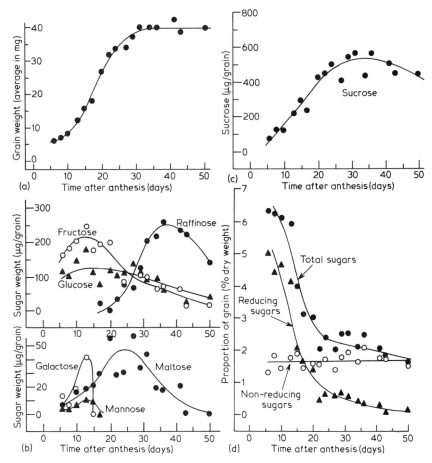

Fig. 6.10 Changes in the weight and the simpler carbohydrate components of growing Betzes barley grains (after La Berge et al. [231]).

238 Barley

fructose', and sucrose are the major alcohol-soluble sugars in the grain. Total fructosan, fibre, pentosan and glucosan levels have also been determined. The starch content of the grain rises in a sigmoid pattern with time, 95% being deposited in an initial short period of about 11–28 days after ear emergence in Canada, compared to 35–40 days in Britain (Fig. 6.11) [260]. At maturity starch is the major grain component. During starch deposition the ratio of amylose to amylopectin increases to a constant final value [24,163,260,270]. There is a dramatic increase in the proportion of small starch granules with time; in a high-amylose barley

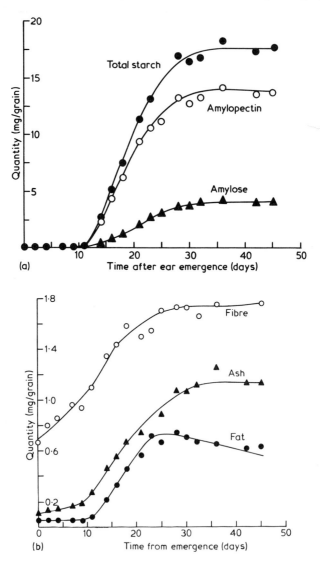

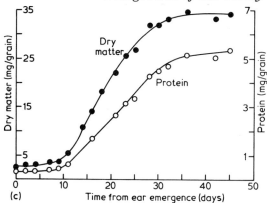

Fig. 6.11 Alterations in the dry weight, starch, amylose, amylopectin, protein, ash, fibre and fat of developing Conquest barley grains, growing in Manitoba, Canada. Grain was sown on 24 April; ears emerged, 30 June. The grain was mature by 14 August. The moisture content of the grain began to decline steadily about 15 days after anthesis (after MacGregor *et al.* [260]).

small grains predominate [66,158,260,270]. There is a transient appearance of starch in the pericarp during development, but this has disappeared at maturity. Chlorophyll and various enzyme activities associated with photosynthesis are found in the pericarp [93].

The fat content of the grain also increases in a sigmoid fashion and then may decline slightly [260,270]. There are slight variations between different cultivars. Alterations in the weight, carbohydrate, protein and nucleic acid contents of developing embryos are recorded [94], as are some changes in the levels of various enzymes [35,36,93,335]. β-Amylase is laid down in the developing grain in soluble and insoluble forms. There are wide variations in the levels attained and the ratio of 'bound' to 'free' enzyme [163]. The free form gradually becomes bound during grain development [259–261,387]. α-Amylase has a transient existence in maturing grain [260]. Possibly this enzyme, which has been purified, helps to remove the starch from the pericarp [259,260]. The enhanced level of enzyme seems to be controlled by gibberellins [92].

During grain formation a period of maximal respiration coincides with the maximum rate of dry weight increase [3,255]. During the formation of the grain its respiration varies in its resistance to cyanide [3]. The gibberellin level passes through two maxima, 9–15 days and 21–30 days after anthesis, at the respective times of rapid embryo growth and differentiation, and rapid accumulation of substances in the endosperm. An inhibitor, possibly abscisic acid, rose to a small extent in days 18–21 and, after a slight fall rose from days 30 to a very high level at maturity [345]. High gibberellin levels were associated with stages of rapid development, and high levels of inhibitor with the termination of these stages [285,345].

240 Barley

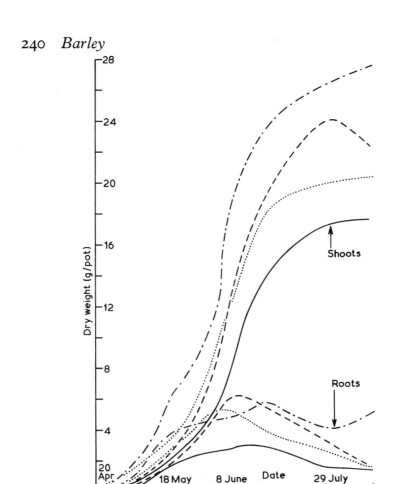

Fig. 6.12 The growth of the roots and shoots of barley plants grown in pots on soil alone, or with one of three different fertilizer applications. One plant, from a 50–60 mg grain, was allowed to survive in a pot containing 10.4 kg (23 lb) of soil. Anthesis occurred near to the time of maximal root yield (after Brenchley and Jackson [58]). ———, No fertilizer; ----, sodium nitrate added; —·—·—, sodium nitrate and superphosphate added; ......, superphosphate added.

6.5. Root growth

Roots growing in soil are difficult to study. Information is needed to enable estimates to be made of total dry matter production (above ground *plus* below ground) in the field, to clarify effects of fertilizer placement, to understand competition between the crop and weeds, and to check if there are varietal differences in root morphology that should be exploited in breeding programmes. Many variations in the patterns of root growth are due to differences in growing conditions [156,413]. Estimates

Table 6.7 The extent of root growth at various times of Hannchen barley plants growing in rows. The plants, which were growing in Saskatchewan, matured in 80 days (from Pavlychenko [314]).

Days after emergence	Roots without branches			Greatest penetration		Branch roots, 1st order			Total root	
	No.	Length (in)	(cm)	(in)	(cm)	No.	Length (in)	(cm)	Length (in)	(cm)
Seminal roots										
5	6.6	57.8	146.8	11.0	27.9	391.4	155.0	393.7	212.8	540.5
22	6.6	175.0	444.5	28.0	71.1	1 582	5 233	13 292	5 408	13 736
40	6.6	257.2	653.3	39.0	99.1	4 176	8 798	22 347	9 055	22 300
80	6.6	283.0	718.8	46.0	116.8	4 720	11 370	28 880	11 653	29 599
Crown roots										
5	0	0	0	0	0	0	0	0	0	0
22	11.2	58.6	148.8	5.8	14.7	592.6	183.6	466.3	242.0	614.7
40	11.2	70.6	179.3	13.2	33.5	1 349	849.2	2 157	919.8	2 236
80	13.0	231.0	586.7	31.0	78.7	3 084	4 080	10 363	4 311	10 950
Entire root system										
5									212.5	540.5
22									5 650	14 351
40									9 975	25 337
80									15 964	40 549

of root/shoot dry matter ratios vary from 0.18 to 0.26 [52]. Seminal roots develop first, followed by the coronal roots. Length and branching increase, but after a certain point parts of the root system begin to die (Fig. 6.12; Table 6.1; 6.7; Chapter 1) [58,314]. The differences in the extensive growth of isolated plants and the reduced growth of plants placed in rows, where size is curtailed by inter-plant competition, are striking [315]. Root growth has been studied in water culture, by digging or washing roots from the soil, and by taking cores from the soil and searching for roots at various depths either visually or by following the distribution of radioactivity from rubidium, ^{86}Rb, introduced into the aerial parts [222,432,434]. The seminal roots generally penetrate the soil more deeply than the adventitious roots, but there are exceptions [130,433].

The supply of soil moisture affects root growth. Barley roots will not grow into dry soil, i.e. soils at or below the permanent wilting percentage nor will they appreciably penetrate a stationary water table, presumably because of a lack of oxygen [364,433]. In dry conditions the root/shoot ratio is increased as it is when shoots are in full sunlight compared to those grown in the shade [415].

Root growth, like shoot growth, varies with the nutrient status of the soil as does the root/shoot dry matter ratio (Fig. 6.12) [58,415]. Root weight is maximal and survival is best with adequate levels of nutrients. Root weight begins to fall at, or a little before, the time the ear emerges from the boot. The loss in root weight is presumably due to death and decay and perhaps to migration of substances into the ear. However pot experiments on which such conclusions are based and in which root growth must be restricted should be interpreted with caution when roots penetrating to 2.04 metre (6.7ft) have been observed in deep soils [432,433]. The nutrient status of soil alters the extent of the root system and the degree of branching [58]. High concentrations of nitrate enhance branching, but reduce length [78,432,433]. The effect of nitrogen and other nutrients on root growth has been strikingly demonstrated in water and sand cultures (Fig. 6.13) [90]. Generally all roots, whether shallow or deep, whether of seminal or nodal origin, are able to take up nitrate and presumably other nutrient salts (78,227].

Deficiency of phosphorus or potassium reduces the size of the root system, especially the nodal component, the number of axes and lengths of the primary laterals especially being reduced. Potassium deficiency can completely inhibit the formation of secondary laterals [150]. Under a range of conditions the relationships between the dimensions of the root systems are fairly constant (Fig. 6.14) [151,266]. Thus any of several measurements may be related to nutrient uptake. The root profile (change in lengths of laterals along each axis) varies with variety and nutrient supply as does the frequency of the laterals [153,266]. Models have been

The growth of the barley plant 243

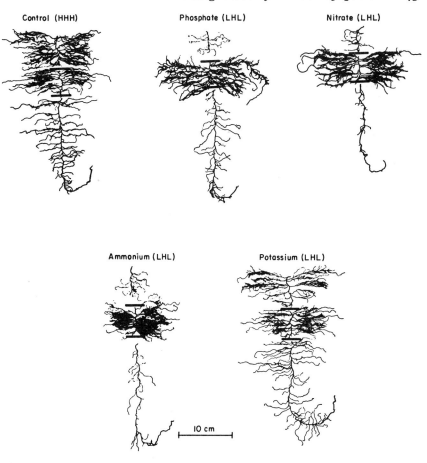

Fig. 6.13 The effects of localized differences in nutrient concentrations on the form of the main seminal root axes of 21 day old barley plants growing into sand divided horizontally into three compartments and separately irrigated with low (L) or high (H) concentrations of the nutrients indicated. All other nutrients were supplied at high concentrations (from Drew, 1975 [90]).

proposed to describe the growth pattern of roots, or to denote their distribution in the soil [154]. Removal of root tips greatly altered the branching pattern of barley roots but some dimensional relationships are little altered, indicating that root growth is a highly coordinated developmental process (Fig. 6.14) [152]. Probably the starch grains situated in the root cap allow roots to perceive gravity. Removal or other damage to the root caps destroys gravity perception; the ability to perceive gravity is restored as the cap regenerates [70].

Barley roots, in contrast to rice roots, will not grow under anaerobic conditions or in the presence of high concentrations of carbon dioxide,

244 *Barley*

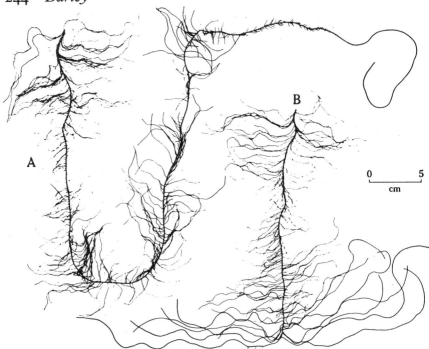

Fig. 6.14 Intact (a) and decapitated (on day 8, b) roots of 23 day old barley plants, after fixing and staining (from Hackett, 1971 [152]).

nor will they survive long periods of anaerobiosis [142]. However, under conditions of low oxygen tension some concentrations of carbon dioxide may enhance root growth [124,169]. In poòrly aerated water culture cereal roots grow poorly, have increased diameters, of about 15%, and develop large intercellular air-spaces in the cortex (Chapter 1). These spaces allow some oxygen to reach the roots from the upper parts of the plant. Such roots contain 100% more soluble sugars, 4–25% more nitrogen but much less polysaccharide cell wall material than roots from well-aerated cultures [165,321,445]. The movement of oxygen from the shoot to the root has been demonstrated in barley and in rice where the resistance to diffusion is lower [27,197]. Soil aeration fluctuates, and the growth of roots is probably controlled by the local oxygen supply.

The growth of roots into a soil is also regulated by its physical structure. Roots do not grow through hard pans, and they grow better in porous soils [169]. Experiments in which roots grew through rigidly packed glass beads indicated that while the mass of roots was little altered the morphology was greatly changed when the spaces between the beads were small [136]. Sterile roots exude increased quantities of amino acids and carbohydrates when growing between packed beads, rather than free in solution culture [28]. Root distribution may be influenced by

The growth of the barley plant

hydrotropism and geotropism. The influence of allelopathy on root distribution, and competition with other plants, is hard to assess. Barley roots release phenolic substances, and give off biologically active unsaturated substances, notably ethylene [326,379]. Root extension is checked by increasing concentrations of ethylene (0.1–100 ppm); inhibiting concentrations of ethylene can occur in anaerobic soil [379]. Barley roots release quantities of organic materials into the soil which normally help support the growth of the micro-organisms of the rhizosphere [31]. In roots growing free in solution culture, or between glass beads the quantity of amino acids *plus* carbohydrates released were equivalent to 5% and 9% of the root dry matter increments, respectively [28]. The growth and survival of roots is also influenced by soil temperatures.

6.6. Water supplies

Plants draw practically all their water from the soil, through the roots. When the soil spaces become full of water the soil is waterlogged and roots soon stop growing. Freshly drained soil is at 'field capacity'; the actual *amount* of water present varies greatly with the soil type. As the soil moisture falls below field capacity its availability to plants decreases until eventually the roots cannot replace their evaporative losses and the plants wilt. The degree and extent of root ramification controls the volume of soil from which water is withdrawn. Below field capacity moisture movement in the soil is slow [138,218]. Barley varieties differ in the efficiency with which they utilize soil moisture [2]. Water is continually evaporating from plants into the atmosphere. The quantity used is critical, since the supply of soil water must be sufficient to produce a useful crop. Some effects of varying soil moisture are illustrated by Table 6.8. The ratio of weight of water transpired to the weight of crop grown or to the weight of grain obtained varies between varieties, and between seasons, and with fluctuations in the incoming solar radiation, wind speed, temperature, and relative humidity [351,374]. Transpiration declines with decreasing moisture availability, or with increasing osmotic tensions in the soil water [296,363]. After a period of water lack (stress) transpiration may be less, so some change favouring 'drought-hardiness' has occurred [363]. Transpiration is a physical process modulated by the plant's ability to collect moisture from the soil, regulate its movement

Table 6.8 The effect of varying soil moisture on the growth of barley plants grown in pots (from Hellriegel, quoted Russell [356]).

Water (% soil saturation)	0	5	10	20	30	40	60	80
Grain (g dry weight)	0	0	0.72	7.75	9.73	10.51	9.96	8.77
Straw (g dry weight)	0	0.12	1.80	5.50	8.70	9.64	11.00	9.47

through the plant, and control its evaporation. It is dependent on the plant's microclimate. The pattern of water removal from soil is similar to the profile of root distribution, being greater in the upper soil levels [130, 433].

Water may enter the roots either via the hydrated root mucilage and the cell walls (apparent free space) of the cortex, or it may enter the protoplasm of the cells. It can enter root hairs directly, and rates of 0.07 to 0.44 $\mu m^3/\mu m^2$ root hair/min have been recorded [354]. The extent of free space in roots and leaves is uncertain [323,324,403]. Water must pass through the cytoplasm of the endodermal cells of the older roots, since the walls are probably sealed by the thickening substances. The waterflow, 'transpiration stream', enters the xylem vessels and moves upwards to the aerial parts of the plants. It is, at least approximately, in osmotic equilibrium with the adjacent tissues. Some water evaporates directly from the epidermis, via the cuticle, but most evaporates into inter-cellular spaces, and reaches the exterior via the stomata.

Under humid conditions drops of liquid, 'guttation fluid', may collect at the apices of coleoptiles or leaves. This liquid contains various substances in solution, including minerals, vitamins, amino acids and sugars [131]. Drying can lead to the deposition of these substances on the leaf tip, and tissue scorch and damage. Guttation fluid is driven up the plant by root pressure, which depends on the metabolism of the roots, supported by a supply of sugars. Probably the roots actively 'pump' salts into the xylem, and water flows in to maintain osmotic equilibrium. The rate of guttation varies with the root medium (which must contain salts), altering with pH and temperature, and being checked or prevented by high osmotic pressures, anaerobiosis or toxic substances [61,132]. When the roots are severed, or their metabolism is checked so that 'root pressure' is not operative, the flow of water to the aerial parts is maintained by evaporation into the atmosphere unless this is saturated with water vapour. The transpiration is mainly driven by a potential gradient from the soil to the atmosphere through the plant. The moving water must overcome various resistances on the way. Leaf water potential increases basipetally with plant leaf position. In one case it reached about 16.5 bars when the soil was at field capacity, but was reduced to 5.6 bars at a particular water stress [275]. The energy required to evaporate the water from the aerial parts, and so maintain the potential gradient, is mainly supplied by the incoming solar radiation, which also warms the plant above the temperature of the surrounding air.

The relative importance of the routes of water loss from leaves, via the cuticle or the stomata, is disputed [403]. However, when stomata are caused to open or close by changes in the gas atmosphere or light supply, cutting stems, or supplying abscisic acid or xanthoxin (both of which cause closure), kinetin or gibberellic acid (which cause opening), or by other means, or when backcross plant populations having different stomatal frequencies are compared, then it is seen that stomata have a major regulatory function [5,76,79,203,252,277,279]. The findings imply that hormones may regulate transpiration, and indicate that 'anti-transpirant' sprays might be used to check water losses. However sprays of abscisic acid have proved ineffective 'anti-transpirants' in the field. The awns are major sites of water loss. Intact, awned ears transpire 4–5 times as much as when awns are clipped [147,453]. The passage of water or other substances through the cuticle is of importance in transpiration, the possible uptake of moisture from saturated atmospheres, fogs and water droplets, and the uptake of growth regulators and systemic insecticides or fungicides from agricultural sprays [114]. In general uncharged organic molecules penetrate cuticles more readily than ionic species, suggesting that the waxy nature of the cuticle limits penetration. In leaves the surface wax, when undamaged by wind or friction with other plant parts, provides a water-repellant surface giving contact angles with water droplets variously reported as (i) $165° \pm 1.5°$ (upper and lower surfaces); (ii) $132° \pm 8.6°$ (upper surface) and $134° \pm 8.8°$ (lower surfaces) [114,177,346]. As different wax types occur in different genotypes or on different parts of one plant these differences are not surprising.

6.7. Water stress

When transpiration exceeds water uptake the plant's moisture content falls, and it is stressed. Consequently its metabolism changes and its root and shoot growth slows or ceases. (Fig. 6.15). It may wilt and, if the stress is not relieved, it may die. Stress may be due to a reduction in soil moisture content, or to the soil-water having a high osmotic pressure, or to the air being excessively dry. In saline soils there may be additional toxic effects due to particular ions. Of the substances used to induce experimental osmotic stress purified polythylene glycol seems acceptable, whereas small molecules like mannitol are not. Mannitol permeates the plant and is excreted in the guttation water, causes leaf burn, and may even crystallize on the leaf [146]. Even slight water stress affects growth. The fastest growing tissues are checked most, possibly because cell expansion is impeded; meristems may survive water stress longest. Roots grow less and mature leaves undergo hastened senescence under stress [256]. Water stress reduces a plant's relative turgidity and the vapour pressure at its surface, increases the osmotic pressure of the sap and alters its freezing point and refractive index, characteristics which may be measured in field studies [228]. Arid conditions depress grain yield, the relationship between the yield and the osmotic pressure of the sap being approximately inversely logarithmic. Within limits fertilizers can compensate for such losses [228]. However nitrogenous fertilizers may

248 Barley

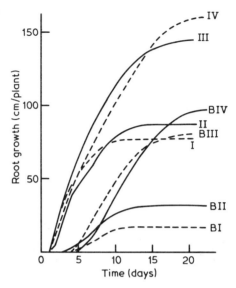

Fig. 6.15 The growth of entire root systems (I-IV) and the roots in the lower layer of soil (BI-BIV) of plants of Roger's barley, subjected to various degrees of water stress (after Salim *et al.* [364]: by permission of the American Society of Agronomy).

Plants were grown individually in boxes 10 × 11 × 60 cm, (3.9 × 4.3 × 23.6 in) containing 6.58 kg (14.5 lb) soil, in a greenhouse. The initial moisture tension at 15 cm (5.9 in) depth was less than 0.1 atm. At the lower depth, 45 cm, (17.7 in) the water tensions were I, about 10 atm; II, 6.5 atm; III, 1 atm; and IV, 0.45 atm.

cause leafy growth and so excessive transpiration, resulting in premature utilization of the soil water, and reduction in grain yield [256]. In the osmotic pressure range of 10–17 atmospheres in the soil water, work in Canada and Iraq indicates that each 1 atm increase is associated with a yield decrease of 10–13% [228]. In E. England a lack of soil moisture limits yields in many years (Chapter 7; Table 6.9) [339]. Late rainfall or irrigation after a dry period induces the production of tillers which is counter-productive since nutrients are directed away from the developing grains, and the new tillers are immature, green and without grain at harvest.

Drought may reduce plant height, tiller number, tiller fertility, the number of spikelets/ear, the extent of tiller mortality, the number of grains/ear, or grain size. Which effects occur depends on the timing and severity of the stress [366]. Relief of moisture stress is often followed by a spurt of rapid growth, possibly at the expense of photosynthate accumulated in the plant and nitrate accumulated in the soil. Water stress is more harmful at some stages of development than others. The concept of 'critical periods' of growth is helpful in understanding the effects of rainfall or irrigation [160,187,298,366]. While stress early in the growth

Table 6.9 Characteristics of barley sown late in E. England in 1970 when early rainfall was deficient. G – 30–60% of the rainfall intercepted by gutters between rows. GI – as G, but with one late irrigation to within 5 cm of field capacity in July. I – Irrigated regularly, so soil maintained within 5 cm of field capacity (from Rackham [339]: copyright by Academic Press).

	G	G + I	I
Plants/m^2	260	270	262
Stems (at harvest)/m^2	661	681	648
Ears/m^2	321	302	512
Spikelets/m^2	7 004	6 384	11 142
Grains/m^2	4 977	4 291	8 711
Grain dry weight, (g/m^2)	180	152	300
Total, above ground dry weight (g/m^2)	390	428	623
Green tillers dry weight (g/m^2)	7	85	2

period may check tillering and root establishment it may, if not too severe, be beneficial in preventing lodging and in 'hardening' the plant. Grain yield need not be reduced. Later mild stress may slow the initiation of floret initials and – if extended – result in ears with fewer spikelets [298]. Formation of apical primordia is checked at low levels of soil water stress (– 0.8 bar or more) that have little effect on plant growth; differentiation ceases at about – 2 to – 2.5 bars. Spikelet growth is less sensitive to stress than differentiation of primordia whereas root growth is more resistant still [187]. The barley spikelet, being determinate, cannot make good this loss. Later stress, particularly during meiosis, prevents the formation of fertile pollen, leading to sterile spikelets. Water stress also checks grain filling and stops growth early [160]. While best yields are obtained by maintaining the water supply near field capacity, this is usually impractical. The most critical period for water supply is from the onset of shooting until anthesis is complete.

Barley varieties vary in their resistance to moisture stress. It has been claimed that various presowing treatments involving wetting and drying cycles sometimes enhance drought resistance; various studies show that these do not always work. Some of the growth changes observed in stressed plants may be due to alterations in nutrient distribution. Stressed plants tend to become chlorotic, photosynthesis is checked and chlorophyll levels decline. Flag leaves with water potentials of – 28 bars were found to stop transpiring; photosynthesis ceased at – 31 to – 33 bars [199]. Nitrate reductase and phospho*enol*pyruvate carboxylase activities are suppressed, but phosphoribulokinase and ribulose diphosphate carboxylase activities are not [185]. The stomata close, presumably checking photosynthesis, as well as transpiration. This may be due to abscisic acid (ABA) being released from the wilting tissues. Like water stress, ABA alters the uptake of nutrients by roots and hastens leaf senescence [20,324].

250 Barley

Water stress alters the nitrogen metabolism of barley, and causes the accumulation of proline in the leaves, by blocking its oxidation [47]. This accumulation is altered by ABA. The extent of proline accumulation in response to water stress varies between cultivars [20,72]. While choline levels remain approximately constant betaine levels are increased, like proline, in stressed barley [391].

Irrigation frequently leads to soil salination and the need for crops tolerant to these conditions. Barley is among the more salt-tolerant crops. The ability of varieties to germinate in the presence of different salt concentrations varies widely [125,343]. At high osmotic pressures grain respiration is also suppressed [125]. The roots of salinized plants may not supply cytokinins to keep the stomata open [223], or perhaps the wilting-induced ABA release causes stomatal closure. As salinity is often unequally distributed in soil, it is interesting that plants with roots having access to solutions of differing osmotic strength behave as though growing in an 'averaged' solution. Salinity causes some changes in composition of the root lipids [110]. Starch and sucrose contents rise, but reducing sugars are little altered by increasing salinity. Chloride and sulphate in the water supply may lead to leaf scorch [97,122].

6.8 Mineral requirements [80,353,358]

Roots take up minerals that are in solution in, or are released from the soil materials into, the soil water. The soil–soil water-root system is complex, and for simplicity water cultures and sand cultures are often used to study plant nutrient requirements. Salts obtained from soil washings can support barley growth. Complications include the ability of humus substances to accelerate growth, possibly by chelating iron and keeping it in solution. It is difficult to exclude all 'trace elements' needed by plants from experimental equipment so that identifying essential minor nutrients has been difficult.

The essential major or 'macro'-elements needed for barley growth are carbon, hydrogen and oxygen (supplied by photosynthesis and soil water), nitrogen, phosphorus, potassium, magnesium, calcium and sulphur. The status of silicon is uncertain. Essential trace elements are iron, manganese, copper, zinc, boron, molybdenum, chlorine and possibly aluminium. Sodium additions often enhance growth. Iodine may or may not be essential for the plant but, like cobalt, selenium and chromium, it is essential in animal feeds and so should be present in the plant. Nitrogen can be supplied to the plants as nitrate or ammonium ions. In the soil ammonia is oxidized to nitrate in a few weeks. However these ions (NO_3^- and NH_4^+) are at different oxidation levels and are not equivalent; thus they are used best at different pH values and root aeration is less critical with nitrate-grown plants [13]. Phosphorus is taken up as inorganic

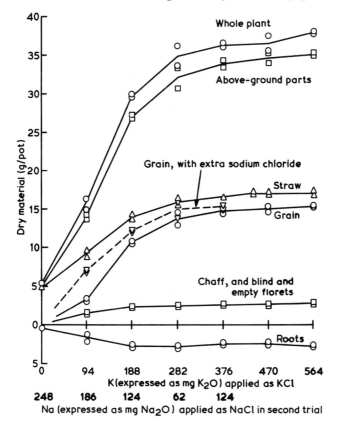

Fig. 6.16 The yields of parts of barley plants grown in pots with increasing supplements of potassium (as KCl) and, in a second trial, with variable additions of sodium chloride (from Hellriegel *et al.* [167]). Solid lines – KCl experiment. Dotted line – experiment with supplemental sodium chloride. Note the irregular additions of the sodium salt.

phosphate. Unlike nitrate, which can be leached by water, phosphate tends to bind to the soil. Sodium can replace potassium to a limited extent, when the latter is in short supply (Fig. 6.16) [167,289]. Calcium and magnesium are taken up as ions as are chlorine, iodine and most other minerals. In water culture calcium helps to maintain root integrity; it also has a marked effect on the uptake of other ions [306]. Sulphur is normally taken up as sulphate. When the soil becomes anaerobic the sulphate present may be reduced to sulphide, which is toxic.

The availabilities of 'trace elements' to barley are greatly affected by soil conditions, pH for example. Only tiny quantities of these substances are required, and in excess they are frequently toxic [57,200,249]. Barley needs less than 0.1 ppm of boron in the soil, excess being toxic; the plant contains about 2.3 ppm in the aerial parts [39]. At very low levels (0.5 ppm) iodide stimulates barley growth; at higher levels it is toxic, but it is not clear if it is *essential* [418].

The role of silicon (either as silica, SiO_2, or as silicates) in barley growth is puzzling. Soluble silicates enhance silica uptake, which is large, improve plant growth and reduce the plant's susceptibility to wilting and mildew [358]. In the field the beneficial effects of added silicates may be due to the enhanced availability of soil phosphates [237,249]. This can only be a partial explanation, since in water cultures silicates protect plants against toxic levels of manganese [443,444]. In rye, competition studies with germanic acid suggest that silicic acid is essential for normal growth. Silica, SiO_2, is deposited in barley in the form of opal [238]. Its location in various tissues has been studied, e.g. in spodograms, or by electron-probe micro-analysis (Chapter 1).

Barley varieties vary in the soil pH values they will tolerate. Few will tolerate soil acidity (one has been reported to yield well at pH 3.5) yet optimal pH values of 5–7 are usual. In the U.K. barley fields are normally limed to a pH of 6–7 [104,388]. The optimal pH for root growth rises with increasing calcium ion concentration *in vitro* [65]. pH has indirect effects on plant growth. Thus in acid conditions toxic levels of aluminium or manganese may be held in solution. In alkaline conditions iron is precipitated from solution and plants become chlorotic (lime chlorosis).

Animals require traces of selenium. However on seleniferous soils plants may accumulate so much that they become toxic, and their growth is checked. The selenium partially replaces sulphur within the plant [414]. Selenium uptake is reduced by increasing supplies of sulphate ions, which are absorbed in preference to the selenate ions.

Abnormal growth in plants may be due to many factors, including mineral deficiencies (Table 6.10). In addition to the appearance of the

Table 6.10 Symptoms in barley caused by some mineral deficiencies or excesses [118,171,248,302,383,424]. Mineral deficiencies probably cannot be reliably diagnosed from plant appearance alone. Symptoms of N, P and Mg deficiency are first detected in older basal leaves. Symptoms of Ca, B and Cu deficiency are first seen in young leaves, but spread back to the older parts.

Nitrogen, N. Generally a limiting factor in agriculture. Deficiency restricts root and shoot growth. Tillering reduced, shoots small and thin, with sparse, pale green leaves. Chlorophyll and carotenoid levels are reduced. Ears are small. Grain ripening may be premature. Excess nitrogen favours lush, leafy growth, lodging, and delayed maturity.

Phosphorus, P. Deficiency much restricts growth. Leaves are bluish-green and eventually the tips brown and die. In severe deficiency purple anthocyanin pigments may colour the leaves, stems and ears. (Some cultivars are always coloured). The grain may ripen late. Large excesses may depress growth. Young plants are most susceptible to this deficiency.

Table 6.10 (Contd.)

Calcium, Ca. Lack results in stunted growth and high seedling mortality. Growing points are defective, both in roots and shoots. Leaf tips die; there may be chlorosis. Calcium effects are closely linked with differing soil pH. Calcium deficiency is rare.

Magnesium, Mg. In deficiency growth may appear normal, but leaves may be a pale green (chlorophyll is a Mg tetrapyrrole). Older leaves may yellow and die. Necrotic areas can occur near the leaf margins. Plants are dwarfed in extreme deficiency.

Potassium, K. A lack causes dwarfed shoots, with short internodes. An excessive number of tillers, but with few ears, may occur. The small heads have few grains. Leaves bluish-green, slightly chlorotic with tip and marginal scorch preceded by general mottling with brown-purple spots. The dead leaf ends become ragged. In acute deficiency the chlorotic leaves have white, blotchy lesions. In mild deficiency symptoms appear first in old leaves as the element is withdrawn and moved to young tissues.

Iron, Fe. Deficient plants, especially the young parts, are chlorotic. Continuous interveinal stripes or totally bleached leaves may occur. This pattern of chlorosis may be 'lime-induced', but cereals are reasonably resistant. Applications of iron salts are not economic, perhaps iron chelates would be more effective. Excessive quantities of P, Cu, Mn, or Zn may cause iron deficiency.

Manganese, Mn. Leaves of deficient plants, particularly the distal parts, are chlorotic, with mottled brown spots and stripes. (Oats are markedly more susceptible than barley; they show 'grey-speck' disease). Mn deficiency often occurs on calcareous soils. Requirements for Mn increase with increasing supplies of Cu.

Boron, B. Boron deficiency has not been detected in the field. Experimentally deficient plants have thick stems, growing points die, and the ears are malformed. Elongated, dark brown blotches may occur, appearing first at the tips of older leaves.

Sulphur, S. Deficiency occurs in some few soils and becomes apparent when extra NPK fertilizer is supplied. Leaves are pale yellow, growth is stunted and maturity is delayed. In contrast to N-deficiency, leaf yellowing is most noticeable in younger leaves.

Copper, Cu. Deficiency causes 'reclamation disease'; yellow tip; wither tip; whiteheads. Occurs frequently on polders and reclaimed heaths. Tips of younger leaves wither with marginal chlorosis. Young leaves may not unroll and the foliage tends to wilt. Ears are small and malformed and may be chlorotic (giving one name to the deficiency syndrome). Grain formation is depressed proportionately more than vegetative growth.

Zinc, Zn. A deficiency mainly found on sandy or calcareous soils. Young plants show purple colours, older leaves die, and mature straw has a grey tinge. Diagnosis can be aided by analysing foliage for Zn or noting response to a spray application of a solution of a Zn compound.

Molybdenum, Mo. Deficiency has been produced experimentally. Growth droops and the stems become limp. The leaves become dull, pale green developing chlorotic mottling and, when nitrate is the source of N, with necrosis. They often have a 'papery' appearance and a tendency to dry. Youngest leaves stay rolled and chlorotic, with a constriction near the tip. Stems collapse after the youngest leaves die, before emergence. This deficiency is rare.

Acid soil – Manganese toxicity Leaves have brown spots between the veins.

Acid soil – Aluminium toxicity Leaves and shoots show symptoms akin to those of P deficiency. Root growth is poor and individual roots are stubby in appearance.

254 Barley

Table 6.11 The mineral contents (means of 4 locations; dry matter basis) of different parts of barley plants (from Goodall [133]).

Mineral (Units)	Lower leaf blades	Lower leaf sheaths	Upper leaves	Stems	Ears
Ca (%)	0.440	0.204	0.211	0.121	0.095
Fe (p.p.m.)	228	127	147	90	133
Mg (%)	0.225	0.133	0.165	0.133	0.165
Mn (p.p.m.)	140	100	65	53	32
K (%)	0.76	0.72	1.58	0.93	2.23
Na (%)	0.766	0.978	0.255	0.757	0.086

Table 6.12 The effect of increasing applications of potassium fertilizer on the potassium and sodium contents of parts of barley plants (% dry matter basis; from Goodall [133]).

Fertiliser, K_2SO_4 (lb/acre)	Lower leaf blades		Lower leaf sheaths		Upper leaves		Stems		Ears	
	K	Na	K	Na	K	Na	K	Na	K	Na
0	0.51	0.908	0.46	1.294	1.09	0.405	0.68	0.863	2.23	0.118
56	0.80	0.755	0.73	0.917	1.82	0.228	0.96	0.694	2.20	0.088
112	1.07	0.655	1.09	0.787	1.99	0.179	1.22	0.725	2.26	0.062

crop, aids to recognizing mineral deficiencies include the appearance of particular weeds, and of 'tester' plants known to be particularly vulnerable to a lack of particular elements, observation of the effects of applications of nutrients in pot or field trials, and leaf and soil analyses. Some deficiencies may be diagnosed by responses to solutions of salts supplied as foliar sprays [133–135,257,383,424]. Plant analyses are difficult to interpret because the concentrations of minerals vary in different parts of the same healthy plant with age, with the supplies of other elements, between varieties, and with changing growth conditions (Table 6.11; 6.12) [297]. The mineral content of plants does not increase linearly with increasing availability. 'Minimal contents' of elements have been suggested (e.g. in mature plants) but the values do not agree. Mineral ratios (e.g. P:N or K:P) may be more valuable in detecting deficiencies [135].

Increasing the supply of a mineral, when other conditions are fixed, results in improved growth up to a maximum (Fig. 6.16). Beyond this point growth may decline, when an excess is harmful, or may 'plateau', in which case other factors are limiting. At intermediate supplies also the response will vary with other conditions. An 'optimal' concentration of a nutrient is only constant when all other cultural conditions are defined (Table 6.13). In some greenhouse trials extra growth in response to added nitrogeneous fertiliser only occurred above a certain level when extra water and phosphate were supplied, indicating that all these factors

The growth of the barley plant

Table 6.13 The variation in grain weight with varying combinations of P and N fertilizer, in a pot experiment (three plants/pot; from Goodall and Gregory [135]). By permission of The Commonwealth Agricultural Bureau.

N (mg/pot)	Grain yield (g dry weight)				
	P_2O_5 (mg/pot)				
	405	135	45	15	5
1215	33.0	18.9	2.7	1.3	0.7
405	10.5	10.9	4.9	3.6	2.5
135	4.6	4.4	4.2	3.5	3.5
45	2.0	1.9	2.2	2.0	1.3
15	1.2	1.2	1.0	1.2	1.0

Table 6.14 The growth of Plumage barley plants in water cultures in which the culture media, diluted to different extents, were changed with various frequencies. Growth period, 5 April – 24 May (from Brenchley [54]).

Weights of parts (dry weight, g/10 plants)

		Culture solution changed								
		Every 4 days			Once			Never		
		Root	Shoot	Total	Root	Shoot	Total	Root	Shoot	Total
Medium undiluted		4.44	1.33	5.77	3.75	1.32	5.07	3.05	1.16	4.21
Diluted to	0.2	3.37	1.14	4.50	1.62	0.71	2.33	1.14	0.63	1.77
	0.1	1.90	0.84	2.79	0.84	0.47	1.31	0.53	0.41	0.93
	0.05	1.28	0.68	1.97	0.36	0.32	0.68	0.25	0.26	0.51

limited growth [91]. In solution cultures various 'optimal' concentrations have been reported (e.g. Fe, 11.2ppm; < 0.7ppm Fe, deficiency; Ca, 10μM for roots, 1000μM for shoots, yield levels above 10μM; K, 95μM – yield).

Many attempts have been made to relate growth to nutrient supply in quantitative terms [358]. In water culture experiments not only the initial nutrient concentration, but also the continued availability as effected e.g. by the ratio of root to liquid volume, and the frequency of liquid renewal limit growth (Table 6.14) [54,193]. Requirements for a particular nutrient vary with the stage of growth of the plant [56].

Variations in plant composition with changing nutrient supply give some indications of the role of individual nutrients. In water and sand culture experiments deficiencies in nitrogen, potassium and phosphorus interfere with photosynthetic assimilation and respiration to an extent depending on the age of the plant and the light intensity. Leaf water content is also altered [348,350]. Potassium deficiency seems to allow an abnormally high concentration of cellular inorganic phosphate, which may be the stimulant causing the excessive respiration which occurs [349]. Nitrogen deficiency causes a reduction in the rate of leaf expansion, the final dry weight, area, and chlorophyll content [145,292]. The carbohydrate and nitrogenous leaf fractions also alter with nutrient status [12,348,357]. Many effects are due to imbalances of nutrients rather than to inadequate quantities being supplied [149,348]. Optimal ratios for nitrogen, phosphorus and potassium have been suggested. When adequate supplies of one nutrient are available the needs for all other nutrients, to achieve maximal growth, are enhanced (Table 6.13, Chapter 7). The alterations in nitrogenous substances that follow from nutritional imbalances are striking. Nitrogen is translocated from older, senescing leaves, especially in nitrogen-deficient plants. In potassium-deficient plants there is a rapid decline in protein and rise in leaf basic substances especially agmatine and putrescine (Chapter 4) [75,378]. Feeding putrescine induces necrotic lesions on leaves similar to those occurring in potassium deficiency. Alterations in the relative levels of potassium and phosphorus also alter agmatine and putrescine levels [155]. Potassium may be partly replaced as a basic counter-ion in cells by agmatine and putrescine, since these bases increase when barley plants are fed inorganic acids [380]. Barley cultivars differ in their responses to fertiliser levels [144].

6.9. The uptake and release of substances by roots

Work on the uptake of substances by barley roots has been very extensive, studies on substances released by roots less so. Substances released include sugars, amino acids, organic acids and phenols. Release is particularly noticeable after a period of desiccation. This encourages alterations in the soil microflora, members of which may even form a mycorrhiza around the roots; these alter the uptake of nutrients [31,175,209,419]. Roots do not take up nutrients from dry soil; uptake is modulated by the soil water supply and by the temperature [263,272]. Nutrients are not merely swept into the plants in the transpiration stream. This is shown, for example, by the independent increases of minerals, of dry matter and total transpiration during plant growth (Fig. 6.6) [351,431]. The uptake of ions is selective; e.g. potassium is taken up in preference to sodium [321]. The part of mineral uptake that is against a concentration (or more properly an activity) gradient is energy – and hence respiration – dependant [29]. However, when the external salt concentration is high (so there is little or no gradient) and transpiration is rapid, the salt uptake may follow transpiration quite closely [62,180,288]. Confusion is caused by difficulties of interpretation. For example it is difficult to decide how to make allowance for the pumping of ions from the cortex into the stele and their leakage back into the experimental vessel [322]. In intact plants changing rates of transpiration may alter rates of salt uptake, and all aspects of the metabolic state of the plant, (e.g. oxygenation of the roots, availability of carbohydrates to support respiration and the temperature) are important [61,62,74,164,422].

Substances may be dissolved in the film of surface moisture around the root and in the 'apparent free space' which is located, at least mainly, in the cell walls of the cortex. Penetration into the free space may be modulated by Donnan equilibria established with bound charged groups in the wall [33,116]. Ions in the free space may exchange with ions within the cell without the expenditure of energy, or they may be actively accumulated. Active accumulation may occur at the plasmalemma and into the vacuole at the tonoplast. Ions can be moved from one tissue to another, for example from the root cortex into the stele, creating a high osmotic pressure there and the motive power for guttation. Ions accumulate rapidly in root apices; however it is not necessarily from these regions that most export to the rest of the plant occurs [438]. Roots seem to deplete soil nutrients in any zone they reach, but different substances are not necessarily removed in the same proportions, or in proportion to root density [78,294]. The stream of ions from outside the root to the shoot need not equilibrate with the 'pools' retained in the root cells.

The energy-dependent aspects of ion uptake are indirectly inhibited when respiration mediated by the cytochromes is blocked, by lack of nutrients, or by adverse temperatures [181,262,268]. Uptake is also dependent on the concentration of the ion. To maintain electrical neutrality ions taken up by the roots must be mixed so that their charges balance, or they must be exchanged for others. For example potassium ions may be exchanged for hydrogen ions causing a fall in the pH of the medium [174]. Sometimes roots take up the potassium ions faster than the counter ion of the potassium salt. When this occurs the roots fix carbon dioxide by reversible carboxylation reactions forming organic acids, particularly malic acid, which act as counter ion to the accumulated K^+ within the root [172,173,194,195,417].

The uptake of organic acids, herbicides, phenols, proteins and other macromolecules have all been studied. Of more general interest are the essential components of the soil solution. At neutral pH soluble silica occurs as the essentially uncharged silicic acid $(Si(OH)_4)$ or its dimer. This can apparently penetrate into plants by a passive mechanism, but as it can reach the upper parts of the plant at rates exceeding that of the transpiration stream, active accumulation must also occur [32]. The mechanism by which silica is deposited in particular cells and structures is not understood. The quantitative effects of micro-organisms on uptake processes are in dispute [25,103]. However as micro-organisms have been largely ignored in salt uptake studies much 'quantitative' work must be suspect. Roots (and leaves) can accumulate choline sulphate, and it is claimed that infecting micro-organisms transfer at least part of this capacity (permease) to the plant [301]. Micro-organisms alter the apparent rate of uptake of e.g. phosphate, and also its distribution in the root [25,30]. At a fixed temperature isolated roots take up ions from a bathing medium at rates that increase with increasing concentration up to a maximum at about 0.5–1 mM; the rates increase again very rapidly at still higher concentrations. The two phases of such 'dual absorption isotherms' may be described approximately by Michaelis–Menten type kinetics, and are seen whether or not roots are sterile [26,304,341]. However the existence of 'multiphasic' isotherms has been claimed [178]. These apparent discrepancies need to be resolved, and decisions must be made regarding the validity of various proposed accumulation mechanisms, such as: (i) uptake at a single carrier site modulated by phase changes in the cell membrane; (ii) the successive involvement of two uptake sites, either both in the plasmalemma, or one at the plasmalemma and one at the tonoplast; (iii) the existence of an active, energy-dependent accumulating mechanism, and a more passive, inward leak (diffusion) operating at abnormally high ionic concentrations [23,26, 126,245,300,304]. When the uptake of ions by the *whole* plant is considered other active steps are involved.

The rate of uptake of an ion is frequently altered by other ions. Thus potassium ions limit the uptake of sodium ions, and their addition to a root culture medium may even cause a transient efflux of sodium ions against a concentration gradient [198]. A sodium extrusion mechanism has been detected in roots [290]. Various ionic uptake patterns show interactions between H^+, Na^+, K^+ and NH_4^+ for example. These interactions may be altered by other ions, such as Ca^{2+} [340]. Calcium ions maintain root integrity in water cultures. The addition of increasing quantities of calcium ions to a culture medium at first enhances and later depresses the uptake of potassium ions and alters the uptake pattern of many ions, and the responses to pH alterations [307,421]. Phosphate may be taken up as $H_2PO_4^-$ or HPO_4^{2-}, hydroxyl ions may compete with them [157]. Phosphate differs from some other ions in its alternative levels of ionization and in its being rapidly metabolized within the root [192]. Sometimes low temperatures seem to check growth in the field by limiting phosphate uptake, a situation that may be remedied by giving phosphate fertilizer. The route(s) by which ions penetrate the stele and are carried about the plant are in doubt. The Casparian band in the older parts of the roots seems to seal the cell walls of the endodermis and so prevent passage of at least the large molecules through the cell wall free space and into the stele [352]. However, the Casparian band does not extend to the young root tips. Once in the stele, in the cytoplasm of the symplast, ions may be pumped by the cells of the xylem parenchyma into the liquid in the xylem vessels (Chapter 1) [232]. Abscisic acid

(10^{-5} M) may enhance or reduce ion transport from the roots, depending on other conditions [323,324].

6.10. Coleoptile growth and gravity perception

Data on the effects of exogenous hormones, light and gravity on barley coleoptiles are comparatively meagre in contrast to information on oat and wheat coleoptiles. When seedlings are grown in a 'weightless' or gravity-compensated environment, obtained by simultaneous rotation around two axes, responses to light are altered, coleoptiles are aligned along the longitudinal axes of grains, and the roots grow apart into the moist atmosphere, apparently repelled by the other roots [184]. Probably gravity is normally perceived by a central group of starch-containing cells in the root cap which have a characteristic form of endoplasmic reticulum, and may use the starch grains as statoliths [205]. Strong magnetic fields alter the growth pattern of barley seedlings [281]. Abscisic acid reduces the growth of barley coleoptiles [48]. Sections of barley coleoptiles respond to auxin by extending. At the same time cellulase and glucanase activities increase – perhaps allowing turgor-powered extension by weakening the walls [401]. Coleoptile growth varies between grains of one variety. It is independent of cell number since, as in other cereals, coleoptiles of γ-irradiated seedlings elongate, although cell multiplication is blocked. Applications of kinetin increase the proportion of long coleoptiles found in a seedlot [250].

6.11. Leaf unrolling and greening

The first leaf is rolled when it emerges from the coleoptile. Subsequent unrolling is enhanced by light and is reversibly controlled by red (stimulates) and far-red light (inhibits). The red–far red response occurs in albino and etiolated as well as green leaves suggesting control by phytochrome. Applications of gibberellic acid or kinetin stimulate unrolling, while abscisic acid inhibits it. As in wheat, red light increases the 'extractable-gibberellin' content of leaves, and enhances the rate of metabolism of exogenous gibberellin. Leaves of albino mutants respond similarly, so photosynthesis is not involved [208,254,269,330,344].

As coleoptiles reach the light the leaf within normally becomes green. For convenience, to study greening and the onset of photosynthesis, seedlings are normally grown in the dark, and the changes that occur when the colourless (etiolated) leaves are exposed to light are followed. In response to light the etioplasts undergo characteristic morphological and biochemical changes as they become photosynthetically functional chloroplasts (Chapters 2 and 4). In darkness or in light δ-amino laevulinate is incorporated into protochlorophyll(ide); in light this is changed into chlorophyll(ide). Spectroscopic studies indicate that a range of molecular forms is involved. In light succinyl CoA synthetase activity rises – possibly allowing a faster synthesis of δ-amino laevulinate [386], (but see Chapter 4). Fatty acids, particularly linoleate (18 : 3), increase in amount [295]. Chlorophylls *a* and *b*, various carotenoids and vitamin K are synthesized in parallel with the development of the thylakoids, followed later by tocopherol, tocoquinone and plastoquinone; abscisic acid can partly block these changes [247]. Photophosphorylation and photoreduction of $NADP^+$ build up, and photosystems I and II develop, but at different rates. Electron transferring factors including plastocyanins, cytochromes *f*, *b*-559 (2-forms), *b*-563 and pigments P546 and P700 also appear [99,168, 320,325]. The light quality and intensity and the ambient temperature alter the rate of greening; different steps in the formation of chlorophyll have different action spectra [276]. Enzymes of the Calvin carbon-fixation cycle increase in greening leaves [303]. Ribulose–1,5-diphosphate (RuDP) carboxylase is synthesized *de novo* in light and is degraded in darkness [318]. The levels of enzyme activity attained and the carbon dioxide compensation point observed depends on many cultural factors [105]. The subunits of RuDP carboxylase are synthesized an appreciable time before they are assembled as an active entity [377]. Dark carboxylation reactions take place in etiolated leaves and in greening leaves in the dark [41]. As leaves green so photosynthetic carbon dioxide fixation becomes dominant in the light [398]. The rates of development of the different parts of the photosynthetic mechanisms are not exactly 'in step' so that, for example, chlorophyll accumulation does not exactly parallel increasing oxygen evolving power [375]. Nitrate reaching the leaf is reduced and incorporated into proteins. Glutamic dehydrogenase activity alters during greening [60]. Chlorocholine chloride (CCC, 1 mM) prevents greening; the effect is not reversed by gibberellic acid [40]. Isolated barley roots can be induced to 'green' in light [123].

6.12. Leaf senescence

When mature barley leaves senesce they gradually lose their chlorophyll and photosynthetic ability, and export some of their dry matter to the younger parts of the plant. Studies on senescing leaves provided early evidence for protein-amino acid turnover; extra supplies of nitrogen to the plants checked the decline of amino acids – and to a lesser extent of proteins [423]. Mineral elements are retranslocated, young leaves accumulating potassium and phosphate exported by older leaves [141]. Interactions between the mineral levels within the plant are probably common. For example calcium movement is dependent on the copper and phosphate status of the plant [59]. Detached leaves are often studied on the doubtful premise that the changes that occur are comparable to those that happen in old, senescing leaves attached to the plant. In detached leaves respiration (carbon dioxide output) first rises and later declines. Initially the respiration is at the expense of carbohydrates, glucose, fructose, sucrose, fructosans and sometimes starch. With time, progressively more carbon dioxide comes from other sources and the respiratory quotient, R.Q., falls from 1.0 to about 0.8, then rises again as the leaf turns brown. As the R.Q. reaches 0.8, protein is being broken down and amino acids are being respired. The ammonia released may kill the leaf [448]. Glutamine and especially asparagine levels change markedly during leaf starvation [267,449].

Materials from sunflower root exudates delay the senescence of detached barley leaves, as does the synthetic cytokinin kinetin [211]. This, and subsequent work, suggests that natural cytokinins regulate senility. Natural cytokinins, possibly zeatin and zeatin riboside, have been detected in barley [269,308,385]. However, detached oat leaves respond to the synthetic compounds kinetin and benzyladenine but *not* the naturally occurring zeatin or isopentenyl adenine [420]. In contrast chlorophyll retention in detached barley leaves is enhanced by zeatin, zeatin riboside and possibly zeatin mononucleotide [101]. Wheat leaf senescence is also delayed by nickel and cobalt salts, or yeast extract. Benzimidazole checks the decline of photosynthetic capacity in pieces of barley leaves [291]. Most studies have been made with the synthetic substance kinetin so their significance is uncertain. In detached leaves kinetin delays the fall in levels of chlorophyll, proteins (including ribulose diphosphate carboxylase), nucleic acids, ribosomes and polysomes, while the rises in levels in deoxyribonuclease and ribonuclease (including those associated with the chromatin) and peptidase are delayed [21,127,183]. Kinetin also enhances the turnover of nucleic acids and protein in detached leaves [21]. Cycloheximide prevents the senescent rise in proteolytic activity and fall in chlorophyll and ribulose diphosphate carboxylase. Kinetin retards the decline of other enzymes and enhances others. In the presence of light reducing sugars accumulate in detached leaves; this accumulation is enhanced by kinetin [84]. Removal of the second seedling leaf delays senescence in the 1st leaf, but kinetin is ineffective [384]. An application of kinetin to a cereal leaf creates a metabolic 'sink' inducing a local retardation of senescence and an accumulation of metabolites, possibly analagous to the 'sink' created by a focus of fungal infection in a leaf. Some of the enzymic changes noted in detached leaves are similar to those occurring in leaves attacked by pathogens [106]. Other unexplained effects, such as leaf bleaching by light in the present of EDTA (ethylene diamine tetra-acetate), coumarin or other agents that seem to retard pigment loss in the dark, also occur.

6.13. Growth regulation

Plants grow in a regular fashion and growth is altered by changing conditions. Regularity in dimensions is seen in the mature culms of some of the Gramineae in that mean log internode length is linearly related to log internode number [333]. There are numerous suggestions that tillering is regulated by hormone levels. Damage to the main shoot apex by pests may induce extensive tillering reminiscent of the effects of removing the dominant apex of many dicotyledons (Chapter 9). The striking effects on subsequent growth of removing root tips have been noted. However, studies of endogenous hormone levels are rare, and the exogenous application of a hormone often alters levels of other endogenous hormones; e.g. auxin applications are reported to lead to enhanced ethylene production and gibberellin applications to increased auxin levels. Single sprays of gibberellic acid accelerate the growth of barley plants, but eventually untreated plants grow to the same final size; repeated doses of gibberellin can lead to larger plants [370,392]. The chlorophyll content of treated plants is reduced, causing a pale appearance and plant enzyme levels change [396]. Barley ears may be distorted and be partly sterilized by applications of gibberellin (Chapter 2). Synthetic growth regulators, used as weedkillers, and many experimental alterations in growth conditions, can also cause ear distortion or sterility (Chapters 2, 9 and 12). Auxin may play a role in regulating barley growth since semi-brachytic barley *(uz, uz)* apparently lacks an enzyme for converting tryptophan to tryptamine, but responds to added tryptamine as to auxin [229].

Allelopathic substances may play a part in interplant competition. Alkaloids from barley roots may check the growth of weeds [311], and possibly ethylene produced by barley will do the same. Ethylene increases top growth and reduces root growth in barley, making the roots short, broad and curled [77]. Barley residues decomposing in the soil give rise to phytotoxins, including aromatic organic acids, which can reduce the establishment of a subsequent crop [411].

Extensive removal of roots, a supposed source of cytokinins, from barley grown in water culture, did not produce a marked disorganization of growth. However, as root regeneration was rapid, this may have masked any effects [186]. Undoubtedly a major control system of growth is exerted by the endogenous levels of nutrients, photosynthate, water, and minerals. Mineral levels in the plants can be altered by changing the supplies to the roots; carbon dioxide levels and light intensity can be altered to change the available supplies of photosynthate. The effects on tillering of growing plants from different sized seeds, of sowing at different densities, of applications of fertilizers, or irrigation water, and of shading or supplementary light support this view (Chapter 7) [96,119,140]. Competition between barley and neighbouring plants

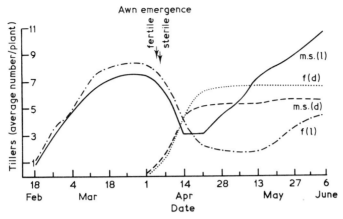

Fig. 6.17 The number of living (l) and dead (d) tillers carried by fully fertile (f) and male-sterile (m.s.) plants of Atlas 57 barley, grown in a greenhouse (after Laude et al. [241]: by permission of the American Society of Agronomy).

may be for nutrients, water or light (Chapter 12) [15,55,216,244]. Tillers of one plant partly resemble closely competing, separate plants. Separated tillers can survive and grow [400]. In general tiller-formation is little altered by increasing plant density, presumably because it precedes the onset of competition, but the development and survival of these tillers is altered, as is floral initiation and development [220,221,243]. Tiller senescence during heading cannot be entirely due to a drain of nutrients from the plant into the swelling grain since it occurs in infertile, male-sterile plants. Nevertheless the extensive growth of new tillers that occurs in male-sterile plants about two weeks after heading is partially prevented by the presence of grains in fertile plants (Fig. 6.17) [241]. Decapitating normal plants causes a marked increase in tillering [17]. Senescence may be accelerated by the removal of leaves from the main shoots, and retarded by supplying nitrogenous fertilizer, reducing main shoot growth with maleic hydrazide, or by removing old shoots [405,406]. Removing young leaves from uniculm barley shoots causes photosynthate from the older leaves to be diverted from the roots to support the regrowth of young leaves [360]. Lack of inter-tiller competition for assimilates may explain the greater leaf and internode growth and eventual size in a uniculm, non-tillering mutant compared to the parent form [219]. If nutrients are continually renewed, tillering is continuous [16]. A lack of minerals is often the predominant factor in limiting tillering; insufficient light can also have an effect [17,69]. The growth correlations observed in the developing spike may well be controlled, in part, by endogenous gibberellins as well as by nutrient supplies. [92,298]. Feeding plants different nutrient mixtures alters their growth rates and relative dimensions (e.g. the top/root ratio of the plants) [283,415,416].

264 Barley

Large grains, which have greater reserves of nutrients, give rise to more robust seedlings [210]. The initial conditions for seedling growth have notable effects on subsequent plant development. Thus, withholding nitrogen (NO_3^-) from seedlings beyond day 4 results in a smaller, less photosynthetically active first leaf [81]. Increasing supplies of nitrogen can increase the levels of some photosynthetic enzymes present in leaves [105]. Shading the first leaf reduces photosynthesis, delays the development of subsequent leaves and reduces root growth; tiller appearance is delayed and yield is reduced and reduces the uptake and assimilation of nitrate [109,113]. Photosynthate from the expanded first leaf travels throughout the seedling, including the expanding second leaf [108]. Thus ephemeral checks to seedling growth can have adverse long-term effects on plant development.

6.14. Temperature and growth

Late sowing or high temperatures readily induce abnormalities in the ear of uniculm barley, similar to those induced by 2,4-dichlorophenoxy acetic acid (Chapters 2 and 9) [219]. The mechanisms by which these dissimilar treatments produce their effects are unknown. Grain yield can be reduced or prevented by growing plants in warm greenhouses [189]. This effect is variously ascribed to 'physiological imbalance', effects on nutrient uptake, the induction of sterility (particularly male sterility), or the failure to meet the cold requirements of a vernalizable, winter variety (see below). Floret numbers are reduced in barley grown at a high temperature, 24°C (75.2°F) compared to 18°C (64.4°F). Growth is improved by alternating day and night temperatures and the selection of the best photoperiods (duration of light and dark periods) [410]. Presumably high night temperatures enhance 'wasteful' respiration in the absence of compensating photosynthesis, but grain yield is variously increased or decreased by warm nights [137,143]. Phytotron or growth cabinet experiments can be used to vary light intensity and temperature separately with clear-cut results; for a given photoperiod increasing temperatures decrease grain weight [87]. In the field higher temperatures are associated with greater radiation intensities and potential photosynthesis. In England positive correlations occur between temperature and grain yield in field-grown plants [435]. The tillering patterns of different barley varieties alter in different ways when the day/night temperature regimes are altered [69]. Comparative trials with different varieties show that maximal photosynthetic rates differ (7.8–14.1 mg CO_2/h/g) and that the maximal rates occur at different temperatures (6°–25°C; 42.8–77.0°F); photosynthesis varies with temperature in different ways in different cultivars [310]. The temperature responses of different varieties were related to the climate of the place of origin. It is suggested that for vegetative growth (in vernalized plants in winter forms) about 15°C (59°F) is best, while at heading the optimum is about 17–18°C (63°F). Plants grown at elevated temperatures and then planted into the field soon show signs of injury, while plants initially grown cool grow vigorously [427]. The damaging effects of high temperatures (enhanced shoot/root ratios; reduction in dry matter; weak, floppy growth) may be partly overcome by increasing the supplies of light and carbon dioxide [14]. Barley roots grow better at 20°C (68°F) than at 10°C (50°F) or 30°C (86°F) [342]. Soil temperatures of about 15°C (59°F) are optimal for barley growth; the range of temperatures allowing maximal growth is increased by supplying more phosphate [331]. At excessive soil temperatures, (27°C; 80.6°F) plants may not respond to applied nutrients [263]. One would expect varietal differences in responses towards high temperatures, e.g. between plants from India and the U.K.; such differences exist [397].

6.15. Cold hardiness

Proline accumulates in heat stressed plants only when this coincides with a water stress [72]. Proline also accumulates in chilled plants, and the levels of other amino acids alter [72,89]. Plants grown under cool conditions are 'hardened', they are less prone to damage by freezing temperatures. Hardening is important in preparing autumn sown cereals for winter survival. The physiological and chemical alterations observed in cold grown plants may contribute to vernalization or cold hardiness or both or neither; proof of the relevance of the changes is lacking. For hardening cool-grown plants must be in light [212]. Mitochondria from cool-grown barley have enhanced respiration rates, while protein and nucleic acid patterns are altered [212,213,404]. The moisture content of seedlings declines during cold-hardening and the content of solids in the sap, including sugars, increases. A period of warm growth rapidly reverses these changes and increases susceptibility to cold damage (Table 6.15) [240].

Glucose and fructose concentrations rise during cold treatment, and the carbon/nitrogen ratio may double [382]. Changes in the levels of glucose, fructose and sucrose follow different time courses from the autumn to the spring; peak values for a winter variety were (as % dry weight), fructose, 16; glucose, 2; sucrose, 5–8. Drier soils probably favour winter survival [451]. Coleoptiles from cold-treated grains emit less carbon dioxide than controls and they have a reduced sugar content [287]. Applications of gibberellins *reduce* frost resistance [286]. The more concentrated the sap is, the lower its freezing point, so the accumulation of sugars may be protective. Polysaccharides that modify the course of ice formation have been isolated from cereals. The critical freezing

Table 6.15 The water content, volume of expressed sap, and sap refractive index of hardened Tennessee barley during subsequent warm growth (a), and susceptibility to freezing injury (b) (from Laude [240]).

(a)	Time in greenhouse (days)					
	0, Hardened	1	2	3	5	7
Water (%)	77.7	80.1	82.6	84.3	84.4	86.0
Sap expressed (%)	23.2	24.1	25.0	29.5	31.8	36.8
Refractive index	1.3475	1.3421	1.3415	1.3404	1.3366	1.3372

(b)	Time in greenhouse (days)				
	0, Hardened	0.5	1.0	1.5	2.0
Freezing injury (%)	8	14	61	71	85

stresses occur in the barley crown, and vascular transition zones where ice crystals may damage the protoplasts [305]. The moisture content of the crown is closely related to susceptibility to freezing injury [271]. In hardy plants ice formation may be delayed and, when it occurs, it tends to occur in the free liquid, withdrawing liquid from the protoplasts. Different freezing patterns have been distinguished.

Plant hardiness varies with the time of planting [451]. Barley crowns grow a different depths, depending on the variety and the growing temperatures (e.g. 3.2–49.1 mm (0.13–1.93 in) deep at 10°C (50°F) 20.6–65 mm (0.81–2.56 in) for the same varieties at 27°C (80.6°F)) [206, 365]. Deeper crowns seem to favour survival through spring frosts, and the soil provides protection from the wind. Barleys vary widely in their response to hardening conditions and their subsequent resistance to winter killing [82,439]. Summer frosts can result in pollen sterility, sterile heads and damaged, non-viable grain (Chapter 7).

6.16. Vernalization

There are varieties of barley that will not head and form grain unless the germinating grain or young seedling has received an adequate cold treatment, which is usually called vernalization [121,258,437]. Requirement for vernalization is independent of the morphological form of the plant, or of daylength (photoperiodic) requirements and is apparently not linked to cold hardiness. Without a vernalization treatment, a cold-requiring 'winter' plant will tiller freely, but will not head. Within limits, more chilling will cause it to head sooner. The recognition of vernalization was originally exploited by wetting seed grain, allowing it to begin to germinate, then exposing it to cold to satisfy the chilling requirement.

The growth of the barley plant 267

Subsequently it was sown in the spring when the danger of winter-killing was past. Warming may easily reverse this effect [230,258]. In England, where winter-killing is usually not a problem, the cold requirements of barleys may be met by autumn or early spring sowing [37]. Most studies of cereal vernalization have been made with winter rye, but winter barleys are apparently similar [337]. The damp embryo (which may be isolated from the grain) or seedling must be cooled for an adequate period, in air, and not be warmed until the risk of devernalization is past. The required duration of chilling depends on the variety and the temperature used, e.g. 2–10 weeks may be suitable. In some trials 9°C (48.2°F) was superior to 4°C (39.2°F) or 12°C (53.6°F) (Fig. 6.18) [412]. Three weeks at 22°C (71.6°F) can devernalize barley seedlings [190]. Cold treated barley plants develop a different morphology

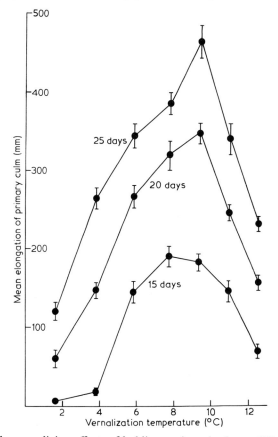

Fig. 6.18 The vernalizing effects of holding a winter barley at different temperatures, for the periods shown. The vernalizing effect is indicated by the mean elongation of the main stems (after Trione and Metzger [412]: by permission of the American Society of Agronomy).

268   Barley

to those without but these changes are *not* a reliable guide to flower induction, which must be checked by inspection of the apex of the stem [156,337,382]. The mechanism of vernalization is unknown. Gibberellin applications can accelerate ear formation and shoot growth, but cannot replace vernalization [196,225,337]. The effects of added auxins are distinct from vernalization and abscisic acid is not involved [19,188]. Vernalization is clearly distinct from (a) cold hardiness, or (b) the breaking of dormancy by stratification (Chapter 5).

6.17. Some effects of light

Light has numerous effects on plant development. Since barley grows from the equator to 60°N, in Europe and Canada, it has forms adapted to a range of light intensities, photoperiods and perhaps spectral compositions [217]. Some effects of altering light intensity may be due to varying quantities of photosynthate and/or altering the plant's temperature, but other effects are consequent on less obvious mechanism(s). When plants are grown at high temperatures (25°C; 77°F) with a large supply of nitrogen, short days and at low light intensities, weak stems and prostrate growth result [427]. With supplementary light, and increased ambient concentrations of carbon dioxide, plants grow more sturdily, their carbohydrate concentrations are increased, their nitrogen contents decline, and grain yields are enhanced by for example, five times [14]. Such results are reasonably explained as adjustments to the altered balance of nitrogen and carbohydrate supplies. Competition for light between plants fully supplied with nutrients in water cultures reduces the number of ears/plant, causes irregularity in tiller production, reduces dry matter production, enhances the shoot/root ratio and decreases nitrogen uptake by the roots [55].

Initial studies on the induction of flowering were hindered by confusion between the effects of vernalization (chilling), nutrient supply, and light. Winter and spring barleys can respond differently to supplementary light. Extra light (longer photoperiods) reduced the time from germination to heading of a spring barley (from 46 to 36 days) and its height (from 117 cm to 102 cm; 46 to 40 in) but it had no effect on the winter barley tested [120]. In the absence of nitrogen barleys tiller poorly or not at all and produce small plants, but even on short days they will produce an ear. When adequate amounts of nitrogen are supplied robust plants result, but these will only ear in long days [83,265]. Incidentally, daylength changes alter the response of barley to boron [428]. Barleys are often regarded as needing long days to hasten spikelet differentiation and anthesis. Short days extend the vegetative period; however there are wide differences between varieties, and even between a mutant and its parental form [100,148]. Other factors controlling responses to changing light are the ambient temperature, the spectral composition of the light, the duration of the light exposure and its intensity. Changes in the combinations of growth conditions can alter the relative duration of developmental stages [309]. In some trials at 13°C (55.4°F) the first node reached ground level (NGL) as the stamen initials were formed, while at 24°C (75.2°F) NGL coincided with progressively earlier stages in floral organogenesis as the photoperiods were shortened. Some varieties appear to be 'daylength neutral'. The stage of development at which photoperiods are varied also has an effect on development [4,18,201,410]. Photoperiods and temperatures interact in controlling development. At fixed long photoperiods of 17–24h increasing temperatures hasten the development of the apex [201,397]. Day-neutral varieties are, in contrast, quite resistant to alterations in daylength and temperature [397]. As in other types of studies floral initiation can only be reliably determined by inspecting the apical primordia, and not by leaf number or other changes in 'external' morphology. While day-neutral varieties produce only 6–8 leaves before earing, varieties that are highly responsive to short days may produce up to 20 leaves [217]. In general, in spring and vernalized winter forms, lengthening photoperiods hasten flowering, but reduce the numbers of tillers and leaves that are formed, enhance the rate of leaf emergence, and change the pattern of leaf size on the stem. Wintex 'semi-winter' barley produced no grains, but the heaviest plants, in 12h photoperiods; grain yields were best in 16h rather than 24h photoperiods [50]. Night interruption, with light too weak to contribute significantly to photosynthesis, is effective in controlling floral initiation, particularly when applied for a 2h period, 6.5h after the onset of darkness. An astonishing finding, frequently confirmed, is that the least effective wavelength (about 480nm) and most effective wavelength (about 620nm) for inducing flowering in long day plants coincides with those that prevent floral initiation in short-day plants. These and other results implicate phytochrome as the light detector in the photoperiodic control of floral initiation [51]. However phytochrome is unlikely to be acting merely as a no flowering/flowering switch, since there is a far-red (730nm) light requirement and a red/(red + far-red) light intensity ratio of 0.3 is most effective for night breaks [233,312]. Following brief illumination with red light there is the expected change in the molecular form of the pigment (P_{FR} to P_R) and a *decrease* in the total quantity of pigment detectable [182].

These considerations, and the low intensities of artificial light generally achieved, make it difficult to choose supplementary greenhouse lighting for barley. Often combinations of high intensity fluorescent lamps and low intensity incandescent lamps are used. Floodlighting field-grown barley has a range of effects on plant morphology. Laser light of the correct composition is able to accelerate apical development [313]. Growing barley in unusual photoperiods, (e.g. transferring to short days for a period after growth under long days) or with light of unusual spectral composition can induce male sterility, abnormal spike morphology or even ear death [34,86,394].

Phytochrome has been found in etiolated seedlings, and its distribution and levels have been measured, together with the changes undergone with age and when exposed to red or far-red light [182,334]. Phytochrome occurs in higher concentrations in the leaf bases, at the apex of the coleoptile, and at the stem apex. The leaves are generally regarded as the major photoreceptors.

Light from one side causes barley coleoptiles to grow unequally and bend towards it. The suggested asymmetric photolysis of auxin is doubted, at least as a complete explanation. An alternative or additional explanation is that photolysis of the xanthophylls gives rise to the neutral inhibitor xanthoxin which in turn may be converted to abscisic acid. This would check coleoptile growth, mainly on the lightward side, and cause the observed curvature.

Light affects barley in other ways. Barley root tips adhere to glass when exposed to red light for 30 sec when under water containing indole acetic acid and some other substances. The tips become detached when exposed to far-red light for 30 sec – a striking, rapid, and reversible 'phytochrome' effect [399]. Abscisic acid rapidly reverses this effect, but gibberellins, cytokinins and ethylene do not. Roots of mung beans behave differently. Red light stimulates the respiration of barley leaves, and increases gibberellin levels [246]. In the long term exposure to red, and especially blue light during greening depresses catalase levels [11]. Light modulates nitrate reductase activity and can alter the levels of niacin, riboflavin, pantothenic acid and biotin in leaves [204,236]. Red light induces rises in phenylalanine ammonia lyase and the flavonoid saponarin in etiolated shoots; acetylcholine depresses the response [367]. Treating plants with some synthetic dyestuffs (e.g. polyhalogenated fluorescein) photosensitises the roots [278].

6.18. Some factors that control yield

Records of experiments designed to test the effects of changing agricultural practices on yield go back several centuries. Many efforts are made to eliminate the effects of adverse factors, such as diseases, and to supply inorganic nutrients. Efforts are also being made to understand characters which control dry-matter production – such as the efficiency at which incoming light is utilized.

Early studies were confined to the 'components of yield', i.e. the mean grain weight (thousand corn weight), the number of grains per ear, the number of ears per plant and the number of plants per unit soil area [235]. Very great changes in these values occur between varieties, and in one

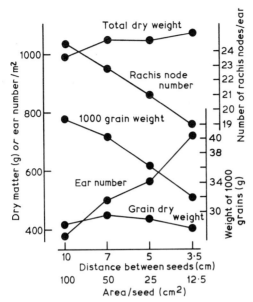

Fig. 6.19 The effects on the component parts of barley plants, both total dry matter and grain, of growing at different spacings (after Kirby [214]): by permission of the Cambridge University Press).

variety under different cultural conditions, with relatively small alterations in grain yield per unit of area cultivated [49,332]. This 'growth compensation' occurs when barley is sown at different densities (Figs. 6.19, 6.20) [220,221,243,442]. In one trial, in which plants were grown at densities of 100 and 400 per square metre, some values were respectively – yield of grain 880 and 830 g/m^2; shoot yield 405 g/m^2 in each case; numbers of ears 470 and 610/m^2; numbers of grains 23.5 and 17.9/ear; mean grain weights 42.5 and 38.5 mg [215]. The effect on yield of large numbers of ears under one set of conditions is partly 'compensated' by each ear having fewer, smaller grains – an inverse relationship that complicates attempts to select for higher yields (Chapter 12). Thus yield is defined by the size of its components, and their product, but the components are not physiologically independent. The existence of 'compensation' shows that one or more factors become limiting during the life of the plant – either limiting the number of fertile (ear-bearing) tillers, the number of florets formed, their fertility or survival, or grain filling. Likely limiting factors are the supplies of water, soil nutrients and photosynthate. The possibility exists of selecting plant forms in which the resources are distributed more advantageously within plants (Chapter 12). Strains that produce the fewest infertile tillers should be more efficient than others; semi-dwarf varieties having a greater grain/straw ratio should be potentially higher yielding than tall varieties because a higher proportion

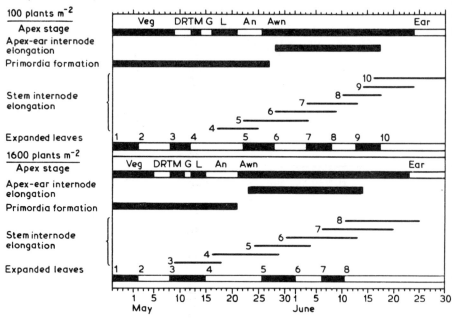

Fig. 6.20 A chart summarizing the temporal relationships between the developing parts of barley plants growing at two densities (from Kirby and Faris [220]). The lowest, chequered line in each set of data represents the time of full expansion of the numbered leaf. Veg: apex in vegetative state; DR: double ridges formed on apex; TM: triple mound; G: glumes detectable;. L: lemma forming; An: anther forming.

of the available resources should be available for use in forming grain; semi-dwarf plants are also less prone to lodging. Theoretically, but possibly not to a significant extent in practice, uniculm (single stemmed) varieties planted at optimal distances, with no gaps, should give the best yields.

When relating yield to photosynthesis it is the performance of the whole crop, rather than individual plants, that is important. If a single plant is illuminated with increasing intensity its net assimilation rate (NAR, photosynthetic gain *minus* respiratory loss) rises to a plateau dictated by the available carbon dioxide, a level that is reached much below the intensity of full sunlight. In the field the NAR of the crop only behaves in this way if the plants are growing far apart. Where they have grown to form a closed canopy, and mutual shading occurs then, within limits, the NAR rises with increasing light intensity as the light penetrates deeper into the canopy and causes the lower, shaded parts to photosynthesize faster. Before the canopy is 'closed' sunlight misses the plants, strikes the soil and is 'wasted'. For maximum rates of photosynthesis the leaf density must be sufficient to intercept all the useful light but not

The growth of the barley plant 273

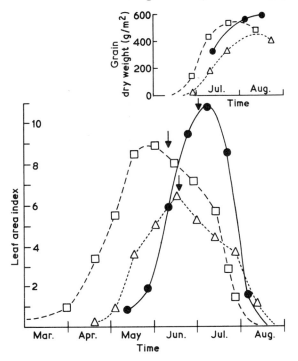

Fig. 6.21 The alterations with time of the leaf area index and the weights of the useful parts of the crops indicated (after Watson [430]): ●——●: Spring barley; △---△: Spring wheat; □---□: Winter wheat. Time of 50% ear emergence, ↓. The crops were grown in eastern England. The spring wheat was sown in early March, the spring barley in late March.

so great that there is excessive shading, so that the lower leaves respire more than they photosynthesize. For maximum dry matter yield this leaf area (often expressed as the leaf area index, LAI, the ratio of the leaf area to the soil area) must last for as long as possible in the growing season [429]. With spring barley grown in Britain maximal LAI lasts only a relatively short time for a period in June – July when the net assimilation rate is maximal (Fig. 6.21). Optimal LAI is about 9, giving about 95% interception of the incoming radiant energy, but varies in different trials [409]. These facts apply to total dry matter production, roots and shoots + grain; but it is the grain yield that is of prime interest. Maximum grain numbers are determined before anthesis when fertile tiller numbers are being decided and when the spikelet initials are being formed and the number of spikelets per ear is being determined. Grain filling is mainly dependent on photosynthesis subsequent to anthesis. The idea that grain yield is *only* dependent on photosynthesis after ear emergence is clearly wrong [409,442]. Of the solar radiation reaching the earth's surface about half that which occurs in the visible region (400–700nm) is useful for photosynthesis. Some of the radiation is direct sunlight arriving at an angle and intensity that depends on the latitude, the weather and time of year. Some is diffuse light, amounting to 50–60% of the total in England [409]. Of this visible light only 2–3% is retained trapped in the plant substance *under optimal, practical* conditions during the growing season. This low value is partly due to transmission and reflection of the light; light exceeding the saturation values of the leaves and being wasted and partly due to respiration by the plant causing a loss of photosynthate. Respiration goes on in all parts of the plants, including the roots and may 'lose' about 40% of the photosynthate. The upper limit of light use has been computed as 12% of the incident, visible radiation [117,207,253,280,446]. Barley has achieved comparable values for short periods [207]. Shading and other experiments indicate that in N. Europe sunlight is a major factor limiting the growth of the barley crops. Shading also alters plant morphology [207,242]. Increasing light intensity at first increases then reduces plant height; increases tillering and early tiller-death; initially increases the number of living leaves and later the number of dead ones; decreases leaf length and breadth, but increases leaf thickness; increases the root/shoot ratio; increases the thousand grain weight; induces the formation of more spikes/plant and earlier ear emergence, greater ear length, and more grains per ear and per plant [207,442].

Within limits, increasing the carbon dioxide concentration of the air (from 0.03% to 0.5–1%) increases barley growth, although supplying carbonates to the roots does not [115,282,390]. The atmospheric carbon dioxide concentration and the radiation intensity cannot be regulated in the field, so the problem is to design and breed plants that utilize these resources as well as possible. This involves producing plants with a high net photosynthetic capacity, that intercept and utilize the sunlight in as near an optimal way through the greatest possible part of the year. Thus autumn-sown cereals should, in the absence of winter-killing, have a higher yield potential than those sown in early spring, which in turn should be superior to those sown in late spring (Chapter 7) [176]. Sowing grain thickly quickly establishes a canopy that intercepts most of the incoming radiation and often increased the yield of total dry-matter. However this does not necessarily give the best or most economic grain yield.

The carbon dioxide content of field air is about 0.03%, but it may fall a little in the day, and be higher at night in the crop canopy, as photosynthesis and respiration proceed. Turbulent mixing and even slight air movements maintain the pCO$_2$ within crop canopies [280]. The rate of photosynthesis is also changed by the numerous factors which cause the stomata to open and close [9]. As light penetrates leaf canopies its intensity falls according 'on average' to Beer's law. However, the leaf area index at which the best net assimilation rate (photosynthesis *minus* respiration) occurs varies with the leaf disposition and angle, being about 3–4 for broad, horizontally disposed leaves, and 8–9 for small, nearly vertical leaves spaced widely on stems. Plants arranged in north to south oriented drill rows should have better light interception than those in rows running east to west. The comparison of the results of field studies and with physical models show that the interception of light by crops is very complex [253,373]. Narrow, upright leaves are characteristic of high-yielding barley varieties in Ontario unless weed competition is strong when broad, floppy leaves are advantageous in smothering the weeds [402]. Experiments with barley seedlings planted at different densities and illuminated from different angles confirm that as leaves become more nearly vertical higher leaf area indices (LAI) are needed to achieve 95% light interception (e.g. 18°, LAI 4.5; 53°, LAI, 7; 90°, vertical, LAI, 11). At higher light intensities greater values of LAI are needed, at any leaf angle, to intercept most of the light [316]. Thus more closely spaced tillers, potentially carrying grain-filled ears, should be tolerated by erect-leaved cultivars; within limits this has been found. Crowded conditions that favoured enhanced yield in an erect-leaved variety tended to depress yield in a lax-leaved form (Table 6.16) [7].

As cereal stems elongate, increasing the vertical separation of the leaf-blades but exposing more of the leaf sheaths (which are also capable of photosynthesis), more light penetrates the canopy, and light intensity at ground level increases. As well as finding a plant type that allows efficient light distribution in the canopy, it may be possible to find types that have, for example, more chloroplasts per cell, or which have the intracellular spaces arranged to reduce diffusion resistance to carbon dioxide, in the way that genotypes with differing stomatal frequencies are available. Accumulating carbohydrates check photosynthesis in barley leaves only slightly [293]. Carbohydrate can be transported about the

Table 6.16 The effects of doubling the seeding rate on the yields of an erect-leaved cultivar, Lenta, and lax-leaved variety, Research. At Deniliquin yields were limited by water-stress (from Angus *et al.* [7]).

Variety	Seeding rate	Grain yield (Total yield) at 10% water content (kg/ha).	
		Site 1. Mt. Derrimont	Site 2. Deniliquin
Lenta	low density	3860 (8440)	3460 (7730)
	high density	4460 (9950)	3480 (8550)
Research	low density	4690 (9860)	3550 (7940)
	high density	4150 (8800)	3450 (8320)

276 Barley

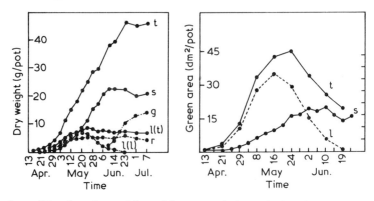

Fig. 6.22 The changing weights of the parts of plants during the growing season (a), and the area of the green photosynthetic parts (b), (after Strebeyko and Góra [393]). Ear length was increasing most rapidly about 7 June. Grains reached the doughy stage on 23 June. t = Total plant. s = Stems. l = Leaves. g = Grain. l(l) = Lining leaves. l(t) = All leaves.

plant, even between tillers, when the need arises [409]. Searches for barley with a 'superior' photosynthetic mechanism, i.e. with superior net assimilation rates, or with less or no photorespiration, and/or with the C-4 carboxylative pathway with its high affinity for carbon dioxide are being made, but as yet none have been found [284]. Sulphate supply and temperature alter the compensation point in barley [170]. Owing to the 'attenuating' effects of many other factors a 'C-4 carbon dioxide fixing barley' could be disappointing even if such a variant could be found [128]. The production of total dry matter is dependent on rates of net photosynthesis and its duration. These are regulated by leaf age, number, and survival and all the parameters, such as nitrogen supply, which alter these. This discussion has been in terms of LAI, the leaf area index; however in barley, with green leaf blades, leaf sheaths, stems, peduncles, ears, and awns, studies involving the whole photosynthetic area would be more appropriate (Fig. 6.22) [393]. When values for the expansion of leaves 5–9 inclusive, assimilation rate, translocation pattern, respiration and other factors were used to predict the growth of a uniculm barley growing under simplified conditions the results agreed with the observed growth well enough to show that probably most of the important factors were recognised, but that detailed understanding is inadequate [359]. The effects of differing canopy structure cannot only be due to alterations in photosynthesis. Thus awns, which transpire freely, reduce canopy temperature significantly [111].

It used to be believed, mistakenly, that barley grains were filled at the expense of preformed carbohydrates stored in the leaves, stem and ear. However the removal of the awns has long been known to give rise to grains of reduced size [453]. Awns photosynthesize as well as transpire, they contain much photosynthetic tissue, and greatly add to the area of the ear. Awns were estimated to contribute 1/6 of the photosynthesis of the whole plant [371]. In whole ears the weight of each grain is roughly related to the length of the subtending awn [159,372]. In the variety Herta awns contribute about 27.4 cm² photosynthetic area/plant; removing 1/3, 1/2 and all of each awn reduced grain yield by 4%, 9% and 19% respectively [389]. Awns carry 72–75% of the stomata of the spikelet [147]. Analysis of yields, photosynthetic and respiration rates of ears of genotypes with awnless, quarter, half and fully awned ears and ears with enlarged awns and glumes show that awn and glume weight is the best measure of the extra tissues [202,224]. However, under adverse conditions better yields may be obtained from half-awned rather than fully awned types, presumably because the development of full awns and their extra transpiration represents a further demand on the developing plant [368]. In *eburata* mutants, having green awns and white paleae and lemmae, awn removal has a more dramatic effect in reducing grain size, than has awn removal on the all-green parent strain. The awn also seems to control the nutation of the barley heads [336]. The glumes also contribute to photosynthesis [362]. Direct measurements indicate that awns perform 70–80% of the photosynthesis in all green ears. In *eburata* ears the value is 85% and this is replaced by net CO_2 output by awn removal [8]. Highest ear photosynthesis is found in long-awned varieties [10]. Awned barleys generally give more yield than hooded or awnless varieties under favourable conditions [159,338]. In some situations this seems to be because awns senesce and cease photosynthesizing after the leaves; this effect is particularly important under arid conditions [44,147]. Assimilate from each awn seems to be utilized mainly by the subtended grain, whereas photosynthate from the flag leaf is used generally throughout each ear [426]. Recent estimates were that awns accounted for 73% of the ear transpiration and 80–115% of the net CO_2 uptake. De-awned ears had a net CO_2 output [44]. However, because of compensatory effects, it cannot be assumed that because awns provide about 75% of the ear photosynthetic surface and awn removal causes a 12% reduction in grain weight that the ear only fixes 16% of its photosynthate requirement.

The progressive senescence of the leaves from the base of the stems upwards means that during heading, and subsequently, most photosynthesis is carried out by the upper parts of the plants. Lower leaves have little effect on grain yield from the time of anthesis onwards [361]. Much effort has been expended in trying to decide what contribution the different parts make to grain-filling [235,407,409,441,450]. Photosynthesis by the ears themselves makes a large contribution to the growing grain that is largely independent of, for example, the arrival of nitrogen. Analysis of control plants and plants in which various parts were shaded or removed showed that preformed sugars were not stored in the roots, and those

Table 6.17 The contributions to final grain weight (W) of ear photosynthesis (P_E), daytime (R_d) and night-time (R_n) ear respiration and carbohydrates from the shoot (S). While barley ears were awned the wheat ears were not; the wheat had large flag leaves and peduncles (from Thorne [408]: Crown copyright, by permission of H.M.S.O.).

	Barley	Wheat
P_E	79	24
R_d	−24	−28
R_n	−10	−11
$(P_E − R_d)$	(55)	(4)
$(P_E − R_d − R_n)$	(45)	(−15)
S	55	115
W	100	100

stored in the stems could not account for more than about 10% of the final grain weight. Sugar stored in the ear before emergence, equivalent to about 5% final grain weight, was roughly equivalent in amount to the sugars of the mature grain. Stem sugars (including fructosans) seemed mainly to support internode respiration. About 80% of the final grain weight could be accounted for by photosynthesis occurring after ear emergence. Estimates from Japan of the final contributions to grain weight by net photosynthesis in the different parts are: ears, 22 or 23%; leaf blades 28 or 25%; stems + leaf sheaths, 43 or 41%; reserve carbohydrates 7 or 11% [395]. An estimate of the photosynthetic contributions to barley and wheat grain yields in Britain are shown in Table 6.17. The experimental methods used most frequently in the past underestimated the contributions made by the ear itself [409]. Direct measurements, over 33 days, on barley growing in the field indicated that on average each ear, which eventually contained 1.09 g grain, fixed 622 mg of carbon dioxide in the 496 h of daylight and emitted 117 mg carbon dioxide during the 296 h of darkness, suggesting that net ear photosynthesis provided 34% of the grain dry weight or 52% of the starch. Of the whole ear 25% of the final dry weight was present at emergence [329]. Together photosynthesis in ears and flag leaves have been estimated to contribute 50–75% of the final grain yield. As expected, and as occurs in wheat, awnless ears contribute less to grain filling than awned or triple awned lines [67]. Competition between grains for the carbohydrate supply can sometimes be demonstrated, but only when the supply is limited, e.g. by stem shading [67], indicating that carbohydrate supply is usually in excess of requirements for grain filling, which is limited by 'sink capacity', the ability of the grain to utilize the carbohydrate available.

This result is supported by the failure of the presence of sterile florets to alter the size of adjacent grains [226]. The differences between the results of different investigators are to be expected. Direct comparisons reveal varietal differences [409,430,431]. Illumination intensities have been very different in different trials and many widely used experimental techniques are open to objection [68]. For example shading an organ reduces its photosynthesis, presumably also its photorespiration, its temperature and therefore its 'normal' respiration. In different climates the foliage will senesce at different rates and so make differing contributions to grain filling. The photosynthetic (and probably the respiratory) contributions of the plant parts alter with age; differences occur between tillers [43]. More sugars are transported from the flag leaf to the ear when the latter is shaded. Removing florets from the ear decreases photosynthesis by the flag leaf – evidently a reduced 'sink size' exerts a feedback control on photosynthesis. Ear photosynthesis seems to be a little dependent on other carbohydrate supplies [42,409]. One might expect that grains with the ability to grow large must have a large individual sink capacity. However, under British conditions the grain size of commercial varieties does not seem to correlate with yield, and it is not known if ears with a few large or numerous small grains have the higher yield potential [409]. Under Australian conditions, with high sunlight intensities, probably 'source' and 'sink' limitations are about equally responsible for limiting grain yields, and both are involved elsewhere [129,409].

The lamina and sheath both contribute to leaf photosynthesis and, within limits, grain yields correlate with leaf area or better with the area of the parts above the flag leaf node after ear emergence [409,447]. Two effects of nitrogenous fertilizers are to increase the leaf area and, by slowing senescence, extend its useful life (Fig. 6.2) [407]. In addition to photosynthesizing, transpiring and controlling nutation, awns may alter the sink capacity of subtended grains [179]. Awns intensify the accumulation of exogenous cytokinins in grains and glumes and larger grains mature more slowly and have larger cytokinin contents than smaller grains [274]. Possibly awns 'collect' cytokinins from the roots via the transpiration stream and direct them to the grains where they may delay senescence, prolong photosynthetic activity, and enhance the sink capacity. Under conditions of high relative humidity (low transpiration) awn removal had little effect on grain weight, while applying kinetin increased grain weight in one trial [273]. Supplying cytokinins to the plant can enhance shoot apex development, produce bigger ears and larger grain numbers [355], but applications of cytokinins have not proved valuable in the field.

Plant morphology relates to yield in a series of complex ways and the limitations imposed on yield by growth compensations, genetic limitations, farming practice, nutrient supplies, climate, disease effects and lodging cannot be ignored (Chapter 7). Still one may hazard that under favourable growing conditions best yields would be achieved by short, stiff-strawed plants with limited numbers of tillers, most of which carry ears. The ears would carry as many grains as possible, each with the ability to grow large when photosynthate was plentiful. To optimize the supply of photosynthate the spikelets might be triple awned, the awns being long, flattened and massive and would have long-awned and massive glumes. The flag leaf would be large, and thick to make best use of the full light to which it is exposed, and should have as long a photosynthetic life as possible. Leaves should be nearly erect. Sufficient photosynthetic capacity should be available in other parts to supply all the requirements needed to provide big ears *before* anthesis, and to supply the rest of the plant, both shoot and roots, during grain filling.

There is a critical need to create a range of locally well-adapted, *near-isogenic* barley lines of various morphologies and basic genotypes to enable physiologists to test the relative values of different plant forms under a variety of different climatic conditions, in the way that tests demonstrated that six-rowed cultivars do not necessarily outyield two-rowed cultivars [436]. More experimental results are needed to guide breeding programmes directed towards producing high-yielding plants of superior morphologies.

References

1 Aberg, E., Wiebe, G.A. and Dickson, A.D. (1945). *J. Am. Soc. Agron.*, **37**, 583–586.
2 Aase, J.K. (1971). *Agron. J.*, **63**, 425–428.
3 Abdul-Baki, A. and Baker, J.E. (1970). *Pl. Physiol., Lancaster*, **45**, 698–702.
4 Aitken, Y. (1966). *Aust. J. agric. Res.*, **17**, 1–15.
5 Akita, S. and Moss, D.N. (1973). *Pl. Physiol., Lancaster*, **52**, 601–603.
6 Andersen, S. (1955). *Physiologia Pl.*, **8**, 404–417.
7 Angus, J.F., Jones, R. and Wilson, J.H. (1972). *Aust. J. agric. Res.*, **23**, 945–957.
8 Apel, P. (1965). *Kulturpflanze*, **13**, 257–265.
9 Apel, P. (1966). *Ber. d. bot. Ges.*, **79**, 279–288.
10 Apel, P. (1966). *Kulturpflanze*, **14**, 163–169.
11 Appleman, D. and Pyfrom, H.T. (1955). *Pl. Physiol., Lancaster*, **30**, 543–549.
12 Archbold, H.K. (1938). *Ann. Bot.*, **2**, 403–435.
13 Arnon, D.I. (1939). *Soil Sci.*, **48**, 295–307.
14 Arthur, J.M., Guthrie, J.D. and Newall, J.M. (1930). *Am. J. Bot.*, **17**, 416–482.
15 Aspinall, D. (1960). *Ann. appl. Biol.*, **48**, 637–654.
16 Aspinall, D. (1961). *Aust. J. biol. Sci.*, **14**, 493–505.
17 Aspinall, D. (1963). *Aust. J. biol. Sci.*, **16**, 285–304.
18 Aspinall, D. (1969). *Aust. J. biol. Sci.*, **22**, 53–67.
19 Aspinall, D., Paleg, L.G. and Addicott, F.T. (1967). *Aust. J. biol. Sci.*, **20**, 869–882.

The growth of the barley plant 281

20 Aspinall, D., Singh, T.N. and Paleg, L.G. (1973). *Aust. J. biol. Sci.*, **26**, 319–327.
21 Atkin, R.K. and Srivastava, B.I.S. (1970). *Physiologia Pl.*, **23**, 304–315.
22 Banerjee, S. and Wienhues, F. (1965). *Z. PflZücht.*, **54**, 130–142.
23 Bange, G.G.J. (1972). *Acta bot. neerl.*, **21**, 145–148.
24 Banks, W., Greenwood, C.T. and Muir, D.D. (1973). *Stärke*, **25**, 153–157.
25 Barber, D.A. (1968). *Ann. Rev. Pl. Physiol.*, **19**, 71–88.
26 Barber, D.A. (1972). *New Phytol.*, **71**, 255–262.
27 Barber, D.A., Ebert, M. and Evans, N.T.S. (1962). *J. expt. Bot.*, **13**, 397–403.
28 Barber, D.A. and Gunn, K.B. (1974). *New Phytol.*, **73**, 39–45.
29 Barber, D.A. and Koontz, N.V. (1963). *Pl. Physiol.*, *Lancaster*, **38**, 60–66.
30 Barber, D.A. and Lee, R.B. (1974). *New Phytol.*, **73**, 97–106.
31 Barber, D.A. and Martin, J.K. (1976). *New Phytol.*, **76**, 69–80.
32 Barber, D.A. and Shone, M.G.T. (1966). *J. exp. Bot.*, **17**, 569–578.
33 Barber, D.A. and Shone, M.G.T. (1967). *J. exp. Bot.*, **18**, 631–643.
34 Batch, J.J. and Morgan, D.G. (1975). *J. exp. Bot.*, **26**, 596–608.
35 Baxter, E.D. and Duffus, C.M. (1973). *Phytochemistry*, **12**, 1923–1928.
36 Baxter, E.D. and Duffus, C.M. (1973). *Planta*, **114**, 195–198.
37 Bell, G.D.H. (1937). *J. agric. Sci. Camb.*, **27**, 377–393.
38 Bergal, P. and Clemencet, M. (1962). In *Barley and Malt*, ed. Cook, A.H., pp. 1–23. New York: Academic Press.
39 Berger, K.C. (1949). *Adv. Agron.*, **1**, 321–351.
40 Berry, D.R. and Smith, H. (1970). *Planta*, **91**, 80–86.
41 Biggins, J. and Park, R.B. (1966). *Pl. Physiol.*, *Lancaster*, **41**, 115–118.
42 Birecka, H. and Skiba, T. (1968). *Bull. Acad. pol. Sci. Cl. II Sér. Sci. biol.*, **16**, 595–601.
43 Birecka, H., Skupinska, J., and Bernstein, I. (1967). *Acta Soc. Bot. Pol.*, **36**, 387–409.
44 Biscoe, P.V., Littleton, E.J. and Scott, R.V. (1973). *Ann. appl. Biol.*, **75**, 285–297.
45 Bishop, L.R. (1928). *J. Inst. Brew.*, **34**, 101–118.
46 Bishop, L.R. and Marx, D. (1934). *J. Inst. Brew.*, **40**, 62–74.
47 Boggess, S.F., Aspinall, D. and Paleg, L.G. (1974 – June). *Pl. Physiol.*, *Lancaster*, abstr., p. 51.
48 Bonnafous, J.C., Mousseron-Canet, M. and Olive, J.L. (1973). *Biochim. biophys. Acta*, **312**, 165–171.
49 Bonnett, O.T. and Woodworth, C.M. (1931). *J. am. Soc. Agron.*, **23**, 311–327.
50 Borthwick, H.A., Parker, M.W. and Heinze, P.H. (1941). *Bot. Gaz.*, **103**, 326–341.
51 Borthwick, H.A. and Hendricks, S.B. (1960). *Science*, **132**, 1223–1228.
52 Bray, J.R. (1963). *Can. J. Bot.*, **41**, 65–72.
53 Brenchley, W.E. (1912). *Ann. Bot.*, **26**, 903–928.
54 Brenchley, W.E. (1916). *Ann. Bot.*, **30**, 77–90.
55 Brenchley, W.E. (1919–1920). *Ann. appl. Biol.*, **6**, 142–170.
56 Brenchley, W.E. (1929). *Ann. Bot.*, **43**, 89–110.
57 Brenchley, W.E. (1943). *Biol. Rev.*, **18**, 159–179.
58 Brenchley, W.E. and Jackson, V.G. (1921). *Ann. Bot.*, **35**, 533–556.
59 Brown, J.C. (1965). *Agron. J.*, **57**, 617–621.
60 Brown, D.H. and Haslett, B. (1972). *Planta*, **103**, 129–133.
61 Broyer, T.C. (1951). *Am. J. Bot.*, **38**, 157–162; 483–495.

62 Broyer, T.C. and Hoagland, D.R. (1943). *Am. J. Bot.*, **30**, 261–273.
63 Buchala, A.J. and Wilkie, K.C.B. (1974). *Phytochemistry*, **13**, 1347–1351.
64 Burd, J.S. (1919). *J. agric. Res.*, **18**, 51–72.
65 Burström, H. (1952). *Physiologia Pl.*, **5**, 391–402.
66 Buttrose, M.S. (1960). *J. Ultrastruct, Res.*, **4**, 231–257.
67 Buttrose, M.S. and May, L.H. (1959). *Aust. J. biol. Sci.*, **12**, 40–52.
68 Buttrose, M.S. and May, L.H. (1965). *Ann. Bot.*, **29**, 79–81.
69 Cannell. R.Q. (1969). *J. agric. Sci., Camb.*, **72**, 405–422; 423–435.
70 Cercek, L. (1970). *Int. J. Radiat. Biol.*, **17**, 187–194.
71 Cerning-Beroard, J., and Guilbot, A. (1975). *Ann. Technol. Agric.*, **24**, 143–170.
72 Chu, T.M., Aspinall, D. and Paleg, L.G. (1974). *Aust. J. pl. Physiol.*, **1**, 87–97.
73 Clark, R.B. (1969). *Crop Sci.*, **9**, 341–343.
74 Clarkson, D.T., Shone, M.G.T. and Wood, A.V. (1974). *Planta*, **121**, 81–92.
75 Coleman, R.G., and Hegarty, M.P. (1957). *Nature*, **179**, 376–377.
76 Cooper, M.J., Digby, J. and Cooper, P.J. (1972). *Planta*, **105**, 43–49.
77 Cornforth, I.S. and Stevens, R.J. (1973). *Plant and Soil*, **38**, 581–587.
78 Crist, J.W. and Weaver, J.E. (1924). *Bot. Gaz.*, **77**, 121–148.
79 Cummins, W.R. (1973). *Planta*, **114**, 159–167.
80 Curtis, O.F. and Clark, D.G. (1950). *An Introduction to Plant Physiology*. London: McGraw-Hill.
81 Dale, J.E., Felippe, G.M. and Marriott, C. (1974). *Ann. Bot.*, **38**, 575–588.
82 Dantuma, G. and Andrews, J.E. (1960). *Can. J. Bot.*, **38**, 133–151.
83 von Denffer, D. (1940). *Planta*, **31**, 418–447.
84 Dézsi, L. and Farkas, G.L. (1968). *Acta biol. hung.*, **19**, 43–48.
85 Dommergues, P. (1951). *Ann. Amélior. Plantes*, **1**, 367–375.
86 Dormling, I., Gustafsson, Å and Ekman, G. (1975). *Hereditas*, **79**, 255–272.
87 Dormling, I., Gustafsson, Å and von Wettstein, D. (1969). *Hereditas*, **63**, 415–428.
88 Dougall, H.W. (1963). *E. Afr. agric. For. J.*, **28**, 182–189.
89 Draper, S.R., Sylvester-Bradley, R. and Keith, D.G. (1972). *J. Sci. Fd. Agric.*, **23**, 1369–1372.
90 Drew, M.C. (1975). *New Phytol.*, **75**, 479–490.
91 Dubetz, S. and Wells, S.A. (1965). *Can. J. Pl. Sci.*, **45**, 437–442.
92 Duffus, C.M. (1969). *Phytochemistry*, **8**, 1205–1209.
93 Duffus, C.M. and Rosie, R. (1973). *Planta*, **114**, 219–226.
94 Duffus, C.M. and Rosie, R. (1975). *Phytochemistry*, **14**, 319–323.
95 Duffus, C.M. and Rosie, R. (1976). *J. Agric. Sci., Camb.*, **87**, 75–79.
96 Early, H.L. and Qualset, C.O. (1971). *Euphytica*, **20**, 400–409.
97 Eaton, F.M. (1942). *J. agric. Res.*, **64**, 357–399.
98 Edgar, K.F. and Draper, S.R. (1974). *Phytochemistry*, **13**, 325–327.
99 Egnéus, H., Reftel, S. and Selldén, G. (1972). *Physiologia Pl.*, **27**, 48–55.
100 Eguchi, T. (1937–1939). *Jap. J. Bot.*, **9**, 100–101.
101 Engelbrecht, L. (1971). *Biochem. Physiol. Pfl.*, **162**, 547–558.
102 Engledow, F.L. (1926). *J. R. agric. Soc.*, **87**, 103–123.
103 Epstein, E. (1972). *New Phytol.*, **71**, 873–874.
104 Essen, A. van and Dantuma, G. (1962). *Euphytica*, **11**, 282–286.
105 Fair, P., Tew, J. and Cresswell, C.F. (1974). *Ann. Bot.*, **38**, 39–43; 45–52.
106 Farkas, G.L., Dézsi, L., Horvath, M., Kisban, K. and Udvardy, J. (1964). *Phytopath. Z.*, **49**, 343–354.
107 Feekes, W. (1941). *Verslagen van de Technische Tarwe Commissie XVII De Tarwe en haar milieu.* pp. 523–588. Groeningen: Gebroeders Hoitsema N.V.

108 Felippe, G.M. and Dale, J.E. (1973). *Ann. Bot.*, **37**, 45–56.
109 Felippe, G.M., Dale, J.E. and Marriott, C. (1975). *Ann. Bot.*, **39**, 43–55.
110 Ferguson, W.S. (1966). *Can. J. Pl. Sci.*, **46**, 639–646.
111 Ferguson, H., Eslick, R.F. and Aase, J.K. (1973). *Agron. J.*, **65**, 425–428.
112 Fischbeck, G. (1965). *Brauwissenschaft*, **18**, 1–7.
113 Fletcher, G.M. and Dale, J.E. (1974). *Ann. Bot.*, **38**, 63–76.
114 Fogg, G.E. (1948). *Discuss. Faraday Soc.*, **3**, 162–169.
115 Ford, M.A., and Thorne, G.N. (1967). *Ann. Bot.*, **31**, 629–644.
116 Franklin, R.E. (1969). *Pl. Physiol., Lancaster*, **44**, 697–700.
117 Gaastra, P. (1959). *Mededel. Landbouwhogesch. Wageningen.* **59** (13) 1–101.
118 Gallagher, P.H. and Walsh, T. (1943). *J. agric. Sci., Camb.*, **33**, 197–203.
119 Gardner, J.L. (1942). *Ecology*, **23**, 162–174.
120 Garner, W.W. and Allard, H.A. (1923). *J. agric. Res.*, **23**, 871–920.
121 Gassner, G. (1918). *Z. Bot.*, **10**, 417–480.
122 Gauch, H.G. and Eaton, F.M. (1942). *Pl. Physiol., Lancaster*, **17**, 347–365.
123 Gautheret, M.R. – J. (1935). *Revue gén. Bot.*, **47**, 401–421; 484–511.
124 Geisler, G. (1967). *Pl. Physiol., Lancaster*, **42**, 305–307.
125 George, L.Y. and Williams, W.A. (1964). *Crop. Sci.*, **4**, 450–452.
126 Gerson, D.F. and Poole, R.J. (1971). *Pl. Physiol., Lancaster*, **48**, 509–511.
127 Giacomelli, M., Donini, M.L.B., and Cervigui, T. (1967). *Radiat. Bot.*, **7**, 375–384.
128 Gifford, R.M. (1974). *Aust. J. Pl. Physiol.*, **1**, 107–117.
129 Gifford, R.M., Bremner, P.M. and Jones, D.B. (1973). *Aust. J. agric. Res.*, **24**, 297–307.
130 Gliemeroth, G. (1957). *Z. Acker- und PflBau*, **103**, 1–21.
131 Goatley, J.L. and Lewis, R.W. (1966). *Pl. Physiol., Lancaster*, **41**, 373–375.
132 Görbing, J. and Munkelt, W. (1928). *Angew. Bot.*, **10**, 79–87.
133 Goodall, D.W. (1948). *Ann. appl. Biol.*, **35**, 605–623.
134 Goodall, D.W. (1949). *Ann. appl. Biol.*, **36**, 352–363.
135 Goodall, D.W. and Gregory, F.G. (1947). *Imperial Bureau of Horticulture and Plantation Crops – Technical Communication* No. **17**.
136 Goss, M.J. and Walter, C.H.S. (1969). *Agric. Res. Council Letcombe Lab. ARCRL 19 (Ann. Rep. 1968)*, pp. 27–38.
137 Grafius, J.E. (1956). *Agron. J.*, **48**, 56–59.
138 Grant, D.R. (1970). *J. agric. Sci., Camb.*, **75**, 433–443.
139 Greaves, J.E. and Hirst, C.T. (1929). *Cereal Chem.*, **6**, 115–120.
140 Green, J.T., Finkner, V.C. and Duncan, W.G. (1971). *Agron. J.*, **63**, 469–472.
141 Greenway, H. and Gunn, A. (1966). *Planta*, **71**, 43–67.
142 Greenwood, D.J. (1968). In *Root Growth. Proc. 15th Easter School in Agric. Sci., Univ. Nottingham*, ed. Whittington, W.J. pp. 202–227. London: Butterworth.
143 Gregory, F.G. (1926). *Ann. Bot.*, **40**, 1–26.
144 Gregory, F.G. and Crowther, F. (1931). *Ann. Bot.*, **45**, 579–592.
145 Gregory, F.G. and Sen, P.K. (1937). *Ann. Bot.*, **1**, 521–561.
146 Groenewegen, H. and Mills, J.A. (1960). *Aust. J. biol. Sci.*, **13**, 1–4.
147 Grundbacher, F.J. (1963). *Bot. Rev.*, **29**, 366–381.
148 Gustafsson, A., Ekman, G. and Dormling, I. (1974). *Hereditas*, **76**, 137–144.
149 Hackett, C. (1966). *Ann. Bot.*, **30**, 321–327.
150 Hackett, C. (1968). In *Root Growth. Proc. 15th Easter School in Agric. Sci., Univ. Nottingham*, ed. Whittington, W.J. pp. 134–147. London: Butterworth.
151 Hackett, C. (1969). *New Phytol.*, **68**, 1023–1030.
152 Hackett, C. (1971). *Aust. J. biol. Sci.*, **24**, 1057–1064.

153 Hackett, C. and Bartlett, B.O. (1971). *New Phytol.*, **70**, 409–413.
154 Hackett, C. and Rose, D.A. (1972). *Aust. J. biol. Sci.*, **25**, 669–679; 681–690.
155 Hackett, C., Sinclair, C. and Richards, F.J. (1965). *Ann. Bot.*, **29**, 331–345.
156 Hänsel, H. (1952). *Z. PflZüch.*, **31**, 359–380.
157 Hagen, C.E. and Hopkins, H.T. (1955). *Pl. Physiol., Lancaster*, **30**, 193–199.
158 Harlan, H.V. (1920). *J. agric. Res.*, **19**, 393–430.
159 Harlan, H.V. and Anthony, S. (1920). *J. agric. Res.*, **19**, 431–472.
160 Harlan, H.V. and Anthony, S. (1921). *J. agric. Res.*, **21**, 29–45.
161 Harlan, H.V. and Pope, M.N. (1921). *J. agric. Res.*, **22**, 433–449.
162 Harlan, H.V. and Pope, M.N. (1923). *J. agric. Res.*, **23**, 333–360.
163 Harris, G. (1962). In *Barley and Malt*, ed. Cook, A.H. pp. 431–582; 583–694. London: Academic Press.
164 Hatrick, A.A. and Bowling, D.J.F. (1973). *J. exp. Bot.*, **24**, 607–613.
165 Heide, H. van der, De Boer-Bolte, B.M., and van Raalte, M.H. (1963). *Acta bot. neerl.*, **12**, 231–247.
166 Heller, V.G. (1948). *Oklahoma Agr. Exp. Sta. Bull.*, **B–319**, 100–104.
167 Helliegel, H., Wilfarth, H., Römer, H. und Wimmer, G. (1898). *Arbeiten der Deutscher Landwirtschafts—Gesellshaft.*, **34**.
168 Henningsen, K.W. and Boardman, N.K. (1973). *Pl. Physiol., Lancaster*, **51**, 1117–1126.
169 Hentrich, W. (1966). *Züchter*, **36**, 25–36.
170 Herath, H.M.W. and Ormrod, D.P. (1972). *Pl. Physiol., Lancaster*, **49**, 443–444.
171 Hewitt, E.H. (1956). *Soil Sci.*, **81**, 159–171.
172 Hiatt, A.J. (1967). *Z. PflPhys.*, **56**, 233–245.
173 Hiatt, A.J. and Lowe, R.H. (1967). *Pl. Physiol., Lancaster*, **42**, 1731–1736.
174 Hoagland, D.R. and Broyer, T.C. (1940). *Am. J. Bot.*, **27**, 173–185.
175 Hofbauer, J. and Minar, J. (1968). *Biol. Pl.*, **10**, 166–176.
176 Holliday, R. (1966). *Agric. Prog.*, **41**, 24–34.
177 Holloway, P.J. (1969). *Ann. appl. Biol.*, **63**, 145–153.
178 Holmern, K., Vange, M.S. and Nissen, P. (1974). *Physiologia Pl.*, **31**, 302–310.
179 Holmes, D.P. (1974). *Nature*, **247**, 297–298.
180 Hooymans, J.J.M. (1969). *Planta*, **88**, 369–371.
181 Hopkins, H.T. (1956). *Pl. Physiol., Lancaster*, **31**, 155–161.
182 Hopkins, W.G. and Hillman, W.S. (1965). *Am. J. Bot.*, **52**, 427–432.
183 Horváth, M. and Lásztity, D. (1967). *Acta agron. hung.*, **16**, 393–397. (via *Field Crop Abstr.*, **21**, No. 779, (1968).
184 Hoshizaki, T., Adey, W.R. and Hamner, K.C. (1966). *Planta*, **69**, 218–229.
185 Huffaker, R.C., Radin, T., Kleinkopf, G.E. and Cox E.L. (1970). *Crop Sci.*, **10**, 471–474.
186 Humphries, E.C. (1958). *Ann. Bot.*, **22**, 251–257; 417–422.
187 Husain, I.C. and Aspinall, D. (1970). *Ann. Bot.*, **34**, 393–407.
188 Hussey, G. and Gregory, F.G. (1954). *Pl. Physiol., Lancaster*, **29**, 292–296.
189 Hutcheson, T.B. and Quantz, K.E. (1917). *J. Am. Soc. Agron.*, **9**, 17–21.
190 Ikeda, H. and Nagamatsu, T. (1958–9). *Proc. Crop Sci. Soc. Japan*, **27**, 284.
191 Ivanko, S.I. (1971). *Biol. Pl.*, **13**, 155–164.
192 Jackson, P.C., Hendricks, S.B. and Vasta, B.M. (1962). *Pl. Physiol., Lancaster*, **37**, 8–17.
193 Jacobson, L., Hannapel, R.J., Moore, D.P. and Schaedle, M. (1961). *Pl. Physiol., Lancaster*, **36**, 58–61; 62–65.
194 Jacobson, L., Schaedle, M., Cooper, B. and Young, L.C.T. (1968). *Physiologia Pl.*, **21**, 119–126.
195 Jacoby, B. and Laties, G.G. (1971). *Pl. Physiol., Lancaster*, **47**, 525–531.

196 James, N.I. and Lund, S. (1965). *Am. J. Bot.*, **52**, 877–882.
197 Jensen, C.R., Stolzy, L.H., and Letey, J. (1967). *Soil Sci.*, **103**, 23–29.
198 Jeschke, W.D. and Stelter, W. (1973). *Planta*, **114**, 251–258.
199 Johnson, R.R., Frey, N.M. and Moss, D.N. (1974). *Crop. Sci.*, **14**, 728–731.
200 Johnson, C.M., Stout, P.R., Broyer, T.C. and Carlton, A.P. (1957). *Pl. Soil*, **8**, 337–353.
201 Johnson, L.P.V. and Taylor, A.R. (1958). *Can. J. Pl. Sci.*, **38**, 122–123.
202 Johnson, R.R., Willmer, C.M. and Moss, D.N. (1975). *Crop. Sci.*, **15**, 217–221.
203 Jones, R.J. and Mansfield, T.A. (1971). *Nature*, **231**, 331.
204 Jordan, W.R. and Huffaker, R.C. (1968). *Pl. Physiol., Lancaster*, **43** (Suppl.) 5–8.
205 Juniper, B.E. and French, A. (1973). *Planta*, **109**, 211–224.
206 Kail, R.M., Kolp, B.J. and Bohenenblust, K.E. (1972). *Crop Sci.*, **12**, 872–874.
207 Kamel, M.S. (1959). *Meded. LandbHoogesch. Wageningen.*, **59** (5), 1–101.
208 Kang, K.B. (1971). *Pl. Physiol., Lancaster*, **47**, 352–356.
209 Katznelson, H., Rouatt, J.W. and Payne, T.M.B. (1954). *Nature*, **174**, 1110.
210 Kaufmann, M.L. and Guitard, A.A. (1967). *Can. J. Pl. Sci.*, **47**, 73–78.
211 Kende, H. (1965). *Proc. natn. Acad. Sci. U.S.A.*, **53**, 1302–1307.
212 Kenefick, D.G. and Swanson, C.R. (1963). *Crop. Sci.*, **3**, 202–205.
213 Kenefick, D.G. and Whitehead, E.I. (1970). In *Barley Genetics II*, ed. Nilan, R.A. pp. 378–387. Washington State University Press.
214 Kirby, E.J.M. (1967). *J. agric. Sci., Camb.*, **68**, 317–324.
215 Kirby, E.J.M. (1969). *Ann. appl. Biol.*, **63**, 513–521.
216 Kirby, E.J.M. (1969). *J. agric. Sci., Camb.*, **72**, 467–474.
217 Kirby, E.J.M. (1969). *Field Crop Abstr.*, **22**, 1–7.
218 Kirby, E.J.M. (1970). *J. agric. Sci., Camb.*, **75**, 445–450.
219 Kirby, E.J.M. (1973). *J. exp. Bot.*, **24**, 567–578; 935–947.
220 Kirby, E.J.M. and Faris, D.G. (1970). *J. exp. Bot.*, **21**, 787–798.
221 Kirby, E.J.M. and Faris, D.G. (1972). *J. agric. Sci., Camb.*, **78**, 281–288.
222 Kirby, E.J.M. and Rackham, O. (1971). *J. appl. Ecol.*, **8**, 919–924.
223 Kirkham, M.B., Gardner, W.R. and Gerloff, G.C. (1974). *Pl. Physiol., Lancaster*, **53**, 241–243.
224 Kjack, J.L. and Witters, R.E. (1974). *Crop Sci.*, **14**, 243–248.
225 Koller, D., Highkin, H.R. and Caso, O.H. (1960). *Am. J. Bot.*, **47**, 518–524.
226 Kramer, H.H. and Veyl, R. (1952). *Agron. J.*, **44**, 156.
227 Krassovsky, I. (1926). *Soil Sci.*, **21**, 307–325.
228 Kreeb, K. (1961). In *The Water Relationships of Plants*, eds. Rutter, A.J. and Whitehead, F.H. pp. 272–288. London: Blackwell Scientific Publications.
229 Kuraishi, S. (1974). *Pl. Cell Physiol.*, **15**, 295–306.
230 Kurth, H. (1954). *Züchter*, **24**, 300–304.
231 LaBerge, D.E., MacGregor, A.W. and Meredith, W.O.S. (1973). *J. Inst. Brew.*, **79**, 471–477.
232 Läuchli, A., Kramer, D., Pitman, M.G. and Lüttge, U. (1974). *Planta*, **119**, 85–99.
233 Lane, H.C., Cathey, H.M. and Evans, L.T. (1965). *Am. J. Bot.*, **52**, 1006–1014.
234 Lang, R.W. and Holmes, J.C. (1969). *Expl. Husb.*, **18**, 1–7.
235 Langer, R.H.M. (1967). *Field Crop Abstr.*, **20**, 101–106.
236 Langston, R. and Leopold, A.C. (1954). *Physiologia Pl.*, **7**, 397–404.
237 Lanning, F.C. (1966). *J. agric. Fd. Chem.*, **14**, 636–638.
238 Lanning, F.C., Ponnaiya, B.W.X. and Crumpton, C.F. (1958). *Pl. Physiol., Lancaster*, **33**, 339–343.

239 Large, E.C. (1954). *Pl. Path.*, **3**, 128–129.
240 Laude, H.H. (1937). *J. agric., Res.*, **54**, 899–917.
241 Laude, H.M., Ridley, J.R. and Suneson, C.A. (1967). *Crop Sci.*, **7**, 230–233.
242 Leach, G.J. and Watson, D.J. (1968). *J. appl. Ecol.*, **5**, 581–408.
243 Leakey, R.R.B. (1971). *J. agric. Sci., Camb.*, **77**, 135–139.
244 Lee J.A. (1960). *Evolution*, **14**, 18–28.
245 Leggett, J.E. (1968). *Ann. Rev. Pl. Physiol.*, **19**, 333–346.
246 Leopold, A.C. and Guernsey, F.S. (1954). *Physiologia Pl.*, **7**, 30–40.
247 Lichtenthaler, H.K. and Becker, K. (1970). *Phytochemistry*, **9**, 2109–2113.
248 Lichtenthaler, H.K. and Verbeek, L. (1973). *Planta*, **112**, 265–271.
249 Lipman, C.B. (1938). *Soil Sci.*, **45**, 189–198.
250 Liptay, A. and Davidson, D. (1973). *Pl. Physiol., Lancaster*, **51**, 405–406.
251 Liu, D.J., Pomeranz, Y. and Robbins, G.S. (1975). *Cereal Chem.*, **52**, 678–686.
252 Livnè, A. and Vaadia, Y. (1965). *Physiologia Pl.*, **18**, 658–664.
253 Loomis, R.S., Williams; W.A. and Hall, A.E. (1971). *Ann. Rev. Pl. Physiol.*, **22**, 431–468.
254 Loveys, B.R. and Wareing, P.E. (1971). *Planta*, **98**, 109–116; 117–127.
255 Lüers, H., Fink, H., and Riedel, W. (1930). *Woch. Brau.*, **47**, 393–397; 405–409.
256 Luebs, R.E. and Laag, A.E. (1967). *Agron. J.*, **59**, 219–222.
257 Lundergårdh, H. (1951). *Leaf Analysis*, transl. Mitchell, R.L., London: Hilger and Watts.
258 Lysenko, T.D. (1954). *Agrobiology.* transl. 4th Russian edition. Moscow: Foreign Languages Publishing House.
259 MacGregor, A.W., Gordon, A.G., Meredith, W.O.S. and Lacroix, L. (1972). *J. Inst. Brew.*, **78**, 174–179.
260 MacGregor, A.W., La Berge, D.E. and Meredith, W.O.S. (1971). *Cereal Chem.*, **48**, 255–269.
261 MacGregor, A.W., Thompson, R.G. and Meredith, W.O.S. (1974). *J. Inst. Brew.*, **80**, 181–187.
262 Machlis, L. (1944). *Am. J. Bot.*, **31**, 183–192; 281–282.
263 Mack, A.R. (1965). *Can. J. Soil Sci.*, **45**, 337–346.
264 Martini, M.L., Harlan, H.V. and Pope, M.N. (1930). *Pl. Physiol., Lancaster*, **5**, 263–272.
265 Maximov, M.A. (1929). *Biol. Zbl.*, **49**, 513–543.
266 May, L.H., Randles, F.H., Aspinall, D. and Paleg, L.G. (1967). *Austr. J. biol. Sci.*, **20**, 273–283.
267 McKee, H.S. (1950). *Austr. J. sci. Res. Ser. B.*, **3**, 474–486.
268 Mengel, K. and Herwig, K. (1969). *Z. PflPhysiol.*, **60**, 147–155.
269 Menhenett, R. and Carr, D.J. (1973). *Aust. J. biol. Sci.*, **26**, 527–537; 1073–1080.
270 Merritt, N.R. and Walker, J.T. (1969). *J. Inst. Brew.*, **75**, 156–164.
271 Metcalf, E.L., Cress, C.E., Olein, C.R. and Everson, E.H. (1970). *Crop. Sci.*, **10**, 362–365.
272 Metwally, S.Y. and Pollard, A.G. (1959). *J. Sci. Fd Agric.*, **10**, 632–636.
273 Michael, G., Allinger, P. and Wilberg, E. (1970). *Z. PflNarung Boden.*, **125**, 24–35.
274 Michael, G. and Seiler-Kelbitsch, H. (1972). *Crop Sci.*, **12**, 162–165.
275 Millar, A.A., Duysen, M.E. and Wilkinson, G.E. (1968). *Pl. Physiol., Lancaster*, **43**, 968–972.
276 Miller, R.A. and Zalik, S. (1965). *Pl. Physiol., Lancaster*, **40**, 569–574.
277 Miskin, K.E., Rasmusson, D.C. and Moss, D.N. (1972). *Crop Sci.*, **12**, 780–783.

278 Mitrovich, G. and Kuehner, C.C. (1967). Can. J. Bot., **45**, 333–338.
279 Mittelheuser, C.J. and van Steveninck, R.F.M. (1969). Nature, **221**, 281–282.
280 Monteith, J.L. and Elston, J.F. (1971). In Potential Crop Production—a Case Study, eds. Wareing, P.F. and Cooper, J.P., pp. 23–42. London: Heinemann.
281 Montgomery, D.J. and Smith, A.E. (1963). In Biomedical Sciences Instrumentation, **1**, ed. Alt, F. New York: Plenum Press.
282 Mortimer, D.C. (1959). Can. J. Bot., **37**, 1191–1201.
283 Morton, A.G. and Watson, D.J. (1948). Ann. Bot., **12**, 281–310.
284 Moss, D.N., Kreuzer, E.G., and Brun, W.A. (1969). Science, **164**, 187–188.
285 Mounla, M.A. and Michael, G. (1973). Physiologia Pl., **29**, 274–276.
286 Müller, F. (1962). In Eigenschaften und Wirkungen der Gibberellins, ed. Knap., R. pp. 173–180. Berlin: Springer Verlag.
287 Müller-Stoll, W.R. and Augsten, H. (1961). Planta, **56**, 97–108.
288 Muenscher, W.C. (1922). Am. J. Bot., **9**, 311–329.
289 Mullison, W.R. and Mullison, E. (1942). Pl. Physiol., Lancaster, **17**, 632–644.
290 Nassery, H. and Baker, D.A. (1974). Ann. Bot., **38**, 141–144.
291 Nátr, L. (1967). Photosynthetica, **1**, 29–36.
292 Nátr, L. and Purš, J. (1969). Photosynthetica, **3**, 320–325.
293 Nátr, L., Watson, B.T. and Weatherley, P.E. (1974). Ann. Bot., **38**, 589–593.
294 Newbould, P. (1968). In Ecological Aspects of the Mineral Nutrition of Plants, ed. Rorison, I.H. pp. 177–190. Oxford: Blackwell.
295 Newman, D.W., Rowell, B.W. and Byrd, K. (1973). Pl. Physiol., Lancaster, **51**, 229–233.
296 Newton, J.D. (1925). Sci. Agric., **5**, 318–320.
297 Nicholas, D.J.D. (1952). J. agric. Sci., **42**, 468–475.
298 Nicholls, P.B. and May, (1963). Aust. J. biol. Sci., **16**, 561–571.
299 Nielsen, N. (1936). C. r. Trav. Lab. Carlsberg (ser physiol), **21**, 247–269.
300 Nissen, P. (1973). Physiologia Pl., **28**, 304–316.
301 Nissen, P. (1974). Physiologia Pl., **30**, 307–316.
302 Nyborg, M. (1970). Can. J. Pl. Sci., **50**, 198–200.
303 Obendorff, R.L. and Huffaker, R.S. (1970). Pl. Physiol., Lancaster, **45**, 579–582.
304 Oertli, J.J. (1967). Physiologia Pl., **20**, 1614–1626.
305 Olien, C.R. (1970). In Barley Genetics II, ed. Nilan, R.A. pp. 356–363. Washington State University Press.
306 Olsen, C. (1939). C. r. Trav. Lab. Carlsberg (Ser. Chem.), **23**(5), 37–44.
307 Olsen, C. (1942). C. r. Trav. Lab. Carlsberg (Ser. Chim.), **24** 69–97.
308 Onckelen, H.A. Van and Verbeek, R. (1972). Phytochemistry, **11**, 1677–1680.
309 Ormrod, D.P. (1963). Can. J. Pl. Sci., **43**, 323–329.
310 Ormrod, D.P., Hubbard, W.F. and Faris, D.G. (1968). Can. J. Pl., Sci., **48**, 363–368.
311 Overland, L. (1966). Am. J. Bot., **53**, 423–432.
312 Paleg, L.G. and Aspinall, D. (1964). Bot. Gaz., **125**, 149–155.
313 Paleg, L.G. and Aspinall, D. (1970). Nature, **228**, 970–973.
314 Pavlychenko, T.K. (1937). Ecology, **18**, 62–79.
315 Pavlychenko, T.K. and Harrington, J.B. (1935). Sci. Agric., **16**, 151–160.
316 Pearce, R.B., Brown, R.H. and Blaser, R.E. (1967). Crop Sci., **7**, 321–324; 545–546.
317 Pedersen, A. (1950). Landbrugets Plantekultur, II Kandrup og Wunsch Bogtrykkeri, København, 32, 34.
318 Peterson, L.W., Kleinkopf, G.E. and Huffaker, R.C. (1973). Pl. Physiol., Lancaster, **51**, 1042–1045.

319 Phillips, M. and Goss, M.J. (1935). *J. agric. Res.*, **51**, 301–319.
320 Phung nhu Hung, S., Hoaran, A. and Moyse, A. (1970). *Z. PflPhysiol.*, **62**, 245–258.
321 Pitman, M.G. (1969). *Pl. Physiol., Lancaster,* **44**, 1233–1240.
322 Pitman, M.G. (1971). *Aust. J. biol. Sci.*, **24**, 407–421.
323 Pitman, M.G., Lüttge, U., Kramer, D. and Ball, E. (1974). *Aust. J. Pl. Physiol.*, **1**, 65–75.
324 Pitman, M.G., Lüttge, U., Läuchli, A. and Ball, E. (1974). *J. exp. Bot.*, **25**, 147–155.
325 Plesnicar, M. and Bendall, D.S. (1973). *Biochem. J.*, **136**, 803–812.
326 Plhák, F. and Urbánková, V. (1969). *Biol. Pl.*, **11**, 226–235.
327 Pomeranz, Y., Robbins, G.S. and Gilbertson, J.T. (1976). *J. Fd. Sci.*, **41**, 283–285.
328 Pope, M.N. (1932). *J. agric. Res.*, **44**, 323–341; 343–355.
329 Porter, H.K., Pal, N. and Martin, R.Y. (1950). *Ann. Bot.*, **14**, 55–68.
330 Poulson, G. and Beevers, L. (1972). In *Plant Growth Substances* (1970), ed. Carr, D.J. pp. 646–653. Berlin: Springer Verlag.
331 Power, J.F., Grunes, D.L., Reichman, G.A. and Willis, W.O. (1970). *Agron. J.*, **62**, 567–571.
332 Power, J.F., Willis, W.O., Grunes, D.L., Reichman, G.A. (1967). *Agron. J.*, **59**, 231–234.
333 Prat, H. (1934). *Can. J. Res.*, **10**, 563–570.
334 Pratt, L.H. and Coleman, R.A. (1974). *Am. J. Bot.*, **61**, 195–202.
335 Prentice, N., Moeller, M. and Pomeranz, Y. (1971). *Cereal Chem.*, **48**, 714–717.
336 von Proskowetz, E. (1893). *Landw. Jbr.*, **22**, 629–717.
337 Purvis, L.O.N. (1961). In *Handbuch der Pflanzephysiologie*, ed. Ruhland, W. **16**, pp. 76–122. Berlin: Springer Verlag.
338 Qualset, C.O., Schaller, C.W. and Williams, J.C. (1965). *Crop. Sci.*, **5**, 489–494.
339 Rackham, O. (1972). In *Crop Processes in Controlled Environments*, eds. Rees, A.R., Cockshall, K.E., Hand, D.W. and Hurd, R.G. pp. 127–138. London: Academic Press.
340 Rains, D.W., Schmid, W.E. and Epstein, E. (1964). *Pl. Physiol., Lancaster*, **39**, 274–278.
341 Rains, D.W., and Epstein, E. (1967). *Pl. Physiol., Lancaster*, **42**, 314–318; 319–323.
342 Ranson, S.L. and Parija, B. (1955). *J. exp. Bot.*, **6**, 80–93.
343 Rauser, W.E. (1967). *Can. J. Pl. Sci.*, **47**, 614.
344 Reid, D.M., Tuing, M.S., Durley, R.C. and Railton, I.D. (1972). *Planta*, **108**, 67–75.
345 Rejowski, A. (1969). *Bull. Acad. pol. Sci. Cl 11 Sér. Sci. biol.*, **17**, 641–644.
346 Rentschler, I. (1971). *Planta*, **96**, 119–135.
347 Rhodes, A.P. and Jenkins, G. (1975). *J. Sci. Fd. Agric.*, **26**, 705–709.
348 Richards, F.J. (1938). *Ann. Bot.*, **2**, 491–534.
349 Richards, F.J. and Rees, A.R. (1962). *Indian J. Pl. Physiol.*, **5**, 33–52.
350 Richards, F.J. and Sheng-Han, S. (1940). *Ann. Bot.*, **4**, 165–175; 403–425.
351 Richardson, A.E.V. and Trumble, H.C. (1928). *J. Dept. Agric. S. Australia*, **32**, 224–244.
352 Robards, A.W. and Robb, M.E. (1974). *Planta*, **120**, 1–12.
353 Rorison, I.H. (1968). (Editor) *Ecological aspects of the Mineral Nutrition of Plants*. Oxford: Blackwell.
354 Rosene, N.F. and Walthall, A.M.J. (1949). *Bot. Gaz.*, **111**, 11–21.

355 Ruckenbauer, P. and Kirby, E.J.M. (1973). *J. agric. Sci., Camb.*, **80**, 211–217.
356 Russell, E.J. (1909). In *The Standard Cyclopedia of Modern Agriculture and Rural Economy*, ed. Wright, R.P. **5**, 222–235.
357 Russell, R.S. (1938). *Ann. Bot.*, **2**, 865–882.
358 Russell, E.W. (1973). *Soil Conditions and Plant Growth* (10th edition). London: Longmans.
359 Ryle, G.J.A., Brockington, N.R., Powell, C.E. and Cross, B. (1973). *Ann. Bot.*, **37**, 233–246.
360 Ryle, G.J.A. and Powell, C.E. (1975). *Ann. Bot.*, **39**, 297–310.
361 Saghir, A.R., Khan, A.R. and Worzella, W.W. (1968). *Agron. J.*, **60**, 95–97.
362 Sagromsky, H. (1954). *Z. PflZücht*, **33**, 267–284.
363 Salim, M.H. and Todd, G.W. (1965). *Agron. J.*, **57**, 593–596.
364 Salim, M.H., Todd, G.W., Schlehuber, A.M. (1965). *Agron. J.*, **57**, 603–607.
365 Sallans, B.J. (1961). *Can. J. Pl. Sci.*, **41**, 493–498.
366 Salter, P.J. and Goode, J.E. (1967). *Crop responses to Water at Different Stages of Growth* (Research Review No. 2 of Commonwealth Bureau of Horticulture and Plantation Crops).
367 Saunders, J.A. and McClure, J.W. (1973). *Pl. Physiol., Lancaster*, **51**, 407–408.
368 Schaller, C.W., Qualset, C.O. and Rutger, J.H. (1972). *Crop Sci.*, **12**, 531–535.
369 Schjerning, H. (1914). *C. r. Trav. Lab. Carlsberg*, **11**, 45–145.
370 Schmalz, H. (1962). *Eigenschaften und Wirkungen der Gibberelline*, ed. Knapp, R. pp. 180–191. Berlin: Springer Verlag.
371 Schmid, B. (1898). *Bot. Cbl.*, **76**, 1–9; 36–41; 70–76; 118–128; 156–166; 212–221; 264–270; 301–307; 328–334.
372 Schulte, H.-K. (1955). *Z. PflZücht.*, **34**, 157–196.
373 Scott, D. and Wells, J.S. (1969). *New Zealand J. Bot.*, **7**, 373–385.
374 Shantz, H.L. and Piemeisel, L.N. (1927). *J. agric. Res.*, **34**, 1093–1190.
375 Smith, J.H.C. (1954). *Pl. Physiol., Lancaster*, **29**, 143–148.
376 Smith, D.B. (1972). *J. agric. Sci., Camb.*, **78**, 265–273.
377 Smith, M.A., Criddle, R.S., Peterson, L. and Huffaker, R.C. (1974). *Arch. biochem. Biophys.*, **165**, 494–504.
378 Smith, T.A. and Richards, F.J. (1962). *Biochem. J.*, **84**, 292–294.
379 Smith, K.A. and Robertson, P.D. (1971). *Nature*, **234**, 148–149.
380 Smith, T.A. and Sinclair, C. (1967). *Ann. Bot.*, **31**, 103–111.
381 Sotola, J. (1937). *J. agric. Res.*, **54**, 399–415.
382 Sparmann, G. (1961). *Planta*, **57**, 176–201; 447–474.
383 Sprague, H.B. (1964; Ed.). *Hunger Signs in Crops – A Symposium*. (3rd Edit.) David McKay, Co., New York.
384 Srivastava, B.I.S. and Atkin, R.K. (1968). *Biochem. J.*, **107**, 361–366.
385 van Staden, J. (1974). *Physiologia Pl.*, **30**, 182–184.
386 Stobart, A.K. and Pinfield, N.J. (1970). *New Phytol.*, **69**, 31–35.
387 Stoddart, J.L. (1971). *Planta*, **97**, 70–82.
388 Stølen, O. (1965). *Arsskr. k. Vet. – Landhøjsk.*, København, 1965; 81–107.
389 Stølen, O. (1967). *Ugeskr. Agron.*, **112**, 643–646.
390 Stolwijk, J.A.J. and Thimann, K.V. (1957). *Pl. Physiol., Lancaster*, **32**, 513–520.
391 Storey, R. and Jones, R.G.W. (1975). *Pl. Sci. Lett.*, **4**, 161–168.
392 Stoy, V. and Hagberg, A. (1967). *Hereditas*, **58**, 359–384.
393 Strebeyko, P. and Góra, B. (1964). *Biol. Pl.*, **6**, 152–157.
394 Stroun, M. (1958). *Bull. Soc. bot., Fr.*, **105**, 1–9.

395 Sugahara, T., Murata, Y. and Kikkawa, M. (1958–9). *Proc. Crop Sci. Soc. Japan*, **27**, 392.
396 Szalai, I. (1969). *Physiologia Pl.*, **22**, 587–593.
397 Takahashi, R. and Yasuda, S. (1960). *Ber. Ohara Inst. Landw. Biol. Okayama Universität.*, **11**, 365–381.
398 Tamàs, I.A., Yemm, E.W. and Bidwell, R.G.S. (1970). *Can. J. Bot.*, **48**, 2313–2317.
399 Tanada, T. (1973). *Pl. Physiol., Lancaster*, **51**, 154–157.
400 Tange, M. (1965). *Sci. Rep. Hyogo Univ. Agric.*, **7**, 13–20; 21–28.
401 Tanimoto, E., and Masuda, Y. (1968). *Physiologia Pl.*, **21**, 820–826.
402 Tanner, J.W., Gardener, C.J., Stoskopf, M.C., Reinbergs, E. (1966). *Can. J. Pl. Sci.*, **46**, 690.
403 Tanton, T.W. and Crowdy, S.H. (1972). *J. exp. Bot.*, **23**, 600–618; 619–625.
404 Thomason, R.C. (1970). *Crop Sci.*, **10**, 618–620.
405 Thorne, G.N. (1962). *J. agric. Sci.*, **58**, 89–96.
406 Thorne, G.N. (1962). *Ann. Bot.*, **26**, 37–54.
407 Thorne, G.N. (1966). In *The Growth of Cereals and Grasses*, eds. Milthorpe, F.L. and Ivins, J.D. 88–105. London: Butterworths.
408 Thorne, G.N. (1969). *NAAS Quart. Rev.*, **85**, 42–46.
409 Thorne, G.N. (1974). *Rothamsted Report for 1973(pt. 2)*, 5–25.
410 Tingle, J.N., Faris, D.G. and Ormrod, D.P. (1970). *Crop Sci.*, **10**, 26–28.
411 Toussoun, T.A., Weinhold, A.R., Linderman, R.G. and Patrick, Z.A. (1967). *Phytopathology*, **57**, 834.
412 Trione, E.J. and Metzger, R.J. (1970). *Crop Sci.*, **10**, 390–392.
413 Troughton, A. (1962). *The Roots of Temperate Cereals (Wheat, Barley, Oats and Rye). (Publication No. 2/1962) Commonwealth Bureau of Pastures and Field Crops*, Hurley, England.
414 Turina, B. (1922). *Biochem. Z.*, **129**, 507–533.
415 Turner, T.W. (1922). *Am. J. Bot.*, **9**, 415–445.
416 Turner, T.W. (1929). *Bot. Gaz.*, **88**, 85–95.
417 Ulrich, A. (1942). *Am. J. Bot.*, **29**, 220–227.
418 Umaly, R.C. and Poel. L.W. (1970). *Ann. Bot.*, **34**, 919–926.
419 Vančura, V. (1964). *Plant and Soil.*, **21**, 231–248.
420 Varga, A. and Bruinsma, J. (1973). *Planta*, **111**, 91–93.
421 Viets, F.G. (1944). *Pl. Physiol., Lancaster*, **19**, 466–480.
422 Vlamis, J. and Davis, A.R. (1944). *Pl. Physiol., Lancaster*, **19**, 33–51.
423 Walkley, J. and Petrie, A.H.K. (1941). *Ann. Bot.*, **5**, 661–673.
424 Wallace, T. (1961). *The diagnosis of Mineral deficiencies in Plants by visual symptoms, A colour Atlas and Guide*, (3rd edition). London: H.M.S.O.
425 Walpole, P.R. and Morgan, D.G. (1971). *Ann. Bot.*, **35**, 301–310.
426 Walpole, P.R. and Morgan, D.G. (1972). *Nature*, **240**, 416–417.
427 Walster, H.L. (1920). *Bot. Gaz.*, **69**, 97–126.
428 Warington, K. (1933). *Ann. Bot.*, **47**, 429–458.
429 Watson, D.J. (1952). *Adv. Agron.*, **4**, 101–145.
430 Watson, D.J. (1971). In *Potential Crop Production, A Case Study*, eds. Wareing, P.F. and Cooper, J.P. pp. 76–88. London: Heinemann.
431 Watson, D.J., Thorne, G.N. and French, S.A.W. (1958). *Ann. Bot.*, **22**, 321–352.
432 Weaver, J.E. (1926). *Root Development of Field Crops*. New York: McGraw-Hill.
433 Weaver, J.E., Jean, F.C. and Crist, J.W. (1922). *Development and Activities of Roots of Crop Plants. A study in Crop Ecology*. Washington: Carnegie Institute.

434 Welbank, P.J. and Williams, E.D. (1968). *J. appl. Ecol.*, **5**, 477–481.
435 Welbank, P.J., Witts, K.J. and Thorne, G.N. (1968). *Ann. Bot.*, **32**, 79–95.
436 Wells, S.A. (1962). *Can. J. Pl. Sci.*, **42**, 169–172.
437 Whyte, R.O. (1960). *Crop Production and Environment* (2nd edition). London: Faber and Faber.
438 Wiebe, H.H. and Kramer, P.J. (1954). *Pl. Physiol., Lancaster*, **29**, 342–348.
439 Wiebe, G.A. and Reid, D.A. (1958). *U.S. Dept. Agric. Tech. Bull.* (*No. 1176*).
440 Wilfarth, H., Römer, H. and Wimmer, G. (1905). *Landw. Vers. Stat.*, **63** (1/2). 1–70; pl. 1–3.
441 Willey, R.W. and Dent, J.D. (1969). *Agric. Prog.*, **44**, 43–55.
442 Willey, R.W. and Holliday, R. (1971). *J. agric. Sci., Camb.*, **77**, 445–452.
443 Williams, D.E. and Vlamis, J. (1957). *Pl. Physiol., Lancaster*, **32**, 404–409.
444 Williams, D.E. and Vlamis, J. (1957). *Plant and Soil*, **8**, 183–193.
445 de Wit, M.C.J. (1969). *Acta bot. neerl.*, **18**, 558–560.
446 Wittwer, S.H. (1968). *Mich. Agr. Exp. Sta. J.*, (*No. 4176*) 98–123.
447 Yap, T.C. and Harvey, B.L. (1972). *Can. J. Pl. Sci.*, **52**, 241–246.
448 Yemm, E.W. (1937). *Proc. R. Soc. Ser. B*. **123**, 243–273.
449 Yemm, E.W. (1950). *Proc. R. Soc. Ser. B*, **136**, 632–649.
450 Yoshida, S. (1972). *Ann. Rev. Pl. Physiol.*, **23**, 437–464.
451 Young, A.L. and Feltner, K.C. (1966). *Crop Sci.*, **6**, 547–551.
452 Zadoks, J.C., Chang, T.T. and Konzak, C.F. (1974). *Weed Res.*, **14**, 415–421.
453 Zöbl, A. and Mikosch, C. (1893). *Bot. Cbl.*, **54**, 240 (abstr.).

Chapter 7
Agricultural practices and yield

7.1. Introduction

Most simply yield is the total weight of material (dry, or at some fixed moisture content – e.g. 15%) produced per unit area cultivated. As estimates of roots and stubble are imprecise a more usual meaning is 'recoverable yield' i.e. the harvested straw and grain. Where immature plants are grazed or clipped the yield of the forage must also be taken into account. Where straw is considered valueless the 'yield' is the quantity of grain harvested. The economic yield, the cash value of the crop to the farmer, is different again. Grain, and straw of different qualities fetch different prices. Malting barley usually receives a premium, as does seed quality corn. However, the feed value of barley grain (nitrogen content; lysine content) is rarely assessed. In the U.K. naked barley is rarely grown because of its 'low grain yield' no note being taken that the husk (10–12%) of covered varieties is equivalent at most only to straw in feed value, and that naked grain may be more palatable to stock and valuable to processors. The profit from a crop is the price received *less* costs such as transport, labour, use of farm machinery, seed, fertilizers, herbicides and insecticides. The best economic yield is below the greatest achievable yield since increasing inputs, e.g. of fertilizers or labour, yield diminishing returns, and ultimately exceed the value of the extra yield. Crops can have less obvious values. Barley may be sown thinly, at the expense of grain yield, when it is used as a 'nurse' for a companion crop. The reduced value of the barley is more than compensated for by the enhanced value of the forage obtained subsequently. Barley may profit from fertilizers applied to a previous crop, and paid for in a previous year. Growing plants in rotation conserves soil fertility and controls pests and weeds – valuable assets that add to yield and quality, but not all crops in a rotation are equally valuable. 'Yield' from a national viewpoint is properly the

Agricultural practices and yield

quantity of material left after storage, when other losses occur (Chapter 10).

7.2. Soil preparation

Waterlogged soil must be drained. Often this is accomplished by a permanent land-drain system, below a mole-drained level (Chapter 8). Applications of lime, to adjust soil pH, should be made before the soil is worked. The object of tillage (working the soil) is to achieve the best growing conditions for the crop. To achieve this weeds, pests and diseases must be suppressed, and a good, damp but well drained and aerated seed bed must be provided. Plant residues, weeds, manures, fertilizers, soil dressings and chopped straw may be incorporated into the soil. However the straw must be fully decomposed before sowing, or the rotting residues will check or prevent seedling growth.

The time and extent of tillage is regulated by the climate, the cropping sequence, and the needs of water conservation, erosion control and weed control. In moist, temperate areas complete clearance of crop and weed residues, and volunteer plants, is desirable. In arid areas trash is left on the surface as a 'mulch' to minimize erosion and evaporation; weeds are destroyed with herbicides to conserve soil moisture [6]. Bare fallows, in which unplanted fields are cultivated at intervals throughout the year, are excellent at reducing infestations of weeds, even of stoloniferous grasses. The land is normally ploughed into large 'clods' in the autumn. Throughout the winter the soil – and weeds – freeze and desiccate. Subsequent ploughings complete the drying and bury seedlings. Trash may be collected, e.g. with chain-harrows, and burnt [112]. Fallows leave the land unproductive for a year, so are now avoided where possible. In addition to controlling weeds in drier areas fallows allow the accumulation of soil moisture, and in drier areas nitrates also accumulate from the activities of nitrogen fixing microbes. With half – or bastard – fallows the land, cleared of a crop in early summer, is ploughed to allow the surface to dry. In the autumn the ground is rapidly broken up to a fine, dry tilth. Weeds are gathered and burnt. The soil inversion and drying kills many weeds and the soil disturbance encourages the seeds of others to germinate so that the later working can kill the seedlings. Cleaning ground in the spring is avoided, since it delays sowing. In Britain it is usual to plough land to a depth of 15–25 cm (6–10 in) in the autumn when preparing for spring barley, care being taken to bury weeds and crop residues [104]. Elsewhere depths of 10–20 cm (4–8 in) are normal [105]. In the spring, when the soil has begun to warm, it is harrowed or disc ploughed and worked well to a depth of 7–10 cm (3–4 in). Other working may be needed to control stoloniferous weeds. The aim is to make a fine tilth in the top 5 cm (2 in) over a well tilled but more compact layer, to promote even seed distribution at a uniform depth, and early seed germination. If the soil is light it may be lightly rolled. Undue soil compaction, as caused by tractor wheels, checks growth badly. Attempts are being made to create and use 'tramlines' in the crops to ensure that the tractors wheels follow the same path for sowing and all subsequent work. This results in more even applications of top-dressings of fertilizers and of sprays, as well as compacting only the soil of the 'tramlines'. In dry weather the soil should be tilled less, to minimize water losses. For autumn-sown barleys the ground should be ploughed as soon as possible after clearing the previous crop [104,105,125]. In arid areas soil may be undercut with 'ducks-foot' blades or broken with chisel ploughs, to leave a surface 'mulch' of trash and stubble to conserve moisture, or trap snow, and minimize erosion by wind and water. Elsewhere the land may be ridged to trap snow, or fallowed to build up soil moisture.

Other methods of cultivation have often been tried. Older studies indicated that ploughed and harrowed soil carried fewer weeds and out-yielded rototilled plots. Usually ploughing below 10 cm (4 in) was not advantageous, but deep tilling or grubbing, to 20 cm (8 in) was. More recently trials with corn (wheat, barley and oats), indicated that on heavier soils deeper working, and subsoiling often increased yield [116]. It was not clear why deep working enhanced yield in one field but not in another. Ploughing at 28 cm (11 in) and 36 cm (14 in) seemed equally effective. Deep ploughing reduced stoloniferous weeds, sometimes produced a drier soil surface and perhaps brought up nitrate from the subsoil. It was essential to avoid bringing up subsoil from below the water table as the newly exposed soil sometimes needed liming. In 'minimal cultivation' husbandry, seeding into poorly worked soil can result in stunted growth [8]. If weeds are adequately controlled by chemical means the method of soil preparation is less critical [4]. Reasons for 'minimal tillage' are the conservation of soil moisture and the reduction of the cost and the acceleration in the rate of working the soil.

7.3. The choice of seed; sowing [104,105,125]

The barley variety chosen must be suitable for local conditions. Adaptation may be quite local [13,66]. For example the short-strawed, upright-leaved variety Midas outyielded Proctor and Zephyr in wetter, low-sunshine areas but not elsewhere [126]. Autumn-sown or early-maturing spring-sown varieties may avoid water shortage, and allow the land to be cleared early. By sowing different varieties maturing at different times it is possible to stagger the harvest. Usually the best can be chosen from a list of varieties thoroughly farm-tested [92]. Evaluation will have included comparative yields, straw strength, other agronomic characteristics, hardiness and disease resistance. The intended use of the crop must also

Agricultural practices and yield 295

be considered – for example only certain varieties are suitable for malting.

The quality of the seed is important. It should be authentic – being only of one named variety, and should have been gently threshed, dried and carefully stored so that it is vigorous and has a high percentage germination, preferably >95%. There are legal quality standards to be met in many countries. The seed should be clean and pure. There are legal limits on the quantities of offensive weed seeds that may be present (Chapter 5) [83]. In Britain poor standards in home-saved seed have led to the spread of wild oats. The seed grain should be free of seed-borne diseases, which may necessitate treatment – e.g. with hot-water or a systemic fungicide (Chapter 9). Before use seed should be dressed to control soil-born fungal pathogens and possibly insect pests. Systemic fungicides, to combat subsequent attacks by mildew for example, will possibly also be applied (Chapter 9). The effectiveness of seed dressings and other types of pest control partly explains the drop in preferred seed rates in Britain from 125–314 kg/ha (2–5 bushels, 112–280 lb, per acre) in about 1940, when only 25–30% of seedlings survived to the 94 kg/ha ($1\frac{1}{2}$ bushels, 84 lb, per acre) usual today when 50–70% establishment is usual [13,108]. Sometimes poisoned bait is also mixed with the seeds to control slugs and prevent damage after sowing.

The relationships between seed sizes and subsequent yields are not simple. Varieties differ in the size range of their seeds, and the seed size varies with growing conditions – raising difficulties in comparative yield trials (Chapter 12). Grown widely spaced, or in isolation, large seeds give rise to larger, more competitive seedlings which grow and tiller more, producing larger plants carrying more grains than plants from small seeds separated from the same grain lot, or even the same ear (Chapter 12) [12,35,71,122]. The removal of 'thin' and 'light' seeds reduces the level of infection with seed-born diseases and perhaps viruses (Chapter 9) [23]. The important question is whether the selected largest seeds, grown closely packed in the field with tillering suppressed by competition, give a crop that outyields a crop from a cleaned but otherwise unselected seed lot. Opinion is divided. When *equal numbers* of large and of unselected seeds are sown then large seeds usually give rise to the better crop. However, if *equal weights* of seeds are sown, the lighter grains are more numerous and may give as good or a better crop [38,72]. In some trials seed from different sites varied in value (up to 16% grain yield) and better crops were obtained from the heavier seeds (Table 7.1) [88,102]. Larger seeds give more vigorous seedlings that establish a closed leaf canopy sooner, and compete more effectively with weeds. On present evidence the use of the 'boldest' grain as seed will often give the best crop, if 1–1.3 million established plants/acre (0.405 ha) are obtained.

The date of sowing has marked effects on cereal yields. The guiding principle is to plant the seed as early in the season as varietal character-

296 Barley

Table 7.1 The characteristics of mature plants of pure varieties grown at about equal densities from large (l) and small (s) seeds (from Pinthus and Osher [102]).

Variety	Seed weight (mg)		Seed N (%)		Main culm length (cm)	
	l	s	l	s	l	s
(a) Glacier	61.5	37.7	1.26	1.27	109	107
(b) Nissani	48.4	31.7	1.51	1.41	101	88
(c) Tunis	52.5	38.3	0.97	0.97	97	93

	Tillers (No./plant)		Spikes (No./plant)		Kernels (No./spike)		Grain (g/plant)	
	l	s	l	s	l	s	l	s
(a)	10.3	8.4	5.70	4.47	46.5	48.0	11.6	8.7
(b)	10.5	10.4	5.29	5.00	46.3	36.8	10.5	7.6
(c)	8.0	7.9	3.65	3.81	42.0	37.0	6.6	6.3

Table 7.2 The effects of sowing date, variety, and fertilizer application on the grain yields of two spring (Proctor, Maris Concord) and two alternative (Maris Puma, Maris Otter) varieties grown in Norfolk, England (from Bell and Kirby [14], 1 cwt/acre = 125.54 kg/ha).

	Yield (cwt/acre)					
	Autumn Sown, (22 Oct. 1962)			Spring Sown (4 April 1963)		
Nitrogen applied (units/acre)	40	60	80	40	60	80
Proctor	13.0	14.5	14.6	33.7	37.8	36.9
Maris Concord	15.2	19.8	20.9	38.9	45.9	43.4
Maris Puma	29.8	33.3	35.7	34.3	32.2	32.4
Maris Otter	33.2	39.9	40.8	32.4	35.3	38.1

istics, the weather and other circumstances permit to give the longest possible growing season (Tables 7.2 and 7.3) [11,14]. In England, soil conditions permitting, spring barleys and 'alternative varieties' (moderately hardy forms not requiring vernalization) are often sown in February or March, but exceptionally seeding may begin in January or may be delayed by bad weather to late April when yields are substantially reduced. It is not admitted that the 'fundamental' test – dropping one's trousers and sitting on the seed bed, to see if it is warm and dry enough for sowing, is still employed. In mild areas spring barleys and alternative varieties may be autumn sown in September or October, as are winter barleys (hardy, with a vernalization requirement). Winter killing can reduce yields

Table 7.3 The effect of sowing date on the subsequent crop of spring barleys, (from Bell and Kirby [14], 1 cwt/acre = 125.5 kg/ha).

	Spratt–Archer			Carlsberg			Maythorpe		
Sowing date	Yield (cwt/acre)	Grain N (%)	TCW* (g)	Yield (cwt/acre)	Grain N (%)	TCW* (g)	Yield (cwt/acre)	Grain N (%)	TCW* (g)
16 March 1955	29.1	1.63	38.2	34.8	1.46	42.5	34.3	1.52	40.5
5 April 1955	25.0	1.69	34.6	28.8	1.61	37.7	29.8	1.57	40.9
27 April 1955	8.7	2.25	33.3	15.2	1.98	35.9	17.5	2.12	34.1

*TCW: Thousand corn weight (dry)

Table 7.4 The effects of different fertilizer (nitrate of lime) applications and seed rates on the yield components of two barley varieties averaged over six years. (from Holm and Pedersen [60], 100 kg/ha = 0.797 cwt/acre).

Seed rate (kg/ha)	200 (normal)			100			50			25		
Nitrate of lime (kg/ha)	Basic	200	400	Basic	200	400	Basic	200	400	Basic	200	400
Grain Yield (Hkg/ha)	40.3	48.0	49.2	39.0	45.2	47.2	36.3	42.1	42.5	32.7	37.0	37.9
Culms (No./m^2)	568	670	733	485	572	616	385	474	512	317	371	409
Grain (No./ear)	15.7	16.3	15.5	17.4	17.6	17.6	19.9	19.1	18.5	21.1	20.9	19.6
1000-grain weight (g)	45.9	45.4	45.0	46.8	45.9	45.4	48.3	47.3	45.9	48.8	47.8	46.8
Plants (No./m^2)	406	—	—	202	—	—	105	—	—	61	—	—
Culms (No./plant)	1.4	1.6	1.8	2.4	2.8	3.0	3.7	4.5	4.9	5.2	6.1	6.7

of insufficiently hardy varieties, and may necessitate re-seeding in the spring (Tables 7.2 and 7.3) [14]. Autumn sown barleys often have a marked yield advantage over spring sown barleys especially if the spring is dry [69]. The lower yield consequent on later sowing goes with less well filled grains, of higher nitrogen content, that are less valuable for malting (Table 7.3) [66].

The density at which seed is sown, the seed rate, depends on the location and season; autumn-sown grain may be planted more thickly where winter killing is expected. Where soil moisture limits growth grain is sown thinly. With late-sown spring barley the seed rate is increased, partly to control, or compensate for, increased levels of weeds and diseases [148]. As the seed rate is increased so the plant density increases, the number of culms/plant decreases, the number of grains/ear declines, the average grain weight declines, but the number of ears/unit area increases (Table 7.4, Chapter 6). Usually the total yield (grain + straw) increases with increasing seed rate, but grain yield levels off, or may show a distinct optimum. Percentage seedling establishment is not altered, but survival to maturity is depressed by increasing plant densities, and senescence is accelerated. The proportion of sterile culms also increases with crowding [52,58,60,75]. The effects of plant density on grain yield are minimal in early-planted crops, but with late sown plants there are marked yield changes with alterations in density [75]. Numerous trials have attempted to establish average optimal seeding rates. British farmers usually sow at rates of about 112–168 kg/ha (100–150 lb/acre). Small-scale trials indicate that grain yields are optimal at 78–101 kg/ha (70–90 lb/acre) when the seed is planted early [144], while the best average seed rate in national trials is about 157 kg/ha (1.25 cwt/acre), for achieving the highest grain yield but the most *economic* rate is likely to be about 125 kg/ha (1 cwt/acre) [58]. An estimate for the best average seed rates in Scotland was 173 kg/ha (about 154 lb/acre). On heavy and light land, in England, optimal rates were estimated to be 141 and 110 kg/ha (126 and 98 lb/acre) respectively [113]. Different cultivars respond differently to altered seeding rates, short varieties with small upright leaves yielding better at high seeding densities, as expected (Chapter 6) [126]. In the U.S.A. seed rates vary from 40–161 kg/ha (36–144 lb/acre) [105]. In Britain barley is generally sown 2.5–3.8 cm (1–1.5 in) deep, greater depths being used in drier conditions. Depths down to 7.6 cm (about 3 in) are used in dry land areas of the world [105]. Deeper sowing carries the risk that more seedlings will not grow up to the soil surface and will die. In Britain barley is usually planted by seed drills, although seed broadcasting machines and 'tail-gate seeders' may be used. In other areas broadcasting may be usual and, in wet seasons in California, the crop may be seeded from the air. The orientation of the seed in the soil (e.g. base up or down, etc) is of no practical importance, but seed drills

Agricultural practices and yield

with hoe (rather than disc) soil openers are sometimes preferred, as these distribute seed in a band, rather than crowded in a slot, so the onset of competition is delayed. Exactly evenly placed plants should grow best, but normal seed drills do not distribute seed like this. Drills with closer coulters, say 9 cm (3.5 in) apart, rather than the usual 18–20 cm (7–8 in), delivering fewer seeds so that the rate per unit area is unaltered give a significantly better distribution. In Britain, when planting at a constant seed rate, increasing row widths above 18 cm (7 in) usually reduces yield, while reducing row spacing increases yields by 2–10% [9,59]. Closer spacing requires the use of 'precision' drills. Recently trials with precision drilling seed in 10 cm (4 in) rows, with 2.5 cm (1 in) between seeds, showed an advantage of 11–13% in yield over conventionally drilled plots. Seedlings were established as 1–1.3 million/acre – a near optimum density for the area. Precision drills are likely to be costly, but the greater yield is attained with the minimal practical amount of expensive seed. Doubts exist about the economic value of closer drilling on heavily fertilized soils. Continuing agricultural trials, taking account of spacing, seed type (unsorted or plump), variety (tall or short stems; floppy or erect leaves), row-direction (N–S or E–W) and soil fertility are needed. Under some conditions, with wide spaces between the coulter rows, corn (small grains) planted in rows running N–S will outyield that in E–W rows owing to better light penetration in to the canopy [7]. In areas where the soil moisture will not support a thick crop, the seed-rate is reduced and the width between the rows is increased to as much as 25–36 cm (10–14 in) [105].

Cereals, used as companion crops, are planted more thinly when undersown with plants being used to establish a ley, for forage. Barley tends to reduce the growth of the undersown crop. The cereal should be cut high (over 20 cm, 8 in) at harvesting to minimize damage to the undersown crop. In dryland conditions results are often better if the forage crop is seeded at right angles to the cereal [73]. Elsewhere undersowing with grasses near to the cereal sowing time can significantly reduce gram yields, but legumes (clovers, trefoil) do not necessarily do so. By sowing clover–grass mixtures a month after the barley, yield reductions are avoided [22]. The risk of reduced cereal yield and enhanced difficulty in harvesting has to be offset against the value of an earlier establishment of the ley. When barley was grown mixed with nitrogen-fixing leguminous crops in sand culture, its nitrogen content was enhanced when grown with peas, but not with red clover or lucerne [94].

7.4. Nutrient supply and barley yield

The need to apply manure to maintain or augment soil fertility has been known at least from the earliest historical times. A scientific understanding of manures – the gradual recognition of the importance of minerals, and the practical difficulties that may arise from their use – has come slowly (Chapter 6) [114,115]. The nature and quantity of manures or fertilizers needed to achieve the best yields depend *inter alia* on the local climate, the cropping sequence and the nature of the soil. Fertilizers need not contain all the essential elements for cereal growth – rather they should supplement the nutrients available in the soil to allow the plants to achieve as near their maximal yields as other factors (water supply, sunlight, lodging, pests), permit.

In various areas applications of sulphate (up to 28 kg S/ha/yr, 25 lb S/acre/year) significantly increase yields and allow greater responses to other added nutrients [95]. In many other areas sulphates have little effect on yield. On sulphur deficient soils application of sulphate gives grain with enhanced levels of β-amylase without altering the nitrogen content; this enzyme is dependent on sulphur-containing thiol groups for its activity [106]. Ammonium sulphate has a tendency to make soil acid, an alteration which reduces grain yields and may show itself by the plants going purple (Chapter 6) [24,84]. Acid soil – considered in Britain as soil with a pH below 6.5, which is higher than in some other areas – is corrected by applications of limestone or chalk (both calcium carbonate, $CaCO_3$) or lime (normally 'slaked lime', $Ca(OH)_2$), which also maintain soil levels of calcium. Dolomitic limestones, and limes derived from them, also contain magnesium and may also be used to correct magnesium deficiencies [25]. In some areas yields are enhanced by applications of magnesium [18], but in other trials applications increased the magnesium concentration of the barley foliage and prevented the appearance of transient deficiency symptoms without increasing yields [61]. The occurrence of magnesium deficiency in Britain may be traced to the decline in use of farmyard manure (FYM).

On some soils yields are increased by applications of micro-elements, of which copper, boron and manganese are the best known [25,127]. Doses required depend on the soil nature and pH, and previous manurial and cropping history, but copper sulphate, 3.4–11.2 kg/ha (3–10 lb/acre) has been successfully used on sandy soils and 11.2–22.4 kg/ha (10–20 lb/acre) were used in N.E. Scotland, where yields of grain were enhanced by 36% and straw by 14% [107]. In various areas applications of other micro-elements increase yields.

Fallowing was important in Britain to 'rest' soil, to control weeds and to achieve increases in soil fertility due to nitrogen fixation. Elsewhere it is still used; in dryland zones a crop may be taken every second or third year. In the absence of added nitrogen a preceding fallow gives a striking increase in yield, which falls off rapidly in succeeding years if no fertilizers are used (Table 7.5) [85,118]. Organic matter in the soil favours the activity of nitrogen-fixing organisms during a fallow. Farmyard

Agricultural practices and yield 301

Table 7.5 Mean yield (lb/acre) of barley after fallowing, with adequate cultivation, after different types of manuring on soil cropped continuously since 1877 (from Mann [85] 1 lb/acre = 1.121 kg/ha). By permission of the Cambridge University Press.

	Yield (lb/acre)		
Period	Unmanured	Artificial manure	Farmyard manure
10 yr before fallow (1917–1926)	404	—	—
After 1st fallow:			
1929	1049	1395	1890
1930	628	745	1064
1931	439	(496)*	887
1932	43	(98)*	447
1933	14	42	146
After 2nd fallow:			
1936	896	1034	1529
1938†	688	834	1006
1939	376	489	668
1940	20	25	135

*Results for some plots were excluded. †In 1937 the barley crop failed.

manure (FYM, dung, stable-manure) is at present comparatively little used on barley in Britain owing to the inconvenience of handling it, the tendency to separate livestock and arable farm projects, and prejudice since the organic nitrogen it contains is released slowly through the growing season giving grain rich in nitrogen, that is unsuited for malting. Nevertheless the faeces and urine of animals mixed with bedding such as litter, straw and sawdust, rotted down in compacted heaps protected from the weather are a valuable source of organic matter and nitrogen, phosphorus and potassium [25,112,136]. The increasing costs of 'artificials' and the need for intensive animal units to dispose of their waste by means that avoid pollution, e.g. of water courses, should encourage the use of FYM. Organic manures, like FYM or ploughed-in legume remains have residual effects over several years (Tables 7.5 and 7.6), whereas nitrogen from 'artificial' fertilizers tends to be leached from the soil by winter rainfall and only contributes to subsequent crops in dry areas. Organic matter in the soil enhances nitrogen fixation and alters mineral uptake by barley seedlings. The use of FYM gave the most reproducible yields in some exhaustion trials [118].

Potassium and phosphate often have prolonged residual effects, e.g. nearly 50 years [47], which may explain the frequent lack of responsiveness of barley to these fertilizers when previous crops have been liberally

302 Barley

Table 7.6 The residual effects of previous leguminous crops, grown for the number of years shown to 1911, on barley. The crop for 1912 was oats. The effect in 1913 approximated to that obtained with an application of 3 cwt ammonium sulphate/acre (376 kg/ha; from Nicol [93]).

Legume, (to 1911)	Yield per acre					
	1913			1914		
	Dressed grain (bu)	Straw (lb)	Total produce (lb)	Dressed grain (bu)	Straw (lb)	Total produce (lb)
Lucerne (7 years)	55.17	2955	6218	32.97	1960	3853
Red Clover (3 years)	38.51	2106	4339	20.26	1187	2347
Alsike Clover (3 years)	33.05	2006	4037	21.94	1266	2522

1 Bushel (bu; British barley) = 25.4 kg; 56 lb nominal; 1 lb = 0.4536 kg

fertilized [142]. On some S. Australian soils initially deficient in phosphate, applications had dramatic stimulatory effects on barley yield for the first five years, but subsequently nitrogenous fertilizers had the major effect [146].

Nitrogenous fertilizers usually have the most marked effects on yield. Under favourable conditions increasing applications of 'N' increase dry matter production and grain production to a maximum, beyond which yield may decline (Chapter 6) [109]. This decline is frequently associated with lodging, and may not occur if lodging is prevented [123]. Older varieties lodge more readily than the new, and varieties differ in their response to nitrogen. Possibly the better yields of newer varieties are mainly due to their resistance to lodging when economically maximum applications of nitrogen are made [40,44]. Newer Danish barleys tiller more than the old, with increasing doses of nitrogen, as well as lodging less [121]. In some trials no lodging or yield increases occur in response to high doses of nitrogen, indicating that some other yield-limiting factor is operating. The increases in grain nitrogen ('protein') that occur above a certain level of fertilizer application make the grain unsuitable for malting, but increase its feed value. The application of nitrogenous fertilizers to barley at different stages has markedly different effects on growth, grain quality, and yield. In general applications made up to the early tillering stage enhance tiller production and survival, so grain yields are increased and grain nitrogen is little altered or may even be depressed (Table 7.7) [65,67]. Thus malting barley should be grown with early applications of moderate amounts of balanced fertilizers. Late applications increase grain nitrogen without greatly increasing yield. To get a maximal yield of high nitrogen grain requires two fertilizer applications, one

Table 7.7 The effects of adding nitrogenous fertilizer to barley at different growth stages (Spratt–Archer, 1937; data of Hunter and Hartley [67]). Plants were sown on 16 April in drills that allowed 77.4 cm² (12 in²) soil/plant, and were harvested on 18 Aug. All results are the means obtained from seven replicated plots, arranged in a Latin square. By permission of the Cambridge University Press.

NaNO₃ applied (dose 125.3 kg/ha; 1 cwt acre) (date)	Grain yield (g/plot)	Tillers/plant, at harvest	Grain N (%)	1000 grain weight (g)	Grain yield (g) per ear	Grain yield (g) per plant	No. plants per plot	No. ears per plot
Control; (no NANO₃)	137.4	0.95	1.66	39.2	0.71	1.41	98.1	190
At sowing (16 April)	206.9	1.50	1.61	39.8	0.86	2.14	97.1	243
At appearance 1st tiller (18 May)	218.4	1.60	1.62	41.1	0.83	2.17	101.6	264
At erection of tillers (4 June)	211.3	1.73	1.60	40.6	0.76	2.07	101.9	278
At flowering (3 July)	145.9	0.91	2.04	41.6	0.78	1.49	97.6	187
3 weeks after flowering (23 July)	134.6	0.81	2.01	40.2	0.75	1.37	98.6	179
At sowing (16 April) and 3 weeks after flowering (23 July)	213.6	1.51	1.91	41.8	0.87	2.16	98.7	247

before and one after tillering or a much larger application early in the life cycle (Table 7.7). The nitrogen content of grain is partly a varietal character, and is dependent on factors besides applied doses of N-fertilizer. Rainfall, particularly on light soils, tends to decrease the nitrogen content of grain, probably by leaching nitrogen either into the subsoil or run off water. Factors that favour growth, such as early sowing, or relieving moisture stress, also tend to reduce it [36,117,151]. Factors which check grain filling favour the formation of thin, steely grains that have high nitrogen contents [65,66]. Responses to added nitrogen are often greater on sandy, readily leached soils [118,119]. Evidently barley plants normally deplete the soil of available nitrogen and use it to make as much growth as possible. Late supplies result in enhanced levels in the grain and also delay ripening. Low tillering, or tiller removal, or thin planting all lead to high nitrogen grains. FYM, unlike artificial fertilizers, is not leached from the soil by rain and supplies nitrogen to the plants throughout the growing season, resulting in nitrogen-rich grain. Similar results occur on soils with high levels of organic matter, such as the Fens of East Anglia, where nitrification proceeds for the whole growing season [119,151].

The relationships between grain size, nitrogen content (% dry matter) and cultural conditions are not simple, but certain trends occur. In a field of a single variety there is a wide range of nitrogen content in individual grains (Chapter 6). Where interplant competition is reduced, or where unequal fertilizer distribution has made high levels of nitrogen available, the grain produced will be larger than average, high in nitrogen and 'coarse, thick-skinned and steely' (in maltsters' terms). Thus the largest grains from one sample often have the highest nitrogen content [42,66,91]. Munro and Beaven [91] give the values for three grain grades (a, b, c) for a particular barley as, thousand corn weight (g), mealy grains (%) and total nitrogen (%) respectively as (a) 51.0, 29, 1.30; (b) 41.6, 69, 1.18; (c) 29.1, 66, 1.19. Within *one ear*, where the nitrogen supply to each grain is potentially the same, the largest grains tend to have the lowest nitrogen content as if the fixed nitrogen supply has been more diluted by photosynthate. Nitrogen is concentrated in the peripheral layers of the true seed, which form a smaller percentage of the whole in plump, starch-gorged grains. There is a high proportion of alcohol-soluble protein, hordein, in immature, steely grains (Chapter 6) [91].

When ears from individual plants are classified by size it is found that the grain and straw associated with smaller ears have the higher nitrogen contents (Table 7.8) [42]. Trials of particular fertilizer applications made each year to barley grown continuously, since 1852, have been reviewed at various times [117–119]. Such trials established the need to maintain supplies of mineral nutrients to barley, and that no mysterious organic

Agricultural practices and yield

Table 7.8 Grain numbers, and the nitrogen contents of the grain and straw associated with tillers bearing ears of different sizes (after Engledow and Wadham [42]).

	Plants with 6 ears			Plants with 7 ears		
Tiller type	Grains (No./ear)	N (%) dry weight		Grains (No./ear)	N (%) dry weight	
		Straw	Grain		Straw	Grain
Large ears	32.0	0.39	2.07	32.8	0.36	1.93
Medium ears	27.5	0.42	2.08	27.9	0.40	2.00
Small ears	16.3	0.64	2.49	21.2	0.54	2.20
Very small ears	14.5	1.10	3.65	14.2	0.59	2.04
Infertile	0	1.26	—	0	1.20	—

factors were involved. However the conclusion, sometimes drawn, that the use of organic fertilizers in conjunction with normal 'artificial' fertilizers is generally valueless is contentious, as it takes no account of supplies of trace elements or the deterioration of soil structure. In many parts of the world failure to use availabe animal fertilizers would be bad economics.

While available, soil potassium, phosphate and other elements may be used up only slowly, although responses to fertilizers containing them can be very striking. Sometimes increased yields are obtained in response to added soluble silicates or sodium chloride. Some grain components seem to vary less than the nitrogen content. For example in a wide range of samples the phosphate contents of grain, as P_2O_5, only ranged between 0.74% and 1.18% with a mean of 0.96% [28]. Soil composition varies between sites and with time. Growth may be limited by factors other than supplies of mineral nutrients. Consequently fertilizer trials are carried out repeatedly, and at many locations.

By increasing the growth of the whole plants fertilizers increase the forage available for grazing animals. However high concentrations of nitrate can occur in plants, especially if excess nitrate is applied or growth is checked shortly before grazing. The nitrate ingested by the animal is reduced to nitrite in the rumen, causing nitrite poisoning and methaemoglobinaemia. Soil deficiencies of molybdenum or manganese, probably involved as co-factors in the reduction of nitrate in the plant, favour nitrate accumulation, as does a deficiency of soluble carbohydrates [149].

'Artificial' fertilizers are sometimes broadcast and harrowed into the seedbed, or are spread among the seedlings at a later time. It is often more efficient to drill the fertilizer with the seed or, to avoid damage due to high local salt concentrations, a band is drilled into the soil slightly to one side of the seeds [55,82]. On some soils only half the quantity of potash or super-phosphate needs to be drilled to get a yield response equal to that obtained by broadcasting [141]. Nitrogenous fertilizers, in contrast, are not used much more efficiently when drilled. Winter cereals require some extra nitrogen in the spring, and this must be supplied as a 'top-dressing', as must second applications of fertilizers to spring-sown seedlings. Excessive amounts of nitrogenous fertilizer drilled with the seed checks growth; it is better to split the application of high doses and give the second part as a top dressing [55,80]. Spraying growing barley with solutions of phosphates may correct deficiencies developing late in the growing season.

The response to added fertilizers is dependent on available moisture levels [81,87]. Ploughing evidently allows more nitrogen fixation than subtilling, chisel-, or disc-ploughing. Fertilizer applications or ploughing only enhanced yields under Californian conditions when rainfall exceeded 33 cm (about 13 in) per year. In drylands barley yields can be reduced by nitrogen applications. The enhanced, early vegetative growth causes the soil to dry faster; this later limits grain formation [81]. Sometimes yields are limited by the availability of nitrogen, phosphorus and water, and the changing optimal supply of phosphate with altered nitrogen and water supplies are well known [87].

The form in which fertilizer elements are supplied alters the cost, ease of handling and efficiency of use by the plant. In field trials nitrogen in the form of nitrate sometimes seems superior to nitrogen in the form of ammonium. In greenhouse trials ammonium chloride was utilized better than ammonium nitrate or ammonium sulphate. Fertilizers are usually mixtures chosen to make good deficiencies of the soil to be dosed. They may be mixed on the farm or purchased ready mixed. The components must not interact in an adverse way – for example to fix a component in an insoluble form. The mixing must be thorough and the mixtures must run smoothly, and be suitable for automatic distribution. More concentrated fertilizers are becoming available, reducing the labour required for their distribution but increasing the need for precision. Ammonium sulphate tends to reduce soil pH; however calcium nitrate maintains it. Where barley is grown in rotation it may be most economic to give heavy doses of potash and phosphorus to a preceding responsive crop, e.g. roots; the cereal will utilize the residues [140].

An understanding of fertilizer practice is complicated by the range of units employed. Doses may be recorded as kg/ha or lb or cwt/acre for example. These may refer to the weight of a salt, e.g. $(NH_4)_2 SO_4$, the weight of an element, e.g. N or P, or the weight of an equivalent of a mineral oxide, e.g. K_2O, or P_2O_5. In Britain 'units' of N, P, and K are 0.01 cwt (1.12 lb; 0.508 kg) of N, P_2O_5 and K_2O respectively. Thus, a 6:15:15 compound fertilizer contains, in 1 cwt, N : P_2O_5 : K_2O in the unit quantities (percentages) shown in the ratio [25]. Recommendations for fertilizer doses only apply to locally evaluated average conditions; individual circumstances should decide the doses actually used [25]. In Britain older recommendations for nitrogen applications to malting barley ranged from none (on rich soil) to 30 units of N/acre, (1 unit/acre = 1.25 kg/ha) applied to the seedbed. With other spring barleys up to 60 units/acre were applied to the seedbed or for winter barley, split 15 units at sowing and 45 units top dressed in the spring. However economic increases in yield can be obtained from an extra 30 units of N, i.e. a total of 90 units/acre [80,141], and higher doses of 100–120 units N/acre (125–150 kg/ha) are needed for maximal yields with continuous barley [90]. Each extra increment of fertilizer produces a smaller increase in yield than the one before. It follows that, in periods of fertilizer shortage, it is often better to use a given quantity of fertilizer at a lower rate, over a greater area to get the best possible increase in overall yield. In Britain modern fertilizer usage has developed from wartime recommendations based on calculated optimal applications [29]. Surveys since then indicate that fertilizer applications have increased. In 1962, depending on district, nitrogen applications averaged about 30–45 units/acre – well below the recommended levels [150]. Since then the average nitrogen application rates have risen to 64–82 units N/acre for spring barley and 74 units/acre for winter barley, doses still too low to achieve maximal yields in some areas [25]. Many fertilizers supply nutrients other than N, P and K – e.g. superphosphate also supplies Ca and S.

7.5. Some chemical treatments

Barley may be treated with sprays, or have poisoned bait spread among the seedlings to control a range of pests and diseases (Chapter 9). There are circumstances where spraying to control weeds is not economically attractive – essentially in conditions where either the weeds present are resistant to the sprays available, or they are not economically important. Weeds can cause large reductions in yield. Early European trials indicated that weed control, by sprays of sulphuric or nitric acids, could increase yields by 20–25% [17]. Weeds compete for soil nutrients and moisture – especially in dryland areas – explaining the importance of preventing development beyond the seedling stage (Chapter 9) [45,96,97]. Pot experiments show that many weeds are less damaging if the barley grows away first. If the weed is well established before the cereal it has a competitive advantage and may compete for light as well as moisture and nutrients [68,86]. No special interactions between the roots of barley and weeds have been found. Weeds vary in their competitive abilities. Chickweed (*Stellaria media*), a fragile-appearing plant, is a damaging competitor, sharply reducing yields even of closely planted barley. Field observation in Manitoba showed that yield reductions of up to 61.5% were attributable to weeds [46].

308 Barley

Lodging reduces grain yield and quality. To minimize this cultural conditions and fertilizer applications are chosen with care, and breeding programmes designed to shorten and strengthen culms are continuously being undertaken. Recently sprays of CCC (chlorocholine chloride) have been used to shorten and thicken the stems of wheat and so increase lodging resistance. High doses regularly applied to barley do reduce plant growth – and increase the soluble protein levels of seedlings. However, in field trials the responses of barley to CCC have generally been marginal and uneconomic [16,64]. Possibly this is because barley degrades CCC faster than wheat, or fails to translocate it within the plant [15]. Other substances have been reported to cause potentially useful effects. Acetylcholine enhances the growth of wheat seedlings and antagonizes the effects of CCC [34]. Experimentally the antimitotic agent maleic hydrazide stunts barley plants [49]. Phenylphosphonic acid diamide enhanced grain yields, but depressed straw yields, in pot experiments with barley [145]. Sub-lethal applications of the herbicide simazine (2-chloro-4,6-bisethylamino-s-triazine) increase the nitrate reductase activity and protein content of various crops – raising the possibility of making them more nutritious to stock [110]. The isopropyl ester of 2,4-dichlorophenoxyacetic acid unexpectedly enhanced yields of barley, particularly when applied with iron salts [62]. Such observations suggest that in time sprays may be found to usefully alter many of the characteristics of cereals.

7.6. Damaging factors

Cultivating or spraying the growing crop necessitates tractors running over a proportion of the plants. The yield reduction may reach 125–188 kg/ha (1–1.5 cwt/acre). However if 'tramlines' of unseeded ground are left these guide the tractor driver, giving more uniform treatments to the field, and yield losses are less. The plants adjacent to the unseeded strips will grow more strongly, taking advantage of the lack of competition.

Barley is damaged by a range of factors other than pests or diseases. It is harmed by levels of sulphur dioxide that occur in smelter fumes and some other industrial waste gases [27]. The gas is readily taken up by barley leaves, particularly when the stomata are open (Chapter 4) [130]. Nevertheless oxides of sulphur, carried to the soil by rain, may prevent sulphur deficiency in some soils near to industrial areas [25].

Winterkilling can be a problem with autumn sown barley, even when the most hardy varieties available are sown. Survival can be enhanced by draining the soil well and planting the grain on the optimal date for that variety [131,152]. Sowing grain at the bottom of furrows 10–13 cm (4–5 in) deep enhances survival – perhaps by shielding the seedlings from wind and accumulating a protective layer of snow [120]. Using 'press wheels' to compact the soil over the seeds sometimes enhances

Agricultural practices and yield 309

survival. Survival is also increased by fertilizers and by reducing the levels of soil pathogens by pre-planting applications of chloropicrin [103].

Frosts may damage barley in various ways besides freezing the plant. When the ground is frozen and 'heaved' by the freezing water autumn-sown seedlings may be pulled apart [26]. When barley is heading frosts may induce sterility, mainly by damaging the anthers (Chapters 6 and 12) [132]. Freezing also damages swelling grain at the milk ripe stage, but not mature grain, resulting in shrivelled kernels with low viability and a high nitrogen content [101,111]. Blindness – infertility, or the production of very small grains – may be caused by frost, heat, or attacks by viruses, fungi, or insects (Chapter 9). Deficiencies of nutrients may induce partial sterility. There are a range of syndromes due to mineral deficiencies, including a combined lack of potassium and magnesium (Chapter 6). Partial sterility often occurs in smooth-awned barleys, e.g. Velvon, perhaps because the genetically linked lack of hairs on the stigma reduces the chances of fertilization [147]. A non-pathogenic leaf-spotting syndrome has been observed, but the cause could not be identified. Colour-banding of seedling leaves has been noted in deep-sown cereals [143].

7.7. Water supplies and yield

Grain will not germinate in waterlogged soil and roots will not grow into a water table, or survive long in waterlogged soil. Wet soil is slow to warm in the spring [112]. Waterlogged soil frequently has a poor structure that improves on drainage. It seems that the water table should be about 275 cm (108 in) down for barley (Chapter 8) [128,135]. Heavy rainfall during the winter tends to leach out soluble soil constituents, especially nitrogen, and reduce yields. Autumn-sown barley makes better use of the winter rainfall than the spring-sown crop. Humid areas are usually the most suitable for the production of high quality barley, at least in the absence of irrigation, and a moist growing season but dry harvest period is favourable [19,50,77]. Rainfall zones clearly limit the areas where barley can be grown without irrigation, in N. America for example [139], and the annual rainfall has a dramatic effect on yields in S. Australia (Chapter 12). In moist areas, where high yields are achieved, the *variabilities* in yields are less than in lower-yielding, drier, but otherwise comparable areas [76]. Cropping and tillage practices may be chosen to conserve soil moisture; fallowing can achieve an accumulation of soil moisture. Run-off of surface moisture may be limited by ridging, stubble mulching, or contour ploughing, that is ploughing around hillsides to check the downward flow of surface water. Breaking up the land helps water penetration from heavy rain, especially if trash remains on the surface, but in dry periods of the year water losses are reduced by flattening and compacting the soil [6,10]. Fallows are sprayed to kill weeds, limiting water-losses caused by their transpiration. Barley varieties vary widely in their ability to withstand arid conditions, and the efficiency with which they use irrigation water [1,6,54,74,124]. Possibly long coleoptiles favour a better percentage emergence from grains planted deeply to reach a moist layer of soil [70]. Growing barley with irrigation is widely used with different levels of sophistication [6]. Probably growth is best when the soil is maintained at a steady moisture level. As only limited quantities of irrigation water are normally available, and this is expensive to distribute, the question is how may the water be used best? In one trial alteration of the time of application of irrigation water altered the components of yield, while the yield itself was little changed [124]. It has been recommended that in Alberta irrigation should be applied when the soil moisture is more than 25% but less than 50% below field capacity. Other experiments suggest that a water content of 62% of field capacity is optimal [53,56]. Frequently water is supplied at the 'critical times' in plant development, rather than when the water potential of the soil reaches a particular value (Chapter 6). In Utah, where the average annual rainfall is about 42 cm (16.5 in) and the soil moisture-holding capacity is good, so that irrigation supplies supplementary water, spreading 20 acre-inches of irrigation water over 4 acres produced three times as much grain as if all the water had been applied to one acre [53]. Irrigation enhances yield, increases plant height, lodging, the time to maturity, the ash content of grain and the proportion of plump grains. Irrigated grain has a reduced nitrogen content and diastatic power (Table 7.9) [51,129]. Irrigation can increase yields at some locations in N. Europe. In Finland one application of 76–94 cm (30–37 in) water up to two weeks before ear emergence increased the yield of barley by up to 50% [41]. In eastern England the potential transpiration of corn often exceeds the summer rainfall, and although practically no

Table 7.9 Average analyses of barley grain grown under different levels of irrigation in Utah over several years (from Greaves and Carter [51]: by permission of the American Society of Biological Chemists).

Irrigation water		Grain composition (%)					
(in)	(cm)	N	Ash	P	K	Ca	Mg
0	0	2.06	2.37	0.309	0.389	0.107	0.179
5	12.7	2.03	2.33	0.302	0.401	0.103	0.177
10	25.4	2.08	2.33	0.300	0.447	0.103	0.178
15	38.1	1.81	2.72	0.316	0.477	0.107	0.186
20	50.8	1.81	2.81	0.335	0.546	0.102	0.195
35	88.9	1.77	2.98	0.402	0.516	0.145	0.171
52.5	133.4	1.74	3.23	0.375	0.444	0.150	0.185

irrigation is carried out there is ample evidence that it could sometimes increase yields [20]. In trials on sandy soil (Woburn) supplemental irrigation was advantageous in about two years out of three [98,99]. In Britain, unless irrigation equipment is available, it may not be profitable to irrigate cereals in most years [99,100].

Irrigation brings its own problems. Late irrigation, after heading, favours lodging, with attendant reductions in yield and quality [30]. Irrigation over porous soils can, like rainfall, leach nutrients. It may be used deliberately to leach out salts either for experimental purposes, or to reduce soil salinity [6]. One of the major hazards of irrigation is that it can cause the build up of soil salinity. This is liable to occur where the salt concentration of the soil is high, and inadequate drainage and/or a high water-table preclude leaching [5,6,39]. The installation of adequate drainage and the use of leaching programmes may allow salinized soil to be reclaimed. High levels of salts check seed germination and plant growth by non-specific osmotic effects and by the toxic effects of specific ions. Varieties of barley are known which are markedly tolerant to saline conditions (Chapter 12) [5,6,33]. The way the soil is worked and the seed is distributed alters plant responses to salinity, because as the water evaporates from the 'high points' of the soil the salts accumulate there, and may even form a crust. Planting barley on the top of raised flat-top beds between the irrigation channels has been advocated, as has planting the seed on the sides of the ridges raised between the irrigation channels, since the salts accumulate along the top of the ridge [6,33].

In Arizona winter barley, grown for forage, has been successfully irrigated with partially treated sewerage effluent. The effluent had a fertilizer equivalent of 65 lb N, 50 lb P_2O_5, and 32 lb K_2O/acre/foot. Some comparative yields of forage, with different water supplies, were: (a) pump water, 100% (13.21 tonne/ha; 5.26 tons/acre); (b) pump water with added fertilizers, 161%; (c) pump water and synthetic sewerage, 234%; (d) sewerage effluent, 212% [32]. The effluent as used was unobjectional, but the risk of toxic agents, excess salts and detergents occurring in harmful amounts was noted.

7.8. Barley as forage

In many areas barley is grown for forage. Often this is autumn-sown since in the warmer areas growth continues through the winter. The forage may be taken by grazing, or in a more controlled fashion by clipping. The clippings may be fed direct, or dried to form hay, or used to make silage. Usually a grain crop is taken subsequently. It is advantageous to clip 'high' above the growing points of the barley or to graze varieties with low placed growing points to minimize damage to the plants and obtain the best grain yields (Table 7.10) [138]. Sheep which graze the plants

Table 7.10 Forage from autumn-sown barley in Tennessee, and the effect of close grazing by sheep, twice, in autumn and spring, on grain yields (from Washko [138], 1 in = 2.54 cm, 1 lb/acre = 1.121 kg/ha; data by permission of the American Society of Agronomy).

Variety	Season	Growth-stage when grazed		Composition of the forage					Character of the grain and the mature plants			
		Plant height (in)	Air-dry forage (lb/acre)	Dry matter (%)	Crude protein (%)	CaO (%)	P_2O_5 (%)			Grain yield (lb/acre)	Mature plant height (in)	Delay in ripening (days)	Number of productive tillers
Missouri Early Beardless (erect variety)	Autumn (Fall)	9.5	638	16.0	27.9	1.03	0.70	Missouri Early Beardless				
	Spring	7.0	933	23.2	16.6	0.45	0.56	(a) ungrazed	1683	43.5	—	26
		(total)	1571					(a) grazed	897	36.0	8	22
Jackson strains (prostrate variety)	Autumn (Fall)	12.0	733	16.1	23.9	1.17	0.68	Jackson strains				
								(b) ungrazed	2013	41.5	—	28
	Spring	8.0	927	22.0	12.8	0.58	0.48	(b) grazed	1237	35.0	5	23
		(total)	1660									

Agricultural practices and yield 313

close to the ground, are particularly damaging. The frequency of grazing or clipping must be controlled. Grain yields are usually reduced, particularly by excessive or late grazing [134]. Sometimes clipping or grazing has prevented lodging as the regrown plants are shorter, and yields have been unaltered or increased [31,57]. In Britain clipping or grazing the young plants is not much used. However, as grain and straw are increasingly fed to livestock the idea of harvesting the whole plant has been evaluated [21]. The yield of nutrients is highest before maturity in barley; it can readily be exceeded by grasses, which can be taken at an equivalent yield four weeks sooner (Chapter 14). Thus whole-barley forage is not generally useful in England, unless the ground must be cleared early. Estimates of the yields of essential amino acids (lysine, methionine, tryptophan, phenylalanine, threonine, valine, leucine, isoleucine) that might be recovered as leaf protein indicate that under some conditions barley might provide 56 kg/ha (50 lb/acre) compared to 336 kg/ha (about 300 lb/acre) from alfalfa, or 168 kg/ha (150 lb/acre) from clover [2]. The legumes have the added advantage of obtaining their own nitrogen supplies by fixation.

7.9. Harvesting the grain

Barley is being harvested somewhere in the world at every time in the year [43]. The importance of careful harvesting and threshing have been noted (Chapter 5). Barley harvested with a binder, or presumably with sickles or scythes, should be taken about a week before dead-ripeness, to reduce the chances of grain or ear shedding. Maturation can be completed before the sheaves are threshed. Combine harvesting standing corn involves simultaneous threshing and for this grain should be fully mature, preferably with a moisture content below 20%. In Britain the moisture content at harvest is normally 14–21%; in S. Australia the value may be 12% or less. Grain moisture shows a diurnal variation, rising at night [79]. In humid climates the ideal practice is to harvest only from noon until just before the evening dew forms. In practice pressure to complete the work usually precludes this. Combine harvesting is rapid and needs comparatively little manpower. However as the plants are dead-ripe grain losses may occur through shattering. To minimize this plants may be swathed, that is mown, while unripe and laid in a windrow on the stubble to dry out. Drying is slightly accelerated by this. Later the combine can raise the plants, using a pick-up attachment, and thresh them in the usual way. Entangled green weeds dry out, thus simplifying threshing. However, periods of rain can reduce the quality of the grain [105]. The effects on grain quality and yield of swathing at different times have been investigated. Different estimates suggest that barley may be safely swathed if head or grain moisture are 40% or less, 35% or less, or

314 Barley

are in the range 18–25% [37,78]. Probably the best moisture content for swathing varies with the local climate.

7.10. Actual and potential yields

Theoretical yields of barley are substantially greater than those actually attained. However uncertain the numerical value of the theoretical yield, the best yields *actually attained* each season in Britain are very much greater than the national average. In Wales the average yield of barley grain is about 3635 kg/ha (29 cwt/acre), but 6278–7533 kg/ha (2.5–3 tons/acre) is immediately attainable with best yields of 12555–15066 kg/ha (5–6 tons/acre) probably becoming attainable in the near future [3]. Surveys in Scotland show that mean yields are often 30–60% below the best – which are usually about 7500 kg/ha (60 cwt/acre). Analysis of the complex of factors that reduce yield show that they act at each stage in production. With Golden Promise barley, seedling establishment averaged about 70%; the potential yield from these was 7370 kg/ha (58.0 cwt/acre). Failure to produce ears reduced the potential yield by 12%, which was reduced further by mildew, 5.6%; *Septoria*, 0.5%; rusts, 0.05%; *Rhynchosporium*, 1.5% other leaf damage 4.1%; other factors, including diseases of plant base, 10.4%; non-productive ears, 0.5%; and harvest losses, about 5%. Thus the final mean yield was 4826 kg/ha (38.5 cwt/acre) [108]. Some believe that with limited population growth, food rationing and a change in diet the next generation of British could feed themselves from their own resources [89]. Others regard this as optimistic. Starvation is already widespread around the world and as the world population is increasing faster than food supplies the situation is deteriorating, and could precipitate a world food crisis by A.D. 2000 [137]. While barley is only one crop among many, it seems intolerable that yields less than the economic optimum should be regarded as acceptable.

Some of the causes of low yields may be unavoidable, but for others there are immediate remedies. Avoidable causes include the consequences of variations in understanding and skill between farmers and the willingness or ability to commit extra labour, fertilizers, pesticides, herbicides, machinery, farm buildings, or other resources as required, and to make sure that top-quality seed is planted into a good seed bed, and that other operations are carried out at the best possible times of year. Farms tend to be short of labour and equipment. An increase in yield must be valuable enough to pay for the 'capacity' to allow the farmer to sow, harvest or otherwise treat a large area at exactly the correct time, as the weather allows. The highest yields are most often attained on experimental farms, or by enthusiasts who pay detailed attention to all aspects of crop production. To simplify farm management, and to allow the

Agricultural practices and yield

economic use of single-purpose equipment and facilities, some farms have specialized in growing barley. The decision to grow cereals continuously usually means a lower yield for a given fertilizer input, coupled with extra problems of weed control and disease build-up (Chapter 9). On the whole yields have been maintained better than expected, where good husbandry has been used, and disease has been less of a problem, but the control of some weeds has been difficult [63,90]. Growing 'break crops' such as beans, to interrupt the barley sequence, allows the land to be cleared. The break-crops themselves must have an adequate cash value, since they have only limited improving effects on subsequent barley crops [63]. An early method for keeping as much land as possible under a corn crop, and in an adequately fertile state, was the Chamberlain system [48,133]. Fertilizers were used and the barley was undersown with trefoil (*Medicago sativa*, a legume, at 11.2 kg/ha; 10 lb/acre), and Italian ryegrass also at 11.2 kg/ha (10 lb/acre). The undersown crop grew away after harvest, used the available soil nitrogen and starved out soil pathogens such as the take-all fungus *(Ophiobolus graminis)*. The undersown crop was ploughed in November–January as the weather allowed – adding to the soil organic matter and releasing nutrients for the next barley crop as it decomposed. Ryegrass may 'carry' fungal pathogens through the winter, and is a strong competitor. Undersowing with trefoil alone can still usefully be used to minimize take-all, although with higher applications of nitrogenous fertilizers its value as a source of nitrogen is reduced [90]. 'Continuous barley' needs high applications of nitrogenous fertilizers, while after clover less nitrogenous fertilizer is needed.

Best yields are obtainable when barley is grown in rotation. The different opportunities for cultivation, choice of herbicide sprays, and so on allow the best possible control of pests and diseases. The simplest 'rotation' under dryland conditions – barley alternating with fallow – has been noted. In mediaeval England, with the three-field system the sequence was autumn corn (wheat), spring corn (barley), and bare fallow. Increasing pressure on land, greater flexibility in the way it could be worked, and the introduction of new crops led through the Norfolk four-course rotation (roots – barley – seeds – wheat) to the many sequences now employed. In England barley is often taken after roots or wheat, sometimes after sugar beet or potatoes [112]. Elsewhere many different cropping sequences are used. In some suitable areas, in parts of California, summer crops can be taken as well as autumn-sown barley each year [105].

References

1 Aase, J.K. (1971). *Agron. J.*, **63**, 425–428.
2 Akeson, W.R. and Stahmann, M.A. (1966). *Econ. Bot.*, **20**, 244–250.
3 Alcock, M.B. (1971). *J. natn. Inst. agric. Bot.*, **12**, 314–330.
4 Allen, E.J. and Barker, M.G. (1972). *J.-agric. Sci., Camb.*, **78**, 57–64.

5 Allison, L.E. (1964). *Adv. Agron.*, **16**, 139–180.
6 Arnon, I. (1971). *Crop Production in Dry Region*. London: Leonard Hill Books.
7 Austenson, H.M. and Larter, E.N. (1969). *Can. J. Pl. Sci.*, **49**, 417–420.
8 Baeumer, K. and Bakermans, W.A.P. (1973). *Adv. Agron.*, **25**, 77–123.
9 Baldwin, J.H. (1963). *Agriculture*, **70**, 414–417.
10 Barnes, O.K. and Bohmont, D.W. (1958). *Agric. exp. Sta. Univ. Wyoming, Bull.*, **358**.
11 Beard, B.H. (1961). *Crop Sci.*, **1**, 300–303.
12 Beaven, E.S. (1920). *J. Frmrs' Club*, **VI**, 107–121.
13 Beaven, E.S. (1947). *Barley – Fifty years of observation and experiment* London: Duckworth.
14 Bell, G.D.H. and Kirby, E.J.M. (1966). In *The Growth of Cereals and Grasses*, eds. Milthorpe, F.L. and Ivins, J.D., pp. 308–319. London: Butterworths.
15 Belzile, L., Paquin, R. and Willemot, C. (1972). *Can. J. Bot.*, **50**, 2665–2672.
16 Blackett, G.A. and Martin, G.I.M. (1970). *Exp. Husb.*, **19**, 95–100.
17 Blackman, G.E. and Templeman, W.G. (1938). *J. agric. Sci., Camb.*, **28**, 247–271.
18 Bolton, J. and Penny, A. (1968). *J. agric. Sci., Camb.*, **70**, 303–311.
19 Buck, S.F. (1961). *J. agric. Sci., Camb.*, **57**, 355–365.
20 Bullen, E.R. (1971). In *Potential Crop Production*, eds. Wareing, P.F. and Cooper, J.P. eds. pp. 250–259. London: Heinemann.
21 Cannell, R.Q. and Jobson, H.T. (1968). *J. agric. Sci., Camb.*, **71**, 337–341.
22 Charles, A.H. (1958). *Field Crop Abst.*, **11**, 233–239.
23 Chiang, L.-Y., and Robertson, J.A. (1968). *Can. J. Pl. Sci.*, **48**, 57–66.
24 Chiasson, T.C. (1964). *Can. J. Pl. Sci.*, **44**, 525–530.
25 Cooke, G.W. (1972). *Fertilising for Maximum Yield*. London: Crosby Lockwood & Son (2nd edition, 1975).
26 Cox, A.E. and Mason, H.J. (1955). *Pl. Path.*, **4**, 148.
27 Crocker, W. (1948). *Growth of Plants – Twenty Years Research at the Boyce Thompson Institute*. New York: Reinhold.
28 Crowther, E.M. (1930). *J. Inst. Brew.*, **36**, 349–351.
29 Crowther, E.M. and Yates, F. (1941). *Emp. J. exp. Agric.*, **9**, 77–97.
30 Day, A.D. (1957). *Agron. J.*, **49**, 536–539.
31 Day, A.D., Thompson, R.K. and McCaughey, W.F. (1968). *Agron. J.*, **60**, 11–12.
32 Day, A.D. and Tucker, T.C. (1960). *Agron. J.*, **52**, 238–239.
33 Day, A.D., Turner, F. and Kirkpatrick, R.M. (1971). *Agron. J.*, **63**, 768–769.
34 Dekhuijzen, H.M. (1973). *Planta*, **111**, 149–156.
35 Demirlicakmak, A., Kaufmann, M.L. and Johnson, L.P.V. (1963). *Can. J. Pl. Sci.*, **43**, 330–337.
36 Dent, J.W., Elliott, C.S. and Silvey, V. (1968). *J. natn. Inst. agric. Bot.*, **11**, 285–292.
37 Dew, D.A. and Bendelow, V.M. (1963). *Can. J. Pl. Sci.*, **43**, 534–541.
38 van Dobben, W.H. (1966). In *The Growth of Cereals and Grasses*, eds. Milthorpe, F.L. and Ivins, J.D. pp. 320–334. London: Butterworths.
39 Dregne, H.E. (1960). *Agric. expl. Sta. Univ. Wyoming, Bull.*, **367**, 3–10.
40 Dubetz, S. and Wells, S.A. (1968). *J. agric. Sci., Camb.*, **70**, 253–256.
41 Elonen, P., Nieminen, L., and Kara, O. (1968). *Field. Crop. Abst.*, **21**, No. 251. (1967). *Maataloust, Aikakansk.*, **39**, 67–98.
42 Engledow, F.L. and Wadham, S.M. (1924). *J. agric. Soc.*, **14**, 325–345.
43 F.A.O. (1959). *World Crop Harvest Calendar* Part *II*, pp. 173–174. Rome: Food and Agriculture Organisation of the U.N.

44 Fiddian, W.E.H. (1970). *J. natn. Inst. agric. Bot.*, **12**, 57–64.
45 Friesen, H.A. and Dew, D.A. (1967). *Can. J. Pl. Sci.*, **47**, 533–537.
46 Friesen, G., Shebeski, L.H. and Robinson, A.D. (1960). *Can. J. Pl. Sci.*, **40**, 652–658.
47 Garner, H.V. (1952). *J. Inst. Brew.*, **58**, 214–215.
48 Garrett, S.D. and Buddin, W. (1947). *Agriculture*, **54**, 425–426.
49 Gifford, E.M. (1956). *Am. J. Bot.*, **43**, 72–80.
50 Grainger, J., Sneddon, J.L., Chisholm, E.C. and Hastie, A. (1955). *Q. Jl. Roy. met. Soc.*, **81**, 108–109.
51 Greaves, J.E. and Carter, E.G. (1923). *J. biol. Chem.*, **58**, 531–541.
52 Guitard, A.A., Newman, J.A. and Hoyt, P.B. (1961). *Can. J. Pl. Sci.*, **41**, 751–758.
53 Harris, F.S. and Pitman, D.W. (1922). *Utah agric. expl. Sta., Bull.*, **178**.
54 Hartmann, R.W. and Allard, R.W. (1964). *Crop. Sci.*, **4**, 424–426.
55 Harvey, P.N. (1964). *Expl. Agriculture*, **11**, 14–21.
56 Hobbs, E.C., Krogman, K.K. and Sonmor, L.G. (1963). *Can. J. Pl. Sci.*, **43**, 441–446.
57 Holliday, R. (1956). *Field Crop Abst.*, **9**, 129–135; 207–213.
58 Holliday, R. (1960). *Field Crop Abst.*, **13**, 1–16; 247–254.
59 Holliday, R. (1963). *Field Crop Abst.*, **16**, 1–11.
60 Holm, S.N. and Pedersen, A. (1962). *Den Kongelige Veterinaer – og. Landbohøjskole Arsskrift*, pp. 62–93.
61 Holmes, M.R.J. (1962). *J. Sci. Fd Agric.*, **13**, 553–556.
62 Huffaker, R.C., Miller, M.D., Baghott, K.G., Smith, F.L. and Schaller, C.W. (1967). *Crop Sci.*, **7**, 17–19.
63 Hughes, R.G. Roebuck, J.F., Collingwood, C.A., Helme, W.H., Gammon, F.C., Lester, E., Evans, S.G., Hooper, L.J. and Campbell, S. (1969). *N.A.A.S. Q. Rev.*, **83**, 129–137.
64 Humphries, E.C. (1968). *Field Crop Abst.*, **21**, 91–99.
65 Hunter, H. (1953). *Agriculture J. Min. Agric.*, England **59**, 536–540.
66 Hunter, H. (1962). In *Barley and Malt*, ed. Cook, A.H. pp. 25–44. New York: Academic Press.
67 Hunter, H. and Hartley, H.O. (1938). *J. agric. Sci.*, **28**, 472–502.
68 Idris, H. and Milthorpe, F.C. (1966). *Oecol. Pl.*, **1**, 143–164.
69 Jenkins, G. (1970). *N.A.A.S. Q. Rev.*, **90**, 74–83.
70 Kaufmann, M.L. (1968). *Can. J. Pl. Sci.*, **48**, 357–361.
71 Kaufmann, M.L. and Guitard, A.A. (1967). *Can. J. Pl. Sci.*, **47**, 73–78.
72 Kiesselbach, T.A. (1924). *J. Am. Soc. Agron.*, **16**, 670–682.
73 Kilcher, M.R. and Heinrichs, D.H. (1960). *Can. J. Pl. Sci.*, **40**, 81–93.
74 Kirby, E.J.M. (1968). *J. agric. Sci., Camb.*, **71**, 47–52.
75 Kirby, E.J.M. (1969). *Ann. appl. Biol.*, **63**, 513–521.
76 Klages, K.H. (1934). *Ecology*, **12**, 334–345.
77 Klages, K.H. (1960). *Wyoming agr. expl. Sta. Bull.*, **367**, 11–32.
78 Koenig, R.F., Robertson, D.W. and Dickson, A.D. (1965). *Crop Sci.*, **5**, 159–161.
79 Larmour, R.K., Geddes, W.F., Malloch, J.G., and McCalla, A.G. (1935). *Can. J. Res.*, **13C**, 134–159.
80 Lessells, W.J. and Webber, J. (1965). *Expl. Husb.*, **12**, 62–73.
81 Luebs, R.E. and Laag, A.E. (1967). *Agron. J.*, **59**, 219–222.
82 Lutz, J.A. Terman, G.L. and Anthony, J.L. (1961). *Agron. J.*, **83**, 303–305.
83 M.A.F.F. (1966). *Advisory Leaflet 247, Quality in Seeds*.
84 Mann, H.H. (1937). *J. agric. Sci., Camb.*, **27**, 108–122.
85 Mann, H.H. (1943). *J. Agric. Sci., Camb.*, **33**, 207–212.
86 Mann, H.H. and Barnes, T.W. (1952). *Ann. appl. Biol.*, **39**, 111–119.

87 Martin, W.E. and Mikkelsen, D.S. (1960). *Bull. Californian agr. expl. Sta,* 775.
88 McFadden, A.D. (1963). *Can. J. Pl. Sci.,* **43**, 295–300.
89 Mellanby, K. (1975). *Can Britain Feed Itself?* London: Merlin Press.
90 Mundy, E.J. (1969). *Expl. Husb.,* **18**, 91–101.
91 Munro, J.M.H. and Beaven, E.S. (1900). *J. r. agric. Soc.,* **11**, 185–251.
92 N.I.A.B. (current year), *National Institute of Agricultural Botany – Farmer's leaflet No. 8 – Recommended varieties of Cereals.*
93 Nicol, H. (1933). *Emp. J. exp. Agric.,* **1**, 22–32.
94 Nowotnówna, A. (1937). *J. agric. Sci., Camb.,* **27**, 503–510.
95 Nyborg, M. (1968). *Can. J. Soil Sci.,* **48**, 37–41.
96 Pavlychenko, T.K. (1937). *Ecology,* **18**, 62–79.
97 Pavlychenko, T.K. and Harrington, J.B. (1935). *Sci. Agric.,* **16**, 151–166.
98 Penman, H.L. (1962). *J. agric. Sci., Camb.,* **58**, 343–348; 349–364; 365–379.
99 Penman, H.L. (1970). *J. agric. Sci., Camb.,* **75**, 69–73; 75–88; 89–102.
100 Penman, H.L. (1971). In *Potential Crop Production,* eds. Wareing, P.F. and Cooper, J.P. pp. 89–99. London: Heinemann.
101 Pessi, Y. and Kivinen, P. (1957). *Valt. Maatalousk Julk.* **161.** via (1959) *Field Crop Abst.,* **12**, 116, No. 702.
102 Pinthus, J.J. and Osher, R. (1966). *Israel J. agric. Res.,* **16**, 53–58.
103 Price, P.B. and Pederson, V.D. (1965). *Agron. J.,* **57**, 199–201.
104 Rayns, F. (1959). *Barley* – (MAFF publication), London: H.M.S.O.
105 Reid, D.A., Shands, R.G. and Suneson, C.A. (1968). In *U.S. dept. Agric., A.R.S. – Handbook, No. 338. Barley* pp. 32–38.
106 Reisenauer, H.M. and Dickson, A.D. (1961). *Agron. J.,* **53**, 192–195.
107 Reith, J.W.S. (1968). *J. agric. Sci., Camb.,* **70**, 39–45.
108 Richardson, M.J. (1975). *ADAS Q. Rev.,* **16**, 152–163.
109 Richardson, A.E.V. and Gurney, H.C. (1933). *Emp. J. exp. Agric.,* **1**, 325–332.
110 Ries, S.K., Chmiel, H.C., Dilley, D.R. and Filner, P. (1967). *Proc. natn. Acad. Sci. U.S.A.,* **58**, 526–532.
111 Robertson, D.W., Haus, T.E. and Hoff, J.C. (1959). *Agron. J.,* **51**, 658–660.
112 Robinson, D.H. (ed. 1972). *Fream's Elements of Agriculture* (15th edition), London: John Murray.
113 Rule, J.S. and Fiddian, W.E.H. (1974). *Expl. Husb.,* **27**, 79–98.
114 Russell, E.J. (1909). In *The Standard Cyclopedia of Modern Agriculture and Rural Economy,* ed. Wright, R.P. **V**, 222–235. London: Gresham Publishing Co.
115 Russell, E.W. (1973). *Soil Conditions and Plant Growth* (10th edition), London: Longmans.
116 Russell, E.W. (1956). *J. agric. Sci., Camb.,* **48**, 129–144.
117 Russell, E.J. and Bishop, L.R. (1933). *J. Inst. Brew.,* **39**, 287–421.
118 Russell, E.J. and Watson, D.J. (1938). *Emp. J. expl. Agric.,* **6**, 268–292; 293–314.
119 Russell, E.J. and Watson, D.J. (1939). *Emp. J. expl. Agric.,* **7**, 193–220.
120 Salmon, S.C. (1916). *J. Am. Soc. Agron.,* **8**, 176–188.
121 Sandfær, J. (1953). *Danish Royal Veterinary and Agricultural College Yearbook,* pp. 1–14.
122 Sandfær, J. (1970). *Risø report No. 230* Danish Atomic Energy Research Commission.
123 Sandfær, J., Jørgensen, J.H. and Haahr, V. (1965). *Danish Royal Veterinary and Agricultural College Yearbook,* pp. 153–180.
124 Schreiber, H.A. and Stanberry, C.O. (1965). *Agron. J.,* **57**, 442–445.
125 Shands, H.L. and Dickson, A.D. (1953). *Econ. Bot.,* **7**, 3–26.

126 Silvey, V. and Fiddian, W.E.H. (1972). *J. natn. Inst. agric. Bot.*, **12**, 477–485.
127 Smilde, K.W. and Henkens, C.H. (1967). *Neth. J. agric. Sci.*, **15**, 249–258.
128 Smith, L.P. (1972). *Outlook on Agriculture*, **7**, 79–83.
129 Sosulski, F.W. and Bendelow, V.M. (1964). *Can. J. Pl. Sci.*, **44**, 509–514.
130 Spedding, D.J. (1969). *Nature*, **224**, 1229–1231.
131 Stickler, F.C. and Pauli, A.W. (1964). *Crop Sci.*, **4**, 487–489.
132 Suneson, C.A. (1941). *J. Am. Soc. Agron.*, **33**, 829–834.
133 Sylvester (1947). *Agriculture*, **54**, 9, 422–424.
134 Thakur, C. and Shands, H.L. (1954). *Agron. J.*, **46**, 15–19.
135 Visser, W.C. (1961). In *The Water Relations of Plants*, eds. Rutter, A.J. and Whitehead, F.H. pp. 326–336; 356–365. London: Blackwell.
136 Voelcker, J.A. (1909). In *The Standard Cyclopedia of Modern Agriculture and Rural Economy*, ed. Wright, R.P., **5**, 163–168. London: Gresham Publishing Co.
137 Wareing, P.F. (1971). In *Potential Crop Production*, eds. Wareing, P.F. and Cooper, J.P. pp. 362–378. London: Heinemann.
138 Washko, J.B. (1947). *J. Am. Soc. Agron.*, **39**, 659–666.
139 Weaver, J.C. (1950). *American Barley Production – A Study in Agricultural Geography*. pp. 54–65, Minneapolis: Burgess.
140 Webber, J., Boyd, D.A. and Hill, J.R. (1968). *Expl. Husb.*, **17**, 36–44.
141 Widdowson, F.V. and Penny, A. (1966). *Expl. Husb.*, **14**, 83–97.
142 Widdowson, F.V. and Penny, A. (1968). *J. agric. Sci., Camb.*, **70**, 53–58.
143 Wilcox, H.J. (1959). *Pl. Path.*, **8**, 34–35.
144 Willey, R.W. and Holliday, R. (1971). *J. agric. Sci., Camb.*, **77**, 445–452.
145 Williams, E.C. (1970). *Nature*, **227**, 84.
146 Woodroffe, K. and Williams, C.H. (1953). *Aust. J. agric. Res.*, **4**, 127–150.
147 Woodward, R.W. (1949). *Agron. J.*, **41**, 430–434.
148 Woodward, R.W. (1956). *Agron. J.*, **48**, 160–162.
149 Wright, M.J. and Davison, K.L. (1964). *Adv. Agron.*, **16**, 197–247.
150 Yates, F. and Boyd, D.A. (1965). *Outlook on Agriculture*, **4**, 203–210.
151 Yates, F. and Watson, D.J. (1939). *J. agric. Sci.*, **29**, 452–458.
152 Young, A.L. and Feltner, K.C. (1966). *Crop Sci.*, **6**, 547–551.

Chapter 8
Production and harvesting machinery

8.1. Introduction

In various places barley is grown and harvested with equipment that ranges from the ultramodern to that which would have been familiar 3000 years ago. Consequently some note of the less advanced techniques and machines will be taken here. A 'stepwise' modernization programme in a backward area, following the guidelines of historical development, might be one way of increasing world food production. For example the replacement of scratch-ploughs with single furrow mouldboard ploughs, also animal-drawn, would achieve improved tillage without the displacement of men or organizational disruption caused by the introduction of tractor-drawn ploughs, which need fuels, lubricants, and 'back-up facilities' and which are used most economically on large arable units having appropriate grain-handling, storage, and marketing facilities.

To clear the ground in Neolithic times trees were ring-barked with stone axes and, when dead, they and the undergrowth were burnt. Then cereals, including barley, were grown in the cleared and ash-fertilized land. In modern farming practice 'clearance' usually means killing weeds and removing or burying unwanted residues of previous crops (Chapter 7). However grubbing out trees and hedges, killing undergrowth, disposing of the trash and levelling the ground may all be involved.

8.2. Irrigation and drainage

For healthy growth barley requires soil that is damp, but not waterlogged. Thus water may need to be brought to the fields and arrangements must be made to carry away any excess. Irrigation is an ancient practice that carries the risk of soil salination, and ultimately the decline in the value of the land. Water has been obtained in many ways from the heaps

Production and harvesting machinery 321

of stones used as 'dew traps' by the Nabataeans in Palestine [5], to the distribution of water from rivers and lakes by canals, and aqueducts and the pumping of water from wells and boreholes. Water has been raised into irrigation ditches by shadoofs, Archemidean screws, by wheels with buckets at the periphery, or many types of pumps operated by human, animal, wind, electrical or engine power. The supply of artesian water has allowed barley to be grown in the Sahara. To spread irrigation water the soil must be carefully graded; this is now often carried out by machines (land-planes). The water may be spread by flooding the soil before sowing, or by leading it along channels between small ridges on which the plants are growing [5].

Neolithic agriculture in Britain was concentrated on quick draining soils. The Romans extended cereal farming to lower-lying, heavier soils with the help of drainage systems. Ditches were cut across fields, the bottoms were filled with faggots of wood which were covered with straw and stones, then the soil was returned. Excess water percolated downwards then drained through the channels in the faggots into collecting ditches, and ultimately into streams and rivers. Now permanent land-drains are normally of cylindrical tile-pipes (8–15 cm, 3–6 in) internal diameter and usually about 30 cm. (12 in) long, or of continuous porous plastic pipe. The drainage system used depends on whether it is required to deal with rain water, spring-water, a fluctuating water table, or drainage from adjacent land. Tile-pipes are usually laid in ditches and covered with porous back-fill (often gravel or granular industrial waste), but some plastic pipes may be 'ploughed in' without first digging a ditch. Drains and ditches are made by tractor-mounted power arm diggers or more specialized machines, usually owned by contractors. Although most drainage water flow is under gravity it is sometimes necessary to pump water to a higher level. Pumps may be powered in various ways – e.g. by electrical or diesel motors. In many areas the pumps used to be driven by windmills.

In the English Midlands surface drainage was encouraged by ploughing the soil into 'lands', a series of ridges with small water-furrows between, arranged to lead downhill to a ditch. As modern drainage systems occur some feet down in the soil, certain cultivation practices are needed to allow moisture to reach them. Heavy farm equipment must not be allowed on wet land, otherwise the soil is compressed and becomes impervious. The effects of soil compaction are often seen near field-gates, where plant growth is reduced or may even cease. By using multi-wheel tractors or by bolting metalwork 'cages' to the sides of tractor wheels the load is spread over a wider area and less soil compression occurs. Repeated ploughing to the same depth may form a compressed 'pan' some 20–30 cm (8–12 in) down, which prevents drainage. Pans may be broken up by deep-working chisel-ploughs, or subsoiling ploughs

322 Barley

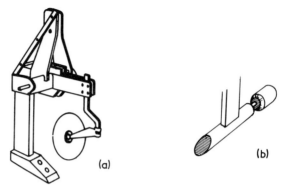

Fig. 8.1 (a) Diagram of a subsoiler with a disc-coulter. The subsoiling attachment might be replaced by a mole, (b), (after Lovegrove [11]).

which break pans and shatter the subsoil without bringing it to the surface (Fig. 8.1). To prevent the vertical slots in the soil closing quickly, organic materials, such as chopped cereal stubble, have been used experimentally to fill them [23]. A widely used device is the mole-drainer or plough. A cylindrical plug, sometimes with an 'expander' behind, is mounted at the base of a vertical blade, which is preceded by a coulter blade. This 'mole' is forced through the subsoil forming an unlined, cylindrical drain 7.5–10 cm (3–4 in) in diameter (Fig. 8.1). This may stay open for some years. Mole drains are often 51–76 cm (20–30 in) below the surface, and run parallel at 2.7–15.1 m (3–16.5 yd) intervals, the frequency depending on soil type. A field may contain land-drains sloping downhill, a system of mole-drains above the land drains and then the ploughed topsoil. Ditches at the sides catch and divert water from higher elevations and receive water from the land drains.

8.3. Tillage

Tools used to prepare seedbeds have included stone-weighted digging sticks, mattocks, hoes, spades, forks, breast-ploughs, Irish loys, Skye caschroms and numerous types of cultivators and ploughs. Some Roman ards (scratch ploughs) had replaceable metal points and projecting metal wings or 'ground-wrests' mounted each side behind the point. Ground was probably 'scratched' at least twice at right-angles. Coulter-ploughs, as used early this century, might be steadied by wheels, (Fig. 8.2) or have no wheels ('swing', or foot ploughs). A modern plough, like the older ones, has an upright blade or coulter, which makes a vertical cut in the soil (Figs. 8.3 and 8.4). Behind and below it another blade, the share, makes a horizontal cut, so that as the plough advances a furrow slice is cut. A curved metal surface, the mould-board twists the cut furrow-slice over

Production and harvesting machinery 323

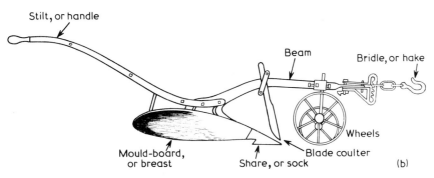

Fig. 8.2 Older types of plough (a) From a Mesopotamian clay-seal, about B.C. 2000. It is an ard, with a tube down which seed is being delivered. (b) A horse-drawn, single-furrow wheel-plough of about 1910. The bridle or hake connected with the harness pulled by the horse. (a) after Hodges [10a] (b) after Tyler and Haining [26]).

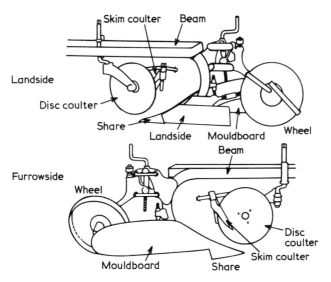

Fig. 8.3 The body of a tractor-trailed, single-furrow, mouldboard plough (after Robinson [21]).

324 *Barley*

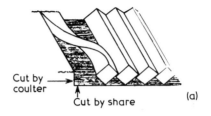

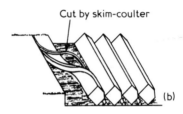

Fig. 8.4 Diagram of the way in which (a) a single mouldboard plough cuts and partially inverts a furrow-slice. (b) The effect of a skim-coulter; by paring off the top, inner corner of the furrow slice this part (and any weeds on it), are more deeply buried.

through about 130°, so that it is nearly inverted, and weeds and crop residues are buried, as they rest mainly against the base of a previous slice (Fig. 8.4). The flat 'landside' of the plough bears against the side of the soil cut by the coulter. After ploughing a field lies in parallel ridges formed by the partly inverted furrow-slices. With tractors or big animal teams, and especially on light soils, each plough body may support several sets of blades, staggered behind each other. Modern tractor-ploughs may carry 4, 5 or even up to 12 sets of blades. When steam-ploughing, steerable six-furrow ploughs were pulled to and fro across the field by reversing the pull on steel cables. To avoid the formation of gulleys, 'one-way' or 'reversible' ploughs were devised. In these one plough set would work the soil in one pass, while another was lifted clear; on the return pass the plough was 'tipped over' and the second set worked. As the two sets of mould-boards were twisted in opposite senses the furrow slices from all passes were turned one way [26,27]. Many modern tractor-ploughs are also reversible; right- and left-handed plough bodies are mounted on each side of a beam which can be rotated to bring either set into operation [7,11,25].

Modern tractor-drawn ploughs are of all-metal construction with readily replaced, interchangeable parts. Adjustments may be made – e.g. to control the depth of ploughing and the alignment of the parts. Sub-soiling tines may be fitted. The type of parts fitted, e.g. the shares, may be altered for different soil types or to achieve different results. The beam may be steadied by land-wheels. On heavier soils or with ploughs with higher numbers of blade-sets greater tractor power is needed. This in turn means larger, heavier tractors with greater risks of soil compaction. Some ploughs are tractor-mounted, and so provision is made to lift them clear, into the air, to allow for backing or travelling along roads. Other ploughs are separate units. Blade coulters may be preferred on hard soil or in some other difficult circumstances. Usually the coulter is a cutting disc, either plain or wavy-edged. A skim-coulter, or skimmer, may be mounted to achieve more efficient burial of surface trash (Fig. 8.4). Ploughing to a depth of about 15 cm (6 in) is often considered adequate but local norms vary [19–21], and subsoiling is often practiced.

Disc ploughs have saucer-shaped steel discs, 60–90 cm (2–3 ft) in diameter, set to dig into the soil at an angle. These leave a low profile of well-pulverized small ridges and furrows (Fig. 8.5). They leave some surface trash unburied. Chisel ploughs resemble heavy-duty, rigid line cultivators (Fig. 8.6). Tines penetrate to 36 cm (14 in) or more and break up and shatter the soil without inverting it, or bringing subsoil to the surface. These allow fast working but they do not bury surface trash.

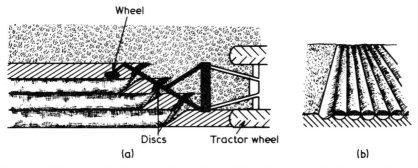

Fig. 8.5 Diagram of a disc plough working, (a) in plan view, (b) the type of soil-disturbance produced (after Lovegrove [11]).

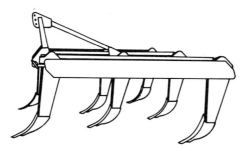

Fig. 8.6 A chisel plough.

They can conveniently be used after applications of total weedkillers. Chisel ploughs, wide 'ducks-foot' blades, and other devices may be used to cut along in the soil and disturb it, and yet leave the surface protected by plant remains.

After ploughing, furrow slices are usually left to weather, and then are reduced by further cultivation to a tilth suitable for a seed bed, often by harrowing. The soil and clods are broken into small pieces, levelling the soil surface, and firming it so that it is not 'fluffy'. There are many types of tractor-drawn harrows ranging from simple metal frames often of a zig-zag pattern, of varying weights, carrying spikes, to types that resemble small chisel-ploughs, or those with spring-loaded, spring-steel tines which vibrate and help disintegrate clods as they are pulled through the soil [7,11,25]. Some harrows have tines that are vibrated mechanically by a drive from the tractor. In some rigid-frame harrows the angle of the spikes to the soil can be adjusted. Chain-link harrows are essentially nets of metal links carrying downward-directed spikes; they readily follow irregularities in the ground surface. Disc-harrows have sets of saucer-shaped steel discs, 30–50 cm, (12–20 in) in diameter mounted in sets in frames which hold them at an angle to the direction of advance to obtain soil penetration (Fig. 8.7). Successive sets of discs move the soil in opposite directions. These harrows 'firm' or lightly compact the soil. Wide areas may be worked by pulling 'gangs' of harrows, possibly followed by rollers. Rollers, which firm and level the soil and break up clods, may be hollow or solid and concrete-filled; they may be divided into independent sections or rings, and may have smooth, ridged or spiked surfaces. Other mechanical cultivators have been devised to break up the soil in place of, or as an adjunct to, ploughing. Tractor-

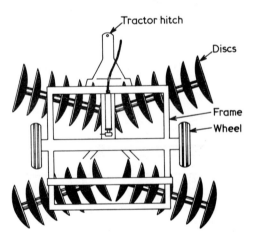

Fig. 8.7 A wheel-mounted, four-gang disc-harrow in plan view (after Lovegrove [11]).

Production and harvesting machinery 327

powered cultivators (trailed or mounted) which break up the soil by sets of rapidly rotating (100–250 r.p.m.) horizontally or vertically mounted sets of blades or tines are used from time to time. These chop and vigorously stir the soil. Opinions are divided on the suitability of these implements for the preparation of cereal seed-beds. Sometimes they produce too fluffy a tilth.

8.4. Sowing

Until recently seed was usually sown 'broadcast', the sower walking across the field carrying the seed and casting it onto the seedbed, using both hands if he was skilled [28]. The seed was harrowed into the soil quickly to minimize losses to birds. Such techniques are still in use in the Near East, as are 'scratch ploughs' with a seed-tube down which grain is dropped into the furrow, a primitive seed drill known in Babylonian times and at a similar period in China (Fig. 8.2) [5,8]. The history of seeding devices is complex [1,9]. In England, from about 1600 onwards, there was a limited interest in 'setting' seed. A man with 'dibbling-irons' (spiked metal rods swollen and pointed at the tip, and fitted with handles) in each hand, walked backwards prodding regularly spaced holes into each furrow-slice. Behind came 2–3 women or children, the 'droppers', dropping three or four seeds into each hole. Each group 'set' about 0.4 ha (1 acre) every two days. The technique was used in this century for planting grain on small, private allotments and has been used recently to sow small experimental plots. For precision seed spacing can be dictated by setting-boards, containing a regular pattern of guide holes. Grain may be sown 'broadcast' from machines or dropped from tail-gate seeders. Usually, however, it is 'drilled' in the U.K. except on very wet land. Many types of seed drills are used. The grain is held in hoppers which are divided vertically to prevent lateral movement. The machines are mounted on wheels and are trailed. The seeding mechanism may be driven from the land-wheels or by power take-off (p.t.o.) from the tractor (7,11,14,21,25]. A set of coulters opens slots in the soil to a chosen depth to receive the regulated streams of grain. The coulters may be discs, hoes, double-discs or Suffolk types. Ideally the seeds should be evenly spaced and covered to an even depth to obtain uniform growth. Grain is fed at a predetermined rate into seed tubes which drop it behind the coulters (Fig. 8.8). The metering devices in use are usually based on cup-feed, internal force-feed, or external force-feed mechanisms. Controls are provided to regulate the rate of seeding. When the coulters and tubes are raised (e.g. to turn the drill) the seed delivery system stops working. Uniform seeding rates must be obtained despite machine vibration, angle or slope of the soil, or other disturbing factors. Each seed drill has many grain delivery sets. Usually it will be

328 Barley

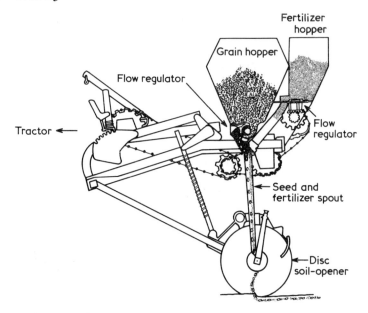

Fig. 8.8 One unit of a combined seed and fertilizer drill in which the grain and fertilizer, held in separate hoppers, are metered into a spout which discharges into a slot in the soil created by a disc-opener (after Robinson [21]).

followed by a harrow. At sowing the seed-bed should be level and of the correct tilth, with a fine surface and firm base so that the seed-depth is regular. When crops are 'tramlined' seed-spouts exactly in line with the tractor-wheels are blocked off. As the crop grows, the 'wheel tracks' remain bare, and guide the tractors over exactly the correct path in subsequent operations. It is usual to employ combine drills, which dispense measured amounts of fertilizer either with the seed down the seed tube, or preferably 2.5–3.8 cm (1–1.5 in) to one side and slightly below the grain, where it will be used more efficiently than if it is broadcast (Fig. 8.8). Combine drills work more slowly than simple seed drills, when these are used with spreaders that broadcast fertilizers, but their use involves heavy machinery on the field only once, so soil compaction is reduced. Placing fertilizer as deep as 9 cm (3.5 in) with spring-sown barley may be advantageous, especially in a dry season. The fertilizer is metered into the delivery tubes from a hopper, often by a star-wheel dispenser [7].

The direct drilling of seed into the stubble of a previous crop is a recent development. With the use of effective 'total' weed-killers the ground is cleared, and then the seed is drilled. Labour and cultivation requirements are reduced. However the surface trash tends to clog conventional seed drills, and so new types are needed to make this technique reliable.

Acceptable yields can be obtained from direct-drilled barley. A 'compromise' approach is the idea of minimal cultivation. A shallow primary cultivation of the 'weed-free' land is followed later by a second cultivation combined with seed-drilling. A rig used to test this approach, with apparent success, has the tractor pulling a cultivator, and a seed drill following immediately behind linked to the tractor by a metal frame bridge. Various other cultivating devices are also being tested [18].

8.5. Post-sowing treatments

To firm the soil and re-bed the seedlings after heaving, caused by water freezing in the soil, fields may be lightly rolled.

Weeds have always been a problem in cereal fields. Once they were removed by hand, but now they are controlled by cultivations and applications of herbicides. Systemic fungicides have also become available and some of these, and insecticides, are applied in sprays; others are applied as seed-dressings (Chapter 9). Other chemical sprays may become available e.g. to limit lodging. Sprays are usually applied from tractors which carry liquid reservoirs, pumps, and spray-nozzles (usually mounted on booms). The cones of spray from the jets must give an even coverage, and avoid overlapping or missing areas. The tractor driver must drive accurately along chosen tracks ('tramlines' assist in this), and the height of the spray nozzles and the spread of the cones of sprayed liquid droplets must be set correctly. Sprays may be applied in high volumes, by convention 670–1130 l/ha, (60–100 gal/acre); medium volume 340–670 l/ha (30–60 gal/acre); low volumes, up to 340 l/ha (30 gal/acre) [25]. With low-volume sprays a sprayer can cover a larger area before needing to refill. Wide areas can be quickly covered by sprays from light aircraft or helicopters.

A second, 'top' dressing of chemical fertilizer is often applied to the growing crop. Usually concentrated granular material is spread from tractor-mounted or trailed hoppers either by oscillating 'pendulum' spouts, or is broadcast by spinning vane-carrying discs. Otherwise the fertilizer may be dropped from beneath a wide hopper, for example being measured and dispensed by a plate and flicker mechanism. Alternatively, reciprocating-plate distributors may be mounted beneath the hopper. Fertilizers should be spread uniformly for the best results. However, the results of recent trials show that even distribution is exceptional.

Before the soil is ploughed, lime, marl, slurry or farmyard manure may be spread. Lime is usually broadcast by spinning discs fed from hoppers via endless belts. Farmyard manure (FYM) is bulky, for example it may comprise cattle-droppings mixed with bedding straw, or semi-liquid slurry. FYM is usually spread from tractor-pulled trailers, from which it may be thrown by rapidly moving prongs or chain flails. Liquid farmyard manure may be spread from tankers or may be comminuted, then pumped through pipework, to be dispensed as a coarse spray. Ammonia liquid, the gaseous material liquefied under pressure, or ammonia solutions (some under pressure) may be injected directly into ploughed soil, at least 15 cm (6 in) below the surface, where it is rapidly fixed and is available for plant use [21].

8.6. Harvesting and threshing barley

Hordeum spontaneum was probably originally harvested by pulling the ears, 'brushing' the fragile ripe ears into baskets, or by reaping the unripe plants with sickles. In 1855 it was estimated that in a 10-hour day a man could reap 1.6 ha (4 acres) of barley with a scythe, or 0.8 ha (2 acres), with a smooth-edged or toothed sickle. Harvesting on a larger scale was carried out by teams of men using scythes often fitted with wooden cradles or 'barley bales'. Stems were cut 10–15 cm (4–6 in) above the soil and the severed plants were laid on the stubble in a swathe. Others tied bundles of plants into sheaves with twisted straw [8,9]. The Bell reaping machine of 1828 was pushed by horses and it cut the plants which were laid in a swathe [28]. In later reaper-binders a cutter bar with fixed and oscillating teeth cut the plants at a chosen distance above the soil. Rotating vanes steadied the plants and as they were cut swept them back onto an endless belt. This carried them into the machine where they were automatically gathered, tied into sheaves and unloaded. The sheaves were later stooked by hand, i.e. arranged in threes or fours, resting on the cut ends with ears in the air and left to dry. Tractor-drawn reaper-binders have mainly been replaced by combine harvesters [22]. Reaper-binders were used to cut the plants some days before they were dead-ripe, to minimize grain shedding. After drying the stooked sheaves were gathered, carted and built into stacks.

These corn ricks were often built on a layer of branches to lift them clear of the ground. The sheaves were laid ears inwards, the severed butts facing the outside. The rick was built up to a ridge or point and capped with thatching. Here the barley remained, to be threshed as needed. It was liable to rodent depredations and, if the rick was wet and built without ventilation spaces, the rotting straw would heat and damage the grain, or even ignite.

Originally grain was probably rubbed from the ears by hand. If the grain was for food the process could be made easier by scorching away the awns by 'parching' the ears in a small straw fire. In the damp Scottish northern islands and Norway the harvested plants used to be dried in kilns before they were threshed. Simple methods of threshing include beating the sheaves against a sieve, by hand, or stripping the heads from sheaves by drawing them back through metal rippling combs. In the Near East corn may still be trampled out by draught animals, walking round on the broken sheaves spread out on a threshing floor [8,28]. Sometimes a wooden threshing sledge, or tribulum of a type used since Roman times is dragged by the animals. In the U.K. the flail was still in occasional use for small parcels of grain in 1939. A flail (a stick-and-a-half) was made of two pieces of wood, e.g. a handle about 1.5 m (5 ft) long, and a striking 'swingle' about 0.9 m (3 ft) long, joined by two flexible loops of leather or eelskin. The flail was swung over the head and the swingle struck the plants, which were spread on a flat surface, breaking open the ears. In England threshing was normally carried out under cover, on the 'thresholds' of barns between two pairs of doors, an arrangement which encouraged a breeze to carry away dust. Separating the beards or awns was often a problem, and these were sometimes broken up by pounding ears with hummelers, or 'awners', metal grids mounted on the ends of rods fitted with handles [28]. After threshing and removing straw by forking the grain was 'winnowed' from dust and chaff. The mixture of grain and chaff was tossed in a breeze; the chaff was blown aside while the dense grain fell back to earth.

Early threshing drums had a rapidly rotating shaft covered with metal pegs, which 'scutched' and beat the grain from the heads and allowed it to fall through sieves in the wall of the threshing chamber, so separating it from the straw. Some incorporated a small fan which blew away dust and chaff. Later, in steam-driven threshing and winnowing machines, sheaves were undone, and were fed into the top. The plants were beaten against perforated 'concaves' by rapidly rotating beaters, fitted with bars or pegs. Some grain and chaff fell through the concave. The remaining grain, with the straw, was moved along with violent agitation, over sieves. The grain remaining was shaken free and fell through the sieves; the straw was discharged at the rear. The grain was passed over and through shaking sieves to remove stones and small impurities, was aspirated to remove dust, and was spouted into sacks.

Combine harvesters, 'combines', are combined reaping and threshing machines [7,11,13,25]. They may be small machines towed by tractors, and actuated by power take-off, harvesting a strip as little as 1.8 m (6 ft) wide or they may be self-propelled vehicles with cut-widths of 5.5 m (18 ft) or more. The rate of advance, typically 4.8–6.4 km/h (3–4 m.p.h.) depends on the machine and the harvesting conditions. They may deliver grain into sacks, which are dropped off at intervals around the field, or they may accumulate it in grain tanks which are unloaded when convenient by a built-in auger. As the machine advances it deposits the straw and cavings (awns and chaff) on the stubble behind it. Grain is normally combine-harvested when it is 'dead-ripe' and moderately dry, 16–22% moisture being best under British conditions [17]. Sometimes the plants

332 *Barley*

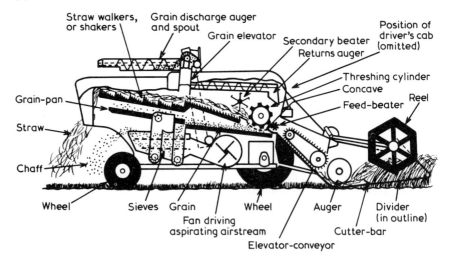

Fig. 8.9 A simplified diagram of a combine harvester. The drivers cab is omitted for simplicity. (Simplified from Turner [25]).

are cut earlier and laid in a swathe or windrow with self-propelled 'windrowers', machines with reels and cutter bars [7]. Here they dry. The grain is then harvested with a combine fitted with a 'pick-up' attachment. As the combine harvester works the bars of the rotating reel, or the pick-up reel used on lodged corn, steady the plants while the stems catch between the 'fingers' of the cutter-bar, and are severed by the triangular, oscillating blades (Fig. 8.9). The blades may have smooth or serrated edges; the latter are said to cut straws more cleanly. Projecting 'dividers' stand forward each side of the reel and cutter-bar and direct the plants to the side of the machine or to the cutter-bar. In some machines the cut plants are elevated 'ears-first' up endless belts, to the threshing mechanism; in others they are moved towards the centre of the machine by augers, situated immediately behind the cutter-bar, and are then elevated butts-first. A stone-trap is incorporated at one end of the elevator. Machines may have preliminary beaters to start ear disintegration and grain separation. The major threshing operation is carried out next. The rapidly rotating threshing-cylinder, or drum (about 1000 r.p.m.; peripheral speed about 1700 m/min, 5500 ft/min) carries longitudinally-mounted beater bars or rasps close to a curved metal grating, the concave. Here some 80–90% of the grain is separated and falls through the perforations in the concave and together with chaff and fragments of ears, is directed to the sieving and winnowing system. The straw, with unbroken, attached ears and entrapped grain now passes another beater which 'fluffs' it, stops it wrapping round the drum, shakes out grain, and directs the straw to the straw-walkers. While the straw is carried up the oscillating straw-walkers, to be deposited behind the vehicle, remaining grain is shaken loose and drops through perforations in the walkers and is directed to the sieving and winnowing system. Some machines have extra flail-beaters behind the walkers, that free the last of the grain before the straw is discarded. The streams of grain, chaff and separated ears combine, and are sieved and aspirated by a fan-driven airstream.

The sieves collect unthreshed heads, which are returned to the threshing mechanism. The sieved grain is delivered to a sump, from which an auger elevates it to a grain tank or bagging spout. Chaff and dust are blown from the back of the machine. Threshing mechanisms differ in detail but in all cases the exact settings are important. The gap between the cylinder and the concave, and the working speed are both critical. If the gap is too wide threshing is incomplete, while if it is too narrow, or the drum speed is too high, grain is likely to be split, bruised or skinned. More than 35% of the grain (w/w) can be mechanically damaged in combining [17]. Barley, with its husk, is much less prone to harvest damage than wheat [4]. The best working settings and speeds vary with the state of the crop. Combines often collect a good proportion of the grain from lodged plants but the rate of working and efficiency is reduced by the presence of lush green weeds. The working height of the cutter-bar is usually variable. In weedy conditions, or with an undersown crop, it may be set to cut above the green material. Some grain losses occur with the straw, and accurate estimates must be made to determine the efficiency of the machine. One grain per square foot ($0.093\,m^2$) indicates a loss of approximately 4.5 kg grain per ha (4 lb/acre) [2,3,6,13]. Other major grain losses occur at the front, at the cutter bar, due to shedding grains and heads, or bent-over plants and can amount to 5% of the potential yield. In some combine trials losses of barley, as kg/ha (lb/acre) averaged:

1969 – front 94, (84); rear 57, (51);
1970 – front 96, (86); rear 30, (27);
1971 – front 155, (138); rear 52, (46) [2,3].

Removing 'crop lifter' fittings doubled grain losses. It was hard to reduce rear-end losses below 56 kg/ha (50 lb/acre) and maintain a reasonable rate of working. On one machine altering a concave setting by only one place reduced threshing losses from 121 to 75 kg/ha (108 to 67 lb/acre). Devices for automatically monitoring grain losses are being tested. Grain losses rise with increased working speeds; a balance must be struck between the quantity of the crop recovered and the rate of working. The rate of working is mainly limited by the quantity of straw. 'Semi-dwarf' varieties, with shorter, stiff straw, should therefore show advantages in harvesting even if overall grain yields are not increased.

The tractors and trailers, or lorries, used to cart grain should be clean, have an adequate capacity, and unloading and turn-round speeds, and

334 Barley

be sufficiently numerous so that combines only stop to unload or refuel. The intake facilities of the store are also important. The rate of combine working must be matched to the other factors so that no delays occur particularly in an uncertain climate. Combine harvesters are expensive and only work in one area for a short period each year. To increase their usage it may be economic to move them north as the harvest ripens, as in N. America, or from England to Scotland.

8.7. Straw

Straw left by the combine is often burnt in the field. Because of the dangers, burning is usually regulated by law. The ash produced is of marginal nutritional value to the subsequent crop but disease and weed-seed carry-over is minimized. Unburnt straw and trash may be broken up with disc harrows before ploughing. Rarely the straw is chopped and ploughed under. Some few combine harvesters are fitted with devices that chop and spread the straw as it is discharged from the machine. Complete burial of straw is difficult to achieve and unrotted or partially rotted residues can harm a subsequent crop. The rotting straw consumes soil nitrogen.

Straw to be collected is baled. Stationary balers used to be used next to stationary threshing machines. Rarely some combines have built-in balers. Modern balers are usually towed and powered by tractors (Fig. 8.10). They pick up the straw from the rows left by the combines with finger-like tines and cut and compress it to various extents, sometimes

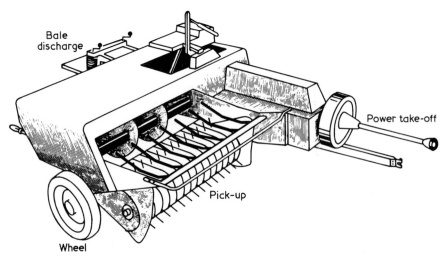

Fig. 8.10 A baler to be trailed by a tractor, powered by the tractors power take-off, for producing rectangular bales.

Production and harvesting machinery 335

by winding or rolling it into a cylindrical mass, sometimes by ramming and cutting it into a rectangular block [7,11,12]. The compressed material is tied or wired and dropped from the machine. Often the pick-up device feeds an auger which moves the straw sideways into the path of a ram and cutter. Bales 36 cm (14 in) × 46 cm (18 in) × 76–107 cm (30–42 in) of densities 130–240 kg/m^3 (8 and 15 lb/ft^3) are common, but some very large bales may be produced weighing e.g. 400–500 kg (8–10 cwt). Some cylindrical 'large' bales may be more than 150 cm (5 ft) diameter × 150 cm (5 ft) wide. Straw is made into large bales in a shorter time than small bales, but small bales can be handled by one man. When a farm is fully mechanized handling large rectangular or square bales is more efficient. For straw to become an industrial raw material the cost of weatherproof storage and cartage must be reduced. This can be achieved by making high-density bales. Bales 60 cm × 50 cm × 120 cm (24 in × 20 in × 48 in) of 77 kg (170 lb), about twice the normal British bale density, are used in the U.S.A. and experiments show that it is feasible to compress conventional bales to about 0.3 their usual volume [10]. Bales may be gathered at once, for example by dropping them onto a sledge towed behind the baler. When sufficient (often 10–20) have accumulated the tail-gate of a simple sledge is opened, and the bales are dumped, being dragged from the sledge by the friction of the stubble, through the slotted bottom. Sledges or wheeled trailers with built-in power-driven stacking devices may hold up to 150 bales.

Bales may be moved conveniently by fork-lifts or by other, more specialized, mechanical handling equipment. They may be stored in stacks in the fields, when they are frequently covered with waterproof sheeting, or they may be stored in Dutch barns. Straw must be kept dry to retain its value.

8.8 Harvesting the whole plant [7,11,15,16,21,24,25]

Whole crop barley is used for animal feed, as grazing, hay, and silage. The choice of equipment for harvesting the entire plant partly depends on the intended use. With foresight the ground, and especially stones will have been levelled by rolling at the start of the season. Less machine damage, and a cleaner, better quality product will result from this precaution. Crops may be cut by trailed or tractor-mounted cutter-bar mowers in which reciprocating triangular knives move between guide fingers and sever the stems, and guide the cut material into a swathe. Rotary mowers may be used in which blades, moving horizontally, mounted on one or two pairs of rapidly contra-rotating drums or discs, slash through the stems. A mower should be followed promptly by a machine for crushing, crimping or lacerating the plants to allow them to wilt or 'condition' quickly to a suitable moisture content for conversion to

336 *Barley*

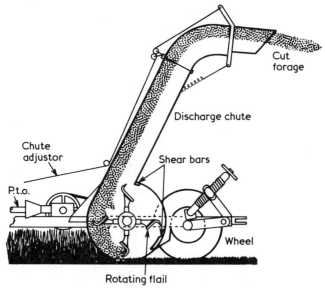

Fig. 8.11 A tractor-trailed, power take-off driven, flail-type, forage harvester, with the chute in a 'high discharge' position (after MAFF [16]: Crown copyright; by permission of HMSO).

silage, or for storage as hay (Chapter 10). Some self-propelled mowing machines cut and crush the crop in one operation. Mown crops used to be baled or even cut and bound into sheaves before ensiling. To allow compaction in the silo, and to allow silage to be fed mechanically the crop should be loose and well chopped. Flail mowers are tractor-trailed and actuated by p.t.o. (Fig. 8.11). They have a rapidly rotating horizontal shaft carrying rows of loosely swinging, sharp cutting-tools, or flails encased in a metal hood which move in the opposite direction to the wheels. The flails are used to cut and move the crop, which moves past a shear-bar, narrowly missed by the flails, chopping it to a variable extent. The cut material is blown into a chute from which it may discharge 'low' to the ground, to form a windrow, or 'high', into a high-sided trailer. The chute may also swivel, to allow discharge to the side, or the rear. Double-chop forage harvesters break up the crop more thoroughly. In these a bank of flails cuts the crop and blows it back into the machine. Here it is gathered and fed by an auger cross-conveyor to a chopping mechanism, which may be a rotating fly-wheel carrying cutters moving past a shear-plate. A paddle-fan blows the chopped material from the discharge chute. Even with double-chop machines the chopping length is not uniform. More complex metered chop harvesters give a more uniform product. The cut crop is elevated into the machine through a metering device – e.g. of rollers, and into a chopping mechanism of a cylinder or flywheel type which divides it into even lengths.

Production and harvesting machinery

To aid the conditioning (wilting, drying) of the cut and windrowed crops the swathes, preferably of lacerated, crushed or crimped material, may be moved about by rakes and 'tedders', to aid ventilation The crop may be collected from the windrow with various harvesters, tractor-mounted buckrakes, green-crop loaders, pick-up balers, or self-loading pick-up trailers. As material is delivered to a trailer for ensilage, whether it has just been cut, or is being picked up, it may be sprayed with silage additives such as molasses or propionic acid. Some machines are fitted with magnets which abstract tramp iron from the crop before it goes to store, thus saving wear on machinery, and protecting livestock.

8.9. Conclusions

In agricultural areas using simple, 'labour-intensive' farming techniques the sudden introduction of mechanization might be socially destructive. Suitable 'small-steps' for the third world, may be suggested by the history of improvements in agriculture which would enhance the well being of a country populace without massive disruption.

In more advanced areas further developments of farming equipment and resources are desirable. Cheaper methods of irrigation and of storing supplies of irrigation water, more exact seed drills, better fertilizer-spreaders and liquid sprayers, and combine-harvesters with more complete grain recoveries under all conditions are obvious examples. The choice of machinery, its maintenance and careful planning of its use are also of critical importance. Individual farmers do not have to buy many machines. Their choice is simplified by reference to independent advisory bodies and to farming societies which exchange information, carry out comparative trials, and report their findings. In Britain, the N.I.A.E. (National Institute for Agricultural Engineering) performs some of these functions.

References

1. Anderson, R.M. (1936, 10 Oct.) *Agricultural History* 157–205.
2. Anon., (1972, 19 May) *Farmer's Weekly*, 68–69.
3. Anon., (1972, 12 May) *Farmer's Weekly*, (Extra) Corn Harvesting.
4. Arnold, R.E. and Jones, M.P. (1963). *J. agric. Engng. Res.*, **8**, 178–184.
5. Arnon, I. (1972). *Crop Production in Dry Regions. 1, Background and Principles, 2. Systematic treatment of the principle crops.* London: Leonard Hill (Books).
6. Catt, W.R. (1967). *N.A.A.S. Q. Rev.*, **76**, 153–160.
7. Culpin, C. (1969). *Farm Machinery*, (8th edition) London: Crosby, Lockwood & Son Ltd.
8. Fussell, G.E. (1952). *The Farmer's Tools, 1500–1900.* London: Andrew Melrose.
9. Fussell, G.E. (1966). *Farming Technique from Prehistoric to Modern Times.* Oxford: Pergamon Press.

10 Hansford, R.J., Hughes, R.G., Truman, A.B., Wilson, P.H. and Woods, R.S. (1975). *Report of Working Conference on Straw utilization at Oxford, 5th Dec. 1974*, ed. Staniforth, A.R. Oxford: ADAS.
10a Hodges, H. (1971). *Technology in the Ancient World*. Harmondsworth: Penguin.
11 Lovegrove, H.T. (1968). *Crop Production Equipment*. London: Hutchinson.
12 MAFF (Ministry of Agriculture, Fisheries and Food; U.K.) (1966). *Mechanisation Leaflet No. 10, Pick-up balers*.
13 MAFF (1968). *Mechanisation Leaflet No. 18, Combine Harvesters*.
14 MAFF (1968). *Mechanisation Leaflet No. 22, Corn Drills*.
15 MAFF (1970). *Bulletin No. 37 Silage*. London: HMSO.
16 MAFF (1971). *Mechanisation Leaflet No. 13, Forage Harvesters*.
17 Mitchell, F.S., Arnold, R.E., Caldwell, F.Y.K. and Davies, A.C.W. (1955). *Rep. natn. Inst. agric. Engng.*, **56**.
18 Patterson, D.E. (1975). *Outlook on Agriculture*, **8**, 236–239.
19 Rayns, F. (1959). *Barley*. London: MAFF, HMSO.
20 Reid, D.A., Shands, R.G. and Suneson, C.A. (1968). In *U.S. Dept. Agric. – Agric. Res. Service, Handbook*, **338**, Barley, 32–38.
21 Robinson, D.H. (Ed. 1972). *Fream's Elements of Agriculture*, 15th edition, London: John Murray.
22 Shands, H.L. and Dickson, A.D. (1953). *Econ. Bot.*, **7**, 3–26.
23 Spain, J.M. and McCune, D.L. (1956). *Agron. J.*, **48**, 192–193.
24 Squire, M.J.E. (1958). *Agric. Gaz., New South Wales*, **69**, 505–511; 543.
25 Turner, J.C. (1964). *Farm Machinery—Operation and Care*. London: Cassell & Co.,
26 Tyler, C. and Haining, J. (1970). *Ploughing by Steam*. Hemel Hempstead: Model and Allied Publications.
27 Wik, R.M. (1953). *Steam Power on the American Farm*. Philadelphia: University of Pennsylvania Press.
28 Wright, P.A. (1961). *Old Farm Implements*. London: A. & C. Black.

Chapter 9
Weeds, pests and diseases in the growing crop

9.1. Weeds and the need to control them [10,38]

Weeds are wild plants growing in inconvenient situations, and also 'volunteer' plants from previous crops. By competing for space, water, mineral nutrients and light, they reduce yields. They may help pests and diseases to overwinter. They can make the crop ripen unevenly, and can also slow the rate and efficiency at which combine harvesters work. Weed residues in the straw or grain may reduce their value or acceptability. The reproductive bodies of species like wild onion *(Allium vineale)* and corncockle *(Agrostemma githago)* are poisonous and make grain unsaleable. All the treatments used to control weeds represent extra work and expense, so instead of eradication attempts are made to achieve control, so that the hoped-for extra yield and quality will at least pay for the costs incurred. Adequate control must be attempted every year, since a build-up of weeds may take several years and be laborious and expensive to control.

The number of weed species that may be encountered is large. To combat infestation the weeds, or their seedlings, must be correctly identified, so that their importance may be assessed and the correct treatment applied [15,38]. In each of the main weed classes (the broad-leaved dicotyledonous types and the narrow leaved monocotyledonous species) annual weeds occur. The variable dormancy and different preferred seasons of germination make some of them difficult to control [56]. Perennial weeds may or may not reproduce by seed. Their underground stems or swollen roots are often hard to kill, and when broken up during cultivation each fragment may give rise to a new plant. Over the years the relative importance of different weeds changes. Thus mechanical threshing and seed cleaning has nearly eliminated the poisonous corncockle *(Agrostemma githago)* and darnel *(Lolium termulentum*). The use of contaminated seed has increased the British acreage infested with wild oats *(Avena* spp.*)* [56]. Certified seed has maximum limits of seed impurities. Grain grown on infested land cannot meet the purity standards. Legislation regulates the nature of field inspections and the tests that must be made on the harvested grain to detect dead or damaged corns, weed seeds, some other impurities and what statement must be made regarding the quality of the grain (Chapter 5) [10,38]. Even one plant/acre (0.4 ha) of wild onion *(Allium vineale)* will cause the crop to be rejected.

Some broad-leaved weeds that are important in British barley crops are spear- and creeping-thistles *(Cirsium vulgare, C. arvense)*, various docks and sorrels *(Rumex* spp.*)*, cleavers or goosegrass *(Galium oparine* L*)*, black bindweed *(Polygonum convolvulus)*, and common persicaria *(Polygonum persicaria)*. Among the important monocotyledonous weeds are the wild onion *(Allium vineale)*, annual grasses such as black grass *(Alopecurus myosuroides)*, and wild oats *(Avena fatua* L. and *A. ludoviciana* Dur.*)* and the perennial grasses couchgrass *(syn* twitch, quackgrass; *Agropyron repens (*L*)* Beauv*)* and bent grasses *(syn*. couch; *Agrostis* spp.*)*.

9.2. Weed control

Weed control is an integral part of good farming. The use of clean seed housed, handled and drilled with clean equipment is essential. Keeping the edges of the fields, hedge-bases, and headlands clean reduces infestation. Straw infested with wild oats should not be moved about the farm. Rotating crops discourages build-ups of particular weeds. Ploughing to a depth of 20 cm (8 in) or more and thoroughly inverting the furrow slice and repeatedly cultivating as it becomes friable kills many seedlings (Chapter 8). In the past the land was allowed to lie fallow, bearing no crop, but being tilled at intervals to kill seedlings as they appeared. Fallows are now rare. It is now common for corn, often wheat and barley, or barley only, to be grown on land for several years in succession, a sequence requiring careful weed-control.

The earliest selective weedkillers sprayed onto cereal crops were copper sulphate (3–5% solution) used to control charlock, *Sinapis arvensis*, and sulphuric acid or vitriol, which controls various annual weeds [63]. Many chemicals are now in use, with widely varying properties. Only a few are mentioned here. They are often applied as mixtures to achieve control over a wider range of weeds. Formulations often contain wetting agents, spreaders and other substances to increase their effectiveness. Doses are varied, depending on the weed to be attacked. Total weedkillers may be used to clear stubble or to kill weed seedlings before drilling the grain. The use of substances such as metham-sodium (sodium *N*-methyl dithiocarbamate) or dazomet is probably confined to experimental or horticultural soils. Their decomposition gives rise to

Weeds, pests and diseases in the growing crop

methyl isothiocyanate, which kills weed seeds, bacteria, fungi and insects. Residues from metham-sodium disperse in about two weeks, while dazomet disperses in about three months. To combat perennial and other weeds, especially couch, an area may be sprayed with a non-selective herbicide such as aminotriazole, dalapon (2,2-dichloropropionic acid) (or a mixture of these two), sodium trichloroacetate, or paraquat. Crops may be sown a few weeks after an application of aminotriazole. The chlorinated aliphatic acids are slow to disperse. Other 'pre-emergence' weedkillers include terbutryne, and substituted ureas such as methabenzthiazuron. The bipyridyl compound paraquat is rapidly absorbed by and kills the green aerial parts of plants. Paraquat is completely adsorbed by most soils, leaving no toxic residues, so it may be applied to control weeds immediately before or immediately after sowing, or to kill grassland leys before ploughing. It, and some other 'total weed-killers', provide the basis for various minimal- or zero-cultivation techniques.

Wild oats are controlled by incorporating triallate into the top 4–5 cm (1.5–2 in) of soil and sowing the barley 6.4–7.6 cm (2.5–3 in) deep so that there is 2.5 cm (1 in) of untreated soil between the seed and the herbicide. The herbicide is best applied 2–21 days before drilling, but it may be applied immediately after seeding. In practice seed and herbicide applications will be at irregular depths and so some thinning of the crop will occur, and must be allowed for by increasing the seeding rate. As the wild oats grow the mesocotyl extends and carries the sensitive part of the weed into the triallate zone, where it is killed. With barley the mesocotyl does not extend, and the sensitive growing point is protected by the leaf bases as it passes through the treated soil [10]. Triallate treatment prevents undersowing with grass, but not with peas, beans or vetches. Nitrafen is also used for 'pre-emergence' wild oat control. Barban is used as a post-emergence spray to control wild oats in wheat, but some varieties of barley, including many that are grown in the U.K., are killed by it.

Selective herbicides may be applied as post-emergence sprays to the growing crop. They are toxic to various weeds but cause little or no permanent damage to barley if correctly used. The basis for selectivity is unknown. Evidently the small, upright leaves of the cereal will intercept and retain less spray than the horizontal leaves of many broad-leaved weeds. Also the axillary and apical buds of the cereal are well protected by the enclosing leaf-sheaths. Selectivity and effective doses may be altered by additives, the nature of the spray formulation, the volume of spray applied, the droplet size, the weather, the age of the crop, and so on. Increasing the seed rate establishes ground cover sooner, weeds tend to be smothered, so smaller doses of herbicide are needed to achieve control. When herbicides are incorrectly used the crop itself may be badly damaged.

Barley

The bipyridyl morfamquat is a selective weedkiller that kills mayweed for example, but leaves the cereal unharmed. However, like paraquat, it is inactivated by contact with the soil and so the crop may be undersown shortly after spraying. The substituted nitrophenols DNOC (3,5-dinitro-o-cresol) and dinoseb are contact herbicides. They control many broad-leaved weeds. However, they are highly toxic by ingestion, inhalation or penetration through the skin. Phenoxyaliphatic acids are widely used as herbicides. The activities of the acetate derivatives MCPA (4-chloro-2-methyl phenoxyacetic acid) and 2, 4-D (2,4-dichlorophenoxyacetic acid) are attributed to excessive, long-lasting, auxin-like effects. MCPA and 2, 4-D damage crops and reduce yields if applied at the wrong growth stage [2,19,30]. If applied before the five-leaf stage, malformations of the leaves and ears may occur in up to 85% of cases [1]; spraying slightly later results in malformations of the head; if applied after jointing, when the ear is in boot, sterile or 'blasted' spikelets may result [37]. The sprays often cause mitotic abnormalities, but plants derived from seed taken from damaged ears are normal, so the agents are not mutagenic [86]. The malformations are due to damage to the apical primordium, which is often visibly distorted. Common leaf abnormalities include fused leaves, leaves with two 'mid-ribs', and tubular leaves. Sometimes the flag leaf grows like a tube, with an indistinct transition between the blade and the sheath. As the ear forms and grows the awns stick in the fused ring of tissue that occurs at about the level of the auricles; the head may not emerge, or may burst from the side of the sheath and grow 'bowed out' with the stem retained in the bottom and the awns in the top. Occasionally tiller ripening is delayed. Ear abnormalities include 'opposite spikelets', where two kernels arise on opposite sides of one node, whorls or clusters of spikelets, three to six, around the stem at one node, or other supernumary spikelets. Unilateral spikelets may occur – grain forming on only one side of the ear. Twisted and twinned kernels and fused awns and lemmae also occur. Occasionally stems with two ears are found, or 'branched ears', where a rachilla has lengthened and taken on the function of a rachis. Frequently ears are 'tweaked', that is parts of the ear looks 'bald' as the internodes are lengthened (cf. Chapter 12).

The phenoxy-propionic acid derivatives mecoprop and dichloroprop are less toxic to young cereals than the acetate analogues, and control some different weeds. The butyric acid derivatives MCPB and 2, 4-DB are also less toxic to young cereals than MCPA or 2,4-D and are more selective. They may only become active after conversion to the acetate derivatives within sensitive plants. Other agents such as the nitrile ioxynil, the substituted benzoic acids dicamba and 2,3,6-TBA, benazolin, and picloram are used in mixtures with other agents to control particular groups of weeds. 2,3,6-TBA and picloram leave residues which are persistent and which may remain in straw, making it unfit for greenhouse and perhaps other uses.

Weeds, pests and diseases in the growing crop

Every year a number of people die from the effects of accidents with agricultural chemicals. Instructions and safety regulations must be obeyed and protective devices or clothing must be used as appropriate. New herbicides are continuously being marketed [4,11,63].

9.3. The economics of weed control

Of the barley grown in Britain 80–90% is treated with herbicides, Of these 35% are likely to be formulations of MCPA or 2,4-D. Early reports indicated that the use of herbicides increased crop yields by 5–23%. However, recent results suggest that the increased yield obtained from applying herbicides is often negligible [31]. Probably routine spraying has eliminated susceptible weeds, and the species that now grow are either resistant to the herbicides used, or grow after the spray applications. Annual spraying may be wasted unless the herbicide is chosen and applied to control troublesome weeds that are actually present.

9.4. Nematode pests

The cereal cyst eelworm, *Heterodera avenae*, has a world-wide distribution and often attacks barley. In 1963 about 0.5% of the crop in England and Wales was lost, but local losses were much higher [25,83]. The use of resistant barleys increased yields by 9% and 20% over susceptible controls in trials on heavily infested land; on clean land the yields were equal [16]. Nematode varieties vary in their ability to attack particular barley cultivars [81,92]. Experimental soil sterilization with chemicals which, *inter alia*, reduced nematode populations and attack by the take-all fungus, significantly increased yields of grain even after three years [93]. Attacked plants occur in patches, and are stunted and yellowish. Tillering is reduced, and ears may be shrivelled. Attacked plants have many dense, short, thick, highly-branched roots, matted together. The pest attacks each root near the growing point, growth is arrested and laterals are formed to be checked and to form more branches in turn. The reduced root system allows plants to be pulled up whole. Tiny cysts about 1 mm (0.04 in) long, round or lemon-shaped, varying in colour from white to dark brown may be present. Usually the presence of 80 or more cysts/100 g soil indicates that damage will occur [81]. However, there is a lack of correlation between pre-sowing cyst-count and depression of grain yield [25]. Eggs hatch when the soil is damp, whether or not a host is present [28]. Most damage occurs in May-September. Survival can occur for several years. Control by fumigants is generally uneconomic; to be effective a 20cm (8 in) depth of soil must be treated. Fumigants effective against nematodes, and other soil organisms, include methyl bromide, 'D-D' soil fumigant (a mixture of 1,2-dichloropropane and

344 Barley

1,3-dichloropropene), chloropicrin (trichloronitromethane), metham-sodium, dazomet, ethylene dibromide (EDB), 1,2-dibromo-3-chloropropane (DBCP) and methyl isothiocyanate (MIC). Control is achieved by growing resistant crops, rotating crops, avoiding oats (which allow a rapid nematode population increase) and in severe cases grassing the land for three or more years [28,55].

In north America other nematodes may attack barley [20]. Root lesion nematodes *(Pratylenchus* spp.*)* invade roots, retard plant development and may be very damaging when combined with a fungal attack. Dagger nematodes *(Longidorus* spp.*)* also invade roots, and occasionally the wheat gall nematode, *Anguina tritici*, forms galls on leaves or kernels giving rise to twisted leaves and irregular spikes.

9.5. Molluscs

Slugs (three spp. in the U.K.), and probably snails, eat barley plants [28]. Attacks are worst on heavy, wet soils with a high humus content. They occur throughout the year, but are worst in October–December and March–May. The animals may rasp away the ends of grains; sever seedlings at or below soil level; sever the ends of the leaves, leaving a ragged stump; eat holes in the leaves, or shred the leaf into longitudinal strips, leaving only the veins (Fig. 9.1). Damage may be so bad that a field has to be resown. Rolling the soil to compact it and restrict slug movements is helpful. Fertilizer hastens the recovery of damaged seedlings. Chemical control is effective if the weather is warm and damp, and the slugs are near the surface. Spraying with copper sulphate solution kills slugs on the surface. Alternatively, poisoned baits made with bran mixed with Paris Green or metaldehyde are used successfully. The bait with the arsenical Paris Green is also effective against leatherjackets.

Fig. 9.1 Barley seedlings damaged by slugs (after Robinson [75]).

9.6. Birds and mammals

These groups attack barley at all stages of growth, but the control methods available are unsatisfactory [28,44]. Attempts have always been made to scare birds away from corn, employing clappers, slings, scarecrows, and latterly automatic bird scarers. Deliberate poisoning of birds is not a legal control measure in the U.K. Damage varies with the season and the species of bird. Some will dig up seed, others snip seedlings in two, or pull them up and discard them. Grain may be crushed by small birds when milk ripe or may be eaten. Geese and swans may graze young plants.

Rabbits and hares eat plants, and may cause extensive damage. They are shot, and rabbit burrows may be gassed. Poisoning rabbits is illegal in England, but appears to be efficient in New Zealand [44]. Myxomatosis, a virus disease, sharply reduced rabbit populations in many areas but increasingly they are recovering. Damage caused by rabbits is severe locally. Before myxomatosis, about 0.7% overall of the barley crop of England and Wales was taken by rabbits and hares [83]. Some 30% of fields were moderately or severely grazed, with an estimated average reduction in grain yield of 200kg/ha (1.6 cwt/acre). Rodents damage growing corn. Rats and mice dig up and eat freshly sown grain or seedlings. No attempts are made to control rodents in the field, although they are vigorously controlled in and near grain stores. Other mammals also cause damage occasionally.

9.7. Insect and some other pests

Numerous other creatures attack barley. Occasionally millipedes (var. spp.) eat parts of seedlings, causing damage that may resemble that caused by slugs [28]. The grass and cereal mite *(Siteroptes graminum)* attacks the stem within the leaf sheath, just above a node. Often the damaged stem rots and is easily pulled apart; the upper parts of the plant whiten and die [28]. Many insects occasionally attack barley. Often no control measures are taken. In the U.S.A. grasshoppers *(Melanoplus spp.)* are destructive in some years, and are controlled by the use of poisoned baits or sprays or dusts of chlordane or toxaphene (camphechlor) [17]. Also the cinch bug, *(Blisus leucopterus)*, a sucking insect, frequently causes damage [17]. Outbreaks may so reduce a crop that it is uneconomic to harvest it. The occurrence of the cinch bug limits barley production in some parts of Kansas, and adjoining areas.

In parts of the U.S.A., and some other areas, the greenbug *(Toxoptera (Schizaphis) graminum)*, is a severe pest which sucks the plant's sap and weakens or kills it [17,21]. It also transmits barley yellow dwarf virus. Swarms of greenbugs may infest whole fields. Control may be achieved by insecticides and the use of resistant or tolerant barley varieties [17,21].

Resistant strains of barley contain benzyl alcohol. Susceptible strains become phenotypically resistant if 100 p.p.m. benzyl alcohol is supplied to the roots [45]. Aphids (var. spp.) occur around the world. A heavy attack may significantly reduce yields [17,21,28,44,72]. Infested plants are weakly and poorly coloured, and may develop reddish or purple blotches. Attacked grains may shrivel. Some species attack the roots. The distribution of species within a field varies with the degree of shelter. Aphids transmit barley yellow dwarf virus. They vary in their efficiency as vectors, and one species of aphid is variably efficient as a vector of different pathotypes of the virus [90]. Systemic insecticides may be employed for control purposes; e.g. dimethoate or dimeton methyl.

Barley thrips (e.g. *Limothrips cerealium*) cause white or silvery markings on green parts of barley. The adults are tiny 'thunder-flies'. Their economic importance is uncertain, but they may cause blind spikelets or shrivelled grain [28]. The tiny maggots of the minute wheat blossom midges (e.g. *Contarinia tritici*) feed on spikelets which do not then form grain. Later the larvae fall to the soil to pupate [28]. In areas where the pest is important DDT smokes will control it in barleys tolerant towards this insecticide. The larvae of other midges, e.g. the saddle-gall midge *(Haplodiplosis equestris)* occur in depressions of the stem, beneath the leaf sheaths. They seem to be unimportant economically [28].

Caterpillars and cutworms, the larvae of various moths, may attack plants eating the leaves, leaving ragged holes, or biting off shoots. They can be controlled with insecticides [28].

The Hessian fly, *(Mayetiola destructor)* is not sufficiently important in the U.K. to warrant control [28]. Elsewhere, e.g. North America and mainland Europe, it is an important pest. Barleys vary in their susceptibility to it. In each area attempts are made to determine a 'fly-free' date after which sowing may safely take place. Eggs are laid on early-autumn sown grain. The small white maggots eat the stems within the leaf sheaths. Attacked plants lack vigour and may lodge. Ears are often thin. The pupae occur in the base of the stems as 'flax-seed' [28]. Other control measures include crop rotation, killing all cereal 'volunteers' and ploughing under stubble [17].

Leatherjackets are the larvae of crane-flies, or daddy-long-legs (e.g., *Tipula paludosa*). Eggs are laid preferably in grass in leys or field verges. The pest is 'carried over' when a ley is ploughed. The larvae are tough-skinned, elongated, 2.5–3.8 cm (1–1.5 in) long, grey-brown to black, and are without obvious heads. They feed on roots, and on warm nights come to the surface and bite off plants at ground level, or eat ragged holes in the leaves. Probably about 0.2% of the U.K. barley crop is lost to this pest [83]. Populations of 247 000/ha (100 000/acre) can occur. Control may be by insecticide spray or poison baits containing insecticide or Paris Green.

Weeds, pests and diseases in the growing crop 347

Fig. 9.2 Pupating larva of corn sawfly in the base of a cereal stem (after Smith [79a]).

Various flies have maggots called cereal leaf miners, which tunnel between the leaf surfaces, leaving pale, wavy 'mines' [28]. These pests are not economically important in the U.K. Cereal leaf beetles *(Lema melanopa)* and their larvae eat thin strips from the leaves. In North America fields may show a white appearance due to a heavy infestation [17,28]. They may be controlled with insecticides.

The corn sawfly (wheat stem sawfly; *Cephus pygmaeus*), occasionally attacks barley. This pest is relatively unimportant in the U.K. but elsewhere attempts are made to control it. Eggs are laid on the stems, and the maggot eats its way down the inside. When ready to pupate it turns and eats a notch around the inside of the stem, making a weak line (Fig. 9.2). It pupates under a plug of 'sawdust' and frass. The stem breaks cleanly at the notch, giving a distinctive form of lodging. Ploughing in the stubble before the adults emerge aids control. Some related species cause similar harm, while others damage leaves.

The gout fly *(Chlorops pumilionis)* has two or three generations each year. It lays an egg on a shoot and the cream-coloured maggot, about 6mm (0.25 in) long, eats its way to the centre. Seedlings attacked in the autumn have a dead heart, while the outer parts become swollen or 'gouty', the leaves being broader and a deeper green and often with ragged tips (Fig. 9.3). Subsequently, many die. When plants are attacked during ear formation the ear may not emerge, or may emerge 'one-sided' with a groove eaten down the stem, or may be distorted and dwarfed [28,36].

348 Barley

Fig. 9.3 A winter barley plant in which the main stem (swollen) is attacked by gout-fly (after Frew [36]).

The wheat-bulb fly *(Leptohylemyia coarctata)* is locally important. Eggs are laid on bare soil, and maggots bore into the base of shoots leaving a small, round hole. The central leaf is killed by the attack; each maggot may attack three shoots in succession. Plants die or suffer a severe check [28]. The symptoms are similar to those caused by many ley pests, but the infestation does not transfer from ploughed grass. Systemic insecticides, such as dimethoate will control damage, or, more usually, autumn-sown seeds are dressed with persistent insecticides.

Frit fly *(Oscinella* or *Oscinis frit)* damages barley worst when it is planted on a freshly ploughed grass ley. The crop may be thinned or may fail. Some damage occurs in most years. Three or four generations occur each year. Eggs are laid at the base of the shoot, often under the edge of the leaf sheath. The maggot eats its way in to the growing point, which it destroys. Later it pupates. In young plants tillering is induced, and the tillers in turn may be killed, so a short, bushy 'grassy' plant, often with ragged leaf-tips, is produced [28]. Rarely a late attack may result in white and empty ears. The only control measures used are to sow spring grain early to avoid attack, and to plough grass leys at least six weeks before sowing winter corn.

Wireworms, the larvae of Elaterid or 'click-beetles' *(Agriotes* spp. and *Althous haemorrhoidalis)* are destructive. They are numerous in grassland, and as they take 3–5 years to reach maturity they are persistent. Adults vary in appearance but *Agriotes* spp. are 1.0–1.6 cm (0.4–0.6 in) long. Eggs hatch to larvae which are initially less than 1 mm (0.04 in) long, but which grow to 2.5 cm (1 in) and are elongated, hard and bright yellow-brown. Eventually they pupate, and adults emerge from the pupae. Wireworms are always active but are particularly damaging in autumn and spring. Seeds may be holed, seedlings may have the middle leaf killed or have ragged holes eaten in the lower leaf sheath at the stem base; or the seedlings may be severed at or below ground level. About 1% of the U.K. barley crop was lost to wireworm in 1963 [83]. Populations of more than 2.47×10^6 wireworms/ha (10^6/acre) probably warrant control by seed dressings or applications of insecticides to the soil [28,44]. Care must be taken as γ-BHC (lindane, benzene hexachloride), for example, is phytotoxic, and the use of more persistent insecticides, such as aldrin and dieldrin is regulated by law.

Chemically, pesticides are very diverse [63]. Mishandling these toxic substances puts the operators at risk, can be fatal to wildlife, can leave harmful residues on the crops, and can reduce yields. In the United States to avoid exceeding the legal limits of insecticides that may occur in grain, it is forbidden to spray parathion within 15 days of harvest, malathion within 7 days, and so on [17].

9.8. Virus diseases of barley

The disease caused by barley stripe mosaic virus (BSMV) was, until recently, of unknown aetiology, and was called false stripe. The disease occurs in other grasses [20,21]. In England BSMV is not very damaging, but in North America it reduced barley yields by more than any other disease in the period 1951–1960 [66,67]. In Denmark several older barley varieties carried BSMV. In pot experiments, in which lodging was prevented, BSMV-free lines outyielded infected lines by about 25%. Differences of about 16% were obtained in field-trials where applications of N fertilizer potentiated the lodging of the old varieties [77]. The reduction in yield approximately equals the percentage of infected seed sown. Infected plants carry stripes, especially near the base, which may be bleached yellow or light green. Internode lengths may be reduced and tillering enhanced, or the plants may be severely stunted. Leaves may be mottled or nearly completely yellow. The differences are partly due to strains of virus differing in virulence, and differences in barley varieties, from very susceptible to highly resistant or tolerant. The disease is seedborn, and can be transmitted by pollen or by contact between leaves. The only control is to sow clean seed [20,21,66,67].

The barley yellow dwarf virus, (BYDV) is widespread and reduces yield significantly. Estimates of average losses are about 5–10%, while field losses of 5–30% occur [24,66,67]. Artificial inoculation of plants can reduce yields by 40% to 80% [24]. Many aphids are vectors [79]. Different aphids transmit different isolates of the virus with varying degrees of efficiency. Yellowing of infected plants begins at the leaf tips and spreads down the leaf, often with a stage when the leaf is alternately striped golden yellow and dark green. Ultimately the whole leaf will be golden yellow. The disease is associated with degeneration and necrosis of the phloem [29]. Plants infected early are severely stunted, but only moderately so with later infections. Root growth and tillering are reduced, as are the size and fertility of the ears, kernel weight and total yield. Control is by sowing in late autumn or early spring and to get plants well established before aphids are active. The use of tolerant or resistant barley varieties is advisable. The spread of the disease can be checked by using systemic insecticides to kill the aphids. This disease is *not* transmitted by seeds or soil. A scale of visible damage to plants allows estimates of losses to be made in advance of harvest [20,24,66,67].

Other viruses have been detected in barley and probably more will be found. Their economic importance is usually uncertain, but they may cause substantial losses, at least locally [79]. Vectors spreading the infective agents vary widely. Barley yellow mosaic virus and wheat mosaic virus are transmitted via the soil; bromegrass mosaic virus is transmitted by nematodes. Mites transmit wheat streak mosaic virus and *Agropyron* mosaic virus, and biting leaf hoppers transmit the Aster yellows (the symptoms of which resemble BYDV), cereal enanismo, oat blue dwarf, and wheat striate mosaic viruses. In other cases, e.g. *Hordeum* mosaic virus, the vectors are not known [79].

9.9. Bacterial diseases

The number of significant bacterial diseases of barley is small. In N. America ears and rarely other parts of the plants are attacked by *Pseudomonas atrofaciens*. The lower parts of attacked glumes appear charred and discoloured (basal glume rot). Occasionally small light spots occur on the leaves. Bacteria are abundant in the lesions [66].

Xanthomonas translucens (var. synonyms) causes diseases in several parts of the plant called, for example, bacterial blight, bacterial leaf blight and black chaff disease [20,66]. Black chaff (not detected in the U.K.), occurs when the disease is conspicuous on the glumes or upper parts of the stem, and may not be noted until the ears emerge. Brown or black stripes may form on the stems, especially just below the ears, and sunken dark stripes may occur on the glumes. Bacteria may reach the surface of the grain and penetrate it, usually at the base, so that it is shrunken. Sowing infected grain may result in a massive disease outbreak. Other syndromes caused by *X. translucens* include a leaf blight, in which irregular, narrow, glossy surfaced yellow or brown stripes or blotches appear on the leaves following the appearance of small water-soaked patches. A heavy infection may cause

Weeds, pests and diseases in the growing crop

a slow yellowing and death of the leaf. Lesions may show small grey-white blotches or drops of sticky gelatinous exudate, rich in bacteria, which dry to thin, flaky scales or to little granules. Bacteria remain viable a long time in this dried state. The disease is favoured by warm, wet weather, and is spread by rain, wind, insects, from crop residues, with infected seed, and possibly from the soil. Attacks may cause severe losses. Organo-mercurial seed dressings reduce the incidence of the disease, which is otherwise best controlled by the use of clean seed, thorough soil cultivation, and crop rotations [20,21,66,67].

Organomercurial seed dressings may contain many different active agents in different formulations [63,67]. Very similar names may be used for different compounds, even in the same country. In recent American practice Ceresan M contained 7.7% N-(ethylmercuri)-p-toluenesulphonanilide, Ceresan 100 contained 3.1% ethylmercuri-2,3-dihydroxypropylmercaptide and 1.3% ethylmercuriacetate, Chipcotes 25 and 75 contained respectively 5.41% and 1.85% methylmercurinitrile and Panogen 42 contained 6.3% cyano-(methylmercuri)-guanidine. Such substances should be handled with care.

9.10. Fungal diseases

Fungal pathogens occur widely, and some are ubiquitous. Some of the diseases they cause are of great economic importance. Nomenclature of the fungi is confused. Thus different synonyms are often used when a pathogen attacks a plant producing different symptoms or when it occurs in perfect and imperfect forms [20].

Powdery Mildew, caused by *Erysiphe graminis*, occurs in humid and subhumid regions. It causes large yield reductions, partly by reducing kernel weights but more by reducing kernel numbers and ear numbers. Inhibition of root growth is also noticeable [50,67]. In the U.K. the disease is common, and may cause losses of 20% or more [52]. A scale is available that relates the square root of the leaf infection (%) at a particular stage of growth to the ultimate yield reduction [49]. In mild climates the fungus may overwinter as perithecia (cleistothecia), or as mycelial mats on volunteer plants, winter cereals, and grasses. Conidiospores are the main source of infection in spring barley. In arid winter-barley regions, e.g. Israel, sexual cleistothecia carry the pathogen through the rainless summer [48]. Conidiospores land on the epidermis. A mycelial strand grows out and swells to form an appressorium. From this an infection peg forces its way directly through the cuticle and into a cuticular cell (Fig. 9.4). This may occur in 10 h under favourable conditions. The infection peg probably releases carbohydrases which weaken the epidermal cell wall [65]. Within the epidermal cell the fungal peg swells and branches to form a haustorium. Haustoria withdraw nutrients from the host and

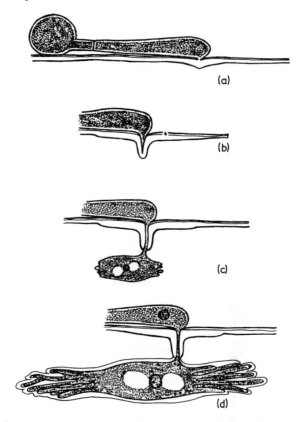

Fig. 9.4 Stages in the penetration of the hosts cuticle, and the formation of a haustorium, by the germ-tube from a spore of mildew (*Erysiphe graminis*, after Butler and Jones [13]). (a) The germ-tube attached to the cuticle; the cell wall of the host beginning to thicken. (b) The germ-tube growing into the host cell and (c) having penetrated into the cell extending into a haustorium. (d) An enlarged, branching haustorium. By permission of the Estate of the Late Edwin Butler.

support the other fungal parts, which are extra-cuticular. Many conidia are formed and are dispersed by air currents, rapidly spreading the disease. Normally the first signs of infection are the appearance of small white/grey fluffy spots of mycelium on leaf sheaths, leaves, or on the ears. These grow, darken and become powdery with millions of spores. Later small, round, black reproductive bodies, the perithecia, may be formed on the mycelial mats. Disease development is favoured by rapid lush growth of the host. Feeding lithium salts is said to increase resistance to mildew. Ways to breed resistant strains of barley are continually sought; genes for resistance have even been incorporated from wild *Hordeum spontaneum* [23], or induced by mutagenesis (Chapter 11 and 12). Experiments with near isogenic lines, differing by a gene for resistance, showed that when the disease was severe the resistant strain had a 30% greater grain yield [20,21]. However, in a few years strains of fungus arise that parasitise previously resistant varieties. Numerous pathotypes have been distinguished. Control consists of sowing thinly, using resistant varieties, using balanced fertilizers, and using seed dressings and sprays of systemic fungicides. Winter hosts should be destroyed before spring to minimize the initial spore inoculum. It has been proposed that growing winter cereals should be banned to reduce 'carry-over' of disease, as early inoculation of the spring crop causes high losses. Small plot trials tend to underestimate the value of fungicides, since unsprayed control plots provide a continuing heavy inoculum of mildew spores [43]. Systemic fungicides give yield increases. Milstem, (ethirimol, 5-n-butyl-2-ethyl-amino-4-hydroxy-6-methyl pyrimidine) is used as a seed dressing or foliar spray. It is persistent, and is translocated in the plant. At high doses it can cause cytological abnormalities in barley [6]. Calixin (tridemorph; mainly 2,6-dimethyl-4-tridecylmorpholine) also gives control for periods of 3–4 weeks. Applications have given increases in yield in a mildew-resistant variety so perhaps it has some other effect on the plant [78]. Among other systemic fungicides may be mentioned lucel (tetrachlorinated quinoxaline), and milfaron (1–3,4-dichloroanilino-1-formyl-amin-2,2,2-trichloroethane). Non-systemic fungicides need to be applied repeatedly to control mildew and this is often uneconomic.

When mildew infection is well advanced, local chlorosis of the host tissue occurs within the centre of the colonies and photosynthesis declines. Chlorophyll develops again at the centre of older colonies. The numbers of chloroplasts decline in the infected leaves, there is a rise in the RNA/DNA ratio, and other changes occur [26]. The parasite accumulates substances at the expense of the host tissue. The parasite/host complex respires rapidly, increases in wet and dry weight, has a high water-demand, and is an effective metabolic sink 'draining' organic and inorganic substances from the host tissues [27]. The enzymatic changes occurring in diseased leaves resemble those of 'senescence' when leaves are detached [34]. There are several mechanisms of resistance to mildew. In resistant plants epidermal penetration may not occur, or it may occur normally but haustorial development may be delayed or the haustoria may be checked or killed. In 'hypersensitive' plants the host mesophyll cells collapse and degenerate when several haustoria are formed, and this local necrosis is associated with a failure of the pathogen to spread further. Resistance parallels the rapidity of mesophyll death [89]. However, it is not clear whether this accelerated cell death is the cause of resistance. Studies of a cultivar containing the 'Algerian gene' for mildew resistance show two types of resistance. First, infection pegs tend to be killed. Secondly, after haustoria are established fungal growth is retarded and ultimately ceases, and the pathogen dies [65].

354 Barley

Downy Mildew (*Sclerophthora* (*Sclerospora*) *macrospora*; conidial state, *Phytophthora macrospora*) attacks cereals and grasses in N. America, S. Europe, Australia and Japan. It has not been detected in Britain. Occasionally it causes severe damage in low-lying, badly drained fields [20,66]. Some infected plants produce numerous small tillers which rapidly brown and die. Others survive but remain stunted, distorted and discoloured. A partial control measure is to improve soil drainage.

Disease caused by *Rhynchosporium secalis* is called leaf blotch, or Rhynchosporium in the U.K., and scald or leaf scald in N. America. In some areas it is a very serious disease. In S.W. England the disease has rapidly increased since 1963, causing losses of up to 30% in autumn-sown and 35% in spring-sown crops [23,42,51,66,67]. A key for rating infection intensity allows grain yield reductions to be assessed in advance of harvest [41]. The loss in yield roughly equals 0.67 of the flag leaf area visibly affected or 0.5 of the infected area of the second leaf, at growth stage 11.1 on the Feekes scale. The primary attack causes oval or lens-shaped blotches on the leaves or leaf-sheaths. At first these are water-soaked or grey-green, but later the central area becomes pale or white, and is surrounded by a brown ring, and may be 1.9 cm (0.75 in) long by 0.6 cm (0.25 in) wide. Sometimes a series of concentric zones is formed. In the growing season the disease is spread by conidiospores. When a spore lands on a host a short germ tube grows and forms an appressorium from beneath which a hypha penetrates the cuticle. Extensive mycelial pads are formed between the cuticle and the cell wall. Epidermal cell walls beneath these mats become swollen, separate into lamellae and collapse. Mesophyll cells also collapse and die before the hyphae reach them. The hyphae colonize the dead cells. Hyphae will grow across the leaf, then grow into mats in the substomatal cavities the other side. The stomata open to a greater extent in epidermis over the fungus [5]. The pest over-winters (or over-summers in winter cereal areas) on volunteer plants and grasses, on crop residues and on seed. Control measures include the use of crop rotations, burning and thoroughly ploughing under crop residues, the destruction of grasses and volunteer cereal plants, growing resistant varieties and using organo-mercurial seed dressings [32,51,66].

Leaf spot, caused by *Septoria passerinii*, is a minor disease in Britain, but is important in some areas of the U.S.A. Linear lesions and spots, initially bleached or yellow-brown in colour, occur on leaves and leaf-sheaths. Later they become brown, with rows of minute dark brown-black dots (pycnidia) embedded in the tissue. Defoliation may occur. The disease is spread by spores and overwinters on crop residues. Its incidence is worst in areas of intensive cultivation. Genes for resistance are available. Control measures include rotating crops and the destruction of crop residues [20,66,67].

Yellow rust (stripe-rust; *Puccinia striiformis* syn. *P. glumarum*) attacks many grasses and wheat as well as *Hordeum* spp. As with all rusts the life cycle is complex. Summer or uredospores spread the disease throughout the crop. Parallel rows of pale to orange-yellow pustules occur in rows on the foliage, stems or ears and may be close enough to appear as continuous stripes. Leaves may wither early; grains are small and shrivelled. At the end of summer brown or black pustules form teleutospores, and so the stripes become dark. The teleutospores germinate in the spring to produce sporidia. These cannot infect cereals but no secondary hosts that can be infected are known. Thus effective overwintering is by mycelium or uredospores on volunteer plants or winter cereals. Races of the pathogen are known, as are resistant barley varieties. The best control seems to be to breed resistant varieties. For special purposes the disease can be controlled by applications of sulphur dust [20,21,51,66,67].

Stem rust, (black stem rust, red rust, *Puccinia graminis*) can cause serious losses [51,66,67]. The dormant teleutospores overwinter on straw or stubble, germinating in the spring to produce sporidia which infest the secondary host, the European barberry *(Berberis vulgaris)*. The fungus grows on the barberry leaves as yellow or orange patches which produce yellow aecidiospores, which infect cereals. The consequent 'rust-red' stage occurs on the cereals as reddish brown rows of spots on stems and leaves. Throughout the summer uredospores spread the disease. In the autumn the streaks darken due to the formation of resting teleutospores, which overwinter. Control is by the eradication of the barberry, the use of resistant cereals, the destruction of crop residues, and the use of cultural practices to hasten crop maturity. Numerous races of the pathogen are known. In warm areas the pathogen overwinters as the rust red stage. As spring comes a wave of rust infection advances from Mexico up into Canada with the mild weather [66,67]. Frequent applications of sulphur dust can give limited control. Scales exist relating the extent of rust infection to subsequent reductions in grain yield [71].

Brown rust of barley (Fig. 9.5; leaf rust; dwarf rust; *Puccinia hordei*; *P. anomala*) is widely distributed, and has become important in Britain where it began spreading in 1966. Infections may reduce yields by 30%, and reduce grain quality [14,51]. In the U.S.A., where barley is sown early in the autumn to provide grazing, infection may reach epiphytotic proportions and cause severe losses of fodder and of grain. The secondary hosts of this fungus are *Ornithogalum* spp. (including Star-of-Bethlehem) but overwintering is often in the uredial form. Numerous small round yellow to orange-brown pustules of the summer, uredial stage, occur scattered over leaf blades and sheaths. The uredospores spread the disease. In late summer the black telial stage occurs, when lead-grey pustules form, still covered by leaf epidermis [51]. Many races of this fungus occur. Some varieties of barley show wide resistance. When the

Barley

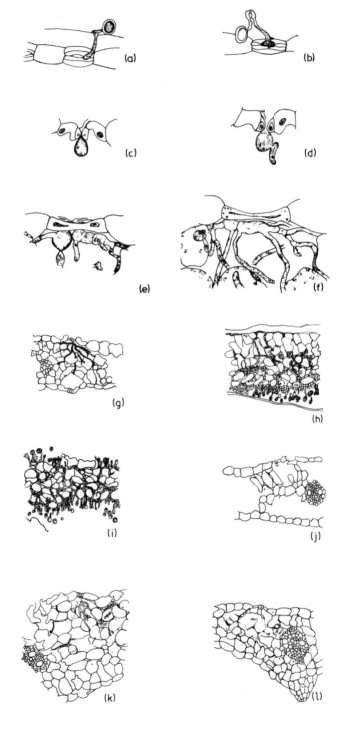

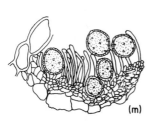 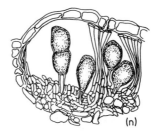

Fig. 9.5 Attack on barley by *Puccinia anomala* (after Butler and Jones [13]). a-f. Uredospore germination and penetration of a stoma. a, A germ tube has grown to a stoma. b, An appressorium has formed over the stoma, c, A vesicle has formed below the stoma and, d, is beginning to extend a hypha. e, f, successive stages of ramification of mycelium in the sub-stomatal space. g-i. Successive stages of disease spread across the leaf of a susceptible barley variety, and the formation and release of uredospores. j-l. Stages in the resistant variety, Quinn; j formation of the sub-stomatal vesicle. k, Rapid, 'hyper-sensitive' necrosis of the barley tissue in contact with the parasite. l, complete destruction of the mycelium; infection checked. Two spore types, m, Uredospores in sorus; n, Teleutospores in sorus. By permission of the Estate of the Late Edwin Butler.

spores germinate a short mycelial strand extends to a stoma where it swells and attaches to form an appressorium (Fig. 9.5). A slender infection hypha then enters the host through the stoma and swells in the sub-stomatal cavity. Branches arise and, in susceptible plants, quickly penetrate cells and form haustoria. In uncongenial hosts haustoria may fail to develop, or are checked. Often the fungus damages cells which die, leaving necrotic spots around infection sites [13]. Hypersensitivity is often associated with resistance but, as with mildew, it is not clear if the swift local necrosis is the cause of the resistance. Systemic fungicides, e.g. benodanil, can provide useful control of brown and yellow rusts.

Ergot *(Claviceps purpurea)* is not serious on barley in Britain but it may occur in up to 1% of grain samples (2.5% in 1941). In the U.K. its incidence rose with the introduction of open-flowering Scandinavian barleys, in 1950–1956 [60]. It occurs frequently in N. America [20,67]. The introduction of outcrossing 'hybrid' barleys, in which parents must be open-flowering, may increase the frequency of this disease. Sclerotia, purple-black bodies with whitish interiors, are formed of compacted mycelia in the place of grains. They become 0.25–5.1 cm (0.1–2 in) long, and may be enclosed within the husk or protrude well out from the ear. They may be harvested with the grain. They are frequently rich in alkaloids which are highly toxic to men and animals. When eaten they cause hallucinations, 'St. Anthony's Fire', convulsions, gangrene, abortions and death. Several of the alkaloids are derivatives of lysergic acid. Ergots may be separated from grain by cleaning machinery or (after a 3 h pre-soak) by floatation in a brine solution [66]. In the autumn

358   *Barley*

Fig. 9.6 (a) The head of a six-rowed barley with covered smut *(Ustilago hordei)*. (b) Head of barley infected with loose smut *(Ustilago nuda)*, beginning to break up (after Tapke [84]).

sclerotia fall to the ground, where they overwinter, or they may be spread with seed. They germinate in springtime elevating long, slender threads which produce spherical heads which release ascospores. These infect barley flowers, which subsequently exude a sticky yellow fluid, 'honeydew' rich in another type of spore. These infect other flowers. Control is achieved by planting closed-flowering barleys (that exclude spores), sowing clean, fungicide-dressed seed, ploughing in crop remains, and by rotating crops [20,21,60,66,67].

Three smut diseases occur in barley. Covered smut (Fig. 9.6; *Ustilago hordei*) is widely distributed, and causes considerable losses. Numerous pathotypes are known, and barley varieties vary in their susceptibility. Spores and hyphae are borne superficially on seeds, and may be killed with organomercurial seed dressings. Seedlings are infected between germination and emergence from the soil, but the disease is noticeable only when the heads emerge from the boot when it is seen that the grains are replaced by hard black spore masses, each covered with a thin membrane. Membranes begin to split after emergence and spores are blown to healthy ears. Others break up and spread spores during threshing, contaminating the grain [13,20,51,66,67]. To infect a high percentage of seedlings experimentally, seeds may be decorticated with sulphuric acid before inoculation [9]. Black loose smut, (semi-loose smut, nigra loose smut, *Ustilago nigra*) is widespread in the U.S.A., but is unknown in the U.K. Its life history resembles that of covered smut, but the spore-masses are dark brown-black, and awns may remain on the ear. The millions of spores are spread by the wind. Ultimately the ears appear 'broken up' [20,51,66,67]. Loose smut (Fig. 9.6, 'nuda'-loose-smut, 'deep-born' loose smut, *Ustilago nuda*), occurs in all barley areas and is an important disease. It differs significantly from the other smuts. The entire ear, except the rachis, is replaced by brown, enclosed spore-masses. Ultimately the spores are liberated and blow away and the naked rachis remains. When the spores reach new ears they infect the developing grain. The fungus grows within the grain, mostly in the scutellum, where it can be detected by staining and microscopic examination of separated embryos (Chapter 5) [74]. Infection occurs only in open florets; varieties with small lodicules are not necessarily closed-flowering and immune [35]. Spore suspensions may be inoculated into the florets. Such artificial inoculation tests give higher 'susceptibility' estimates than are obtained in field trials because they by-pass the 'pseudo-resistance' given by the closed-flowering habit [59,70]. Infection may occur at least 150 m (165 yd) down wind of smutted plants [39]. Upper spikelets of two-rowed barleys and side florets of some six-rowed varieties open more than other florets on the ear, and grain from these is more likely to be smutted [85]. Screening seed to remove small grains reduces the percentage infection in the next crop. The fungus resumes growth when the seed germinates, and the resulting plant usually produces smutted heads. In susceptible varieties crop losses approximately equal the percentage infection of the seed sown. Some varieties show physiological resistance; even if ears are inoculated the crop grown from the infected grain has a low percentage of infected ears. Superficial seed dressings give no control. This is achieved by sowing clean seed, preferably of resistant barley varieties. Various treatments are available for killing the fungus in the seed, but they tend to reduce the viability of the grain. In one type of anaerobic treatment seed is soaked for 4 h, drained for 0.5 h, and held in full, airtight metal drums for times ranging from 30 h at $32°C$ ($90°F$) to 80 h at $18°C$ ($65°F$) [51]. Hot-water treatments vary in detail and difficulty of execution, as a range of exact steeping times and temperatures of grain–water mixtures are involved. Simpler, single-temperature treatments can be effective [22].

360 Barley

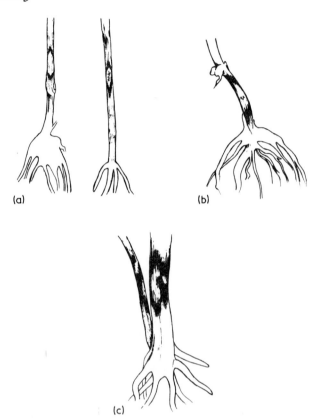

Fig. 9.7 The bases of cereal stems attacked (a) by sharp eyespot *(Rhizoctonia solani)* and (b, c) eyespot *(Cercosporella herpotrichoides)*; after MAFF [57]; Crown copyright: by permission of HMSO).

Recently treating seeds with the systemic fungicide Vitavax (carboxin) has given satisfactory control [64,73]. The fungicide has small adverse effects on healthy plants. Metabolites include a phenol, and insoluble materials bound to lignin [8].

Two pathogens cause eye-like lesions at the base of barley stems (Fig. 9.7). Sharp eyespot is caused by *(Corticium (Rhizoctonia) solani)*, which also causes root rot. This pathogen is only locally important. The roots of young plants are killed and although many plants survive an infection they remain stunted. Leaves may remain rolled, erect and purplish, giving 'purple patch' in the field. Plants may ripen slowly and still be green at harvest [66]. The common symptom is 'sharp eyespot'. Superficial, roughly symmetrical, clearly defined lesions occur from a few up to 38 cm (15 in) up the stem, with brownish-purple mycelial patches at the centre (Fig. 9.7) [57,66]. The fungus is soilborne and is common on light, dry soils.

Eyespot, a much more serious disease, is caused by *Cercosporella herpotrichoides*. The fungus, which occurs in pathotypes of varying aggressiveness, is soil-borne and can live either as a parasite or as a saprophyte on dead stubble [57,66]. Primary inoculation comes from the soil, where the fungus can survive 4–5 years. Levels of infection often build up in continuous cereal cultivation. Fungal spores infect the outer leaf sheaths at soil level. Rapid spread of the fungus through tissues may cause the death of tillers or whole plants, and consequently a thin stand. When the fungus attacks more slowly typical oval, yellowish brown-bordered spots are formed on the base of the stem, inside the leaf sheath, and sometimes on the sheaths (Fig. 9.7). The lesion may have a black fungal 'dot' in the middle and grey mycelium may be found in the stem cavity. The disease is spread by fresh spore crops. Infected stems often bend and twist near the lesion and irregular lodging (straggling) may occur. Yields may be reduced by 30%, with a high percentage of small grains. The disease is discouraged by good drainage (especially on heavy land), thin sowing, preferably in the spring, the choice of lodging-resistant varieties with low susceptibility to disease damage, the use of crop rotations involving two year breaks from cereals, and the adequate control of weeds. The systemic fungicide benomyl reduces the severity of this disease [18]. Cercobin, another systemic fungicide, is also used to control it.

Take-all, caused by *Ophiobolus graminis*, is a serious disease. The causative fungus occurs in most parts of the world. It survives on root and stubble residues, weed grasses, and volunteer cereals and spreads readily. It attacks the roots, which are blackened, and break easily. The stem bases are blackened, growth may be retarded, ears may die or be empty and whole plants may be killed. Black spore-cases (perithecia) may be found on the stem bases. The disease is most serious where cereals are sown frequently. It is favoured by light, alkaline soils. Eradication of perennial weed grasses, maintenance of soil fertility, and thorough cultivation and firming the seed bed (to minimize the rate of spread along roots), all help to control the disease. No resistant barleys are known. The best control is to 'starve out' the fungus by using crop rotations having one or two year breaks between cereals. Where two cereal crops are grown in succession, the second should be spring-sown to allow as much breakdown of infected crop residues as possible. In one season take-all can spread up to 1.5 m (5 ft) along and across the drill row by root contact in wheat [91]. In the Chamberlain continuous cereal-growing system, spring-sown barley is undersown with trefoil *(Medicago lupulina)*, or ryegrass or a mixture. This grows vigorously after harvest, and in late winter is ploughed in ready for the next spring-sown cereal, so acting as a green manure. Perennial weed grasses and volunteers are suppressed, and so is the fungus [66,68].

In N. America considerable losses are caused by browning root rot (*Pythium* root rot, *Pythium* spp. including *P. arrhenomanes*). These fungi live in the soil and attack the roots of seedlings, causing necrosis and browning of the root tips. Attacked plants may be yellow-pale green and short. Although plants may survive they tiller poorly, and remain stunted [20,66]

A species complex of fungi, placed in the genera *Fusarium* or *Gibberella*, (*F. culmorum, F. nivale, F. avenaceum, G. avenacea, F. graminearum, G. zeae, G. saubinetii, F. roseum*) attacks the roots, seedlings or ears of barley at different times giving rise to various diseases, such as brown root rot, seedling blight, *Fusarium* blight, ear-blight, ear-scab and scab [20]. These diseases can be extremely harmful, reducing yields and grain quality. The fungi attack all the cereals and many grasses. Fungi overwinter on seed grain, plants or plant residues. Infected seeds or seedlings develop brown foot rot and seedling blight; roots become discoloured and rotten and plants may be killed outright or may be stunted, with yellowish leaves. The crown can be invaded by the fungal mycelium; plants may fail to produce ears, may lodge, or may produce dwarfed or distorted heads, which may contain small grains or be empty. White or pinkish spore masses may occur. The spores spread the disease through the crop. In warm, humid weather florets may be infected, showing the scab or ear-blight syndrome. Diseased parts may be light brown. Pink mould growth may occur around the bases of infected flowers, 'red mould' and black perithecia (fruiting bodies) may be detectable. Kernels that are scabbed are often shrunken, discoloured grey-brown, with irregularly discoloured surfaces and greyish, floury endosperms [20,51]. The fungus can produce toxins which are emetics and cause vomiting in pigs, dogs and humans. Cattle, horses, sheep and adult poultry are not affected. Control is difficult. Clean seed dressed with organomercurial compounds should be sown early in well worked and drained soil, preferably not in the year following a cereal crop [20,51,66,67].

Spot blotch, foot-rot and (on grains) black point or kernel smudge are diseases caused by *Helminthosporium sativum*, (*H. sorokinianum, Cochliobolus sativus, Bipolaris sorokiniana, Ophiobolus sativus*). This fungus is troublesome in N. America. Initially infection is from the seed or residues in the soil. Dark brown, rotting lesions form on the roots and seedlings may die. Often the plants grow, and the fungus spreads slowly through the tissues forming dark brown, eliptical spots and blotches on the leaf blades, stalks, and glumes. Diseased plants may be stunted and excessively tillered; leaves tend to dry and wither early. Symptoms may be complicated by the presence of other parasites. Infected grain is often discoloured (brown to nearly black) on the base, together with local grain shrivelling, black point or kernel smudge. Infected grain is unsuitable for seed or many other purposes [13,20,51,66,67]. *H. sativum*

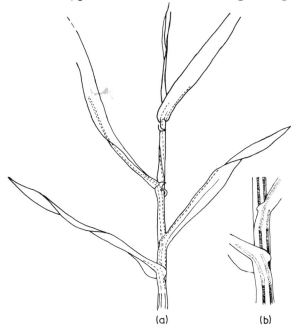

Fig. 9.8 Barley Helminthosporium leaf-stripe, caused by *Helminthosporium gramineum*. (a) A general view of a young infected plant. (b) A detailed view showing the 'continuity' of the stripes due to the fungus penetrating several leaf-sheaths and extending with the growing leaves (after Smith [80]).

produces a phytotoxin, helminthosporal, a terpenoid dialdehyde, which inhibits coleoptile and root growth, rendering them prone to fungal attack. The related compounds helminthosporol, helminthosporic acid and other synthetic compounds have weak gibberellin-like activity [7,46,76].

Barley leaf stripe, *(Helminthosporium* stripe; *Helminthosporium gramineum*; *Pyrenophora graminea)*, seems to occur wherever barley is grown. It can cause 10–25% grain losses. The fungus is carried on infected seed. At germination it penetrates the seedling through the coleoptile and successive leaf-sheaths. As the leaves grow the infection is carried upward, so long, narrow, brown or yellow stripes, as many as seven, may run nearly the length of the plant (Fig. 9.8). These carry spore-forming bodies which spread the disease throughout the crop. Leaves may split, fray and collapse [80]. Even if the plant is not killed vigour is reduced. The infected ear may not emerge from the boot and may carry little or no grain. Grain that is formed is generally discoloured and infected. Crop rotations, the use of organomercurial seed dressings, and the use of resistant barley cultivars are the control methods of choice [13,20,51,66, 67,80].

364 Barley

Net blotch is caused by the related fungus *Helminthosporium (Pyrenophora, Drechslera) teres*. It is widely distributed. Considerable damage may occur when cool, humid conditions favour the fungus. Primary infection may occur from plant residues, infected seed or wild grasses. Sexual ascospores are produced on plant remains, and these and conidia spread infection throughout the crop. Leaf lesions usually appear as brown, net-like blotches of dark brown lines in lighter brown zones or long, narrow rectangular or irregular stripes. Infected stems have brown discolourations, light brown blotches occur on the kernels. Grains are reduced in size. Organomercurial seed dressings, burning infected residues and crop rotations help to control this disease. Resistant barley lines are known, but high post-inoculation temperatures break down resistance [20,32,47,51,66,67].

In some areas winter cereals may become yellowed or rotted under the snow. Such snow-rot (snow scald) is caused by *Typhula incarnata*, and can be serious [66]. Reddish-brown, small sclerotia occur on the infected plants, and these carry the fungus over to infect the next crop.

Cephalosporium leaf stripe, *(Cephalosporium gramineum)* is caused by a soil-borne fungus that is important locally. The fungus causes striping and yellowing of the leaves. Probably it spreads rapidly through the plant as spores are liberated into the transpiration stream [66]. Other fungal diseases of barley are known. For example, Halo spot *(Selenophoma donacis)* has been found on barley in Ireland [66]. Anthracnose *(Colletotrichum graminicolum)* is common but generally unimportant in the U.S.A. [20]. Ascochyta leaf spot *(Ascochyta hordei)* occurs in N. America and Japan [21].

9.11. Some general considerations

To some extent cereal diseases are 'man made', since they are associated with large stands of one species, grown in 'monoculture'. The barley crop is becoming more uniform genetically on a regional and even a world scale. In the U.K. cereals are increasingly being grown for runs of a dozen years in one field. Further, winter barley is being grown more widely, leaving only a small time-gap between harvesting and subsequent seeding. All these changes increase the likelihood of extensive disease outbreaks. The disease pattern in the crop is continually altering [23,88]. The success of 'continuous barley' is surprising, and break crops do not always increase the yield over several subsequent years as anticipated [40, 52]. Possibly micro-organisms such as *Streptomycetes* spp. and *Trichoderma* spp., which are antagonistic to some pathogens, accumulate in the soil and antibiosis regulates the levels of soil-borne pathogens [68]. The mechanisms of resistance to pathogens are little understood. Any plant is resistant to most fungi, possibly due to the tough cuticle which

Weeds, pests and diseases in the growing crop

may have toxic constituents, or the release of toxic chemicals, such as some organic acids, or phenolic substances from roots [61,94]. When pathogens penetrate plants browning often occurs, and it is suggested that phenolic glycosides in the damaged tissues are degraded to phenols which, in turn, are oxidized to quinones which are fungitoxic. Phenolic glycosides (possibly methoxy-quinone β-D-glucoside) decline during warm-water steeping with the appearance of the toxic, free aglycone which may account for the control of loose smut achieved by this process [54].

Aqueous extracts of barley coleoptiles contain thermostable, strongly basic factors toxic to *Helminthosporium sativum*, whose toxicity is counteracted by calcium ions [53]. The active substances are glucosides of hordatines A and B (Chapter 4). They are absent from other cereals, which contain antifungal agents such as avenacin and cyclic hydroxamate glucosides which are absent from barley [82]. Extracts of barley grains show antibiotic properties against gram positive organisms, but the effective agents have not been identified [3].

Breeding for disease resistance is difficult (Chapter 12). When suitable screening processes for detecting resistant plants have been developed genes for disease resistance are sought in collections of genetic stocks or in populations previously exposed to mutagens. If genes able to confer resistance are found they have to be incorporated into locally adapted varieties, of commercial value. As a new resistant variety is often widely sown, strong selection pressure is created for the multiplication of new mutant or recombinant strains of the pathogen (pathotypes). Frequently some years after the release of the resistant variety new pathotypes appear which are able to overcome the resistance. This use of 'major genes' gives so called 'vertical' resistance which may be nearly complete towards the pathotypes used in the screening process. However this 'all-or-none' approach can only be used as long as new genes for resistance can be found; when populations of new pathotypes build up its value is much diminished. A different approach relying on 'field resistance' or 'horizontal' resistance has been debated [12,87]. Rather than sowing pure line barley cultivars the concept is to sow mixtures containing many 'minor' factors for disease-resistance. In these circumstances a pathogen inoculum would meet few plants to which it was fully adapted, and disease build up would be slow. This radically different approach has yet to be used on a wide scale. The increased use of 'pure lines' may have lost much of the field resistance that was probably present in the old, mixed 'land' varieties.

Much more work is needed to learn to combat pests and diseases. Even the estimates of losses are unreliable [33], and yet without these data it is uncertain, for example how the limited research and development resources available should best be spent, or when, on economic grounds, a farmer should or should not take corrective measures. Rapid and accurate diagnosis in the early stages of a disease in the field is desirable. One idea is to use the gas liquid chromatographic analysis pattern of the pyrolysis products from diseased leaves as an aid in diagnosis [69]. A continuing flood of new pesticides is being produced; those available are frequently reviewed [4,58,62,63].

References

1 Åberg, E. (1954). *Ann. R. agric. Coll. Sweden*, **21**, 213–233.
2 Andersen, S. (1954). *Physiologia Pl.*, **7**, 517–522.
3 Ark, P.A. and Thompson, J.P. (1958). *Pl. Dis. Reptr.*, **42**, 959–962.
4 Ashton, F.M. and Crafts, A.S. (1973). *Mode of Action of Herbicides*. London: Wiley-Interscience.
5 Ayres, P.G. (1972). *J. exp. Bot.*, **23**, 683–691.
6 Bennett, M.D. (1971). *Nature*, **230**, 406.
7 Briggs, D.E. (1966). *Nature*, **210**, 418–419.
8 Briggs, D.E., Waring, R.H. and Hackett, A.M. (1974). *Pesticide Sci.*, **5**, 599–607.
9 Briggs, F.N. (1927). *J. agric. Res.*, **35**, 907–914.
10 B.C.P.C., British Crop Protection Council (1968). *Weed Control Handbook* 5th edition, eds. Frayer, J.D. and Evans, S.A. 2 vols. Oxford: Blackwell Scientific Publications.
11 B.C.P.C. (1974). *Pesticide Manual* 4th edition, updated regularly, Nottingham: Boots Co. Ltd.
12 Browning, J.A. and Frey, K.J. (1969). *Ann. Rev. Phytopath.*, **7**, 355–382.
13 Butler, E.J. and Jones, S.G. (1949). *Plant Pathology*. London: Macmillan.
14 Chamberlain, N.H. and Doodson, J.K. (1972). *Agriculture*, **79**, 302–305.
15 Chancellor, R.J. (1966). *The Identification of Weed Seedlings of Farm and Garden*, Oxford: Blackwell Scientific Publications.
16 Cotten, J. (1970). *Ann. appl. Biol.*, **65**, 163–168.
17 Dahms, R.G. (1968). In *U.S. Dept. Agric. Handbook, No. 338 – Barley.* pp. 43–44.
18 Davies, J.M.L. and Jones, D.G. (1971). *J. agric. Sci., Camb.*, **77**, 525–529.
19 Derscheid, L.A. (1952). *Pl. Physiol., Lancaster*, **27**, 121–134.
20 Dickson, J.G. (1956). *Diseases of Field Crops*. New York: McGraw-Hill.
21 Dickson, J.G. (1962). In *Barley and Malt*, ed. Cook, A.H. pp. 161–206. London: Academic Press.
22 Doling, D.A. (1965). *Ann. appl. Biol.*, **55**, 295–301.
23 Doling, D.A. (1967). *J. natn. Inst. agric. Bot.*, **10** (Suppl.) 12–15.
24 Doodson, J.K. and Saunders, P.J.W. (1970). *Ann. appl. Biol.*, **66**, 361–374.
25 Duthoit, C.M.G. (1964). *Pl. Path.*, **13**, 25–31; 73–78.
26 Dyer, T.A. and Scott, K.J. (1972). *Nature*, **236**, 237–238.
27 Edwards, H.H. (1971). *Pl. Physiol., Lancaster*, **47**, 324–328.
28 Empson, D.W. (1965). *M.A.F.F. Bull. No. 186. Cereal Pests*. London: HMSO.
29 Esau, K. (1957). *Am. J. Bot.*, **44**, 245–251.
30 Evans, S.A. (1961). *N.A.A.S. Q. Rev.*, **53**, 16–18.
31 Evans, S.A. (1969). *Expl. Husb.*, **18**, 102–109.
32 Evans, S.G. (1969). *Pl. Path.*, **18**, 116–118.
33 FAO, Plant Production & Protection Division, (1967). *Report on the FAO Symposium on Crop losses*, Rome.

34 Farkas, G.L., Dézsi, L., Horváth, M. Kisbán, K. and Udvardy, J. (1964). *Phytopath. Z.*, **49**, 343–354.
35 Fehr, W.R., Lambert, J.W. and Rasmusson, D.C. (1964). *Crop Sci.*, **4**, 306–307.
36 Frew, J.G.H. (1924). *Ann. appl. Biol.*, **11**, 175–219.
37 Friesen, G. and Olson, P.J. (1953). *Can. J. agric. Sci.*, **33**, 315–329.
38 Gill, N.T. and Vear, K.C. (1966). *Agricultural Botany*, (2nd edition), London: Duckworth.
39 Hewett, P.D. (1970). *Agriculture*, **77**, 20–24.
40 Hughes, R.G., Roebuck, J.F., Collingwood, C.A., Helme, W.H., Gammon, F.C., Lester, E., Evans, S.G., Hooper, L.J. and Campbell, S. (1969). *N.A.A.S. Q. Rev.*, **83**, 129–137.
41 James, W.C., Jenkins, J.E. and Jemmett, J.L. (1968). *Ann. appl. Biol.*, **62**, 273–288.
42 Jenkins, J.E. and Jemmett, J.L. (1967). *N.A.A.S. Q. Rev.*, **75**, 127–132.
43 Jenkyn, J.F. and Bainbridge, A. (1974). *Ann. appl. Biol.*, **76**, 269–279.
44 Jones, F.G.W. and Jones, M.G. (1964). *Pests of Field Crops*. London: Edward Arnold.
45 Juneja, P.S., Gholson, R.K., Burton, R.L. and Starks, K.J. (1972). *Ann. ent. Soc. Am.*, **65**, 961–964.
46 Kato, J., Shiotani, Y., Tamura, S. and Sakurai, A. (1966). *Planta*, **68**, 353–359.
47 Khan, T.N. and Boyd, W.J.R. (1969). *Aust. J. biol. Sci.*, **22**, 1237–1244.
48 Koltin, Y. and Kenneth, R. (1970). *Ann. appl. Biol.*, **65**, 263–268.
49 Large, E.C. and Doling, D.A. (1962). *Pl. Path.*, **11**, 47–57.
50 Last, F.T. (1962). *Ann. Bot.*, **26**, 279–289.
51 Leonard, W.H. and Martin, J.H. (1963). *Cereal Crops* pp. 478–543. London: Collier-MacMillan.
52 Lester, E. (1969). *Agric. Progr.* **44**, 78–84.
53 Ludwig, R.A., Spencer, E.Y. and Unwin, C.H. (1960). *Can. J. Bot.*, **38**, 21–29.
54 Mace, M.E. and Hebert, T.T. (1963). *Phytopathology*, **53**, 692–700.
55 M.A.F.F. (Ministry of Agriculture, Fisheries and Food, U.K.; 1967). *Advisory Leaflet No. 421. Cereal Cyst Eelworm.*
56 M.A.F.F. (1968). *Advisory Leaflet No. 452. Wild Oats.*
57 M.A.F.F. *Advisory Leaflet No. 321. Eyespot of Wheat and Barley.*
58 Makepeace, R.J. (1975). In *Materials and Technology – a Systematic Encyclopedia VII – Vegetable Food Products and Luxuries.* pp. 89–186. London: Longman-DeBussy.
59 Malik, M.M.S. and Batts, C.C.V. (1966). *Ann. appl. Biol.*, **48**, 39–50.
60 Marshall, G.M. (1960). *Ann. appl. Biol.*, **48**, 19–26.
61 Martin, H. (1964). *The Scientific Principles of Plant Protection.* (5th edition), London: Edward Arnold.
62 Martin, H. (ed. 1972). *Insecticide and Fungicide Handbook for Crop Protection* (4th edition), Oxford: Blackwell Scientific Publications.
63 Martin, H. (1973). *The Scientific Principles of Crop Protection* (6th edition), London: Edward Arnold.
64 Maude, R.B. and Shuring, C.G. (1969). *Ann. appl. Biol.*, **64**, 259–263.
65 McKeen, W.E., and Bhattacharya, P.D. (1970). *Can. J. Bot.*, **48**, 1109–1113.
66 Moore, W.C. and Moore, F.J. (1961). *M.A.F.F. Bulletin No. 129, Cereal Diseases*, London: H.M.S.O.
67 Moseman, J.G. (1968). In *U.S. Dept. Agric. Agricultural Handbook, No. 338, Barley.* pp. 45–60.

68 Mundy, E.J. (1969). *Expl. Husb.*, **18**, 91–101.
69 Myers, A. and Watson, L. (1969). *Nature*, **223**, 964–965.
70 Pedersen, P.N. (1967). *Acta Agric. scand.*, **17**, 39–42.
71 Peterson, R.F., Campbell, H.B. and Hannah, H.E. (1948). *Can. J. Res. C*, **26**, 496–500.
72 Rautapää, J. (1968). *Acta Agric. scand.*, **18**, 233–241.
73 Reinbergs, E., Edington, L.V., Metcalfe, D.R. and Bendelow, V.M. (1968). *Can. J. Pl. Sci.*, **48**, 31–35.
74 Rennie, W.J. and Seaton, R.D. (1975). *Seed Sci. Technol.*, **3**, 697–709.
75 Robinson, D.H. (ed. 1972). *Fream's Elements of Agriculture* (15th edition), London: John Murray.
76 Sakurai, A. and Tamura, S. (1966). *Agric. Biol. Chem. (Tokyo)*, **30**, 793–799.
77 Sandfær, J. and Haahr, V. (1975). *Z. PflZüch.*, **74**, 211–222.
78 Slootmaker, L.A.J. and van Essen, A. (1969). *Neth. J. agric. Sci.*, **17**, 279–282.
79 Slykhuis, J.T. (1967). *Rev. appl. Mycol*, **46**, 401–429.
79a Smith, K.M. (1948). *A Textbook of Agricultural Entomology*. Cambridge University Press.
80 Smith, N.J.G. (1929). *Ann. appl. Biol.*, **16**, 236–260.
81 Southey, J.F. (1965). *M.A.F.F. Tech. Bull. No. 7, Plant Nematology*, London: HMSO.
82 Stoessl, A. (1970). In *Recent Advances in Phytochemistry*, ed. Steelink, C., and Runeckles, V.C. **3**, pp. 143–180.
83 Strickland, A.H. (1965). *J. R. Soc. Arts.*, **113**, 62–81.
84 Tapke, V.F. (1940). *J. Agric. Res.*, **60**, 787–810.
85 Taylor, J.W. and Harlan, H.V. (1943). *J. Hered.*, **34**, 309–310.
86 Unrau, J. and Larter, E.N. (1952). *Can. J. Bot.*, **30**, 22–27.
87 Van der Planck, J.E. (1968). *Disease Resistance in Plants*. London: Academic Press.
88 Walker, A.G. (1967). *Agriculture*, **74**, 171–175.
89 Walker, J.C. (1969). *Plant Pathology* (3rd edition). London: McGraw-Hill.
90 Watson, M.A. and Mulligan, T. (1960). *Ann. appl. Biol.*, **48**, 711–720.
91 Wehrle, V.M. and Ogilvie, L. (1956). *Pl. Pathol., Lancaster*, **5**, 106–107.
92 Williams, T.D. (1970). *Ann. appl. Biol.*, **66**, 339–346.
93 Williams, T.D. and Salt, G.A. (1970). *Ann. appl. Biol.*, **66**, 329–338.
94 Wood, R.K.S. (1967). *Physiological Plant Pathology*. Oxford: Blackwell Scientific Publications.

Chapter 10

The reception and storage of whole plants and grain. The micro-organisms and pests of stored grain

10.1. Introduction

Rapid and efficient reception, processing and storage of plant materials is essential to prevent deterioration and losses. Grain for seed or malting must retain a high level of germinative power and so standards for its storage are more demanding than for feed grain. More care is needed for longer periods of storage. The provision of good storage conditions is expensive. The most economic storage conditions are often those which just succeed in preserving the plant material for its chosen purpose. When using marginally 'safe' conditions the risk of sudden deterioration is high.

On a world-wide scale the losses incurred through poor handling and storage are enormous. Bad storage allows decreases in bulk, and in quality; seed and malting grain become animal feed; animal feedstuffs are reduced in nutritional value or become unuseable. Contamination of grain, for example with insect pests or rodent faeces, may cause the rejection of samples of otherwise high quality.

In dry areas of the world grain is sometimes stored in heaps in the open, either loose or in sacks; the sacks may be covered with straw or buried in sand [36]. Infestation by insects and rodents is rapid. In damp, unpredictable climates straw and grain should be gathered quickly and stored under carefully controlled conditions since warmth and damp favour deterioration.

10.2. Barley hay

Grass can outyield cereals and can give a better quality feed, yet cereal hay and silage is sometimes made to extend the period in which the drying or silage making equipment is in use, or to clear the land. As the crop is taken before wild oats have a chance to shed their seed this helps to clear this weed. A heavily diseased crop may be of more value as silage than as a source of straw and a little, poor quality grain. For feeding ruminants it seems more rational to harvest the whole plant at the best time rather than harvest the grain and straw separately, particularly as the nutritive value of the straw is comparatively low (Chapter 14).

Animals may be allowed to graze on young plants, or the plants may be cut and dried for hay, 'dried grass' or made into silage. With increasing age the yield of dry matter increases and goes through a maximum, while the digestibility, crude protein content, and moisture contents decline. Thus there is an optimal period for harvesting the whole crop, which is when the grain is milk-ripe or doughy, before the awns harden. Whole barley is suitable for ensiling at the end of July in the U.K. (Chapter 14) [79]. In haymaking the immature crop is dried. Hay made with hooded barley, when the grains were dough-ripe, seemed palatable and digestible to lambs; the grain and leaves were retained better by the immature plants during harvesting. Cereals cut and laid in a swathe may dry slowly and lose considerable amounts of their nutritional value. Passing the plants through a crusher or crimper, or cutting with a flail forage harvester gives a faster drying product. Barley hay is rarely made in the U.K. but there has been interest in artificially drying the whole crop. Sometimes barley is harvested mixed with an undersown legume. The wilted crop may be compacted into rectangular bales and 'artificially' dried further. Hay with a moisture content of 35% will rot, of 30% will mould and heat, at 25% may mould, but at 20% may be stored in a dutch barn or in a well-thatched stack. Moulding reduces the feeding quality and some metabolic products of the micro-organisms may be poisonous. Fungi may cause Aspergillosis, or farmers lung, in humans. Treating hay with 'additives', (such as propionic acid or ammonium propionate) can reduce moulding.

Baled hay may be dried in barns, where it will be stored. Warm air, of low relative humidity, is forced through a stack of bales. Curves relating moisture contents of hay and straw with equilibrium relative humidities are given (Fig. 10.1). Bales may be arranged on a perforated floor stacked over air-channels, so-called 'tunnel drying'. Alternatively they may be dried on a batch dryer before stacking [28]. The fan-driven air stream may be at ambient temperatures or may be heated by a furnace or a radiator. Usually a temperature rise of about $3°C$ ($5°F$) above ambient is sufficient. As the stack dries extra, damp bales can be added until the barn is full.

Hay may also be dried loose at low temperatures in flat-bed tray driers or on endless perforated belts with automatic loading and discharge devices. The product is cooled by blowing cold air through it as it is discharged. Some hay driers can also be used to dry grain. Very high

The reception and storage of whole plants and grain

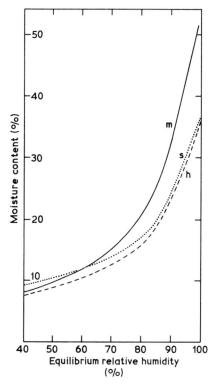

Fig. 10.1 The relationships between the moisture content and equilibrium relative humidity for malt culms, m; straw, s; and hay, h. (after Snow *et al.* [125]).

temperature cylindrical driers may be used for short-chopped grass and might be applied to barley. Two or three stage driers are used and these have a high thermal efficiency [28]. The dried material is heaped for 24 h to allow it to cool and condition. The material can be baled, or milled and stored in thick paper bags. For feeding it must be converted into wafers or cubes [28]. As a means of conservation whole-barley drying may come to compete with silage making. Badly made hay, particularly if it is rained on, loses much of its nutrient value. If dried quickly it retains more of its soluble sugars, minerals and vitamins or vitamin precursors. Cubes of dried barley plants, or barley with peas held together with molasses, seem to give good live-weight gains with wethers.

Bale-handling should be carried out according to an organized system. Bales may be lifted to small heights, or unloaded, with fork-lift trucks. Portable, endless-belt elevators may be used to deliver bales to the top of a stack. Very large barns may have built-in conveyor systems. Stacks may be built up in a regular manner, row upon row, or bales may be allowed to pile up at random, wherever they settle after discharge from the overhead conveyor [28].

10.3. Straw

In the U.K. straw is usually burned or baled, mainly for use as animal litter and roughage for ruminants. Little attention has been paid to ways of maintaining its feed or industrial value. The realization that its nutritional value can be economically increased by spraying with caustic solutions may change this picture, as could the establishment of industries based on straw (Chapters 13 and 14).

Straw is usually baled in the fields and may be stored in the open in stacks that are thatched or covered with plastic sheeting, or that are built in dutch barns. In damp areas provision may be made for drying the bales in a forced-air draught, as with hay. Usually the aim has been to prevent gross deterioration, such as heating in the stack, or extensive mouldiness. Better storage conditions, including the protection of the sides of stacks from rain, and more thorough drying may become important as damp, moulded straw is useless for many purposes. Bales of hay or straw are handled in the same manner. Previous comments on storage largely apply to rectangular bales. Methods for storing giant cylinder-bales have not been perfected.

10.4. Barley silage

Silage is sometimes made from cereals although this is not usual. Silage is preserved by acidity (between pH 3 and 4) and storage under near-anaerobic conditions. Preservation as silage is generally more economic than drying [10,89]. The acidity may be achieved by the addition of mineral acid (the A.I.V. method), propionic acid, or by natural fermentation of the carbohydrates in the crop, with or without the addition of supplemental carbohydrates or other additives.

With grasses maximum yields of digestible matter (not dry matter) occur at the onset of flowering. With barley the best time to harvest is when the plant is beginning to yellow and the grain is at the rubbery stage; the high nutritional value of the grain offsets the low value of the fibrous straw. The dry weight yield is at its maximum, the 'rubbery' grains are relatively easily digested by cattle and the plants are unlikely to shed their grain during harvesting [34,54]. Barley varieties differ in their digestibility and total dry matter yield (Chapter 14) [18,34]. The plants should be chopped, either at harvest or at the silo to allow good compaction. Usually wilting is not needed as at harvest the material has a dry matter content of 27–40% [18,54,79,84].

When the cut and crushed or lacerated crop is ensiled, the respiration of the mass rises and the oxygen in the silo is used up, while the carbon dioxide level rises. At ensilage the crop should have a dry matter content of at least 30% and an available carbohydrate content of 10% [34]. Acetic acid fermentation begins and is succeeded by a lactic fermentation.

The reception and storage of whole plants and grain

The lactic acid content should reach a constant level of about 1–1.5% of the fresh weight, giving a pH of < 4.2 [10]. The wetter the crop the more acid it must be, to prevent undesirable fermentation, for 18% dry matter a pH of about 4.0 is required; for 30% dry matter a pH of about 4.4 and for 40% dry matter a pH of about 5.0 [89]. The fermentation is at the expense of soluble sugars, so the crop should be cut late in the day following a good photosynthetic period when the sugar content is high. The sugar content may be supplemented with sugars such as molasses or with carbohydrate-rich potatoes. Alternatively acids may be added to the mass. The pH may be taken below 4 (but *not* 3) by the addition of hydrochloric acid containing a little sulphuric acid (the A.I.V. process of Virtanen) or phosphoric acid, or organic acids such as lactic, formic or propionic acids. Sodium metabisulphite or calcium propionate additions are also made sometimes to inhibit the activity of undesirable bacteria [10,28,89]. Silos should be filled and sealed rapidly to achieve and maintain anaerobic conditions quickly. The temperature of the fermenting mass should be kept below 32°C (90°F). If the fermentation is inadequate losses due to respiration may be high, or different micro-organisms may cause undesirable changes. The nutritional value of such a product is low, and stock frequently reject it [10,89]. Wet or badly made silage may give rise to a considerable volume of stinking effluent that is unpleasant, wasteful and costly to dispose of. Whole crop barley should not give rise to effluents, but cereal/legume mixtures may. Losses to silage made in clamps should be less than 20% dry matter and in towers < 15% dry matter. Whole crop barley ensiled in sealed towers may lose only 5% dry matter [89].

Silos are chosen with regard to ease of usage, capital costs, and the quality of the final product. Silage may be made in clamps, in unwalled pits, in concrete-walled pits, in containers of strong metal mesh lined with reinforced, waterproof sheeting (e.g. butyl rubber), in silos made of wooden or concrete staves, or in special towers. In general cheaper silos are less convenient to use, more losses are incurred in making the silage in them, and the product is of poorer quality.

Tower silos are best for making top quality silage with minimal effluent. They may be made from concrete, galvanized steel, glass fibre or vitreous enamelled steel. In the U.K. common dimensions range from 4–9 m (14–30 ft) in diameter and 3–21 m (10–70 ft) in height. Towers may be unsealed or sealed. In sealed silos care is taken to exclude air. Inflatable 'breather' bags, open to the outside, are mounted in the top of the silo, which expand and contract thus equalizing the gas pressure inside and outside but reducing the chances of air gaining entry (Fig. 10.2). Pressure-relief valves are mounted in the top of each silo. The silos are filled from the top, chopped material being delivered up an outside duct by a blower. Combined cutter-blowers are often used [28]. Additives may

374 Barley

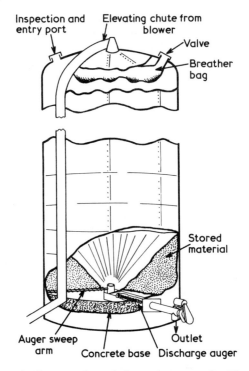

Fig. 10.2 Schematic diagram of a sealed metal storage silo. The breather bag allows the equalization of pressure inside and outside the silo, without the loss of silo gases or the entry of air. Safety pressure-relief valves are also incorporated (after M.A.F.F. [90]).

be sprayed onto the chopped material. Oscillating chutes or rotating deflector plates beneath the inlet spout give the necessary even spread of the material in the tower. The upper surface of the silage should be sealed e.g. with plastic sheeting, when direct access is not required. The presence of asphyxiating gases in towers makes entering hazardous.

Silo unloaders are usually electrically operated, and may unload from the top of the silo or from the bottom [28,89]. Top unloaders are used in unsealed silos. They remove the youngest silage first, and so the tower should be fully emptied before being refilled. Base unloaders are generally employed in sealed silos. They are robust, as access for maintenance is awkward. They cut and remove the oldest silage from the silo and there is no need for personnel to enter. Three main types are employed; (a) the 'coal cutter' type with an endless cutting chain mounted on a sweep arm that rotates round the flat base, cuts the silage and delivers it to the centre, where a chain conveyor, running in a slot in the base, carries it to the outside. (b) Those in which a sweep-auger, fitted with teeth, moves round the cone-shaped base, cutting the silage and delivering it to a

The reception and storage of whole plants and grain

central conveyor. (c) Those in which an endless chain saw reciprocates across the flat base, from a side mounted position, cuts the silage and draws it to the exterior [89].

10.5. Grain reception

Grain is handled in small or large lots coming to temporary or more permanent storage in farms, maltings, feed merchants, feed compounders and some other manufacturers. Malt is handled in a generally similar way. Grain reception and storage may involve stacking a few sacks in an odd, dry corner or receiving thousands of tons of grain which is segregated by type and quality, roughly cleaned, dried, very thoroughly cleaned and stored under rigorously controlled conditions for several years. Grain reception should be rapid, to avoid holding up other deliveries, and the grain should be protected from the weather. It is convenient to discharge loose grain into ground level gratings leading to conveyors, as these can receive from many types of tipping or hopper-bottomed lorries or railway wagons. Alternatively delivery vehicles may be unloaded by pneumatic systems, or may discharge into a pit.

10.6. Handling grain

Sacks may be moved about on wheelbarrows, horizontal conveyor systems, trailers, or fork lift trucks or may be carried up or down on endless belt or chain sack elevators, or hoists. Sometimes the same elevators and conveyors are also used for handling bales. Handling large bulks of grain in sacks is wasteful of manpower [28,31]. Grain bulk handling systems vary according to the quantities to be moved, the rate at which it must be moved, the number of types of grain to be handled, the number of stores to be filled and emptied, and the location and extent of the machinery used to treat and clean the grain [28,31,90,133]. The machinery must be easily cleaned to keep down insect pests and to prevent the admixture of different lots of grain. Physical damage to the grain must be minimized and the dust created by moving it must be trapped. Dust from grain, grain products and malt is rich in bacteria and mould spores. It is unpleasant, constitutes a health hazard as it may cause dermatitis or bronchial problems, and when mixed with air in some proportions forms violently explosive mixtures. Explosions of grain dust have wrecked whole blocks of silos and brewing mill rooms. No smoking or activities involving flames or sparks should be permitted in grain stores. To prevent air-dust explosions travelling along in conveyor casings and triggering explosions throughout an installation it is usual to build in weak spots, often thin metal plates, which burst and allow explosions to 'vent'. Grain dust picks up moisture from the air, becomes 'slack', and provides an ideal breeding ground for many insect pests.

Pneumatic conveyor systems may be either permanent installations or mobile. Grain may be sucked directly into a system through a flexibly mounted nozzle that can be thrust into heaps of grain. Alternatively grain may be introduced to the 'pressure side', into an airstream via an auger or a rotary valve fed by a hopper. Such systems may carry grain upwards, downwards or horizontally. Grain may be discharged through rotary valves or tipping seales. Pneumatic systems are flexible and can work over long distances. They are self-cleaning, and can remove the last traces of grain from stores.

Screw conveyors, augers and 'worm-conveyors' are widely employed. They rely on helical metal screws rotating in cylindrical or U-shaped casings to carry grain along. They may work horizontally or up slight inclines. Simple mobile augers are usually electrically driven, and are mounted at an angle, on a frame with wheels. Discharge is through a spout at the upper end, while intake is through the mesh-guarded open, lower end. They may have helical rubber 'sides-to-middle' sweep-blades which move grain to the middle, at the inlet; these are used for emptying flat stores. Conveyors may be loaded directly, e.g. from the outlet of a silo, or they may be fed by other conveyors. Discharge may be at the end of a run, or by opening ports in the casing at intermediate points. Such conveyor systems are simple to maintain, but they are rarely 'self-emptying', so they must be cleaned regularly. Sometimes, when being used to load a store, or a malt kiln, conveyors discharge onto rapidly moving, steerable, endless canvas belts which throw the grain through the air onto the grain-heap. Other loading devices use rotating high-speed paddles with rubber tips to throw the grain.

Chain-and-flight conveyors move grain horizontally. An endless chain is fitted with projecting vanes, called 'flights', which rest on the bottom of a casing and sweep the grain along. Loading and discharge methods are as for screw-conveyor systems. Rarely skeleton-link chains are used without flights, but these are not so suitable for moving grain. Bucket-elevators lift grain vertically, in conjunction with horizontal screw- or chain-and-flight conveyors. These comprise a series of tipping buckets fastened to an endless belt moving round upper and lower pulleys. The whole is encased, with an enlarged space, called a trough, sump, or 'boot', at the bottom and a discharge chute or pipe at the top. The buckets scoop the grain from the boot, carry it upwards on the ascending side and tip it out into the discharge opening as they go over the top pulley to begin their descent. These devices readily accumulate dust, and are never fully self-emptying, so they must be cleaned regularly.

Endless-belt conveyors are also used to carry small amounts of grain rapidly along horizontally or up slight inclines. They are normally of reinforced cloth, and may be smooth or have ribbed surfaces. Belts run over rollers and pulleys that 'dish' the upper surface by raising the edges

The reception and storage of whole plants and grain 377

so keeping the grain in the centre. They are loaded by meters which feed thin streams of grain to the centre of the belts. Discharge may be by the grain running off the end of a belt, or by a scraper deflecting the grain into a side chute, or by the belt being folded by a special set of rollers to create an artificial 'end' where the grain falls off into a chute, or by a section of the belt being tipped to one side. Such belts may work open inside buildings, or within casings. They generate little dust, are gentle to the grain, and are largely self-cleaning.

Sometimes grain, or green malt, is moved by oscillating jump-conveyors. Each section of the conveyors throws the grain forward in little jumps of 1–8 cm (0.5–3 in) by an upward and forward jerking action. These are very gentle with grain but they are cumbersome and slow.

10.7. Weighers

At each stage of handling barley is weighed [28]. Trucks may be weighed on a weighbridge on entering and leaving a premises, the grain weight being determined by difference. Sacks may be weighed individually and may be filled from a device that fills a hopper to a preset weight then the grain flow is cut off, and the hopper discharges into a sack. Bulk grain may be elevated to a heading hopper. The grain flows in until a certain weight is reached, then the flow is cut off. The grain is then discharged either through the bottom or by tipping. As the pre-chosen weight and the number of emptyings are known, so is the total weight of grain. Sometimes grain is measured by volume.

10.8. Cleaning and grading grain [28,31,88,90,133]

Grain as received is often contaminated with dust, chaff, short straws, leaves, shrivelled and broken grains, the grain of other cereals and wild oats, insects, thistle heads, weed seeds, lush green fragments of weed, soil, stones, snail-shells, string, binder twine and fragments of metal (such as wire and nails). To improve the keeping-quality of the grain, to reduce damage to handling machinery and to increase its value these objects must be removed.

Some grain cleaning machinery is complex, and has a limited rate of working. Often only large seed merchants, maltsters or silo owners can justify the expense of the most sophisticated machines. In humid areas grain from the fields must usually be dried. In some climates artificial drying is unnecessary. The thorough cleaning processes, needed for seed or malting standards, are slow and would hold up the grain intake. Consequently it is usual to quickly and roughly preclean the grain before it is dried, and then to clean the dried grain thoroughly later, as circumstances allow. Precleaning and cleaning are considered together, although they may be separated in time and place.

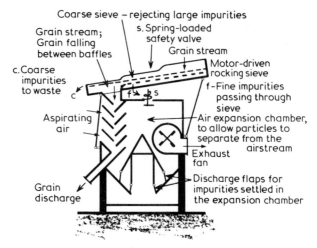

Fig. 10.3 Grain precleaning apparatus. Grain, after passing through an upper sieve and being retained on a lower one, is aspirated by a fan-driven airstream as it falls between baffle-plates (after de Clerck [31]).

Grain is aspirated and roughly sieved by the threshing machinery. It is sometimes necessary to pass it through a 'de-awner' in which rotating paddles rub the grains together, and across a serrated surface and break the awns. Rarely the grain may be 'polished' by passing it down a rotating cylinder where it is stirred by paddles, while dust is aspirated. This scours the surface of the grain, and improves its appearance by dislodging dirt but at the expense of abrading its surface, creating dust and battering it. Precleaners are sets of sieves which separate barley and barley sized impurities from larger and smaller impurities, and dust. As the grain is metered into the machine, dust is carried away by aspiration and the grain is fed to the automatically cleaned screens (sieves). Alternatively the grain may be screened, then aspirated (Fig. 10.3). The slots in the first screen allow the passage of the grain and fine impurities but retain large impurities, which are carried to an outlet. Sometimes this stream is re-sieved to recover any entrained grain. The next sieve retains the grain but allows fine impurities such as sand to pass through. The dust-laden aspiration air is often recirculated through an expansion chamber, where some dust settles, and a textile sleeve filter which retains the very fine particles. The sieves may be decks of slightly inclined, flat beds that oscillate, or rotating inclined cylinders through which the grain moves. Cylindrical screens often contain baffles to mix the grain and ensure that it all meets the sieving surface. On the farm scale small impurities may be removed by moving the grain along an auger with sieve holes in the casing, before it is fed to an aspirator. More complex, slower-working machines work on essentially similar principles (Fig. 10.4). More dust

The reception and storage of whole plants and grain

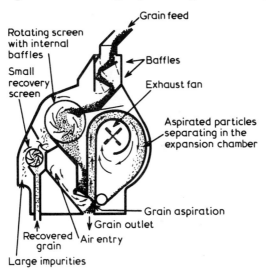

Fig. 10.4 A grain-cleaner with a rotating cylindrical screen (sieve, squirrel-cage scalping reel), a recovery screen, and an aspirator (various sources).

may be removed from the grain stream by aspirators or winnowers. Pieces of iron and steel (tramp iron) are removed early in the handling sequence, to reduce wear to the machinery and the chances of sparks being struck, with the attendant risks of explosions and fires. Magnetic separators are often used on barley intake, (and sometimes on dispatch), on malt on dispatch from maltings, and on malt received in breweries. In the simplest type the grain slides down a wooden or other non-magnetic slide, over a strong permanent magnet, which retains magnetic materials. From time to time the magnet is cleaned. In another type grain is fed in a stream over a cylinder, which rotates in the direction of grain flow (Fig. 10.5). Within the cylinder is an electromagnet. The cylinder is magnetized and so magnetic materials in the grain stream adhere while the grain falls away down a chute. The metal objects are carried round and, as the cylinder becomes demagnetized, they drop into a separate chute. In other equipment the cylinder is replaced by an endless magnetic belt. To remove broken grains and impurities of about the same size as barley grains, sophisticated machinery is needed; this has a low rate of throughput. The machines used are cylinder separators or disc separators. In cylinder separators a regulated stream of grain is fed into the upper end of an inclined, rotating cylinder; the treated grain is drawn off at the lower end. The inner face of the cylinder carries many carefully made small pockets or indentations, say $23\text{--}30\,000\,\text{m}^{-2}$ (Fig. 10.6). Often devices are incorporated to mix the grain, so that as much as possible meets the indented surface. The cylinder may be chosen so

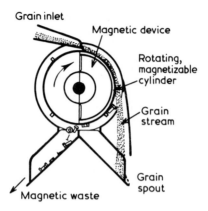

Fig. 10.5 A rotating electro-magnetic cylinder for removing magnetic impurities from a grain-stream (after Vermeylen [133]).

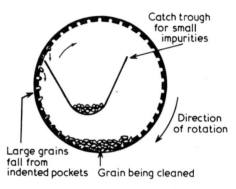

Fig. 10.6 A cross section of a rotating, indented cylindrical grain-cleaner. For clarity the trough-emptying auger and baffles are omitted (after Culpin [28]).

that, for example, half corns and small weed-seeds fit into the indentations and are carried out of the grain mass. When they drop from the indentations they fall into a catch-trough, where an auger moves them away. The reject stream may then pass to a second, smaller, grain-recovery cylinder which picks out any barley and leaves behind the impurities. In high-speed separators the cylinders are rotated at such a rate that objects are partly retained in the indented pockets by centrifugal force. In cylinder-separators only about 20–25% of the indented surface is working at any time and the machinery becomes inefficient if it is overloaded. With disc separators a series of metal discs, carrying indentations on both faces and mounted close together on a revolving shaft, dip into a stream of grain in a trough, lift out impurities, and drop them over the side of the trough into receivers. Because the discs dip deeply

The reception and storage of whole plants and grain 381

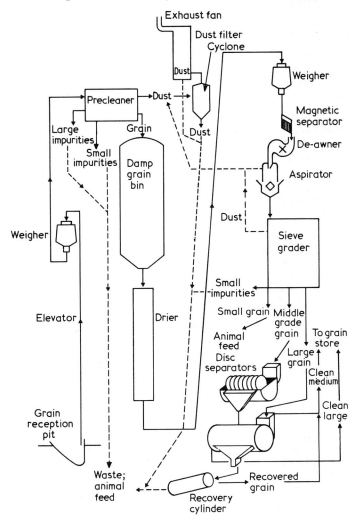

Fig. 10.7 One possible arrangement for grain cleaning.

into the grain, are close together, and are indented on both surfaces they have a high working surface to grain bulk ratio, so they work rapidly. The separated impurities may be re-processed to recover any grain, and may then be graded into mixed cereals, vetch seeds, and other classes so they can be sold. Spiral 'slide' separators may be used for this purpose.

Grain may be separated into grades, differing in width, by screening. This may only involve separating the very thin barley, the 'thins' or 'needles', say < 2.2 mm wide, or it may involve separating it into three or four classes. Usual slot-widths in the screens are 2.8 mm, 2.5 mm, and

2.2 mm, so that the barley grades are > 2.8 mm, 2.8–2.5 mm, 2.5–2.2 mm and < 2.2 mm. However in a bad sample containing a high proportion of thin 'tail corn', the lower limit may be set at 2.1 mm. Medium sized 'bold' grain is preferred for malting, but separating grain into many grades is unusual now. Graders may be in the form of cylindrical screens. However flat-bed screens known as plansifters are more efficient. These consist of sets of flat screens oscillating horizontally and vibrating slightly vertically. Several sets of sieves may be mounted above each other in one shaking frame. Grain is metered onto the top sieve of each set, and the streams of grain from each sieve may be fed down to a lower sieve for reprocessing or withdrawn from the machine. One of the many possible schemes for grain handling is illustrated in Fig. 10.7.

10.9. Drying principles [28,31,56,63,74–76,90,101,102,115,116]

The principles involved in drying hay, straw, barley grain, malt, or spent grains are the same. Most attention has been given to grain drying and malt kilning. Barley drying will be considered in most detail here, malt kilning in Chapter 15. Moisture is removed from materials to enhance their keeping qualities. For example the approximate *maximum* (no 'safety margin') permissible moisture contents (fresh weight basis) for short and for more prolonged storage respectively under average British conditions are barley, 14.8%, 13.6%; malt culms (rootlets), 15.3%, 12.6%; straw, 14.8%, 12.8%; dried grass, 13.7%, 11.1%; hay, 12.6%; 11.0%; distiller's spent grains, 11.0%, 9.8% [125]. In warmer climates these values would have to be reduced. The acceptable moisture level for grain storage is decided with reference to ambient temperatures, the accessibility of oxygen, and whether or not insect pests are likely to be present. Acceptable maximum moisture contents should not be exceeded anywhere in the bulk of material stored. Materials are usually dried to the smallest extent possible, to retain fresh weight and volume, and to minimize costs. This, and the deplorable practice of rewetting grain to increase its weight, are strong arguments for dealing in dry weights, with lower prices given for wetter materials.

The water vapour pressure of a wet material rises with increasing temperature. At a given temperature the water vapour pressure increases with increasing moisture content. In a closed container the moisture in a material comes into equilibrium with the moisture in the surrounding atmosphere. The equilibrium moisture content of the air can conveniently be expressed as the equilibrium relative humidity (r.h.). r.h. is defined as the amount of water vapour present in the air, expressed as a percentage of the amount of water vapour the air would hold at that temperature and pressure if saturated with moisture. This has the advantage that although the moisture-holding capacity of air and vapour pressures increase with

The reception and storage of whole plants and grain 383

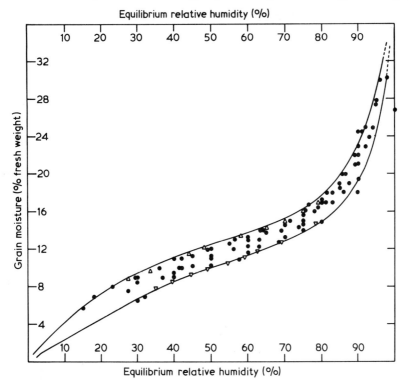

Fig. 10.8 The relationship between the moisture content (fresh weight basis) of barley grain and equilibrium relative humidity, showing hysteresis.
● – other data.
△ – grain initially at 26% moisture.
▽ – initially 4% moisture.
(data of Coleman and Fellows [24]; Finn-Kelsey and Hulbert [37]; Kreyger [76]; M.A.F.F. [90]; Oxley [102]; Robertson et al. [113]; Sijbring [121]; Snow et al. [125]).

temperature, yet the equilibrium r.h. values are relatively independent of temperature in the range 15°–35°C (59–95°F). Curves showing equilibrium relative humidities against moisture content are shown in Figs. 10.1 and 10.8. Nomograms relating equilibrium humidity, temperature, dewpoint and water vapour pressure for barley for adsorptive and desorptive conditions are available [106]. Over certain moisture contents the equilibrium relative humidities of materials are 100%, that is equal to a free water surface. At lower moisture contents the equilibrium r.h. is less as all the water molecules are associated with the structure of the material. As the moisture content is reduced the water remaining is more and more firmly bound as shown by the much reduced equilibrium r.h. Several points should be noticed:

(a) The curves are not *completely* independent of temperatures, for example for a 10°C rise from 25° to 35°C with wheat, which is very similar to barley, the equilibrium moisture content fell by up to 1.3% [61].

(b) Grain moisture may take a considerable time (several months) to come into true equilibrium with the surrounding atmosphere.

(c) Because true equilibrium is rarely reached, different experimental curves relating atmospheric relative humidity to grain moisture content are obtained depending on whether moisture is being gained or lost by the grain, that is the curves exhibit hysteresis (Fig. 10.8) [61].

(d) Different components of stored products (fibre, starch, protein etc.) have different characteristic moisture/equilibrium r.h. curves. Consequently curves for samples of grain, straw etc. with differing chemical compositions will have small but real differences.

(e) Some curves (e.g. for barley and malt) [56,115], appear to be in error. There appear to be no other curves for the moisture/r.h. relationships of malt.

At higher temperatures the rate at which moisture in a material equilibrates with moisture in the air increases. If air with an r.h. below the equilibrium r.h. is passed through a layer of material then it dries until its moisture is in equilibrium with the airstream. Conversely if the r.h. of the air is above the equilibrium r.h., the material picks up moisture. Thus the faster a material is to be dried the higher the temperature of the system and the lower the r.h. of the airstream should be. There are many practical restraints. As the depth of a bed of material is increased, or the rate of airflow through it is increased, the fan power needed to drive the air increases very greatly [63,90]. Then grain or other materials cannot be heated at increasingly higher temperatures and retain their viability or chemical composition.

As hot air passes through a bed of wet grain or malt (> about 30% water), water is withdrawn, initially at approximately a linear rate [116]. Gradually the rate declines until drying ceases. At higher temperatures the final moisture content is reduced since the hotter air has a lower r.h. During drying the distinction has been made between the free-water stage, the intermediate, and the bound-water stages of evaporation [115]. The divisions are not sharp and even in the 'free' water stage evaporation is restricted to some extent. Moisture re-distributes itself more rapidly within grain at higher temperatures. In malt of increasing degrees of 'modification' moisture is increasingly readily removed on kilning. This has been attributed to the moisture being less firmly bound, and to the presence of more roots on well grown malt acting as a big evaporating surface (Chapter 15) [115,116].

To calculate fuel efficiencies and to decide if a drier is being properly managed various pieces of data are required. Thus to calculate the heat needed to warm a given bulk of grain the specific heat is needed. The

The reception and storage of whole plants and grain

specific heat of water is 1, of dry malt is 0.38 (Chapter 15) [55], barley of differing moisture contents has different values [33]. The latent heat of water must be known. The latent heat of steam rises with decreasing temperature, the latent heat of evaporation of water is −539 cal/g; (970 B.T.U./lb) at 100°C (212°F). Many simplified grain-drying calculations may be carried out with the aid of a drying diagram, a modified Mollier diagram [31,56,115]. This diagram relates the relative humidity, the water content and the heat content of air at different temperatures. As air enters a drier it is heated and a small amount of moisture is added from the combustion of the fuel. The net result is that the temperature is increased and the relative humidity is reduced. As the heated air moves through the grain it picks up water, so the r.h. rises, and the temperature drops by an amount controlled by the latent heat of the evaporated water. If the drier or kiln is well insulated no appreciable quantity of heat is lost from the system so the air is in thermal equilibrium with the grain; thus its condition is described by changes occurring along a line of heat content on a Mollier diagram. When the moisture content of the airstream is such that its r.h. equals the equilibrium r.h. of the grain no more evaporation occurs, and the temperature remains constant. If the conditions are badly arranged condensation may occur in the surface layer of the grain, or the air above it, and spoilage occurs. Thus the air must emerge from the grainbed with an r.h. just below the equilibrium r.h. The quantity of moisture removed can be calculated. Comparison with the amount of water that may theoretically be evaporated by the fuel consumed is a measure of the efficiency of the drying unit.

Feed grain may be dried at high temperature e.g. 82°–105°C (180°–220°F) since its viability is immaterial. The wetter barley is, the more readily it is damaged by high temperatures and the longer the exposure, the greater the damage is. No drier treats all grains equally and the moisture contents of grains differ, so to preserve viability for moisture content > 24% the drying temperature should not exceed 43°C (110°F), while < 24% the figure is 49°C (120°F). Some driers have a pre-warming or sweating section where the grain is warmed by passage among radiator bars or in a heated airflow. The airflow is preferable as overheating is unlikely. After grain has been dried with warm air it is cooled before storage to minimize deteriorative changes [28,31,101,102].

10.10. Grain drying in practice

Wet grain must be ventilated to keep it 'in condition'. In the U.K. grain of 25% moisture content (m.c.) can only be kept a short time at ambient temperatures even with ventilation, while at 18–20% m.c. it may stand in sacks for short periods; with 16–18% m.c. grain can be stored in loosely stacked sacks through the winter, if exposed to cold air. At 14–16%

m.c. grain in bulk may be stored temporarily if it is ventilated, for prolonged periods at about 14% m.c. if it is ventilated; at 12% m.c. grain can be stored with minimal ventilation [56,90].

Damage caused during grain drying usually follows from attempts to hasten the process by using too high temperatures, or by stopping the process too soon so that large parts of the bulk are inadequately dried. Samples drawn for analysis may not be representative of the bulk, so the moisture content should be reduced to 1–2% below the value that appears adequate from small-scale trials. Further many electrical moisture meters can easily be 1% in error. The effectiveness of drying is highly dependent on the local weather; drying is easier in the eastern parts of England than the damper western areas [90].

Damp grain has supposedly been 'dried' by mixing it with a dry bulk. This slovenly practice is inadequate, as moisture equilibration is slow and incomplete, and grain mixing is never uniform on a large scale. The usual consequence is a larger bulk of spoiled grain. In a discontinued process grain in a bin with specially designed ducting was slowly dried by passing cold air dried with silica gel [83]. The use of drying drums operating under a partial vacuum has been reported. Barley can withstand higher temperatures under vacuum [138]. Grain and straw are usually dried by forcing a stream of warmed air through the bed. In modern driers or kilns, the pressure is provided by fans. Slight heat may be provided from the radiator of the engine driving the fan, from an electrical heater or from a furnace. In old malt kilns, sometimes used for drying, the convective 'chimney effect' drives the airflow. Generally direct heating is employed, so that the furnace gases go through the product being dried, but occasionally the furnace gases, or hot water, or steam are used to heat a radiator which warms the incoming air. Such indirect heating has the advantage that lower grade fuels may be used without contaminating the material being dried with combustion products. It is possible to dry barley from 24% to 14% m.c. in 35–40 min and fully retain germination, although the husks tend to crack [102]. In large scale practice it is desirable to have a safety margin and take at least 1 h or better 2 h to complete drying.

Many types of drier are in use. Bagged grain (or bales of hay, or straw) may be dried on platform driers (Fig. 10.9). A concrete platform is erected 45–60 cm (1.5–2 ft) above the ground, forming a hot-air chamber, with holes crossed by metal bars across which sacks or bales may be laid, one per hole, or which may be closed with a lid. Generally 3 bushel open-weave sacks containing 55–65 kg (120–140 lb) fresh grain, dried to about 50 kg (1 cwt) are employed. Platforms with 40–50 apertures are usual. Air, at 14°–17°C (25°–30°F) above the ambient temperature is blown under the platform and escapes through the sacks. The sacks may be changed 2 or 3 times in 24 h. Such driers are simple, and relatively

The reception and storage of whole plants and grain

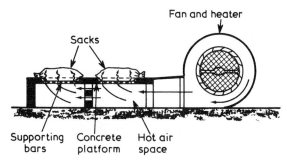

Fig. 10.9 A simple platform drier for drying sacks of grain or bales of straw (after M.A.F.F. [90]).

cheap, but laborious to use and the grain is treated unevenly [90,102].

Grain may be dried slowly in storage bins. This technique is only suitable for reasonably dry grain in districts having suitable weather [90]. A bin or silo may have a perforated floor which allows a vertical flow of air through the grain or a central duct and perforated walls which permit a radial airflow, or a system of inlet and exhaust ducts interlacing throughout the storage space, which keeps the through-grain distance to a minimum and which, although relatively complex, is probably the best. At times of low ambient r.h. cold air, or slightly heated air, is blown through the grain. Using warmed air the drying power of the air is increased but if it is cooled beyond a certain point by the grain then condensation occurs and damp regions form in the grain. Under British conditions with an airflow of about $6.4 m^3/min/m^2$ (21 cu ft/min/sq ft) cross section (at $90°$ to airflow) for a 300 cm (10 ft) bed of grain with the 'air on' warmed to $5.5°C$ ($10°F$) above ambient, it is possible to remove about 0.5% moisture/24 h from grain. At the same time the distribution of the moisture through the grain is made more uniform. The system is unsuitable for the removal of more than 3% water, or for grain of more than 17.5% moisture [76]. However, with an outflow of about $3.6 m^3/min/tonne$ grain (about 130 cu ft/min/ton grain) moisture can be reduced from 21% to 15% in 10–14 days. The maximal practical depth of grain for drying in this way is about 300 cm (10 ft), of which at most about 200 cm (about 7 ft) should be added at one loading to bins where the drying zone spreads up from the bottom. Attempts are being made to economize on fuel by utilizing black-painted surfaces on the bins to 'catch' heat from the sun and pre-heat incoming drying air.

Batch-driers are of many types. In maltings old malt kilns are sometimes used. Simple driers may be horizontal perforated trays on which beds of grain are formed. The beds are ventilated from below. They are charged from a hopper, and may be emptied by a hydraulic tip or through a louvred floor. The cycle may be controlled automatically by a time-clock

388 Barley

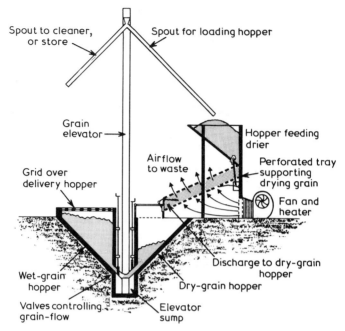

Fig. 10.10 A simple sloping-floor grain tray-drier (after M.A.F.F. [90]).

or a moisture meter and it may incorporate warm up, drying, and cooling (ventilation without heat) periods. The perforated tray, on which the grainbed is formed, may be inclined at 17–20° and be emptied from a gate at the lower and (Fig. 10.10). Some driers are also suitable for drying straw or hay, either loose or baled. Other batch driers are in the form of hoppers with perforated, louvred interior walls. These are filled from above from storage hoppers and empty from a hopper-bottom. Drum driers, which are arranged like malting drums are used as batch driers (Chapter 15) [56,90].

Continuous driers may have inclined shaking trays, or endless perforated belts which may or may not feed lower belts (multiband driers), chain-and-flight conveyors, rotating inclined driers, or towers. Usually they process the grain in 1–2 h. It is usual to have a warming, or 'sweating' section, a drying section and a cooling section. It may be desirable to cool grain before putting it to store since heat is lost slowly from a bulk of grain, and warm grain is more subject to insect attack and other types of deterioration. In maltings grain may be stored warm for a few weeks to overcome dormancy. It is then cooled, often during the cleaning process. The seed trade might usefully adopt this process to improve 'seed vigour'.

In continuous drum-driers grain is fed into the upper end of an inclined, slowly rotating cylinder; as it moves slowly towards the exit it

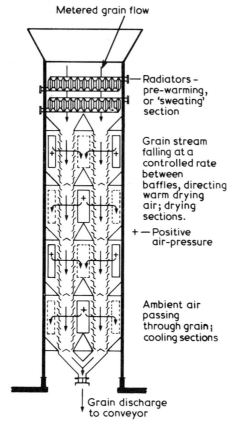

Fig. 10.11 A schematic vertical section of a tower-type of grain drier (after de Clerck [31]).

is mixed and guided by internal baffles. The drying air is introduced at the lower end so that the driest air meets the driest grain and flows counter to the grain stream. Cooling is carried out after the grain has left the drum [74–76]. In tower driers grain is elevated to a feed-hopper at the top of the tower and enters the drier in a metered flow (Fig. 10.11). The grain stream descends at a regulated rate either between perforated walls or ducts which guide the airstream. Water contents may be reduced e.g. from 18% to 12%.

10.11. Grain storage facilities [8,31,56,90,133].

The facilities needed for successful grain storage depend on the local weather, the quantity of grain to be stored, the duration of the storage period and the quality to be maintained. The cost of drying and storage rises sharply with extra provisions to minimize deterioration, so the aim for normal malting grain and top quality seed might be to have less than a 5% decline in germination over two years.

In the Middle-East grain may be stored in simple pits or in heaps in the open with a fence to keep off wandering animals, and perhaps a coating of mineral dust to discourage insects. Such storage is generally unacceptable. Stores should be built to exclude rodents and birds, should keep out rain and draughts of moisture-laden air, should allow the grain to be ventilated in a controlled manner when the humidity is low and should allow fumigation and other control measures to be taken against insect pests. Forced ventilation is essential since the diffusion of oxygen through grain is slow although the volume of intergranular air is about 50% of the bulk volume [9]. Stores should be easy to fill and empty, should keep different grain lots separate, and should permit the early detection of heating or spoilage. The stores discussed first are able to maintain malting-quality barley or seed in excellent condition in temperate, humid climates. The second group of storage facilities to be discussed is used for keeping feed grain by minimizing some forms of deterioration without regard to viability.

Small quantities of dry grain can be stored in sacks. These should be housed in a dry, pest-free and clean place. Draughts which renew damp air from outside should be excluded. Storage is less hazardous in cold weather. Damp grain may be stored for short periods if the sacks are stacked loosely to allow the dissipation of heat. Sacks and the spaces between should be regularly inspected for signs of rodent attack or insect infestation. All sacks should be fumigated or heat-treated before they are allowed near a grain store or are re-used, as old sacks are a frequent source of insect pests.

Many makeshift structures are used especially on farms, but they are usually unsatisfactory. After a short store under adverse conditions deterioration begins and subsequently the grain may deteriorate unexpectedly quickly. Consequently efforts are often made to collect grain from farms soon after harvest. Flat-bed stores are simple and comparatively inexpensive, but they are less easy to empty than silos, and pests are less easy to control. 'Turning the grain over' (mixing it during storage) is nearly impossible. Further, although they can be partitioned into sections, flat-bed stores are inconvenient for keeping different lots of grain separate. Flat-bed stores should have smooth, waterproof floors and smooth side walls that can withstand the lateral thrust of the grain [28,90]. If the grain is loaded from a central overhead conveyor it will form a ridge sloping downward at its natural angle of repose. Depths of about 240–300cm (8–10ft) at the sides are frequent. The whole is enclosed in a weatherproof structure which, since it carries little load, can be of comparatively inexpensive construction. Ducts for forced aeration and cooling may be cast into the floor and covered with perforated material to support the grain, which has the advantage that the floor is flat and so vehicles can move across it. Alternatively the ducts may be of metal, or metal covered with cloth that are erected on the floor before the store is filled and that can be removed and cleaned as the floor is emptied. Ducts are liable to become havens for pests. They should be spaced to ensure adequate ventilation of the whole mass of grain [16,90]. Fans, and sometimes associated heating units, can be coupled to the inlets of the ventilating system. Such stores may be unloaded by augers or pneumatic systems. 'In grain' temperatures should be determined regularly, at different depths and at intervals of a few feet over the entire grain mass to detect any onset of heating. In the U.K. grain, dried to about 12% moisture, is sometimes stored to a depth of 120–180cm (about 4–6ft), without provision for ventilation, on old malting floors. Thorough grain and store cleaning, grain drying, the use of insecticides, the initial acceptance of only uninfested grain, the occurrence of only moderate summer temperatures and regular inspections that allow immediate remedial action should deterioration be detected are the reasons why this risky method is often successful.

Grain is conveniently stored in bins or silos. There is no clear distinction between these structures, although 'silo' is often applied to larger bins, sometimes a whole group of bins or individual silos is called a silo, a silo-block or an elevator [8,28,31,90,102,133]. Bins and silos may be made in many sizes and of many materials including welded mesh lined with butyl rubber, wood and concrete. Steel silos may have individual capacities up to about 3000 tonnes (3000 tons). Wood structures are simple to erect but are difficult to keep sound, clean and uncontaminated. Steel or concrete buildings are preferred. Concrete is worked so that the surface facing the grain is smooth and free of crannies able to hold dust and insects.

Bins or silos may be of circular, square, hexagonal or other cross-section and are often grouped to give mutual support. The containers are filled by delivery from above. They may be emptied by suction, or rarely may be blown free of grain; grain may be withdrawn from below by a built-in auger, or by a 'self-emptying' cone-shaped metal hopper bottom that allows the grain to flow to the conveyor system. Limited ventilation may be achieved by sucking or blowing a stream of air through the grain, or by 'turning the grain over', that is running it from a full bin, elevating it, and running it into an empty one. With damp grain it is necessary to aerate it continuously. Aeration cools the grain, supplies oxygen to maintain its viability, can achieve limited drying, and minimizes the risk of damage by insects and micro-organisms [15]. A silo's contents cannot be adequately inspected, and so probes are incorporated to indicate the grain level and a series of thermocouples, or other distant-reading thermometers are mounted not more than 120 cm (about 4 ft) apart (often on a central 'rope') to detect any temperature rise. Temperature readings should be taken and recorded at least weekly. Grain keeps best when it is cooled. Generally if it is dry (12–13% m.c.) it need only be cooled to about 17°C (63°F) or less to be free from insect attack and other deteriorative changes [90]. When grain has been gathered in a warm weather, or has been dried it often goes to store warm e.g. 21°–27°C (70–85°F). By ventilating the grain with cool night air, especially in winter when the ambient r.h. is low, (75% autumn, or 80% in winter) the grain temperature can be reduced to acceptable levels [13,15,16].

Grain stored cool, ($<15°C$) with a moisture content of about 12% does not require ventilation, and will retain its viability. Dry grain has been stored, even under water, in specially strengthened, laminated water- and gas-proof sealed, bags. It was advantageous to fill the sacks with carbon dioxide gas before they were sealed [95].

Grain for animal feed may be stored damp if the temperature is reduced quickly and is maintained at a low value by the passage of refrigerated air [15,17,28,87]. The grain may retain its viability but although insects will be checked and fungal attack will be slowed the grain will probably become infested with mites, as these are active in damp grain at temperatures below 4.5°C (40°F). Some fungi can grow at $-8°C$ (18°F) and produce toxins. In the U.K. cold stored grain usually contains less than 20% moisture, and pre-drying to about 16% moisture is preferred. Recommendations for *maximum* temperatures of storage and the *maximum* allowable times to reach the temperatures are shown in Table 10.1.

Lots of 30–100 tons may be chilled in insulated bins, preferably not more than 460 cm (about 15 ft) deep, or on floors. One cooler may serve several stores. In freezing weather cool air can be blown directly into the grain. The ducts carrying the cooling air should be sited to ensure

Table 10.1 Maximum permissible temperatures and maximum times allowed to chill grain at different moisture contents to allow storage for the periods shown (M.A.F.F. [87]; Crown copyright, by permission of H.M.S.O.).

Grain Moisture (%)	Permitted time to chill	Storage temperature (maximum)			
		No deterioration, 2 months		Mould-free, but slight fall in germination	
		°F	°C	°F	°C
16	1 week	55	12.8	60	15.6
18	1 week	45	7.1	50	10.0
20	3 days	40	4.4	45	7.1
22	3 days	35	1.6	40	4.4

The reception and storage of whole plants and grain

that the grain is uniformly cooled, and that condensation in the grain is avoided. The acceptance of this method will depend on the relative economics of chilling and drying and the quality of the final products. Mites in the grain will not commend it to any users, and even if conditions are chosen such that viability is maintained maltsters and seedsmen will find it inconvenient as chilling preserves dormancy (Chapter 5).

Grain for animal feed is also stored damp in hermetically sealed containers [28,67,90,102]. The stores, which may be special sealed silos either of metal (for lots up to about 200 tons) or butyl rubber or PVC supported by metal cages (for 20–40 ton lots), or in underground chambers lined with concrete, as are used in Argentina, must be perfectly airtight. Sealed silos must have a sealed-in auger for emptying, a pressure-relief valve, and a breather-bag, as in silage towers. Unloading allows air to get into the store, and then it is desirable to unload it entirely and use its contents as quickly as possible to prevent rapid deterioration. Freshly loaded damp grain, with its associated micro-organisms, uses the oxygen in the store and replaces it with carbon dioxide so that moulds, insects, and the grains all die and there is only a transient temperature rise. The grain dies more rapidly at higher temperatures and higher moisture contents. Fermentative changes go on, and the carbon dioxide content of the intergrain gas may rise to 80% in grain with 19.2% moisture [67]. The stored grain is bright and mould-free, with a distinct sour-sweet smell and a bitter flavour that animals sometimes like. Grain with up to 30% m.c. can be stored in this way, but it deteriorates rapidly when taken out of store, and at water contents of 25% or more the grain may become soft and dark and resemble silage. Faulty sealing leads to spoilage, which in such 'semi-sealed' conditions comprise caking of the grain, the growth of yeasts (e.g. *Candida* and *Hansenula* spp.), thermophilic micro-organisms, and storage moulds such as *Penicillium* and *Aspergillus* spp. Treating the grain as silage and applying preservatives depresses these adverse changes. Suggestions that sealed bins should be purged with carbon dioxide or nitrogen gas immediately after loading have been rejected as uneconomic.

Various chemicals have been tested as preservatives: formic acid, propionic acid and various mixtures are finding acceptance [11,130]. Propionic acid is toxic to moulds and grain. About ten times the theoretical dose needed to control moulds must be applied, possibly due to losses, penetration into the grain and uneven application. For grain of moisture contents of 18–20% propionic acid (PA) equal to 0.8% of the grain weight should be applied; for grain of 25% m.c. the PA dose should be 1%. Trials with stored, acidified barley at high moisture contents (25% or 35%) indicate that at 20° C (68° F), rather than 0° C (32° F), large drops in feeding value may occur [98].

10.12 Seed longevity and grain deterioration

A wide range of survival times have been noted for stored barley [100]. Up to 96% germination has been recorded after 32 years storage [49]. No deterioration was noted in grain stored in closed, dry drawers in a laboratory after 10 years; threshed grain stored in sacks in a cool, dry room lost 50% of its viability in 4 years and when stored in bottles it declined to 6% in 6 years [112]. In other cases germination declined to about 90% in 5 years and had disappeared after 10 years, while in yet another trial barley grain was 80% viable after 15 years storage [113]. Obviously grain survival is strongly dependent on storage conditions, but usually these have not been fully defined. The variety of grain also seems to be important. In one case six-rowed hulled, and awned barley survived better over 22 years than six-rowed, hulled, hooded grain which in turn survived better than two-rowed hulled, awned barley. Naked barley survived least well [114]. However these samples were not adjusted to a constant moisture content (which ranged from 9.5 to 11.5%) nor were they stored at a constant temperature. Furthermore a beetle infestation occurred. Survival after a wet harvest is less good than after a dry one, perhaps due to an initial heavy microbial contamination [86,113]. Storage of 'dry' grains in sealed cans is usually preferable to open storage [100]. Storage in an inert atmosphere is preferable to air, which is preferable to oxygen [110]. Lower temperatures prolong survival. Grain survives exposure to the temperatures of liquid air, liquid nitrogen, or near to absolute zero, 1–4 K [80]. Grain stored in atmospheres of different relative humidities, at $21°C$ (about $70°F$), took 14 to 20 days to approach hygroscopic equilibrium and survived best at lower moisture contents [113]. Deterioration was detected in grain of about 10.5% moisture content when stored for 454 days, and was considerable after 1032 days storage. Deterioration begins slowly and then accelerates [100]. Viability falls faster at higher moisture contents and temperatures [14,74,75]. Aeration slows the loss in viability of damp grain [119]. Graphs relating storage times at different moisture contents and temperatures, above which germination is likely to fall more than a certain extent e.g. below 95% are of value in operating grain stores [75,76]. A 5% fall in germination in a short period represents a large deterioration. It is possible to extrapolate to obtain estimates of the results to be expected from more extended storage.

Cereal grains are practically always infested with fungi, bacteria and yeasts while actinomycetes, nematodes, and slime-moulds occur from time to time [85,105]. It seems impossible to sterilize intact barley grains reliably. Fungi contribute to grain deterioration; grain treated with fungicides spoils more slowly and heats more slowly than untreated grain [29]. Also heat-treated or other grain inoculated with fungi spoils faster than uninoculated grain. When rare samples of seed are found which have little or no fungal contamination after surface-sterilization their survival, under adverse conditions of storage, is surprisingly good. One sample was stored at 20°–25°C (68°–77°F) at 15.2% m.c.; after 71 days viability was 96%, and after 161 days was 93%. When stored at 16.2% m.c. viability was 94% after 46 days and 48% after 71 days [20].

Grain is a good heat insulator, about 0.3 as good as cork, so in bulk it is slow to pick up or lose heat. It respires, and at higher moisture contents and higher temperatures it respires more rapidly [1,7,94]. A large variable part, sometimes more than half, of the respiration and heat output of stored grain is due to micro-organisms. The variable respiration rates observed with different barley lots may well be due to different levels of infestation with micro-organisms. Microbial infestation in wheat has been more thoroughly studied than in barley, but the situation in the two cases is similar. When fungus-free and fungus-contaminated wheats were compared the respiration of the clean grain was low and constant, while in the contaminated grain the respiration increased in a few days storage when the moisture content exceeded 15%. Many fungistatic substances have been experimentally applied to wheat; applications of thiourea, for example, hold down microbial populations and respiration [92].

'Heating' and spoilage may begin at some confined focus in a mass of grain [101]. This may be due to the presence of damp patches, or temperature gradients may have occurred in the grain, with consequent moisture movements and condensation at cool spots such as the edge of the silo or at ventilation ducts. Grain and mould respiration and mould growth occur fast in such places. Also heating may be started by an insect infestation. As the insects breed they generate heat and water, high local temperature gradients are created, convection currents occur, and moisture condenses at the edge, and particularly above the hot spot, and mould heating also begins [58]. Once a hot spot is formed deterioration accelerates, insects spread out into the surrounding grain and the hot damp air slowly rises into the grain above initiating further spoilage. At the surface of the grain the heat dissipates and moisture condenses in such quantities that the top layer may sprout. 'Dry-grain heating', which occurs in grain of less than 15% m.c., is usually initiated by insects, and temperatures do not rise above 38–42°C (100–108°F) [101]. In wetter grain temperatures may become steady at about 50–52°C (122–125°F) and in such 'damp grain heating' insects may or may not be present. To minimize the chances of 'dry-grain heating' insecticide powders are often mixed with the grain. Insects are not easily seen in silos, and the local high temperatures may not be detected until deterioration is far advanced.

When grain deteriorates it loses viability and 'vigour'. After prolonged storage grain gives rise to seedlings with an increased frequency of abnormalities, and mutations, so to maintain genetically pure seed-stocks these alterations should be minimized [5,48,110]. Heat treatments may produce genetic abnormalities in barley [2,5,138]. The percentage of aberrant cells in seedling root tips rises to a plateau with increasingly poor grain storage [2]. Possibly the upper level is set because above a certain level of damage a seed dies. Oxygen at high pressure, 6.9 bar (100 lb/sq in), increases the rate of mutation in stored barley [77]. When seeds, stored so that their viability was reduced to about 50%, were grown barley (in contrast to beans and peas) showed no non-heritable abnormalities. The decrease in seed viability paralleled the decrease in genetic purity.

Attempts have been made to describe in mathematical terms the expected death rate of seeds stored under different conditions [110]. However the trials on which many of these calculations are based were made under favourable conditions, in that small quantities of good quality grain were stored in sealed tubes under controlled conditions, and so give higher estimates of expected viability than would occur in commercial stores. It seems desirable that studies should be extended to cover at least a 15 year period using a variety of grain samples, the samples being ventilated with air of regulated r.h., and stored over the range of temperatures encountered in practice, ($-10°C$ to $+35°C$). Many published results for different lots of grains fit the equations. The most accurate estimates are of the times taken for grain to decline in viability by 50% [107]. For different cereals, including barley, the log of the half-viability period, at a fixed temperature, is inversely related to the moisture content. When results from wheat stored at $15°$ and $25°C$ were plotted two parallel lines were obtained.
Thus:

$$\log p_{50} = K_v - C_1 m - C_2 t \qquad (1)$$

where p_{50} = half-viability period, in weeks; K_v, C_1 and C_2 are constants; m = moisture content (% fresh weight); and t = temperature, °C. When percentage viability is plotted against the cumulative frequency distribution of death the curve is sigmoid. Usually the results fit a normal Gaussian distribution quite well, and so may be plotted either as 'probit % viability' against time or more conveniently as % viability on a probability scale against time on a linear scale. Straight lines fit the results adequately (Fig. 10.12). The time taken for viability to fall from 80% to 50% represents 84% of the standard deviation (σ), and similar values can be calculated for any other interval. But curves relating log half-viability and log 80% viability with time are parallel and so the period between these curves, and therefore σ, increase logarithmically with decrease in moisture content or temperature. As half viability also increases logarithmically then σ is directly proportional to the half-viability period.
Thus:

$$\sigma = K_\sigma p_{50} \qquad (2)$$

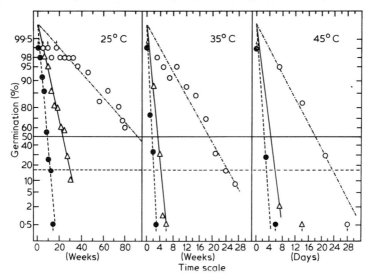

Fig. 10.12 The viability of barley, (plotted on a probability scale), of samples stored at three different temperatures, with moisture contents of o—·—o, 12.1%; △———△, 15.6%, and ●———● 18.0% (after Roberts and Abdalla [111]).

where K_σ is a species constant.

By using appropriate tables and constants the time for viability to decline by any amount can be predicted [108,109]. When barley was stored in sealed ampoules with moisture contents of 12% to 18%, at temperatures from 25° to 45° C (77–113° F) the survival curves were negative cumulative normal distributions except when conditions were very extreme (e.g. 25° C, 77° F; 25.9% m.c., or 60° C, 140° F; 11.3% m.c.), when the distributions became skew (Fig. 10.12). Exclusion of oxygen improved grain-survival, apparently by preventing the activities of micro-organisms, which in any case are inactive at moisture contents below 12%. Thus within limits the behaviour of the barley fitted Equations 1 and 2, the constants determined (± S.E. where possible) being:

$$K_v = 6.745 \pm 0.186$$
$$C_1 = 0.172 \pm 0.0097$$
$$C_2 = 0.075 \pm 0.0031$$
$$K_\sigma = 0.301.$$

The results fitted a slightly more complex equation (Equation 3), almost equally well [111].

$$\text{Log } p_{50} = K_v - C_1 \log m - C_2 t. \qquad (3)$$

398 Barley

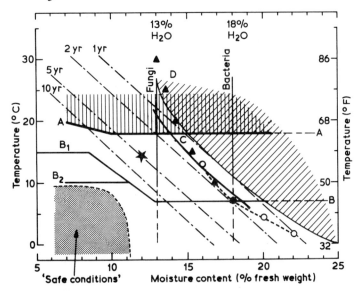

Fig. 10.13 The relationships between storage conditions and the survival of barley grains (Based on data of Burges *et al.* [13]; Burges and Burrell [14]; Kreyger [74]; M.A.F.F. [91]; Roberts and Abdalla [111]). A: Horizontal line limiting vertical shading – conditions under which insect heating is likely to occur. B: Horizontal line indicating the limits at which major insect and mite (B_1) and moth (B_2) pests can multiply. D: Inclined line with diagonal shading – the approximate limit of conditions favouring grain heating by micro-organisms. Vertical lines at 13% and 18% m.c. – the approximate respective lower moisture limits below which fungi and bacteria cannot multiply. Solid, inclined line C, and the line marked o----o, estimates of the conditions at which germination falls to 95% in 35 weeks; ▲: estimates of conditions under which germination falls to 95% in 30 weeks; in each case in open storage. Inclined diagonal lines, ——— –, conditions under which the germination of barley samples, held in hermetically sealed containers, will fall to 95% in the times shown. Note the great survival with storage in sealed containers. Freezing injury may occur to grain of 15% moisture, or more, at temperatures below 0°C (32°F). The star indicates the storage conditions aimed for in many British barley stores. Note that these conditions are not 'safe' if insects and mites are not controlled. Probably fully 'safe' conditions for industrial storage are those within the shaded zone in the lower, left-hand corner.

The results may be used to construct a nomograph for relating storage conditions with viability (Fig. 10.13).

High quality barley is not normally stored for more than two years. Attempts have been made to delimit graphically the conditions under which deterioration may occur in commercial stores [13,110]. Such diagrams are tentative guides, as the data needed for their compilation is incomplete. Thus the conditions under which freezing damage occurs to barley are undefined. Data often relate to levels of deterioration that are

The reception and storage of whole plants and grain 399

not acceptable in commerce. Safety margins must be allowed. Experience suggests that for grain at 17°C (about 63°F) and 14% m.c., storage for a year with adequate aeration is possible, but is certainly risky. For malting purposes it is recommended that grain be dried to about 12% m.c., and be stored at 15°C (59°F) or less [56].

Conditions that allow fungal invasion and insect and mite spread are discussed later. Wheat initially heavily infested with micro-organisms, due to a period of poor storage, continued to deteriorate slowly even when dried and stored at 12% m.c. [45]. At 13.0% m.c. *Aspergillus restrictus* and *A. halophilicus* can slowly invade the germ and cause damage [23]. *Aspergillus glaucus* can grow slowly at 5°C (41°F). The Khapra beetle can grow in malt with an m.c. of only 4% and mites can breed in damp grain at 7°C (44.6°F). Thus a bulk of barley with a moisture content *nowhere* exceeding 12% m.c. or 15°C (59°F) is not 'safe' although from the nomogram a drop in viability of only 5% is predicted to occur in 2000 days storage. The history of the grain, the way it is handled and the chance occurrences of storage all play a part in determining its survival.

For preserving special, relatively small stocks of barley in 'gene banks' with the intention of preserving viability unchanged for ten years or more, while also minimizing genetic deterioration, the problems are different. A few pounds of grain at most, harvested at full maturity in a dry year, carefully threshed to minimize physical damage, (or possibly stored attached to the ear), dried under near-ideal laboratory conditions might be stored under carbon dioxide or nitrogen in sealed containers at low temperatures in the dark. Conditions probably exist such that most samples would be more than 90% viable after 100 years storage. Of the barley grains found sealed in the foundation stone of the Nuremburg Theatre, that had been there 123 years, 3 of the 25 tested germinated [6]. Possibly for really prolonged storage the moisture content should be reduced to about 8% and the sample, sealed under an inert gas, should be stored at about $-20°C$. ($-4°F$). In the Japanese and American seed storage laboratories samples are maintained at 4°C (about 39°F) or lower under conditions that maintain an r.h. of 32% or less [69,70].

10.13 Micro-organisms in grain [19,21,22,32,38,44,73,117].

Over 200 species of micro-organisms, including some pathogens, have been isolated from barley corns, including bacteria, actinomycetes, fungi (mesophilic, thermophilic, thermotolerant), yeasts, nematodes, viruses, protozoa and slime moulds [38,40,73,85,96,105,134,135]. Nematodes and slime-moulds are of rare occurrence. The fungi have been studied most but bacteria, actinomycetes and yeasts are usually present. In the following discussion bacteria, yeasts and actinomycetes are little mentioned because data is lacking, not because they are believed to be unimportant. From 1g of malting barley (about 25 grains) may be obtained tens of thousands of colonies of filamentous fungi, hundreds of thousands of yeasts, and several million bacteria [21]. Grain without a fungal population is very unusual, but may occur in arid areas. In damp climates micro-organisms are universally present, the populations being greatest after a wet harvest. The micro-organisms found and counted in grains depend on the techniques used. It is necessary to utilize selective culture media to detect many types. However it is impossible to know if the media used will allow all the types present to grow. It is usual to study micro-organisms of the grain after washing the grain, or 'surface sterilization' which is usually carried out by shaking with a dilute solution of sodium hypochlorite.

'Field fungi' are found on newly harvested grain. During storage these decline in numbers, being replaced by 'storage fungi'; the rate of replacement and the species of storage fungi appearing are strongly dependent on the conditions. Often storage fungi cannot be detected in newly harvested grain whatever the harvest conditions [21,131]. However, indications have been found that *Aspergilli* and *Penicillia*, 'storage fungi', can invade the grain in the field [81]. Inoculation on the surface during threshing is important. The dust from combine harvesters is rich in micro-organisms. Perhaps the major part of the inoculum of storage fungi comes from the store itself, the walls, the air, the grain handling machinery, and especially the dust, which is rich in bacteria, and fungal spores and mycelium [129].

In wheat most fungi are found in the pericarp; very few occur within the testa. Removal of the pericarp sharply reduces grain respiration without impairing its germination [66]. In barley micro-organisms are found in the husk, particularly between the husk and pericarp, and in the pericarp itself [19,21,39,93,131,134,135]. Many spores become lodged between the husk and the pericarp when the flower 'gapes' at anthesis. Fungi may be present as spores or mycelium. Direct microscopy, the examination of the internal faces of detached paleae and lemmae, showed that in one sample studied mycelium covered 20–93% of the internal area, and was equivalent to a length of about 19–177 cm (7.5–70 in) mycelial thread per grain [134]. A fluorescent antibody demonstrated the presence of one species, *(Penicillium cyclopium)*, *in situ*. The distinction between field fungi and storage fungi is arbitrary, but convenient. In general, field fungi require higher moisture contents (r.h. 90%, grain moisture > 21%) to grow, and they do not proliferate in stored grain. Field fungi frequently present are *Alternaria tenuis*, *Stemphylium* spp., *Cladosporium* spp., *Fusarium* spp., *Helminthosporium* spp., and *Pullularia (Aureobasidium) pullulans* [81,97]. On storage field fungi tend to decline in numbers, being replaced by 'storage fungi' which can grow at lower moisture levels. Notable among these are *Aspergillus* spp., and *Penicillium* spp. When grain of 12, 14 and 16% moisture content (m.c.) was stored at

7°, 20° and 30°C (45°, 68°, 86°F) then, in surface sterilized samples at 7°C (about 45°F) the microbial population remained almost unchanged over many weeks. At 20°C (68°F) and 14 and 16% m.c. the population of field fungi declined, but little invasion by *Aspergilli* occurred. At 30°C (86°F) *Alternaria* spp. had gone in 16 weeks at 14% and 16% m.c. At 12% m.c. and at 30°C the *Helminthosporium* and *Fusarium* spp. had nearly gone in 53 weeks, while storage fungi changed little in amount. In this trial grain germination did not alter much, but germination vigour had declined in those samples where the field fungi had died [82]. Conditions which maintain the viability of the field fungi and prevent invasion by storage fungi are also favourable for grain survival. In well-stored samples of grain storage fungi may remain alive for years; *Aspergilli* and *Penicillia* should not be readily detectable [104,130].

In wet grain, especially if heating has occurred, thermophilic organisms occur frequently [96,97]. These organisms, together with thermophilic actinomycetes, constitute a health risk to farm stock and human beings.

The growth of storage fungi is limited by the relative humidity (r.h.) above the substrate rather than the moisture content (m.c.) of the substrate itself [4,124,125]. Moulds apparently never grow on grain with a moisture content of < 13%. The lower limits of cereal moisture contents that allow the growth of different species are: *Aspergillus halophilicus*, 13.0–13.2%; *A. restrictus*, 13.2–13.5%; *A. glaucus* group (normally *A. amstelodami*, *A. chevalieri*, *A. repens*, and *A. ruber*) 14–14.2%; *A. candidus* and *A. ochraceus*, 15–15.2%; *A. flavus*, 17.5–18%. At 21°C (70°F) and 13.5% m.c., *A. halophilicus* and *A. restrictus* not only grow but can also slowly invade the interior of grains and kill them. Over 15% m.c. other fungi predominate. Members of the *A. glaucus* group only produce perithecia over 15% m.c. at 18.5°C (65°F); over 17% m.c. they are generally overgrown by other fungi. *Aspergillus candidus* occurs in barley stored at 16–19.4% m.c., while *A. glaucus* predominates in the range 16.7–17.4% m.c. *Penicillium* spp. predominate at 19–19.4% m.c. *Aspergillus glaucus* occurs mainly at the ends of the grains [131]. In other trials *Penicillia* predominated and *Aspergillus* spp. grew less at 22.5% m.c. while at 20% m.c. *A. flavus* predominated, at 18.5% m.c. *A. glaucus* and *A. candidus* were dominant and *Penicillia* were sparse [81,104]. Rarely *Mucor* spp. and *Trichothecium* spp. appear with *Penicillia* and *Aspergilli* in damp grain.

Most fungi have temperature optima of about 30°C (85°F) and die a few degrees above this but *Aspergillus flavus* can grow at 45°C (113°F), and *A. candidus* can grow at 55°C (131°F). Some *Aspergilli* do not grow on agar over 35°C (95°F), but will invade grain of 15–16% m.c. at 40°–45°C (104°–113°F). At 12°–15°C (about 53°–59°F) most fungi grow slowly on grain of 15% m.c., and almost cease growth at 5°C (41°F) but some of the *A. glaucus* group can grow slowly at temperatures below freezing. For safe storage grain moisture content and temperature should both be sufficiently low. Fungi are usually strongly dependent on a supply of oxygen; their growth is slowed by 10% carbon dioxide in the atmosphere and strongly checked by 20% carbon dioxide. However some *Fusaria* can grow in atmospheres of less than 1% O_2. Moist grain stored in 'sealed' silos is likely to deteriorate, as in practice conditions are never fully anaerobic. However if an atmosphere of carbon dioxide is maintained spoilage is prevented [99]. The bacteria found on one sample included cocci, coli-aerogenes forms, other gram-negative rods and a few aerobic spore-formers, at about $1-10 \times 10^6$ organism/g grain. During damp storage the total bacterial count declined, the 'coli-aerogenes' group disappeared, but the *Lactobacilli* increased. No strict anaerobes occurred [99]. The total number of yeasts remained about the same, but with an increasing proportion of mycelial forms. Lactic acid and aerogenes-type bacteria occur even on malt [120]. Thermophilic yeasts, moulds and *Streptomyceteceae* were also present. When, during damp storage, grain became bonded together and 'matted' this was due sometimes to fungi, sometimes to mycelialated yeasts [99].

Fungi are important in grain deteriorating with heating, and when no rise in temperature takes place. The heavy fungal inoculum present in cereal dust probably spreads over all grain. Further, insects and mites probably spread infections. Grain that is dead, or skinned, battered, bruised or broken, and is mixed with dust or trash is more likely to spoil than sound grain. Barley in bulk deteriorates rapidly at 18% m.c. while in the shock it keeps well for several months [130]. Barley husks and chaff are known to check grain germination, and reduce mould growth [123]. Injured grain dies more rapidly than sound grain, is more readily damaged by solutions of sterilising agents, and by fungicides. Dead grain is more rapidly invaded by fungi than is sound grain. Fungi invade grain with breaks over the endosperm more rapidly than grain with breaks over the embryo. Grain with a low microbial inoculum survives better than that with a high inoculum [4]. Thus grain should be handled gently and, if it is to be kept viable, the microbial population should never be allowed to rise. Deterioration begun when grain is stored warm and damp can be checked but never reversed. Grain stored unthreshed in the stack was sometimes damaged by heating, the scorched, darkened appearance being called 'mow-burn'. Occasionally it was covered with a mucilaginous slime. When weathered grain is harvested the ends of the grains may be darkened by field fungi. This condition is distinct from black-point or black chaff disease (Chapter 9). When storage fungi invade the embryo, this is discoloured. Mould-spoiled grain undergoes other changes. It often smells, and is tainted, free fatty acids initially increase in amount although they may afterwards decline [21,22,130]. The level of reducing sugars rises, while that of non-reducing sugars

The reception and storage of whole plants and grain 403

declines. Levels of vitamin E and lysine decline, and though protein digestibility remains the same, (until the grain is stored over 18% m.c.), its biological value declines [81,104,130]. Other changes occur also in deteriorating grain (Chapter 5).

Micro-organisms may kill grains, reduce the vigour of seedlings, or may be responsible for 'water sensitivity', (Chapter 5). When heavily infected barley is malted the progress of growth, rootlet production, and modification may be altered [3,44,73,122]. Beer made from such malt may have quality defects. *Fusaria* on malt sometimes lead to gushing (overfoaming) problems in the beer. Poor beer flavours (molasses, burnt, unclean, winey, spicy, harsh, bitter, fruity, astringent) have been associated with infections of various fungi. Haze instability in beer is sometimes linked to microbe-spoiled malt. Probably some storage fungi can make substances that stimulate grain germination and other substances that are toxic to the grain. Fungal culture filtrates frequently check or prevent grain growth.

Mouldy grain, hays, straw and feedstuffs can cause diseases in man and farm animals [78]. Thus in heated grain the spores of thermophilic *Micromonospora (Thermoactinomyces)* spp. may induce the weakening allergy known as farmers lung [97]. Spores inhaled by man, poultry or other animals may give rise to aspergillosis. Some fungi can produce toxins which, when consumed, may be damaging or lethal. Of the plant pathogenic fungi the sclerotia of ergot, *Claviceps purpurea*, are toxic (Chapters 9 and 14). Grain with fusarial scab often has violent emetic effects on farm animals. The water soluble emetic principle may survive in the grain for long periods after the fungi are dead [118]. Feeding barley scabbed with *Gibberella saubinetti* or mouldy with *Aspergillus flavus*, *Pencillium purpurogenum* or *P. rubrum* has often caused sickness or death in pigs and poultry [118]. There are numerous reports of mouldy feed causing illness and death in human beings, horses, pigs, sheep, cattle and poultry [12,21,41,52,78,98]. About 96 fungal spp. are known to cause or are suspected of causing mycotoxicoses. Fungi capable of causing mycotoxicoses regularly occur on stored grain, and many other products. One fungal species can make different toxins, and different strains vary in their toxin-forming ability. When fungal isolates from grain are cultured *in vitro* and the cultures are fed to test animals about 50% of cultures of *Penicillium* spp. are lethal, cause haemorrhages in poultry and damage livers. About 50% of fungal isolates, including *Alternaria* spp., *Aspergillus* spp., *Chaetomium* spp., *Fusarium* spp., *Penicillium* spp. and *Cladosporium* spp. are toxic to rats [21]. Grain moulded with *Fusarium graminearum* (syn. *F. roseum*) may contain zearalenone which, in cattle and pigs, causes disturbances of the reproductive cycle and reproductive organs, reduces litters, and causes abortions. Grain overwintered under snow and ice in the field may be toxic. Various fungi are responsible. Apparently

404 Barley

they have to be on the grain at between $-2°$ and $-10°C$ (29–14°F) to produce toxins [12,41]. Grain and stored foods are not normally harmful, so toxin production must be severely limited under most conditions. In the U.K. the importance of mycotoxicoses is not well defined [52]. In Denmark ergotism can be important, as may cattle sterility due to zearalenone, and nephrosis in pigs caused by feeding grain containing the toxin citrinin produced by *Pencillium viridicatum* [130]. There is a strong supposition that numerous cases of farm animals, or humans, having shortened life spans, being 'off-colour', failing to thrive, or to grow at optimal rates are due to marginal levels of mycotoxicosis. Malt-sprouts that were fatal to dairy cattle were invaded with *Penicillium urticae* that produces patulin, while others that were toxic were infected with *Aspergillus oryzae var. microsporus*, which produced maltoryzine [68,137]. The ready identification of particular micro-organisms in spoiling grain is desirable. By analysing volatile metabolites condensed from air drawn through heated grain, using gas liquid chromatography, this may prove to be feasible [72].

10.14. Insects and mites

Insects and mites often occur in grain, or grain products. Milled products, malt culms, and animal feed are more readily infested than grain, which in turn is more readily infested than malt. Hundreds of species of insects and mites occur in stored grain; some are adventitious, others about 50 spp., are major pests [27,36,42,43,64,101,103,132,136]. At a conservative estimate about 5% of the world's grain is destroyed in store by insects, much more is partially spoiled. Many purchasers will not accept infested grain. Insects harm grain in various ways. Their presence is unattractive and pieces of insect, faeces and webbing are unacceptable in products for human consumption. Mites, and some insects, induce allergies and cause digestive upsets in farm animals. Insects probably spread micro-organisms and create conditions suitable for their multiplication. They also cause a direct loss in grain weight, which can be substantial in hot, humid conditions as are found in ship's holds. At $33°C$ (about $91°F$) a nucleus of 100 grain beetles could give rise to a population of about 12×10^6 in one month, and eat about 51 kg (about 1 cwt) of grain in the process. Insects that bore into grain or eat the embryo may reduce germination percentage or vigour so that it is useless for seed or malting and its nutritional value is reduced. Insect infestations may cause 'hot spots'. The caterpillars of some moths spin webs which bind corns together, preventing the grain from flowing and creating handling problems.

Conditions often favour insect infestation even in soundly designed and carefully operated stores. In barley and malt stores and malting floors situated in the U.K. 134 spp. of insects, mites and related forms were

The reception and storage of whole plants and grain 405

found [42,64,65]. Discounting 'strays', about 86 spp. were probably breeding, of which about 30 spp. were pests. Many others were scavengers, whose presence indicated poor hygiene. Each area of the maltings had its own particular fauna.

Insect populations rise in the spring, with rising temperatures. Their power of multiplication is so great that a small, undetectable winter population may give rise to a massive summer infestation. It is better to maintain a programme to prevent massive infestations than to cope with such infestations as they arise. Only clean loads of grain, malt or grain products, including animal feed, should be accepted. All transport, and grain handling machinery should be thoroughly emptied, cleaned and treated with insecticide after use and sacks should be fumigated or heat sterilized. Dust and spillages of grain should be cleared promptly. Animal feeding-stuffs readily become infested, and should be kept well away from grain stores [27,42,43,103]. Products should be stored only in areas previously emptied, cleaned, and fumigated or sprayed with insecticide. Old and new lots of grain should never be mixed; grain lots should be used in rotation to minimize storage times. Adequately instrumented silos are preferable to flat bed stores for ease of grain handling, infestation detection and control. Store structures should be devoid of crannies where dust can collect and insects shelter. On arrival grain should be well cleaned, dried and cooled. For some purposes grain (but not malt, or many grain products) may be mixed with an approved insecticide. Acceptable insecticides do not persist indefinitely. Stores should be inspected regularly. Insects will not multiply in grain stored cool (usually $< 15°C$) but mites may be a problem (Table 10.2). In sealed stores the level of oxygen must be consistently low to control insects [103]. The first detected sign of infestation is often a temperature rise. Because this often occurs locally and the onset is slow temperature measurements should be taken and recorded frequently and regularly. Control of an outbreak may be by fumigation or with an insecticide. Cleaning, re-drying and sieving is inadequate. Heating grain at $60°C$ ($140°F$) for 15 min is thought to kill most insects [27]. Experimentally grain has been disinfected with nuclear radiations e.g. from ^{60}Co, and high-frequency radio fields have been tested. Different insect strains and different stages in the life cycle have different susceptibilities to γ-irradiation [25,35]. Grain for feed can be disinfested by passing it through an entoleter, which batters it so violently that adult insects, larvae and eggs are killed; the grain is damaged. Insects cannot survive anaerobic storage. Cooling grain with refrigerated air can stop insect multiplication. However many insects survive chilling, and as the temperature declines so in damp grain the mite population may increase.

Purchasers have rejected grain lots in which the insects were adventitious and not grain pests. Owing to the complex life-histories of many

Table 10.2 The appearance, and temperature and humidity requirements of a few common pests of stored grain [59,60,132]. The rate of increase/4 weeks is for favourable conditions. Life cycles vary greatly in length according to the ambient conditions.

Pest	Length and appearance	Optimum temperature	Minimum temperature	Minimum r.h. (%)	Rate of increase in 4 weeks	Comment
Grain weevil (*Sitophilus granarius*)	1.5–3.5 mm. Dark brown, black.	26–30°C (79–86°F)	15°C (59°F)	50	15	Over 35°C (95°F) life cycle not completed. Adults cannot fly.
Rice weevil (*Sitophilus oryzae*)	3.5 mm. Dark. Reddish spots on wing-cases.	27–31°C (81–88°F)	17°C (63°F)	60	25	Adults can fly.
Lesser grain borer (*Rhizopertha dominica*)	2.5–3 mm. Dark brown, black. Cylindrical.	32–35°C (90–95°F)	23°C (73°F)	30	20	Head 'turned down'.
Rust red grain beetle (*Cryptolestes ferrugineus*)	1.6 mm. Flattened, reddish-brown.	32–35°C (90–95°F)	23°C (73°F)	10	60	Adult can fly.
Rust red flour beetle (*Tribolium castaneum*)	4 mm. Reddish-brown.	32–35°C (90–95°F)	22°C (72°F)	1	70	Adults fly.
Confused flour beetle (*Tribolium confusum*)	4 mm. Reddish-brown.	30–33°C (86–91°F)	21°C (70°F)	1	60	Adult does not fly.
Saw toothed grain beetle (*Oryzaephilus surinamensis*)	2.5–3 mm. Dark red-brown, narrow, flattened. Edges of thorax toothed.	31–34°C (88–93°F)	21°C (70°F)	10	50	Difficult to control. Adults rarely fly in U.K.
Khapra beetle (*Trogoderma granarium*)	Adult 2–2.5 mm, light brown. Larva 3.5 mm, hairy.	33–37°C (91–99°F)	24°C (75°F)	1	12.5	Adults hard to detect. Cast skins of hairy larvae more easily seen.
Angoumois grain moth (*Sitotroga cerealella*)	10–14 mm. Yellow-brown.	26–30°C (79–86°F)	16°C (61°F)	30	50	Not established in the U.K.
Warehouse moth (*Ephestia elutella*)	14–23 mm. Variable grey-brown; white bands on forewings.	25°C? (77°F)	10°C (50°F)	30?	15	Larvae (caterpillars) spin webs, bind grains together.
Australian spider beetle (*Ptinus tectus*)	about 2.5 mm. Dark brown, globular.	23–25°C (73–77°F)	10°C (50°F)	50	4	Nocturnal
Flour mite (*Acarus siro*)	0.4–0.5 mm. Whitish, globular, eight brown legs. No division between head and body.	21–27°C (70–81°F)	7°C (45°F)	65	2500	Taint infested grain. Highly allergenic.

pests, the varying growth requirements of temperature and relative humidity (r.h.) of different species, or even of strains of one species, and stages in the life history of one species, and their differing susceptibilities to insecticides and fumigants expert advice is often required to ensure that identification and treatments are correct. The drier the grain, the more difficult it is for insects to multiply. Each insect species has a temperature optimum. The information given for a few pests is in no way comprehensive (Table 10.2). Many major pests are of world-wide distribution [27,132]. Several of the pests important in the U.K. originated in the tropics.

The grain or granary weevil *(Sitophilus granarius*, L; *Calandra granaria* L*)*; and the rice weevil *(Sitophilus oryzae.* L; *Calandra oryzae*, L*)* are major common pests (Table 10.2). They are similar in appearance, but the rice weevil can fly, and is unusual in that in hot areas it can lay its eggs on grain in the field. Minor weevil pests include *Sitophilus zeamays* (Motschulsky) which resembles the rice weevil; the broad-nosed grain weevil, *Caulophilus latinasus* (Say.) and *Araecerus fasciculatus* (Deg.) the coffee bean weevil. Weevils have their heads prolonged into a 'snout' with the mouthparts at the tip [27,59,60,71,132]. Adult grain weevils eat whole grain, but the damage they cause is small compared to that caused by the larvae. In the cold they become inactive but can survive e.g. for 40 days at $-1°C$ (30.2°F). This species overwinters as larvae or adults. The mated female eats small holes in the grain surface, lays an egg, and seals the hole with an inconspicuous mucilaginous plug. About 100 eggs are laid in 8–9 months. The pest is often detected by staining these plugs in grain samples or by sieving out adults. The larva emerges from the egg, eats the endosperm of the grain (leaving the embryo) and after passing through four stages it pupates. Later the adult emerges, bites its way out of the grain, and the cycle re-commences. Different strains show wide variations in their growth requirements. Only the adults are accessible to contact insecticides, of which γ-BHC is the best. Other stages may be killed by fumigation or heat. In the rice weevil the prothorax is slightly indented, and the forewings have four indistinct red-brown spots. Flight wings are present under the elytrae. Each female may lay 300 eggs. One pair can give rise to five broods in 6 months, with a final population of about 152×10^9.

The lesser grain borer *Rhyzopertha (Rhizopertha) dominica* (F) is a widespread, major pest (Table 10.2) [27,57,60,132]. This beetle is slow-moving, with its head 'turned down' so that it is difficult to see from above. Eggs are laid loose and on hatching the larvae wander before they bore into a grain. Adults are voracious feeders, and make large, irregular holes in grains. They require a large bulk of warm grain to survive the British winter.

Various flat beetles occur as pests, but in the U.K. the term is confined to the rust red grain beetle, *Cryptolestes ferrugineus* (Steph, Table 10.2). It is tolerant of dry conditions, but a high mortality of eggs and larvae occurs below 65% r.h. The beetle is only a serious pest in damp grain. The small adult is unusual in having long antennae roughly the length of the body. In warm weather it may fly up to two miles. The adult female can lay 2–3 eggs/day for 6–9 months. The larva eats into the germ and perhaps the endosperm and pupates after four moults. This species is resistant to cold weather.

The rust red flour beetle *(Tribolium castaneum,* Herbst.*)* and confused flour beetle *(Tribolium confusum* Duv.*)* are grouped with other species as bran flour beetles (Table 10.2) [27,42,43,57,60,132]. The adults are similar in appearance. The rust red adult will fly while the confused flour beetle does not. They can secrete pungent material, as a means of defence. Over 9 months the female lays 300–450 sticky eggs loose in the grain. The larva moves about actively and while feeding may spend time within the grain, but the pupa lies loose. These beetles are pests in warm grain or malt stores, where they are established in the U.K. They are often imported with foreign grain. In general these beetles are susceptible to common insecticides.

In recent years the saw-toothed grain beetle *(Oryzaephilus surinamensis* Linn.*)*, so named from the toothed edges of the thorax, has displaced the grain weevil as the most serious pest of farm-stored grain in the U.K. (Table 10.2). It damages the germs of grains directly and causes grain heating with subsequent caking, moulding and sprouting [27,64,71,132]. Outbreaks can easily render barley unsuitable for malting or seed within six months, although the grain can still be fed to stock. The pest damages barley and malt and thrives in dust, broken grain, milled products and animal feeds. The female lays up to 375 eggs at rates of 6–10 per day. The loose laid, white eggs hatch into larvae which move about in the grain and attack the germs, or broken, abraded or otherwise weakened grain. After several moults the larva pupates, the pupa being encased in scraps of food. A population of 500 could reach 5×10^6 in three months. The insect is resistant to extremes of temperature and humidity. It is difficult to control as larvae and adults are highly mobile and shelter in crevices in structures or under dust. The insect shows resistance to some insecticides, and fumigation agents. Control requires that all precautions, namely cleaning, fumigation, and mixing grain with insecticide powder, be rigorously applied, and that the grain be kept cool and dry.

The Khapra beetle *(Trogoderma granarium* Everts*)* is common in hot, dry areas and is an important pest. It came to the U.K. in about 1914–1918 where it is confined to maltings (Table 10.2) [27,42,43,60,132]. The adult lives only 10–12 days and is seldom seen. It does not feed or drink. The larvae, which are very hairy, or masses of their cast skins, are more easily seen. Under unfavourable conditions a special diapause larva

The reception and storage of whole plants and grain 409

occurs which conceals itself and may survive up to 4 years. Larvae stop developing at 8°C (46°F) but can survive for short periods at −10°C (14°F) and can go without food for long periods. The beetle larvae are resistant to various insecticides but pyrethrins seem effective.

Some moths are important pests in stores (Table 10.2) [27,59,60,71, 132]. Adult moths lay their eggs on the grain; the larvae, caterpillars, eat the grain. When they pupate they spin silk which binds grains together, clogging elevators and cleaning machinery. The Angoumois grain moth *(Sitotroga cerealella,* Oliv.*)* is small and yellow-brown with a 10–14 mm (about 0.47 in) wingspan, that arrives in the U.K. on imported grain. In warmer areas it is a major pest. Eggs may be laid on grain in store, or in the field. The larva eats the interior of the grain without touching the germ. Other troublesome moths in warehouses include the Indian meal moth *(Plodia interpunctella* Hbr.*)*, the corn moth (syn. European grain moth, *Tinea* or *Nemapogon granella*) the larva of which eats the germ and the endosperm; the Mediterranean flour moth *(Anagasta kuehniella* Zeller*)*, the brown house moth (syn. false clothes moth, *Hofmannophila pseudospretella,* Staint.) which requires a minimum r.h. of 80%, like the white shouldered house-moth, *(Endrosis sarcitrella* L*)*. These species vary widely in their growth requirements. The warehouse moth (syn. Tobacco or Cocoa moth, *Ephestia elutella* Hbn.) is established in the U.K. and was once a serious pest. In early and mid-summer each female lays 50–200 eggs in the first 4–5 days of her life. The larvae go through 5–6 stages and move from grain to grain eating an average total of 48 embryos each. The caterpillars are most active in the top 30 cm (1 ft) of grain. Suddenly, in late August–September, the mature larvae migrate upwards, mass on the surface of the grain and climb up heaps of grain, walls, pillars, roofs, etc., spinning webbing as they go. At this time many die of diseases. Their decomposing bodies give rise to vile smells. Where possible the surviving larvae find crannies and take a rest (diapause) before pupating in the following spring. The adults emerge from the pupa after 3–4 weeks, appearing in June–July, and surviving about 3.5 weeks.

Some other insects occur that are not directly harmful to stored grain, but their presence indicates inadequate hygiene. Thus in maltings ferment flies *(Drosophila* spp.*)* and fungus beetles (e.g. *Cryptophagus saginatus* or *Mycetea hirta*) are found in the germination areas, but they appear to cause little harm [64]. The yellow mealworm *(Tenebrio molitor* L*)*, like other mealworms, is often found scavenging and breeds in dust and spillages in maltings, grain-stores, and in animal feedstuffs. The adult is a dark brown-black beetle 12–19 mm (0.5–0.75 in) long. The larva, the 'mealworm', is up to 32 mm (1.25 in) in length.

The Australian spider beetle *(Ptinus tectus* Boield*)* is the most common spider beetle in the U.K. It is an important pest (Table 10.2) [27,59,60,

410 Barley

64,132]. It is nocturnal and is only readily detected by night inspection. It is potentially able to multiply in well-kept grain stores. The adult prefers moist, dark crannies. The type of food available may alter the duration of its life cycle between 51 and 300 days. One female can lay up to 600 eggs. The larvae grow most readily on materials like dust and malt culms, but they also bore into faulty grains. The pest is susceptible to several insecticides but, because the larvae and adults hide away, care needs to be taken to make control measures effective.

10.15. The mites of stored grain [27,59,60,62,64,126,132]

Mites are eight-legged, wingless creatures belonging, together with ticks and spiders, to the Arachnida. They are a group distinct from insects. The flour mite, *(Acarus siro* L., *Tyroglyphus farinae* De G*)*, is common in the U.K. and can infest hay, straw, stored grain, flour and compound feedstuffs (Table 10.2). The small adult breaths through the skin, and is very sensitive to changes in humidity. It, and other mites, are active at low temperatures and frequently occur in chilled, damp grain stores [16,17]. A female may lay 500 eggs in a life of 42 days. Eggs hatch to larvae having three pairs of legs, which turn into nymphs with four pairs. The nymph may turn into a slow-moving adult or, rarely in this species, a hypopus, a form capable of resisting adverse conditions. The flour mite produces an oily secretion which has a distinctive, repugnant and penetrating smell and leaves an offensive taint and taste. This is more easily detected than the inconspicuous animal. Animals dislike the taint even after the grain has been dried. The smell usually goes after the mites die. Mites are not known to cause heating in grain, but they eat the germs, reducing viability and quality. They are highly allergenic, causing digestive troubles in men and animals, diarrhoea and scouring and also bronchitis and dermatitis. A range of other mites are found in stored products. In cool, damp grain stores *Glycyphagus* spp. were found in 90% of the samples, and *Acarus* spp. in 60% [16]. In addition *Tarsonemus* spp., *Tydeus* spp., *Chloroglyphus* spp., and *Acaropsis* spp. were found together with gamesid mites and *Cheyletus* spp. These last two groups feed on other mites and may exert some degree of control on their populations. Control is best achieved by drying the grain, and using contact acaricides or fumigants as palliatives.

10.16. Insecticides and fumigants [50]

In part the fact that infestations are not more frequent is due to the use of insecticides and fumigants. The distinction between these is not sharp, some insecticides e.g. γ-BHC and dichlorovos, having high vapour pressures and therefore fumigant action. Purely local applications of pesticides are of little value. Fumigants may be used to disinfect empty

The reception and storage of whole plants and grain 411

silos, or to reduce populations of pests in bulk grain. Measures to prevent reinfestation must be taken if storage is to be prolonged after a fumigation. Fumigants permitted for use on foodstuffs must not leave residues. The choice is controlled by national regulations, the mass and situation of the grain to be treated and the expertise of the operatives [27,53,132]. Some agents, e.g. hydrogen cyanide (prussic acid) and methyl bromide, are so toxic that only trained and well equipped personnel should use them. These are best applied to grain in silos equipped with a closed gas circulation system, but they may be applied to heaps of sacks or loose grain enclosed by plastic sheeting. Ethylene oxide is also used sometimes. Methyl bromide penetrates grain better than hydrogen cyanide, but the latter can be generated throughout a bulk of grain by incorporating calcium cyanide granules with the bulk, as it is run to store. Moisture converts this to hydrogen cyanide. Similarly insects may be controlled by phosphine generated by moisture vapour acting on tablets, pellets, or packets of aluminium phosphide incorporated in the grain. All the phosphine is released in 2–6 days, and penetrates the grain mass well. The aluminium hydroxide residue is unobjectional. Various halogenated hydrocarbons are used as 'liquid fumigants', and in some countries carbon disulphide is also used. They vary in their ability to penetrate masses of grain, their readiness to disperse and rates of adsorption onto the grain. These heavy liquids are poured or sprinkled onto grain masses, where they readily volatilize. In the U.K. carbon tetrachloride (CTC), ethylene dichloride (EDC), and ethylene dibromide (EDB), are frequently used. Elsewhere carbon disulphide and chloropicrin ($CCl_3.NO_2$) are also employed. All fumigation treatments should be thorough, and prolonged. As a first measure liquid fumigants are often sprinkled over, and a good way round a hot spot in a grain heap. The vapours should have dispersed and grain should be well aired before unprotected personnel or animals approach it, particularly in a confined space. Unlike hydrogen cyanide, methyl bromide sometimes reduces grain germination [128].

Insecticides are widely employed. The permitted levels of residues differ in different countries and depend on the purpose for which the grain is intended. In hot, dry climates some control of insects may be achieved by incorporating abrasive mineral dusts in grain. These scratch the water-proofing wax layer of insects cuticles, and they die of desiccation. Chemical insecticides may be applied either by metering a dust formulation from a vibratory feed into a grain stream, or by spraying on a liquid formulation. A few of the numerous insecticides used are mentioned below.

Natural pyrethrins are unstable contact insecticides, with a fast knock-down action, whose activity is often augmented by the addition of synergists such as piperonyl butoxide or sesame oil [46]. Pyrethrins have the advantage of very low mammalian toxicity.

Of the chlorinated hydrocarbons the best known are DDT (dichlorodiphenyl trichloroethane) and γ-BHC *(gamma*-benzene-hexachloride, lindane). Each may be applied as smokes, or sprays, in powder or dispersed in water. There are regulations limiting the residues that may remain in grain.

Malathion, (*S*-[1,2-di(ethoxycarbonyl)ethyl]-dimethyl-phosphorothiolothionate), because of its relatively low mammalian toxicity and its effectiveness against many granary insects, is widely applied to grain as an aqueous dispersion or as dust at 10 p.p.m. It acts as a 'prophylactic' for several months, and does not reduce malting quality. It gradually decomposes, especially in damp or warm conditions, or in contact with alkaline surfaces. Other insecticides, such as fenitrothion do not decompose so rapidly and are effective against different insects and mites at doses of 2 p.p.m. in grain [47]. Dichlorovos (2,2-dichlorovinyl dimethylphosphate), breaks down very rapidly in grain and so has no residual protective action. It readily vapourizes and the vapour can be blown e.g. through bagged grain. The available range of insecticides and their formulations is steadily increasing, creating better opportunities for insect control and greater difficulties of choice and regulation of use. Much remains unknown about insecticides. For example cases have occurred where no residual insecticide was found in grain by chemical analysis, and yet good control of insect infestation occurred [46]. Great efforts are made to treat grain evenly with insecticide, yet from a treated lot (of wheat) only one grain in 20 was toxic to *O. surinamensis* or *S. granarius*.

In a laboratory trial 1 grain in 20 was treated with 40 p.p.m. malathion, at an average dose rate of 2 p.p.m; the control of insects was good and lasted longer than in evenly treated grain. When malathion was sprayed onto grain at a mean dose rate of 9 p.p.m, the dose on individual grains varied between 0.1 p.p.m and 424 p.p.m. Uneven applications of insecticides are probably usual; the consequent danger of high levels of toxic residues even after prolonged storage cannot be overlooked.

10.17. Rodents and their control [30,51,71,90,91]

In the U.K. attention is concentrated on three species, the black or ship rat *(Rattus rattus* L*)*; the brown, common, or Norway rat *(Rattus norvegicus* Berkenhout*)*; and the house mouse *(Mus musculus* L). However at least three other species of mice, together with voles and other rodents are pests. Around the world numerous other species of rodents are troublesome. To obtain adequate control of an infestation it is usually necessary to identify the species involved, and learn its habits before corrective measures can be applied. Grain losses caused by rodents are certainly large. In the 1940s it was estimated that in the U.S.A.

200×10^6 bushels (4.35×10^9 kg) of grain were lost yearly due to rodent depredations. One Norway rat may consume 9.1 kg (20lb) of grain/year [127], and several times this weight of partly eaten and contaminated grain will remain. In addition to direct losses there are others associated with damaged sacks or store structures. Contaminated grain must be cleaned to remove damaged corns, hairs, and rodent faeces. The grain, even after cleaning, is less valuable due to the residues of rodent urine. Rats carry many diseases, including some harmful to man. One may mention bubonic plague, murine typhus and scrubb typhus (rickettsial diseases), leptospiral jaundice, (leptospirosis, or Weil's disease) which is spread by contact with material contaminated with infected rats urine and is sometimes fatal; trichinosis, caused by parasitic nematodes encysting in the muscles; two diseases called rat-bite fever; rickettsial pox; bacterial food poisoning; an incapacitating syndrome with painful swellings due to a rat mite; lymphocytic chorio-meningitis; favus; tularaemia; and some tapeworm infections [26,30,127]. Populations of rodents are never permanently eradicated. Good control may be achieved by the systematic and unremitting application of methods to exclude them from stores and to kill those at least in the vicinity of stores. Stores should be designed to exclude rodents [30,90,91,127]. Buildings should be regularly inspected for signs of rodent activity, and should be kept in good repair. Stores should be kept clean and tidy. Dogs, cats, ferrets and shooting provide more sport than control but they may slow the reinfestation of cleared premises. Trapping, systematically carried out, can be effective. In general an infestation may be cleared by a combination of fumigating grain and gassing holes, before stopping, combined with and followed by a sustained campaign using poisoned water and/or bait. Poison baiting should include the environs, and not just a single store. Trapping and poisoning depend for success on understanding the habits of the rodents involved. Mice are irregular feeders, so that many small lots of poisoned bait must be placed to be sure that lethal doses will be taken. Rats avoid new objects, such as bait. Thus it is essential to 'prebait' with unpoisoned food until the rats are feeding freely before poisoned bait is laid. Rats will then consume lethal doses. Many poisons, but not apparently the anticoagulants, will be avoided by a rat that has received a sublethal dose. Baits are often cereal products, for example pinhead oatmeal mixed with sugar and oil, damp wheat, oatmeal, etc.

For safety, baits should be laid in containers designed to exclude non-rodents. Special drinking stations may also be provided with poisoned water. Dead rodents should be destroyed; some poisons remain in carcasses at lethal levels. The poisons permitted for use are usually controlled by national regulations. They may conveniently be classified into single-dose poisons, that animals will either eat once and die or avoid on another occasion, and anticoagulants which rodents will continue to

Barley

take over long periods. 'Single dose' poisons include formulations of arsenicals, zinc phosphide; barium carbonate; strychnine; yellow phosphorus; alpha-chloralose; red squill (a Mediterranean bulbous plant); thallium sulphate; Antu (α-naphthyl thiourea); sodium fluoroacetate; fluoroacetamide, and norbormide [30,71,127].

The anticoagulants have to be eaten regularly and are relatively specific poisons for rodents. They include such substances as warfarin, coumatetralyl, and chlorophacinone. Rat strains now exist which are resistant to warfarin and so other more dangerous poisons, are being used again.

References

1 Aastveit, K. and Strand, E. (1954–5). *Acta Agric. scand.*, **5**, 76–84.
2 Abdalla, F.H. and Roberts, J. (1969). *Ann. Bot.*, **33**, 153–167; 169–184.
3 Anderson, K. Gjertsen, P. and Trolle, B. (1967, Aug.). *Brewers's Digest*, 76–81.
4 Armolik, N., Dickson, J.G. and Dickson, A.D. (1956). *Phytopathology*, **46**, 457–461.
5 Ashton, T. (1956). In *The Storage of Seeds for Maintenance of Viability* ed. Owen, E.B. pp. 34–38. Bulletin No. 43 Commonwealth Bureau for Pastures and Field Crops. Commonwealth Agricultural Bureau.
6 Aufhammer, G. and Simon, U. (1957). *Z. Acker-u. PflBau.*, **103**, 454–471.
7 Bailey, C.H. (1940). *Pl. Physiol., Lancaster*, **15**, 257–274.
8 Bailey, J.E. (1974). In *Storage of Cereal Grains and Their Products* (2nd edition), ed. Christensen, C.M. pp. 333–360. St. Paul, Minnesota: American Association of Cereal Chemists.
9 Bailey, S.W. (1959). *J. Sci. Fd. Agric.*, **10**, 501–506.
10 Barnett, A.J.G. (1954). *Silage Fermentation*. London: Butterworths Scientific Publications.
11 Bee, R. (1968). *Agriculture*, **75**, 114–118.
12 Brook, P.J. and White, E.P. (1966). *A. Rev. Phytopath.*, **4**, 171–194.
13 Burges, H.D. and Burrell, N.J. (1964). *J. Sci. Fd. Agric.*, **15**, 32–50.
14 Burges, H.D., Edwards, D.M., Burrell, N.J. and Cammell, M.E. (1963). *J. Sci. Fd. Agric.*, **14**, 580–583.
15 Burrell, M.J. (1974). In *Storage of Cereal Grains and their Products* (2nd edition) ed. Christensen C.M. pp. 420–453; 454–480. St. Paul, Minnesota: American Association of Cereal Chemists.
16 Burrell, N.J. and Havers, S.J. (1970). *J. Sci. Fd. Agric.*, **21**, 458–464.
17 Burrell, N.J. and Laundon, J.H.J. (1967). *J. Stored Prod. Res.*, **1**, 125–144.
18 Cannell, R.Q. and Jobson, H.T. (1968). *J. agric. Sci., Camb.*, **71**, 337–341.
19 Christensen, C.M. (1957). *Bot. Rev.*, **23**, 108–134.
20 Christensen, C.M. (1964). *Phytopathology*, **54**, 1464–1466.
21 Christensen, C.M. and Kaufmann, H.H. (1969). *Grain Storage – The Role of Fungi in Quality loss*, Minneapolis: University of Minnesota Press.
22 Christensen, C.M. and Kaufmann, H.H. (1974). In *Storage of Cereal Grains and their Products* (2nd edition) ed. Christensen, C.M. pp. 158–192, St. Paul, Minnesota: American Association of Cereal Chemists.
23 Christensen, C.M. and Linko, P. (1963). *Cereal Chem.*, **40**, 129–137.
24 Coleman, D.A. and Fellows, H.C. (1925). *Cereal Chem.*, **2**, 275–287.

The reception and storage of whole plants and grain 415

25 Cornwell, P.B. (1966). *The Entomology of Radiation Disinfestation of Grain.* Oxford: Pergamon Press.
26 Cotton, R.T. (1954). In *Storage of Cereal Grains and their Products* ed. Anderson, J.A. and Alcock, A.W. pp. 221–274, St. Paul, Minnesota: American Association of Cereal Chemists.
27 Cotton, R.T. and Wilbur, D.A. (1974). In *Storage of Cereal Grains and their Products* (2nd edition) ed. Christensen, C.M. pp. 193–231, St. Paul, Minnesota: American Association of Cereal Chemists.
28 Culpin, C. (1969). *Farm Machinery*, (8th edition) pp. 460–487; 489–536, London: Crosby, Lockwood & Co.
29 Darsie, M.L., Elliott, C. and Peirce, G.J. (1914). *Bot. Gaz.*, **58**, 101–136.
30 Davis, R.A. (1970). *M.A.F.F. Bulletin 181, Control of Rats and Mice* London: H.M.S.O.
31 De Clerck, J. (1958). *A Textbook of Brewing* (trans. K. Barton-Wright) *I* and *II* pp. 587 and 650, London: Chapman and Hall.
32 Dickson, J.G. (1962). In *Barley and Malt*, ed. Cook, A.H. pp. 161–206, London: Academic Press.
33 Disney, R.W. (1954). *Cereal Chem.*, **31**, 229–239.
34 Edwards, R.A., Donaldson, E. and MacGregor, A.W. (1968). *J. Sci. Fd. Agric.*, **19**, 656–660.
35 Erdman, H.E. (1970). *Ann. ent. Soc. Am.*, **63**, 191–197.
36 F.A.O. (1948). *Preservation of Grains in Storage.* Papers presented at the International Meeting on Infestation of Foodstuffs London, 5–12 Aug. 1947, (Ed. Easter, S.S.), Washington.
37 Finn-Kelsey, P. and Hulbert, D.G. (1957). *Tech. Dept. British Elect. Res. Ass. W/T 33 – The relationship between relative humidity and the moisture content of Agricultural Products* – A preliminary report.
38 Flannigan, B. (1969). *Trans. Br. mycol. Soc.*, **53**, 371–379.
39 Flannigan, B. (1974). *Trans. Br. mycol. Soc.*, **62**, 51–58.
40 Flannigan, B. and Dickie, N.A. (1972). *Trans. Br. mycol. Soc.*, **59**, 377–391.
41 Forgacs, J. and Carll, W.T. (1962). *Adv. Vet. Sci.*, **7**, 273–382.
42 Freeman, J.A. (1951). *J. Inst. Brew.*, **57**, 326–337.
43 Freeman, J.A. (1969). *Chem. and Ind.*, 1401–1404.
44 Gjertsen, P. Trolle, B. and Andersen, K. (1965). *Proc. Eur. Brew. Conv.*, Stockholm, 428–438.
45 Golubchuk, M., Sorger-Domenigg, H., Cuendet, L.S. Christensen, C.M. and Geddes, W.F. (1956). *Cereal Chem.*, **33**, 45–52.
46 Green, A.A. (1969). *Chem. and Ind.*, **41**, 1452–1454.
47 Green, A.A. and Tyler, P.S. (1966). *J. Stored Prod. Res.*, **1**, 273–285.
48 Haferkamp, M.E. (1949). *Proc. Ass. off. Seed Anal. N. America*, **39**, 111–114.
49 Haferkamp, M.E., Smith, L. and Nilan, R.A. (1953). *Agron. J.*, **45**, 434–437.
50 Harein, P.K. and de la Casas, E. (1975). In *Storage of Cereal Grains and their Products* (2nd edition) ed. Christensen, C.M. pp. 239–291, St. Paul, Minnesota: American Association of Cereal Chemists.
51 Harris, K.L. (1974). In *Storage of Cereal Grains and their Products* (2nd edition) ed. Christensen, C.M. pp. 292–332, St. Paul, Minnesota: American Association of Cereal Chemists.
52 Harrison, J. (1967). *N.A.A.S. Q. Rev.*, **78**, 78–85.
53 Heseltine, H.K. (1969). *Chem. and Ind.*, 1405–1408.
54 Hopkins, J.R. (1968). *N.A.A.S. Q. Rev.*, **79**, 117–120.
55 Hopkins, R.H. and Carter, W.A. (1933). *J. Inst. Brew.*, **39**, 59–66.
56 Hough, J.S., Briggs, D.E. and Stevens, R. (1971). *Malting and Brewing Science* (Revised 1975), London: Chapman and Hall.

57 Howe, R.W. (1960). *Ann. appl. Biol.*, **48**, 363–376.
58 Howe, R.W. (1962). *Ann. appl. Biol.*, **50**, 137–158.
59 Howe, R.W. (1965). *Nutr. Abst. Rev.*, **35**, 285–303.
60 Howe, R.W. (1965). *Stored Prod. Res.*, **1**, 177–184.
61 Hubbard, J.E., Earle, F.R. and Senti, F.R. (1957). *Cereal Chem.*, **34**, 422–433.
62 Hughes, A.M. (1961). *M.A.F.E. Tech. Bull. No. 9, The Mites Associated with Stored Food Products*. London: H.M.S.O.
63 Hukill, W.V. (1974). In *Storage of Cereal Grains and their Products*, (2nd edition) ed. Christensen, C.M. pp. 481–508, St. Paul, Minnesota: American Association of Cereal Chemists.
64 Hunter, F.A. (1966 Dec.). *Brewers Guild J.*, 591–605.
65 Hunter, F.A., Tullock, J.B.M. and Lambourne, M.G. (1973). *J. Stored Prod. Res.* **9**, 119–141.
66 Hyde, M.B. (1950). *Ann. appl. Biol.*, **37**, 179–186.
67 Hyde, M.B. (1974). In *Storage of Cereal Grains and their Products*. (2nd edition) ed. Christensen, C.M. pp. 383–419, St. Paul, Minnesota: American Association of Cereal Chemists.
68 Iizuka, H. and Iida, M. (1962). *Nature*, **196**, 681–682.
69 Ito, H. (1972). In *Viability of Seeds*, ed. Roberts, E.H. pp. 405–416, London: Chapman and Hall.
70 James, E. (1972). In *Viability of Seeds*, ed. Roberts, E.H. pp. 397–404. London: Chapman and Hall.
71 Jones, F.G.W. and Jones, M.G. (1964). *Pests of Field Crops*. London: Edward Arnold.
72 Kaminski, E., Stawicki, S., Wasowicz, E. and Przybylski, R. (1973). *Ann. Tech. Agric.*, **22**, 401–407.
73 Kotheimer, J.B. and Christensen, C.M. (1961). *Wallerstein Labs. Commun.*, **24**, 21–27.
74 Kreyger, J. (1958). *Petit J. Brass.*, **66**, 811–816; 832–842.
75 Kreyger, J. (1959). *Petit J. Brass.*, **67**, 7–9.
76 Kreyger, J. (1963). *Proc. Intn. Seed. Test. Ass.*, **28**, 753–784; 793–814; 827–836; 861–870.
77 Kronstad, W.E., Nilan, R.A. and Konzak, C.F. (1959). *Science*, **129**, 1618.
78 Lacey, J. (1975). *Trans. Br. mycol. Soc.*, **65**, 171–184.
79 Lawes, D.A. and Jones, D.I.H. (1971). *J. agric. Sci., Camb.*, **76**, 479–485.
80 Lipman, C.B. (1936). *Pl. Physiol., Lancaster*, **11**, 201–205.
81 Lund, A., Pedersen, H. and Sigsgaard, P. (1971). *J. Sci. Fd. Agric.*, **22**, 458–463.
82 Lutey, R.W. and Christensen, C.M. (1963). *Phytopathology*, **53**, 713–717.
83 Macey, A. (1952). *J. Inst. Brew.*, **58**, 25–35.
84 MacGregor, A.W. and Edwards, R.A. (1968). *J. Sci. Fd. Agric.*, **19**, 661–666.
85 Machacek, J.E., Cherewick, W.J., Mead, H.W. and Broadfoot, W.C. (1951). *Sci. Agric.*, **31**, 193–206.
86 MacKay, D.B. and Flood, R.J. (1968). *J. natn. Inst. agric. Bot.*, **11**, 378–403.
87 M.A.F.F. (1965). *Short Term Leaflet No. 42 – Grain Chilling*.
88 M.A.F.F. *Mechanisation Leaflet No. 11, Farm seed cleaning and Grading Machines*.
89 M.A.F.F. (1970). *Bulletin 37, Silage*. London: H.M.S.O.
90 M.A.F.F. (1966). *Bulletin No. 149, Farm Grain Drying and Storage*, London: H.M.S.O.
91 M.A.F.F. (1965). *Tech. Bull. No. 12, Proofing of Buildings Against Rats and Mice*.

92 Matz, S.A. and Milner, M. (1951). *Cereal Chem.*, **28**, 196–207.
93 Mead, H.W. (1942). *Can. J. Res.*, **20, C**, 501–523.
94 Milner, M. and Geddes, W.F. (1954). In *Storage of Cereal Grains and their Products*, ed. Anderson, J.A. and Alcock, A.W. pp. 152–220, St. Paul, Minnesota: American Association of Cereal Chemists.
95 Mitsuda, H., Kawai, F. and Yamamoto, A. (1972). *Fd. Technol.*, (March) 50–56.
96 Mulinge, S.K. and Apinis, A.E. (1969). *Trans. Br. mycol. Soc.*, **53**, 361–370.
97 Mulinge, S.K. and Chesters, C.G.C. (1970). *Ann. appl. Biol.*, **65**, 277–284; 285–292.
98 Munck, L. (1972). *Hereditas*, **72**, 1–128.
99 Nichols, A.A. and Leaver, C.W. (1966). *J. appl. Bact.*, **29**, 566–581.
100 Owen, E.B. (1957). *The Storage of Seeds for Maintenance of Viability. Bureau of Pastures and Field Crops, Bulletin 43*, Commonwealth Agric. Bureau. Hurley: Commonwealth Bureau of Pastures and Field Crops.
101 Oxley, T.A. (1948). *The Scientific Principles of Grain Storage.* Liverpool: Northern Publishing Co.
102 Oxley, T.A. (1959). In *Irish Maltsters Conference – Selected Papers, 1952–1961*. Arthur Guinness, Son & Co. (Dublin) Ltd. pp. 101–114.
103 Oxley, T.A. and Wickenden, G. (1963). *Ann. appl. Biol.*, **51**, 313–324.
104 Pedersen, H., Nørgaard-Pedersen, P.E. and Glahn, P-E. (1971). *J. Sci. Fd. Agric.*, **22**, 451–457.
105 Pepper, E.H. and Kiesling, R.L. (1963). *Proc. Ass. off. Seed. Anal. N. America*, **53**, 199–208.
106 Pixton, S.W. and Warburton, S. (1971). *J. Stored Prod. Res.*, **6**, 283–293.
107 Roberts, E.H. (1960). *Ann. Bot.*, **24**, 12–31.
108 Roberts, E.H. (1961). *Ann. Bot.*, **25**, 373–380; 381–390.
109 Roberts, E.H. (1963). *Ann. Bot.*, **27**, 365–369.
110 Roberts, E.H. (Edit; 1972). *Viability of Seeds.* London: Chapman and Hall.
111 Roberts, E.H. and Abdalla, F.H. (1968). *Ann. Bot.*, **32**, 97–117.
112 Robertson, D.W. and Lute, A.M. (1933). *J. agric. Res.*, **46**, 455–462.
113 Robertson, D.W., Lute, A.M. and Gardner, R. (1939). *J. agric. Res.*, **59**, 281–291.
114 Robertson, D.W., Lute, A.M. and Kroeger, H. (1943). *J. Am. Soc. Agron.*, **35**, 786–795.
115 St. Johnston, J.H. (1954). *J. Inst. Brew.*, **60**, 318–340.
116 Schuster, K. and Grunewald, J. (1957). *Brauwelt*, 1446–1451.
117 Semeniuk, G. (1954). In *Storage of Cereal Grains and their Products* ed. Anderson, J.A. and Alcock, A.W. pp. 77–133. St. Paul, Minnesota: American Association of Cereal Chemists.
118 Shands, R.G. (1937). *Phytopathology*, **27**, 749–762.
119 Shands, H.L., Janisch, D.C. and Dickson, A.D. (1967). *Crop Sci.*, **7**, 444–446.
120 Sheneman, J.M. and Hollenbeck, C.M. (1961). *Proc. A.M. Am. Soc. Brew. Chem.*, 93–97.
121 Sijbring, P.H. (1963). *Proc. Int. Seed Test. Ass.*, **28**, 837–843.
122 Sloey, W. and Prentice, N. (1962). *Proc. A.M. Amer. Soc. Brew. Chem.*, 24–29.
123 Smith, L. (1948). *J. Am. Soc. Agron.*, **40**, 32–44.
124 Snow, D. (1945). *Ann. appl. Biol.*, **32**, 40–44.
125 Snow, D., Crichton, M.H.G. and Wright, N.C. (1944). *Ann. appl. Biol.*, **31**, 102–109, 111–116.
126 Solomon, M.E. (1962). *Ann. appl. Biol.*, **50**, 178–184.

127 Spencer, D. (1954). In *Storage of Cereal Grains and their Products* ed. Anderson, J.A. and Alcock, A.W. pp. 275–307, St. Paul, Minnesota: American Association of Cereal Chemists.
128 Strong, R.G. and Lindgren, D.L. (1959). *J. econ. Ent.*, **52**, 319–322.
129 Swaebly, M.A. and Christensen, C.M. (1952). *Phytopathology*, **42**, 476.
130 Trolle, B. and Pedersen, H. (1971). *J. Inst. Brew.*, **77**, 338–348.
131 Tuite, J.F. and Christensen, C.M. (1955). *Cereal Chem.*, **32**, 1–11.
132 Turtle, E.E. and Freeman, J.A. (1967). In *Modern Cereal Chemistry* (6th edition) ed. Kent-Jones, D.W. and Amos, A.J. pp. 198–247, London: Food Trade Press.
133 Vermeylen, J. (1962). *Traité de la fabrication du Malt et de la Bière*. Assoc. Royale des Anciens Élevès de l'Institut Superieur des Fermentations, Gand *1* and *2*.
134 Warnock, D.W. (1971). *J. gen. Microbiol.*, **67**, 197–205.
135 Warnock, D.W. and Preece, T.F. (1971). *Trans. Br. mycol: Soc.*, **56**, 267–273.
136 Wayman, C. (1969). *Chem. and Ind.*, **41**, 1445–1447.
137 Yamamoto, T. (1954). *J. Pharm. Soc. Japan*, **74**, 797–812 [via. (1954) *Chem. Abst. 48*, 12887b].
138 Zamenhof, S. and Ben-Zeev, N. (1967). In *Physiol. Ökol. Bioch. der Keimung*, **1**, Griefswald, pp. 139–146.

Chapter 11

Barley genetics

11.1. Introduction

The genetics and cytology of barley have been extensively studied because (i) it is an important crop; (ii) it is a convenient experimental material with a wide range of physiological and morphological varieties; (iii) it has a small number of chromosomes ($2n = 14$) which are comparatively large (6–8 µm), and can often be distinguished under the microscope; (iv) thousands of different isolates are readily available; (v) the species is predominantly self-fertilizing, but cross-pollination is easy to perform and a high percentage of seed set is usual. Seed may be stored unchanged for years, or several generations (e.g. 4–6) may be grown, in greenhouses or growth chambers, in one year. The species has been widely used as a 'test' organism in studies on mutagenesis and radiosensitivity [1,58,59,83,98].

11.2. The inheritance of 'distinct' factors

Contrasting traits are inherited in the simplest Mendelian fashion when they are controlled by a single pair of allelic genes. As barley is mainly self fertilizing, plants of established strains or cultivars are nearly homozygous for obvious characters (but see Chapter 12). Crosses between such lines give a uniform, heterozygous F_1 (1st filial) generation. The progeny of the self-fertilised F_1 generation, the F_2, or 2nd filial generation, is of mixed appearance as the characters from the parents begin to segregate. The appearance of the F_1, F_2 and subsequent generations depends upon the degree of dominance of the genes whose phenotypes are being compared and to what extent the phenotypic expression of the genotype is controlled by the environment. Thus when hooded *(KK)* and awned *(kk;* recessive genes*)* lines are crossed the heterozygous

F_1 *(Kk)* is hooded, as the hooded gene is dominant. In the F_2 progeny the characters segregate and when a sufficiently large population is considered, the ratio of approximately 3 hooded to 1 awned is found. When the F_2 plants are allowed to self, and their progeny is investigated, as expected the F_2 awned plants are homozygous *(kk)*, but only a third of the hooded plants are homozygous *(KK)* and give rise exclusively to hooded progeny. The remaining two-thirds are heterozygous *(Kk)* as shown by segregation in the F_3 [65]. Similarly in dihybrid crosses, in which two pairs of allelic genes are involved, segregation in the F_2 gives phenotypic classes in the ratio 9 : 3 : 3 : 1. When plants with white grains *(bb)* and rough-awns *(RR)* are crossed with black-grained *(BB)*, smooth-awned *(rr)* plants the F_1 plants *(Bb.Rr)* are black grained and rough awned. On selfing the genes segregate and recombine at random, so in the F_2 the offspring occur in the ratio 9 black, rough: 3 black, smooth: 3 white, rough: 1 white, smooth.

Such a segregation is achieved only if the genes are not linked, that is if the loci for the two pairs of alleles are in different linkage groups, i.e. different chromosomes, or at least sufficiently far apart on the same chromosome that crossing over makes the effects of linkage minimal. If two genes are close together on the same chromosome they tend to be inherited together and the 'linkage' may be so tight that their effects may be attributed to a single gene with pleiotropic (multiple) expression. With more loosely linked genes a proportion of recombinations will occur due to exchange of genetic material which occurs during the crossing over of chromosomal segments. Thus if two-rowed *(VV)* barley with purple lemmae *(PP)* is crossed with a six-rowed white barley *(vvpp)* the F_1 will be two-rowed and purple *(VvPp)*. If the F_1 plants are test-crossed to the double recessive (six-rowed, white; *vvpp*) the progeny are two-rowed, purple *(VP/vp)* 40.3%; six-rowed white *(vp/vp)* 40.3%; two-rowed white *(Vp/vp)* 9.7%; and six-rowed, purple *(vP/vp)* 9.7%. Thus 80.6% of the progeny look like the parents, indicating that the pairs of genes are linked, while the 19.4% unlike the parents contain recombinants of the parental genes [65]. Crossing over has occurred on 19.4% of the possible occasions. The higher the crossover frequencies between linked genes the further they are apart in genetic terms. Seven linkage groups have been established in barley and have been assigned to each of the seven morphologically distinct, haploid chromosomes. If the F_1 generation is selfed then statistically the proportions of the phenotypes in the progeny, F_2, will be 66.24% two-rowed, purple; 8.76% two-rowed, white; 8.76% six-rowed purple; and 16.24% six-rowed, white [65]. The result can be predicted from the frequency of the gametes (Table 11.1). Massive compilations of linkage data are available [58,75,76,83]. Mutagenic or other agents may be used to enhance crossover frequencies. Crossover frequency is temperature-dependent, being optimal at 15°–16°C (about

Table 11.1 The progeny expected from allowing self fertilization (selfing) in the F_1 of the cross two-rowed (*VV*) purple lemma (*PP*), with six-rowed white lemma (*vvpp*). The frequency of recombinants among the offspring is controlled by the linkage between the genes (see text; after Poehlman [65]).

Gametes	VP 40.3%	vp 40.3%	Vp 9.7%	vP 9.7%
VP 40.3%	VP/VP 16.24%	vp/VP 16.24%	Vp/VP 3.91%	vP/VP 3.91%
vp 40.3%	VP/vp 16.24%	vp/vp 16.24%	Vp/vp 3.91%	vP/vp 3.91%
Vp 9.7%	VP/Vp 3.91%	vp/Vp 3.91%	Vp/Vp 0.94%	vP/Vp 0.94%
vP 9.7%	VP/vP 3.91%	vp/vP 3.91%	Vp/vP 0.94%	vP/vP 0.94%

60° F) [66]. Feeding selenium compounds decreases crossing over, possibly by interfering with spindle-formation [94]. Understanding some segregation patterns is complicated since morphological forms may often be divided into many classes owing to incomplete dominance, or the action of modifying genes, or because the morphological characters under investigation are dependent on groups of non-allelic genes. In studies on the inheritance of awn length disagreements have arisen both because phenotypically similar but genotypically widely-different parents have been used, and because different ways of classifying the progeny have been used. In some cases the F_2 plants from a long-awned × short awned cross may have a frequency of 1 long awned: 2 short-awned: 1 awnless. In other cases two allelic pairs of genes are involved to give the classes long awned (double dominant *LkLk*, Lk_1Lk_1), short awned *(LkLk, lk_1lk_1)*, awnletted *(lklk, Lk_1Lk_1)*, and awnless (double recessive, *lklk,lk_1lk_1*) [55]. In crosses between Nepal (hooded) barley and awnless the pairs of alleles involved are those mentioned above and *KK* (needed for hoods). Thus Nepal is triple dominant, *LkLk,Lk_1Lk_1KK*, and awnless is triple recessive, *lklk,lk_1lk_1,kk* [55]. For a well developed hood all the dominant factors must be present, so the offspring should occur in the proportions 27 hooded; 9 long awned: 12 short awned: 12 awnletted: 4 awnless. However hood morphology is very variable and depends on the 'genetic background' of the plant. In short-awned progeny carrying the hooded factor the short awns are flattened and twisted. Sometimes awnlets carry a small fork at the end, which represent a rudimentary hood [55]. Crosses between hooded and awned phenotypes often give rise to 3 hooded: 1 awned forms in the F_2, but with crosses of short awns × hooded the ratio in F_2 may be 9 hoods: 3 long awns: 4 short awns [100].

The ear-character 'two-rowed' is generally dominant to six-rowed. However the inheritance of fertility in the lateral florets is complex, and variations in grain size and irregularity in fertility are frequent. Harlan and Martini [33], ironically noted that the question of the result of two-row × six-row crosses was repeatedly being settled. The progeny of one cross may be separated into 7 or 8 phenotypic classes while another will give only 4 or 5. The classes found may vary with the growing season. The two- and six-rowed and *deficiens* ear types seem to be controlled by an allelic series of genes, with a second series controlling the fertility of the lateral florets. The complexity detected in the inheritance of grain colour depends very much on the finesse used in colour classification. The more numerous the genes controlling a phenotypic character the more complex the genetic analysis becomes. Other unexpected segregation ratios are known. When two recessive mutants, respectively with narrow and broad leaves, were crossed the F_1 generation had leaves of narrow width. In the F_2 the ratios were 10 normal; 3 narrow leaves: 3 wide leaves, possibly because the double dominants and double recessives had the same phenotype [71]. Anomalous segregation occurs in the F_2 if some genes are lethal to the zygotes (when homozygous), or cause the death of gametes. Cases of cytoplasmic or 'maternal' inheritance are also known. Thus the factor causing the pale-green, chlorophyll deficient *chlorina* form is always inherited via the female parent, never through the pollen. Possibly an abnormal line of plastids is inherited in the maternal, egg cytoplasm.

Genes 'express themselves' differently in plants of differing genetic background, so their effects are most easily studied in lines that are otherwise 'near-isogenic'. In practice one cannot know if lines are truly isogenic, (i.e. differ *only* in the gene to be studied) but near-isogenic lines can be prepared either by backcrossing programmes (Chapter 12) or by obtaining mutants of a particular strain.

11.3. Cytology and chromosome behaviour

Hordeum spontaneum (Koch) and the 'species' of cultivated barley are diploid and, with few exceptions, are fully inter-fertile and have the same number of chromosomes (the haploid number, $n = 7$) which appear similar. On the whole chromosome morphology is rather constant (Chapter 3; Fig. 11.1) [29,58].

Mitosis occurs normally. In the resting nucleus the chromosomes cannot be distinguished. As prophase progresses the 14 ($2n$) chromosomes become visible and become shorter and wider; each is divided into two chromatids, which are held together at the centromere. At metaphase the nuclear membrane disappears, the nuclear spindle is formed and the chromosomes are arranged at the equatorial plate. In anaphase the chromatids move apart the centromeres, which have now divided, leading the way towards the spindle poles. The chromatids have therefore become chromosomes and the two nuclei which reform in telophase have received duplicated sets of genetic information from the parent nucleus. In telophase the nuclear envelopes reform, the chromosomes gradually

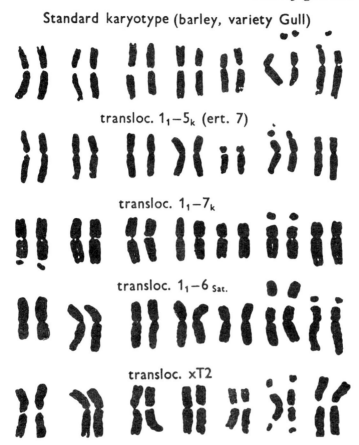

Fig. 11.1 The karyotypes of the variety Gull and of some translocation lines. (from Hagberg and Åkerberg [29]).

become less easily visible until once again resting nuclei are formed.

Meiosis, the division in which the diploid chromosome number is reduced to the haploid number ($n = 7$) during the formation of the germ cells, also follows the usual course. At the onset in prophase I, the leptotene stage, the chromosomes of the mother cell appear as long, slender *single* threads (*i.e.* they are *not* divided into chromatids). In zygotene, the homologous, threadlike chromosomes begin to align with each other, so that the haploid number of bivalents is formed. Chromosomes pairing is complete in pachytene, and the chromosomes, but not the centromeres, divide into chromatids. By diplotene the division of the chromosomes into chromatids is complete, but they remain close together. However, the paired chromosomes move apart, but are held together by chiasmata, usually 1–5, physical joins formed where crossing over has occurred.

The frequency of chiasmata appears to be the same in male and female germ cells [5]. The chromosomes also become shorter and thicker. In the last stage of prophase I, diakinesis, the tetrads (groups of four chromatids) are most easily seen. The nuclear membrane and nucleoli disappear. In metaphase I the tetrads are arranged in a plane. In anaphase I the chromosomes (pairs of chromatids, dyads) separate and move towards the spindle poles, the centromeres leading. As this happens the chiasmata appear to move along the chromosomes to the tips, which separate last. The four arms of each dyad appear to be repelled and spread out. In telophase I the chromosomes contract into small bunches at each spindle pole and a cell plate arises between the new cells. Thus each daughter nucleus has received a haploid set of dyads carrying half the nuclear genetic information of the diploid parent cell. In the resting stage between the first and second meiotic divisions (interphase; interkinesis) the chromosomes elongate and become less easily visible, and the nuclear membranes reform. The interphase chromosome volume ($\pm$ S.E.) is $13.5 \pm 0.8\,\mu m^3$, an estimate that correlates with the 'target area' found in radiosensitivity experiments [85]. At the onset of the second meiotic division, prophase II, the nuclear envelopes go and the chromosomes contract. In metaphase II spindles form and the centromeres of the equatorially arranged dyads are functionally double. In anaphase II the chromatids move to the spindle poles, so each pole now receives a haploid set of single chromatids. In telophase II resting haploid nuclei are formed, and a wall divides the new cells [15]. Thus the nucleus of each gamete mother cell divides twice but the chromosomes only double once, so four haploid cells are formed. The random assortment of chromosomes ensures the random distribution of chromosomal 'sets' of genetic material between the gametes. The formation of chiasmata is a visible expression of crossing over and exchange of genetic material between homologous chromosomes; it allows the re-assortment of partially linked genes. Crossing over occurs most frequently at the ends of the chromosomes, away from the centromeres. The duration of the meiotic stages have been estimated, to assist in cytological studies (Table 11.2) [5,19,51]. Their absolute lengths depend on environmental conditions, but relative durations of the different stages appear to be constant and to be the same during pollen or ovule formation. Under adverse conditions pollen and ovules may not mature simultaneously, leading to poor seed set [5]. In tetraploids the duration of meiosis is reduced. This follows a rule that the meiotic time is inversely related to nuclear DNA content [3]. The DNA content of a root-tip cell nucleus of diploid barley, cv. Sultan is about $20.3 \times 10^{-12}\,g$ [4].

In the formation of the ovule three of the four haploid megaspores degenerate. The remaining haploid cell enlarges, and its nucleus undergoes three successive mitotic divisions, forming eight haploid nuclei, which are retained in an embryo sac (Chapter 2). This process represents

Table 11.2 The duration of meiosis (h) in two diploids and an autotetraploid barley at 20°C, in continuous light (from Finch and Bennett [19]).

	Diploid		Autotetraploid
Stage	Sultan	Ymer	Ymer
Leptotene-telophase II, (inclusive)	39.4	39.1 ± 1	30.9 ± 1
Leptotene	12.0	11.5	9.0
Zygotene	9.0	9.0	7.0
Pachytene	8.8	9.3	7.0
Diplotene	2.2	1.9	1.8
Diakinesis	0.6	0.6	0.5
Metaphase I	1.6	1.6	1.5
Anaphase I	0.5	0.5	0.4
Telophase I	0.5	0.5	0.4
Dyad stage	2.0	1.7	1.5
Metaphase II	1.2	1.5	1.0
Anaphase II	0.5	0.5	0.4
Telophase II	0.5	0.5	0.4
Tetrad stage	8.0	>7.0	>6.0

the development of the female haplophase. Three nuclei, the ovule and the two synergidae form the egg-apparatus. The others are the antipodal cells and the polar nuclei (Chapter 2). In the anther meiosis gives rise to a tetrad of four haploid microspores. These separate and the haplophase develops as the pollen grain. The haploid nucleus divides by mitosis, and one of the daughter nuclei, destined to become the tube nucleus, remains unchanged. The second nucleus divides again, giving rise to the two sperm nuclei. The first sperm nucleus ($n = 7$) fuses with the ovule ($n = 7$) to form the diploid zygote ($2n = 14$). The second sperm nucleus ($n = 7$) fuses with both the polar nuclei (each $n = 7$). The subsequent division of this triploid nucleus ($3n = 21$) gives rise to the cells which form the endosperm. In heterozygotes therefore, the constitution of the endosperm differs in reciprocal crosses. This distinction (A male × B female *or* B male × A female) is also important, for example in trisomics, when extra chromosomes may be transmitted through the egg cell, but not the pollen.

The nuclear processes may malfunction so that binucleate or multinucleate pollen mother cells are formed. Furthermore, a particular recessive gene, called a multiple sporocyte factor, interferes with the meiosis metaphase, and so instead of 7 chromosome pairs at the equator of the spindle, the spindle width is increased and, for example, 14, 21, 28, 56 or 112 chromosome pairs may be found in the 'mixaploid' microsporocytes [82]. The development of the pollen in the central anther lags behind that in the other two [4,15]. Gradients of mitotic development occur along each anther.

The morphology of chromosomes is variable. Usually they are described in metaphase, as seen in root-tip squashes. The karyotypes of cultivated barleys, *H. spontaneum* and *H. agriocrithon* are very uniform. Many reported differences are now attributed to poor cytological techniques [29,58]. However, real karyotype differences do occur, notably in translocation lines (Fig. 11.1). Chromosome 6 has strong and chromosome 7 has weak nucleolar organizing ability. Sometimes at least two others may also show this capacity. In squashes the large chromosomes are 6–8 µm long (Table 11.3). However chromosome 1 may not be the longest in the normal karyotype [92].

The linkage groups of barley have been associated with individual chromosomes. In some cases the order of the genes along a chromosome has been established, together with their approximate locations [75]. However, some of the data are questioned [91]. Desynaptic mutants in which chromosomal pairing, and hence the formation of chiasmata, is reduced, are known in barley. Their expression is conditioned by the

Table 11.3 The relative morphology of somatic metaphase chromosomes (from Nilan [58], Wiebe [98]).

Chromosome Number	Length, relative (6 = 100, excluding satellite)	Ratio length short arm/ long arm	Comments
1	137.2	0.75	Centromere, submedial attachment. Sometimes a constriction in the terminal third of the long arm.
2	132.3	0.86	Centromere submedial.
3	122.3	0.92	Centromere almost medial.
4	119.0	0.77	Centromere submedial.
5	105.0	0.73	Centromere submedial to subterminal.
6 with satellite on	100.0 120.6	0.61 0.94	Chromosome with largest satellite. Centromere almost subterminal.
7 with satellite on	110.7 125.6	0.41 0.60	Centromere subterminal. Long arm probably longest in the karyotype. The short arm sometimes has a tertiary constriction.

plant's environment. The formation of chiasmata evidently involves the formation of breaks in the chromatids and rejoining the 'wrong' broken ends.

11.4. Chromosomal abnormalities [14,29,31,58,69]

New chromosomal types arise as the result of chromatid breakage and repair. Thus inversions occur when a segment of a chromosome is 'switched round' so that the genes in that part run in the reverse order to that originally present (Fig. 11.2). When a heterozygote contains a pair of chromosomes one of which contains an inversion then chromosomal pairing is visibly disturbed, looped chromosomal figures may occur, and fertility is reduced (Fig. 11.3). In principle the genes within the inverted segment should be closely linked in crosses with 'normal' types. In the simplest case when single crossovers occur within an inverted loop

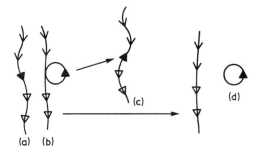

Fig. 11.2 The formation of chromosomal inversions and deletions (schematic). Two parts of a chromosome (a) become joined (b). The join may break either to reform (a), or to form a chromosome with an inverted segment (c), or to allow a small fragment of chromosome to break away, yielding a chromosome with a deletion, and a fragment (d).

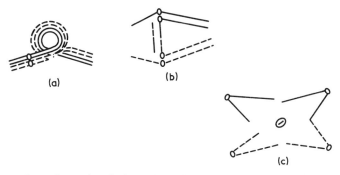

Fig. 11.3 Loop formation during pairing between a normal chromosome, and a chromosome with an inversion (a), illustrating bridging (b), and fragment formation (c). The fragment is eliminated, since it has no centromere.

during meiosis chromosome bridging and fragment formation may occur in anaphase I, and at the end of the second division a micronucleus may be formed, containing a chromosomal fragment. Gametes that should have received the chromosomes that are bridged die. Where more crossovers occur the pattern is more complex [14,67,81]. Plants homozygous for inversions appeared normal, so the sequence of genes is not in itself important. Chromosome breaking and rejoining may also give rise to the loss of a part of a chromosome, a deletion (Fig. 11.2). Chromosomes with deletions often have weakening or lethal effects. However, even large deletions are tolerated in the 'extra' chromosomes of trisomics. A small chromosomal deficiency will behave like a recessive gene; it is unlikely to be detected by cytological examination. Sometimes, e.g. in the progeny of triploids, chromosome fragments behave as extra, tiny chromosomes [79]. When non-homologous chromosomes exchange segments (terminal or otherwise) an interchange (reciprocal translocation; translocation) has occurred, and translocated chromosomes are formed (Fig. 11.1). These chromosomes are of value in genetic studies and may become so in plant breeding. They are used in genetic mapping, in the production of trisomics, and in the duplication of chromosome segments [69]. Many 'stocks' of barley homozygous for translocations between identified chromosomes are available [29,31]. Such translocations occur frequently after mutagenic treatments [28]. Plants homozygous for translocations appear normal so, as with inversions, the movement of genes from one chromosomal location to another has little effect on their expression [62]. When an interchange occurs between chromatids of non-homologous chromosomes the result is an interchange heterozygote. In such a heterozygote chromosome pairing and separation during meiosis gives rise to abnormal figures. At metaphase I the chromosomes may appear as rings or chains. Because of the nature of the abnormality theoretically about the 50% of the pollen will be sterile (Fig. 11.4), since some gametes will not receive all the genetic information. The occurrence of this level of sterility indicates that an interchange (translocation) may be involved [58,69,81]. However in barley about 70% of the pollen is viable. Thus an excess of 'alternate disjunction' of the chromosomes occurs. Crossing over further complicates this picture [69]. Crossing-over is suppressed in the region of the breakpoint of translocated chromosomes, and this is used to associate particular parts of chromosomes with linkage groups. Break-points can be plotted on linkage maps. Translocation lines may also be used in studies of chromosome behaviour, and for moving genes into other linkage relationships. In test crosses between translocated and normal stocks, progeny heterozygous for the translocation are semi-sterile (F_1) due to abnormalities in chromosome pairing. In the F_2 population, from the selfed F_1, homozygotes (with and without the translocation) are fully fertile, the heterozygotes are semi-sterile so the phenotypic

Barley genetics

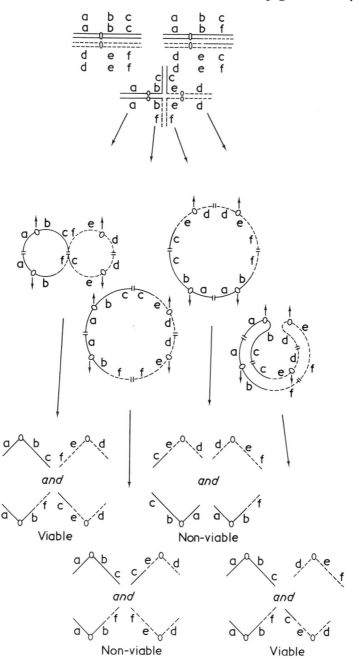

Fig. 11.4 Chromosome pairing and disjunction between normal chromosomes, and those with interchange segments, resulting in the formation of various figures, and the formation of viable and non-viable gametes.

ratio is 1 fertile: 1 semi-sterile. Quadrivalent chromosome groupings in the F_1 indicate that one of the parents contained a translocation (Fig. 11.4). 'Tester sets' of translocations involving all the barley chromosomes are available [28,29]. When a barley with a new translocation is crossed with homozygous tester lines the figures adopted by the chromosomes of the different F_2s in meiosis indicate which chromosomes are involved in the new translocation. Translocation lines may have karyotypes differing from the standard (Fig. 11.1). In plants heterozygous for translocations, abnormal 3:1 (rather than 2:2) disjunction can occur at anaphase I, giving rise to a gamete with an extra chromosome $(n + 1)$ and potentially to a trisomic plant $(2n + 1)$ [68]. Badly paired homologous chromsomes may break and rejoin in such a way that one chromosome, with centromere, is deficient (so that cells receiving it will be eliminated) and the other contains the extra genetic information, and therefore carries duplicated genes. Thus, in principle, translocation lines allow the production of duplications (and deletions) at will by crossing parents having translocations on the same chromosomes, with the relative positions of the breakpoints in critical regions of the chromosome [28,29,31]. In practice, obtaining duplications is difficult but a large part of the short arm of chromosome 6 has been duplicated. Duplications have also been obtained where 'tandem satellites' have been added to chromosomes 6 and 7 [31]. 'Pseudo-isochromosomes', in which interchanges have occurred between opposite arms of homologous chromosomes, have also been found.

Where chromosomes are broken at the centromere two telocentrics ('half chromosomes') are formed, so the chromosome number is increased, e.g. from $2n = 14$ to $2n = 16$, but without an increase in the genetic material [31,90]. Such material gives the chance of producing isochromosomes, and hence duplicating some chromosome arms. Chromosomes have also been obtained from interchanges in which two 'part' 'centromeres or one part and one whole centromeres are present [31].

11.5. Ploidy levels

In 'normal' barley the germ cells are haploid $(n = 7)$, the endosperm cells are triploid $(3n = 21)$ but the embryo and hence the mature plant is diploid $(2n = 14)$. Tetraploid barleys $(2n = 4x = 28)$ have long been known, having arisen from heat-treated seeds, from twin seedlings, and especially following treatments with colchicine, which is also mutagenic [24,58,74,81]. Such autotetraploids usually differ from the diploid forms in being larger and a darker green, with thicker straw, thicker roots, larger and broader leaves, and bigger cells, stomata, kernels and pollen grains. However, the numbers of tillers are reduced and, owing to partial sterility, grain yields are substantially (20–60%) below those of corresponding diploids despite many attempts to improve them [41]. Autotetraploids have lower seed germinability, poorer competitive ability, and often respond to changes in the environment differently to the parental diploid forms. In hexaploid species like *Triticum vulgare* ($2n = 6x = 42$), the chromosome set originates from three different species ($2n = 14$) as they do in the synthetic species *Triticale* (wheat and rye). In such alloploids (amphiploids) homologous chromosomes pair regularly in meiosis. But in autotetraploids, such as tetraploid barley, each chromosome is represented four times, so at meiosis highly irregular pairing and disjunction can occur, gametes with an excess or deficiency of chromosomes are formed, and fertility is reduced. Autotetraploid barleys often appear to be 'physiologically unbalanced'. There is no direct relationship between the formation of multivalents in meiosis and sterility but there is between the disposition of the chromosomes in anaphase I and sterility. No two sets of barley chromosomes differing sufficiently to overcome these 'pairing' problems have been found, although tetraploids derived from the F_1 plants of crosses between unrelated cultivars seem to perform better than tetraploids from pure lines [58]. No commercially acceptable tetraploids have yet been obtained. No other species has been found that can be crossed with barley to give an F_1 that can be converted to an allotetraploid [29,58,98]. The irregular separation of the chromosomes in meiosis in autotetraploids results in gametes with chromosome numbers varying from haploid (7) to diploid (14), and offspring have been found with chromosome numbers ranging between 19 and 42.

Crosses between autotetraploid ($4x$) and diploid ($2x$) barleys gives rise to triploids ($3x = 21$) and hypertetraploids (chromosome numbers 29 or 30) in variable yield. Triploid seed germinates poorly. Special cultural techniques, including the application of gibberellic acid to embryos on the mother plant, increase the yield of triploid seedlings. High levels of spontaneous triploid seeds and other aneuploids, which occur in the small, shrivelled fraction of commercial grain samples, is caused by infection with barley stripe mosaic virus [78]. Trisomic grains are often lighter than related diploid grains [71]. Triploids are genetically unstable and give rise to a high frequency of trisomics ($2n + 1 = 15$ chromosomes) in their progeny. A progressive increase in height has been reported from diploids, through triploids to tetraploids (i.e. $2x$, 67 cm; $3x$, 79cm; $4x$, 86cm) but in other respects – e.g. average tiller number, or pollen diameter or stomatal size the triploid had the largest values [9].

Haploid barley plants are known (chromosome number = 7), but until recently they were very rare [58,83]. In general haploids are weak, short plants with narrow leaves, and small stomata. They are completely sterile, the pollen grains being empty. They have been used in karyotype analysis. Potentially they are important, since by treating a haploid with colchicine to double the chromosome number a perfectly homozygous diploid may be obtained, so saving much time in plant-breeding programmes. In principle haploid plants may be grown from a callus-culture of pollen, but so far such plantlets have been albino and therefore not viable [103]. Not all pollen grains in an anther gives rise to callus. Those few grains that are liable to do so can be detected by their inability to stain with acetocarmine [8]. When the wild barley-grass *Hordeum bulbosum* (female, 2x) is crossed with *H. vulgare* (2x, pollen donor) and embryo culture is used to obtain viable seedlings, the seedlings are haploid. Their vegetative characters are intermediate between the parents but their spikes resemble those of *H. vulgare*. The *H. vulgare* chromosomes are present in *H. bulbosum* cytoplasm [48,88]. When diploid or autotetraploid *H. vulgare* and *H. bulbosum* are crossed mainly haploids (7 chromosomes) and dihaploids (14 chromosomes) are obtained, the *bulbosum* chromosomes usually being eliminated, although hybrids are sometimes obtained [46]. Marker genes on the *H. vulgare* chromosomes are expressed in the haploids, while in diploid and triploid hybrids recessive *vulgare* genes may be masked by *bulbosum* genes [87]. In each case embryo culture is needed to obtain viable plantlets [42,48]. The vigorous efforts made to find ways of raising more haploid seedlings and 'diploidizing' them have been successful, using fed, detached barley shoots, treating with gibberellic acid, multiplying the haploid plants by cloning, and treating with different levels of colchicine [40]. Thus 60% seed set was obtained in pollinated florets, and of these 50% of the embryos were successfully cultivated. For haploids and tetraploids respectively, relative to diploids, some dimensions were: stomatal length, 74% and 163%; pollen diameter, 59%, 147%. For haploids the height of the plants and the length of the awns (relative to diploids) were 70% and 67% respectively [6]. Mitosis and meiosis are irregular in haploids. Multivalents occur as does 'foldback pairing' of single chromosomes and pairing between non-homologous chromosomes. Formations which appear like chiasmata are also seen [77].

Some types of aneuploids, plants with a surplus or deficit of chromosomes, are encountered in barley – e.g. among the offspring of tetraploids having irregular numbers of chromosomes (19–42) [58]. Monosomes (plants lacking a chromosome, $2n - 1$) and nullisomes (plants lacking any of a particular type of chromosome ($2n - 2$), do not occur in barley, in contrast to polyploid species. Complete sets of primary trisomics, that is of plants each carrying one extra chromosome ($2n + 1$) are available in *H. spontaneum*, and in cultivated barleys. Such plants are less vigorous then the parental forms, but viability is adequate for some experimental purposes [29,31,89]. The morphology of trisomics is often abnormal and fertility is poor. Sometimes trisomic grain is appreciably thinner than diploid, and so can be separated from it. Trisomics occur in the progeny of triploids ($3n$) and from crosses between triploids ($3n$) and diploids ($2n$) or tetraploids ($4n$). They also occur in the progeny of interchange heterozygotes when 3:1 disjunction occurs at anaphase I [68,70]. In addition double ($2n + 2$), triple ($2n + 3$) and quadruple ($2n + 4$) trisomics occur. Usually extra chromosomes are not transmitted through pollen, so the progeny of selfed trisomics is a mixture of diploids and trisomics. In the diploid embryo of a trisomic the chromosome number is $2n + 1$, but in the triploid endosperm it is $3n + 2$. Since the extra chromosome carries genetic information which can be expressed particular genes or linkage groups can be assigned to a particular, 'extra' chromosome [58,70].

In tertiary trisomics the extra chromosome is formed by an interchange, and so carries segments from two non-homologous chromosomes. Interchange heterozygotes sometimes undergo irregular disjunction and form ($n + 1$) gametes [68,69]. The progeny may have two normal chromosomes *plus* a translocated chromosome *or* two translocated chromosomes *plus* a normal chromosome. From the progeny of one selfed interchange heterozygote, in the absence of crossing over in the interstitial segments, one should recover eight types of trisomics, two of which are tertiary trisomics and two of which are primary trisomic interchange homozygotes [69]. So-called 'balanced tertiary trisomics' are so arranged that the two chromosomes of the diploid complement of a tertiary trisomic (or primary trisomic interchange marker) carry the recessive allele of a marker gene (closely linked to the interchange breakpoint in the second case) while the extra chromosome carries the dominant allele [69,70,72]. As a consequence of the behaviour of the chromosomes about 60% of the offspring from selfed plants will only have the recessive marker gene. Balanced tertiary trisomics produce three functional types of female gametes but only one type of male gamete, because spores carrying the interchange chromosome alone are deficient in a chromosome segment, so they abort, and $n + 1$ spores rarely or never function as pollen (Fig. 11.5). Thus selfed tertiary trisomics produce three types of progeny of which those showing the dominant marker gene trait must be tertiary trisomics, genetically like the parent. All functional pollen should carry the recessive marker gene. Thus the pollen of such plants can carry a recessive *lethal* or *sterile* gene such as albino or male sterile [72]. Lethal or sterile genes can be maintained in such stocks, which could be used to grow hybrid barley (Chapter 12). Thus all diploid plants produced by selfing would be male-sterile, and could serve as female parents in hybrid formation. Such an approach supposes that the parents can cross freely, and that the genetically undesirable plant type can be eliminated [73,99].

When a chromosome splits through the centromere two functional half-chromosomes (telocentric chromosomes) may be produced, and the haploid number is increased to 8 [31,90]. Such telocentric

434 Barley

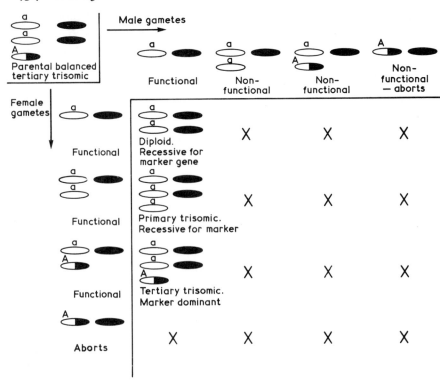

Fig. 11.5 A scheme summarizing the gametes formed by a balanced tertiary trisomic plant, and the genetic constitution of the zygotes expected from selfing (Ramage [69]). In the progeny only those plants which are also balanced tertiary trisomics will carry the dominant marker gene provided that the extra chromosome is stable – i.e. is not involved in crossing over.

chromosomes can be used for mapping genes, 'normally' or when they are present as the extra chromosome in trisomics. The morphologies of telocentric trisomics have been studied and normally fertility is good as the extra part-chromosome is less harmful than an extra whole chromosome [90]. As only one chromosome arm is triplicated the chance of locating a particular gene in that arm is enhanced. For example using telocentric trisomics two recessive mutant genes for 'revoluted leaf' and 'parkland spot' have been assigned to the short arm of chromosome one (1 S) and the long arm of chromosome five (5 L) respectively [17].

11.6. Mutations and mutagenesis [21,26,29,58,60,62]

Many studies are being made on barley: (i) to try and understand mutation processes; (ii) to enhance mutation frequencies, and (iii) to collect mutants and utilize them for genetic and biochemical studies, and for plant breeding. Mutations are sudden changes in the hereditary material of cells that cannot be attributed to simple recombinations of genes. They may be classified as (i) 'point mutations', mutations in the narrowest sense, in which one gene is altered to a new allele; (ii) rearrangements of chromosomal materials, for example to new levels of ploidy, (tetraploids, etc.); (iii) the formation of translocated chromosomes or the occurrence of deletions or duplications. Mutations occur spontaneously at a low frequency, and presumably account for the accumulation of variability in species, and the formation of the species themselves. Mutations are often induced by exposure to agents such as x-rays, γ-rays, ultraviolet light, fast and slow electrons and protons. High centrifugal force is also effective. Chemical mutagens include ethylmethanesulphonate (EMS), diethylsulphate (DES), various epoxides, ethyleneimines, nitrogen mustards, nucleic acid base analogues, nebularine, ethoxycaffeine, nitrous oxide (N_2O) and hydrazoic acid (HN_3). The mechanisms by which the agents act have been extensively discussed [13,34]. Mitotic and meiotic abnormalities are common after mutagenic treatments. Most of these agents are also carcinogens, and so *they are dangerous*. Many agricultural chemicals cause mitotic and morphological abnormalities, but are not mutagens (Chapter 9). Mutagenic treatments are commonly applied to grains, which are easily handled, but male and female gametophytes, zygotes and roots have also been used [10,18]. Generally, dividing cells are the most responsive. When a seed is successfully treated with a mutagen the plant derived from it is a 'chimaera' having mutated and unmutated sectors and 'pure mutants' will appear during successive generations, formed by self fertilization. The size of mutated sectors in barley spikes has been studied by mapping the distribution of 'waxy' pollen grains [50]. Studies of mutated sectors in plants derived from grain treated with mutagens indicate at least nine independent shoot meristems (depending on seed size) give rise to the shoots of normally developed barley, which has 8–9 independent mutation sectors. Presumably each sector comes from one meristem (Chapter 1) [38]. It is expected that treatment of the haploid stages, or zygotes will have the advantage that the plants are genetically uniform in all parts (mutant heterozygotes or non-mutant). Chimaeras should not occur [12]. However this expectation is *not* always fulfilled, for reasons that are unclear [53]. The mutagenic treatments are often harmful to the treated grain or plants, resulting in death or reduced viability, fertility, growth vigour, and sometimes abnormal and stunted growth, leaf-flecking, and extensive mitotic abnormalities in the roots. These 'toxic' effects exhibited in plants grown from treated seeds are not necessarily due to mutations, many of which may be recessive and so will only become manifest in subsequent selfed generations. The observed mutation frequency and the damage to seeds are strongly dependent, in complex ways, upon the exact conditions of pretreatment, treatment (duration and intensity) and storage of the grain after treatment [2,7,21,43,56,58,60].

Mutation frequencies are variously reported as per plant, per spike or per 100 M_2 plants [21]. The highest recorded mutation frequencies appear to have been caused at the rate of 40–57% of M_1 spikes by EMS and 17% per spike by x-rays [58]. The mutation types usually studied are changes in leaf colour, (misleadingly called chlorophyll mutations), because they are easily detected and scored when thousands of plants are grown. In any given population there will be an unknown 'extra' number of undetected mutants. Scoring for subtle physiological and biochemical mutations rarely or never seems to be attempted directly, although utilizing the enhanced variability of irradiated populations may be useful in plant breeding (Chapter 12) [21,23]. Different mutagenic treatments favour particular types of chromosomal damage, e.g. ethoxycaffeine and myleran (1, 4-dimethanesulphonoxybutane) favour chromosome breaks and induce different spectra of mutation types [29]. The induction of mutations is essentially a random process, and cannot be directed. The majority of 'artificial' mutations that have been detected are recessive or partially recessive. Possibly improvements in cultivated barley have been due to the selection for and retention of recessive traits [62]. The interpretation of numerical studies on mutation frequencies is beset with difficulties. What allowance should be made for mutations that are not detected, or for treated plants and seeds that are killed? What allowance should be made for genes that mutate but reveal themselves to various extents, depending on circumstances e.g. the 'labile barley' character, or male sterility [11,22,36]. Probably many mutated cells disappear from chimaeric plants due to an inability to compete with normal cells and so will be 'lost' (diplontic selection).

Some few mutants have been released as new agricultural varieties directly (Chapter 12). Large collections of mutants are held at various plant breeding centres. Artificial treatments rarely cause mutations of types unknown in 'nature'. The supposition that mutated plants will differ from the 'parent' line by only one gene and so will be isogenic with it is not universally true [54]. Numerous morphological mutants illustrate the 'plasticity' of the barley plant (Chapters 1 and 2). Changes in colour include gains or losses of anthocyanin pigments from the lemma veins, a form where the spike and upper part of the shoot is white, *eburata* where the lemma and palea lack chlorophyll, and numerous 'chlorophyll mutations' when the leaves have abnormal pigment patterns in the plastids and are various shades of green, yellow or white. Changes also occur in the nature of the surface wax. In grain waxy starch (high amylopectin) forms occur, and a spontaneous mutant form with an elevated amylose content has been found. Since this gene is expressed by the starch composition of the triploid endosperm it is possible to obtain barleys with four

Barley genetics 437

mutant gene levels namely 0, 1, 2 and 3. With these the amylose contents of the starches were 21.9%, 23.8%, 30.1%, and 43.5% respectively, with decreasing granule size [95]. Genes giving high protein and lysine contents in the grain are also known (Chapter 12). Mutants with high ribulose diphosphate carboxylase activity are available [44]. Numerous mutations lead to partial sterility or are lethal. One lethal mutant gave a seedling with rolled, leek-like leaves. Some mutations confer resistance against pests and diseases, for example to mildew or green bugs.

11.7. The expression of some mutant and other genes

Often although genetic analysis shows that only one gene (or a small chromosome segment) is mutated, this alters many of the plant's morphological and physiological characters. The complex nature of plant metabolism and growth leads one to expect such multiple, or 'pleiotropic' effects, but their occurrence prevents the 'tidy' classification of mutant forms. The occurrence of such diverse and 'natural' varieties, for example those that occur in composite crosses, make the classification of barley nearly impossible in classical taxonomic terms (Chapters 2 and 3) [32,52].

In the *erectoides (ert)* mutations plants usually have dense spikes, stiff straw, and improved lodging resistance. Different mutagenic agents tend to induce *ert* mutations at particular loci [30]. In one collection of about 400 *ert* mutants, 144 were shown to be 'single gene' alterations that had occurred at 26 separate loci. The number of mutants at each locus varied from 1–28 [32,64]. *Ert* mutated genes are usually recessive. However *ert*-14 (locus-*d*) and *ert*-23 (locus-*c*) are recessive regarding ear-length, but dominant with respect to the long first internode character. The combination of different *ert*-genes in one plant results in a denser ear and decreasing internode length for each gene addition. Plants homozygous for 3–4 *ert* genes have the kernels so close together that the very dense spikes are twisted. Extreme combinations may result in dwarf plants with straw only 10–30 cm (about 4–12 in) long [64]. Diallele analysis of internode length in crosses between parental and non-allelic *ert*-mutant forms, an ideal experimental arrangement for this type of analysis, shows no interaction between non-allelic genes [63]. Different mutant alleles, situated at one locus, may give rise to significantly different types of plants. The incorporation of an *ert*-gene into barleys with different genetic make-ups showed that culm length and spike internode length could be varied independently so increasing the possibilities of breeding short, stiff-strawed varieties not having excessively dense spikes [22]. *Erectoides* mutant genes have pleiotropic effects. Often ears are more dense, total plant height is reduced, plants usually have fewer internodes, there is a decrease in the length of lower internodes but an increase in the length of upper internodes, the cross-sectional areas of nodes, internodes and vascular bundles are increased, while root development and cell-size are altered (Chapter 1). However different *ert*-genes vary widely in their expression. In studies of spike density (length in mm of 10 internodes), the 'parental' forms had the values Gull, 29.4; Maja, 29.2; Bonus, 31.2, and Foma, 28.8, while some of the homozygous mutants, grouped by loci, fell into the following ranges: locus-*a*, 17.0–26.3; *b*, 19.1–23.5; *c*, 16.5–28.9; *d*, 15.2–26.9; *e*, 12.4–15.3; *f*, 21.5–25.6; *g*, 20.8–23.9; *m*, 20.3–28.1. The full range was 12.5–31.6 [64]. Clearly the reductions in internode length caused by mutations are partly characteristic for a locus. Gibberellic acid (GA_3) lengthened the internodes of *ert*-mutant or parental forms, if applied to the plants at the correct time, but the effect was not persistent and varied in intensity between different forms [80,86].

Some 876 induced and 4 spontaneous mutations causing abnormal epidermal wax, *(eceriferum* mutants, *cer)* have been shown to occur at 59 different loci. Several of these loci have been located on chromosomes 1, 2, 3, 4, 5 and 7. Different mutagenic agents vary in the frequency with which they induce mutations at different loci [84]. The microscopic morphology of the surface wax is altered in the mutants and its quantity and chemical composition changes so that, for example, wax cover may be reduced by 17–48% of seedling leaf, or β-diketone and hydroxy-β-diketone or alcohol contents of the wax may be reduced [97]. In one mutant the β-diketone content of the wax is enhanced while hydroxy-β-diketone is absent indicating that the former are precursors of the latter. A pleiotropic effect of 12 *cer* mutations (at the *g* locus) is to cause abnormal development of the cells of the stomatal complexes [102].

'Chlorophyll-deficient' mutants, i.e. mutants with abnormal leaf-colours, have been extensively studied [25,37,45]. They have been divided by eye into different colour classes. The leaves of *albina* mutants are white, and lack carotenoids and chlorophyll. They generally have small plastids. *Xantha* forms are yellow, since the carotenoid colour predominates, chlorophyll being nearly or completely absent; in this group plastid morphology varies. *Alboviridis* forms of different types have different rates of plastid development at the base and the tips of each leaf, and so these leaf parts have different colours. *Viridis* forms are light green in colour, while *tigrina* mutants have transverse stripes along the leaf. Generally, green bands alternate with brown or yellow, rather narrow bands, where probably pigment destruction has occurred. In *striata* forms yellow or white longitudinal bands alternate with green while in *maculata* forms chlorophyll and/or carotene is destroyed in spots. Each of these classes is highly heterogeneous. A viable mutant lacking chlorophyll *b* is known [35]. The phenotypic expression of some of these genes is highly dependent on the environment. Weak light and low temperatures often accentuate the deficiencies and the expression of genes in greenhouses (weak light, constant temperatures) and in the field (strong light, fluctuating temperatures) are very different. Some mutants appear normal when fed particular amino acids but their composition is not normal [47]. Single, recessive mutated genes often account for the observed colour deficiencies. However, irregularities in segregating populations from crosses indicate that maternal inheritance, gene complementation, and gamete or zygote elimination occur [37].

Cold treatments or short-day photoperiods, or other treatments at a critical stage of development will cause 'hooded' barley to grow with awns; conditions of growth may dramatically alter the expression of many genes (Chapter 2) [101]. In some chlorophyll 'revertant' mutants, in which the abnormal has changed back to the phenotypically normal, reversion may be due to mutations at widely different loci that suppress the expression of the original 'deficiency' mutation [93]. In a parent wild-type, a yellow virescent mutant, and a revertant carrying a mutated suppressor gene, the wild type and revertant were phenotypically similar but in chemical terms their pigment patterns were distinct [39]. Biochemical and electron micrographic studies are being made in order to locate the plastid morphological and biochemical lesions caused by individual mutations [96]. Some mutations block steps in the biosynthetic sequence leading to the porphyrin ring of chlorophyll. In such mutants the precursor before the block often accumulates and can be identified (Chapter 4).

Many attempts have been made to relate genotype to the enzymes present. This is an attractive approach since the 'one gene – one polypeptide' hypothesis predicts a reasonably direct relationship between a gene mutation and the encoded protein. While in different proteins the occurrence of different polypeptide sub-units, of carbohydrates moieties in glycoproteins, or the occurrence of other cofactors complicate this concept, the relationship between gene and enzyme is more direct than, say, between a gene and spike morphology. Studies have been made on the occurrence of many protein electrophoretic patterns and enzymes and isozymes in barley, including α-amylases, β-amylases, esterases, alcohol dehydrogenases, leucine amino peptidases, and peroxidases. Esterase isozymes can be used to detect some chromosome 3-trisomics [57]. *H. spontaneum* collections from diverse locations show a wide range of esterase patterns, many of which are also found in cultivated barley. The locus for esterase 4 has been located in chromosome 3 by analysing F_1 trisomic plants.

Crosses between closely related mutants sometimes give surprising results. Thus crosses between *albina* and *xantha* mutants and between the mutants and parental forms gave rise to F_1 plants that, in dense stands, were more vigorous than the parental forms [27]. In crosses between 3 allelic mutants (with short heads and culms) and the parental forms the heads and culms of the F_1 plants were larger than those of either parent [61]. Such results are ascribed to 'monohybrid heterosis', 'overdominance', or 'superdominance'.

11.8. The genetics of complex characters

Many important 'quantitative' agricultural characters, such as yield, straw-stiffness, and earliness, are due to the interaction of numerous genes. In these cases the situation is too complex for classical Mendelian genetics to be helpful, and other statistical approaches are used. In a genetically uniform population these characters all exhibit phenotypic variation due to the differing environments of the different plants, and the growing season. A frequency curve for each character can be drawn up when many samples from one population are considered. In a mixed population the observed variability will be due, in addition, to the varied genotypes of the plants. When two pure lines are crossed and the F_1 progeny are selfed in succeeding generations the spread of characters observed in the population will usually coincide with, and overlap the spreads of the parents. However some 'transgressive segregation' may occur, when lines inferior to and lines superior to either of the parents will appear. Clearly this is of outstanding importance in breeding agricultural varieties (Chapter 12). When the F_1 progeny of a cross between pure-line parental forms are superior to either parent then the F_1 generation is said to exhibit heterosis, or hybrid vigour. Heterosis is sometimes less helpfully defined as F_1 superiority to the parental mean (mid-point; mid-parent). Such vigour is used to great effect in enhancing yields in some outcrossing crops, such as maize. Attempts are being made to utilize this heterotic effect on a practical scale and produce 'hybrid barley' (Chapter 12).

When barley lines with widely differing genetic backgrounds are crossed the phenotypic results are unpredictable [52]. Even when genetically 'close' lines are crossed unexpected results may be obtained. Non-allelic gene interactions can have major unpredictable effects in the 'quantitative' characters of progeny. When eight quantitative characters were studied in the progeny of four near-isogenic barley lines, differing only in two short chromosomal segments, it was estimated that 32% of the observed variance was 'unexpected' and was an expression of epistasis [16].

Because of the character's interest as a breeding tool, to eliminate the need for manual emasculation when making crosses, and because of the attempts to develop female parents for 'hybrid' barley production, factors for male sterility, such as the recessive *ms* gene, are being intensively studied. Attempts are being made to obtain cytoplasmic male-sterile barley factors from crosses between wild barley grasses [20], but most genes for male sterility are carried in the nucleus [36]. So far attempts to find a cytoplasmic fertility restorer factor, for use in a breeding scheme like that used with onions, have failed. Numerous non-allelic male-sterility genes are known; they vary in their effectiveness. The recessive gene *ms* produces male-sterility by inducing infertility in the anther. The incorporation of male-sterility in some barleys in a mixed stand facilitates outcrossing and so helps to maintain heterozygosity in a breeding programme (Chapter 12).

Barley is being well characterized in conventional genetic terms, and the genetic data available is correlated on an international basis [1,58,59]. However advances are dependent on the continued expenditure of much time and effort, and the existence of facilities for growing very large numbers of plants. Reports of the wonders of genetic engineering in micro-organisms seem remote from practical work with higher plants. Single-cell cultures of barley may be obtained, but altering their genetic constitution in the ways that are so successful with micro-organisms and then producing normal plants from the genetically modified cells has yet to be achieved. There are indeed reports that microbial DNA fed to barley plants will be associated for a while, in an undegraded state, with nuclei [49]. However, there are no indications that a transfer of genetic information has been achieved. It seems likely that the main advances for agriculture and industrial users will come from the continued application of refined, but otherwise 'conventional' genetic techniques, made in conjunction with biochemical, physiological and agricultural studies.

References

Much new information becomes available in the annuals, *Barley Newsletter* and *Barley Genetics Newsletter*

1. Anon. (1964). *Barley Genetics I. Proceedings of the First International Barley Genetics Symposium, Wageningen, 1963.*
2. Bender, K. and Gaul, H. (1967). *Radiat. Bot.*, **7**, 289–301.
3. Bennett, M.D. (1971). *Proc. R. Soc. Ser. B*, **178**, 277–299.
4. Bennett, M.D. and Finch, R.A. (1971). *Genet. Res.*, **17**, 209–214.
5. Bennett, M.D., Finch, R.A., Smith, J.B. and Rao, M.K. (1973). *Proc. R. Soc. Ser. B*, **183**, 301–319.
6. Clavier, Y. and Cauderon, A. (1951). *Ann. Amèlior. de Plantes*, **1**, 332–335.
7. Conger, B.V. and Constantin, M.J. (1970). *Radiat. Bot.*, **10**, 95–97.
8. Dale, P.J. (1975). *Planta*, **127**, 213–220.
9. Derenne, P. (1967). *Bull. Rech. Agron. Gembloux, N.S.*, **2**, 242–259.
10. Devreux, M. and Donini, B. (1972). *Radiat. Bot.*, **12**, 19–29.
11. Djalali, M., Hoffmann, W. and Plarre, W. (1970). In *Barley Genetics II* ed. Nilan, R.A. pp. 201–207. Washington State University Press.
12. Donini, B., Devreux, M. and Scarascia-Mugnozza, G.T. (1970). *Radiat. Bot.*, **10**, 79–86.
13. Ekberg, I. (1969). *Hereditas*, **63**, 257–278.
14. Ekberg, I. (1974). *Hereditas*, **76**, 1–30.
15. Ekberg, I. and Eriksson, G. (1965). *Hereditas*, **53**, 127–136.
16. Fasoulas, A.C. and Allard, R.W. (1962). *Genetics*, **47**, 899–907.
17. Fedak, G., Tsuchiya, T. and Helgason, S.B. (1972). *Can. J. Genet. Cytol.*, **14**, 949–957.
18. Ferrary, R. (1965). *Radiat. Bot.*, **5**, (Suppl.), 293–297.
19. Finch, R.A. and Bennett, M.D. (1972). *Can. J. Genet. Cytol.*, **14**, 507–515.
20. Foster, A.E. and Schooler, A.B. (1970). In *Barley Genetics II* ed. Nilan, R.A. pp. 316–318. Washington State University Press.

21 Gaul, H. (1964). *Radiat. Bot.*, **4**, 155–232.
22 Gaul, H. and Grunewaldt, J. (1970). In *Barley Genetics II* ed. Nilan, R.A. pp. 106–118. Washington State University Press.
23 Gaul, H.P.K., Ulonska, E., Zum Winkel, C. and Braker, G. (1969). *Induced Mutations in Plant Breeding.* pp. 375–398, Vienna: IAEA.
24 Gilbert, S.K. and Patterson, F.L. (1965). *Crop. Sci.*, **5**, 44–47.
25 Gustafsson, Å. (1942). *Hereditas*, **28**, 483–492.
26 Gustafsson, Å. (1969). *Induced Mutations in Plant Breeding.* pp. 9–31. Vienna: IAEA.
27 Gustafsson, Å., Nybom, N. and von Wettstein, D. (1950). *Hereditas*, **36**, 383–392.
28 Hagberg, A. (1965). *Radiat. Bot.*, **5**, (Suppl.), 741–752.
29 Hagberg, A. and Åkerberg, E. (1962). *Mutations and Polyploidy in Plant Breeding.* London: Heinemann.
30 Hagberg, A., Gustafsson, Å. and Ehrenberg, L. (1958). *Hereditas*, **44**, 523–530.
31 Hagberg, A. and Hagberg, G. (1970). In *Barley Genetics II*, ed. Nilan, R.A. pp. 65–71. Washington State University Press.
32 Hagberg, A. and Persson, G. (1964). In *Barley Genetics I, Proc. 1st Internat. Barley Genetics Symp., Wageningen*, 1963, pp. 55–67.
33 Harlan, H.V. and Martini, M.L. (1935). *J. Heredity*, **26**, 109–113.
34 Heslot, H. (1965). *Radiat. Bot.*, **5** (Suppl.), 3–45.
35 Highkin, H.R. and Frenkel, A.W. (1962). *Pl. Physiol., Lancaster*, **37**, 814–820.
36 Hockett, E.A. and Eslick, R.F. (1970). In *Barley Genetics II*, ed. Nilan, R.A. pp. 298–307. Washington State University Press.
37 Holm, G. (1954). *Acta Agric. scand.*, **4**, 457–471.
38 Jacobsen, P. (1966). *Radiat. Bot.*, **6**, 313–328.
39 Jain, M. (1966). *Genetics*, **54**, 813–818.
40 Jensen, C.J. (1974). In *Haploids in Higher Plants; advances and potential.* ed. Kasha, K.J. pp. 151–190. Guelph: University of Guelph.
41 Kao, K.N., Reinbergs, E. and Harvey, B.L. (1970). *Crop Sci.*, **10**, 491–492.
42 Kasha, K.J. and Reinbergs, E. (1972). *Master Brewer's Assoc. Am. Tech. Quart.*, **9**, 128–130.
43 Kesavan, P.C. and Kamra, O.P. (1970). In *Barley Genetics II*, ed. Nilan, R.A. pp. 127–137. Washington State University Press.
44 Kleinhofs, A. (1969). In *Induced Mutations in Plants.* pp. 101–108. Vienna: IAEA.
45 Kleinhofs, A., Shumway, L.K. and Sideris, E.G. (1970). In *Barley Genetics II*, ed. Nilan, R.A. pp. 194–200. Washington State University Press.
46 Konzak, C.F., Randolph, L.F., and Jensen, N.F. (1951). *J. Hered.*, **42**, 125–134.
47 Land, J.B. and Norton, G. (1973). *New Phytol.*, **72**, 485–492.
48 Lange, W. (1971). *Euphytica*, **20**, 14–29; 181–194.
49 Ledoux, L. and Huart, R. (1970). In *Barley Genetics II*, ed. Nilan, R.A. pp. 254–263. Washington State University Press.
50 Lindgren, D. and Eriksson, G. (1971). *Hereditas*, **69**, 129–134.
51 Lindgren, D., Eriksson, G. and Ekberg, I. (1969). *Hereditas*, **63**, 205–212.
52 Martini, M.L. and Harlan, H.V. (1942). *J. Hered.*, **33**, 339–343.
53 Mericle, L.W. and Mericle, R.P. (1973). *Barley Genetics Newsletter*, **3**, 39–42.
54 Moh, C.C. and Smith, L. (1952). *J. Hered.*, **43**, 183–188.
55 Myler, J.L. (1942). *J. agric. Res.*, **65**, 405–412.

56 Narayanan, K.R. and Konzak, C.F. (1969). In *Induced Mutations in Plants*. pp. 281–304. Vienna: IAEA.
57 Nielsen, G. (1972). *Barley Genetics Newsletter*, **2**, 62–64.
58 Nilan, R.A. (1964). *The Cytology and Genetics of Barley, 1951–1962*. Monographic Supplement No. 3. Research Studies **32**, (1), p. 278. Washington State University.
59 Nilan, R.A. (1970). *Barley Genetics II*, p. 622. Washington State University Press.
60 Nilan, R.A., Konzak, C.F., Heiner, R.E. and Froese-Gertzen, E.E. (1964). In *Barley Genetics I, Proc. 1st Internatn. Barley Genetics Symp. Wageningen, 1963*. pp. 35–54.
61 Nybom, N. (1950). *Hereditas*, **36**, 321–328.
62 Nybom, N. (1954). *Acta Agric. scand.*, **4**, 430–456; 507–514.
63 Persson, G. (1969). *Hereditas*, **63**, 39–47.
64 Persson, G. and Hagberg, A. (1969). *Hereditas*, **61**, 115–178.
65 Poehlman, J.M. (1959). *Breeding Field Crops*. pp. 23–50; 151–171. Henry Holt & Co.
66 Powell, J.B. and Nilan, R.A. (1963). *Crop Sci.*, **3**, 11–13.
67 Powell, J.B. and Nilan, R.A. (1968). *Crop Sci.*, **8**, 114–116.
68 Ramage, R.T. (1960). *Agron. J.*, **52**, 156–159.
69 Ramage, R.T. (1964). In *Barley Genetics I – Proc. 1st Internat. Barley Genetics Symp., Wageningen, 1963*. pp. 99–115.
70 Ramage, R.T. (1970). In *Barley Genetics II*, ed. Nilan, R.A. pp. 89–93. Washington State University Press.
71 Ramage, R.T. and Day, A.D. (1960). *Agron. J.*, **52**, 241; 590–591.
72 Ramage, R.T. and Tuleen, N.A. (1964). *Crop Sci.*, **4**, 81–82.
73 Ramage, R.T. and Wiebe, G.A. (1969). In *Induced Mutations in Plant Breeding*. pp. 655–659. Vienna: IAEA.
74 Reinbergs, E. (1964). *Barley Genetics I, Proc. 1st Internatn. Barley Genetics Symp., Wageningen, 1963*. pp. 151–154.
75 Robertson, D.W. (1970). In *Barley Genetics II*, ed. Nilan, R.A. pp. 220–242. Washington State University Press.
76 Robertson, D.W., Wiebe, G.A., Shands, R.G. and Hagberg, A. (1965). *Crop Sci.*, **5**, 33–43.
77 Sadasivaiah, R.S. and Kasha, K.J. (1971). *Chromosomes*, **35**, 247–263.
78 Sandfær, J. (1973). *Genetics*, **73**, 597–603.
79 Sayed, H.I., Helgason, S.B. and Larter, E.N. (1973). *Can. J. Genet. Cytol.*, **15**, 625–633.
80 Schmalz, H. (1960). *Züchter*, **30**, 81–83.
81 Smith, L. (1941). *J. agric. Res.*, **63**, 741–750.
82 Smith, L. (1942). *Am. J. Bot.*, **29**, 451–456.
83 Smith, L. (1951). *Bot. Rev.*, **17**, pp. 1–51; 133–202; 285–355.
84 Søgaard, B. (1974). *Hereditas*, **76**, 41–47.
85 Sparrow, A.H., Sparrow, R.C., Thompson, K.H. and Schairer, L.A. (1965). *Radiat. Bot.*, **5**, (Suppl.), 101–132.
86 Stoy, V. and Hagberg, A. (1967). *Hereditas*, **58**, 359–384.
87 Subramanyam, N.C. and Kasha, K.J. (1973). *Crop Sci.*, **13**, 749–750.
88 Symko, S. (1969). *Can. J. Genet. Cytol.*, **11**, 602–608.
89 Tsuchiya, T. (1964). In *Barley Genetics I, Proc. 1st Internatn. Barley Genetics Symp., Wageningen, 1963*. pp. 116–150.
90 Tsuchiya, T. (1970). In *Barley Genetics II*, ed. Nilan, R.A. pp. 72–81. Washington State University Press.

91 Tsuchiya, T. and Haus, T.E. (1973). *J. Hered.*, **64**, 282–284.
92 Tuleen, N.A. (1973). *Can. J. Genet.*, **15**, 267–273.
93 Tuleen, N.A., Snyder, L.A., Caldecott, R.S. and Hiatt, V.S. (1968). *Genetics*, **59**, 45–55.
94 Walker, G.W.R. and Ting, K.-P. (1967). *Can. J. Genet. Cytol.*, **9**, 314–320.
95 Walker, J.T. and Merritt, N.R. (1969). *Nature*, **221**, 482–483.
96 von Wettstein, D., Henningsen, K.W., Boynton, J.E., Kannangara, G.C. and Nielsen, O.F. (1971). *Autonomy and Biogenesis of Mitochondria and Chloroplasts.* pp. 205–223. Amsterdam: North-Holland.
97 von Wettstein-Knowles, P. (1970). In *Barley Genetics II*, ed. Nilan, R.A. pp. 146–193. Washington State University Press.
98 Wiebe, G.A. (1968). In *Barley – U.S. Dept. Agric., Agricultural Handbook No. 338*, 105–111.
99 Wiebe, G.A. and Ramage, R.T. (1970). In *Barley Genetics II*, ed. Nilan, R.A. pp. 287–291. Washington State University Press.
100 Woodward, R.W. and Rasmusson, D.C. (1957). *Agron. J.*, **49**, 92–94.
101 Yagil, E. and Stebbins, G.L. (1969). *Genetics*, **62**, 307–319.
102 Zeiger, E. and Stebbins, G.L. (1972). *Am. J. Bot.*, **59**, 143–148.
103 Zenkteler, M. and Misiura, E. (1974). *Biochem. Physiol. Pfl.*, **165**, 337–340.

Chapter 12

Barley improvement

12.1. Introduction

Man's intervention in the development of barley has spread the species around the world and unknown thousands have selected for improved agricultural varieties (Chapter 3). Cultivation and selection have given rise to many thousands of local strains. Meanwhile genes from cultivated strains have probably introgressed into the stands of wild barley, altering it and accelerating its evolution. Improvers always try to achieve a more valuable crop in terms of increased yields and a more desirable product. Improvement techniques have ranged from the magico-religious to patient breeding and selection programmes. Breeder's immediate objectives extend from attempts to find remedies for specific defects, e.g. forms resistant to lodging, shattering, or particular pests or diseases (where relatively few genes may be involved) through to the complex aim of maximizing yield potential, where possibly most of the genome is implicated and the physiological states and morphology of the plant are relevant at all times during the life-cycle. Mendelian genetics is a useful guide for handling a few genes, but when many genes are involved breeding is more empirical, and is guided by 'statistical genetics'. The breeding procedures used depend on individual skill and judgement; no knowledge of plant science or agricultural practices can safely be ignored.

Barley improvement is dependent on the selection of superior strains from among a range of types. Major improvements in the last 200 years were first obtained by selecting outstanding lines from heterogeneous, indigenous 'land' varieties. Later crossing the superior selected lines and, later still in England, crossing English and stiff-strawed Scandinavian varieties led to progressive improvements. Further advances may come from crossing varieties from all parts of the world. Seed of a successful cultivar is distributed for trials around the world. In consequence land races have disappeared from 'developed' countries and the range of types in existence is decreasing rapidly. Thus the 'genetic base' of the cultivated species is being reduced, and with it the chances of further improvements by hybridization. The widespread cultivation of similar varieties also increases the chances of major disease outbreaks. 'Genetic erosion' has been marginally slowed by the establishment of 'world collections' of varieties backed by seed-stores, or 'gene-banks' (Chapter 10). Apparently an unending supply of genes for disease resistance is needed. Inferior plants sometimes prove to be exceptionally good parents, producing many superior offspring [59]. These and other considerations indicate that there is no safe basis on which any 'line' can be discarded. The failure to establish and use more 'gene-banks' for the main world crops is a major crime in a world where the need for higher yielding crops is desperate [17,42,59,160].

Current practice is to insist on using 'pure lines', in which not only undesirable traits but *any detectable variability* is eliminated, reducing the genetic store further. To equate uniformity ('purity'), with high quality is irrational; indeed 'gene buffering' (the presence of a mixture of genotypes) *may* be desirable in a crop [17,150]. Fortunately there is genetic variability to be found in *Hordeum spontaneum*, but locating desirable genes in this wild grass and transferring them to cultivated barley is a difficult undertaking. Two other sources of variation are (i) induced mutations (Chapter 11), and (ii) the extreme variability shown in the offspring of some crosses.

For crop improvement superior plants must be selected from a great mass of average or inferior forms. Screening procedures for making such selections may be a matter of judgement, or of measurements and trials guided by statistical methods. Often plant populations are stressed to aid in selection – e.g. they are chilled if frost resistance is sought, or the ears may be shaken mechanically if shattering resistance is needed. Alternatively, trial plots are artificially inoculated with a pathogen to locate non-susceptible plants when a search for disease-resistance is being made. Screening tests need to be reliable and applicable to large numbers of plants. They are often complex, expensive in space, time and personnel, and are 'bottlenecks' in improvement programmes. Also, as with some 'lodging-resistance' tests, results may not faithfully reflect field performance. Screening for mixtures of attributes, such as 'malting quality' or 'high yielding ability', is difficult indeed.

12.2. Plant introductions, and adapted forms

During the original, slow spread of barley over the Old World, forms adapted to near-arctic, temperate maritime, continental, high mountain, and other climates gradually appeared. Adaptation must involve many little-understood aspects of physiology besides specific adjustments involving photoperiodic, temperature and vernalization responses. The importance of 'adapted' varieties was illustrated when European settlers, carrying their *'local'* cereal varieties, moved to North America, where barley was previously unknown. The N. Europeans – British and Dutch – who settled on the East coast found that two-rowed spring varieties grew badly, although the six-rowed, lax-eared 'bere' from Scotland did better. In contrast the Spaniards carried N. African types to New Mexico, Arizona and California where they grew well, and gave rise to Coast and Atlas types. Mariout types – from Egypt – do well in California, and are tolerant to alkali. Other forms – sometimes of Russian origin – were introduced, often via Germany, and trials showed them to be well adapted to some areas, giving rise to the Manchuria and Oderbrucker strains. Still other varieties came from the mountains of Switzerland and the Balkans [59,111]. Many of the successfully introduced N. American varieties were found by trials embracing as many as possible of the cultivars obtained from around the world. Some less laborious method for making introductions is desirable, but the 'obvious' approach of testing varieties from similar climatic zones has not been improved upon. However successful strains are sometimes introduced from unexpected areas [64]. Introduced varieties may be suitable for immediate release to agriculture, with great savings in time, or they may be used in breeding programmes. Attempts to classify barley types in relation to their climatic site of origin – (latitude, elevation, precipitation, and day-degree or photothermal summations) reinforce commonsense conclusions [106]. Studies in phytotrons or growth chambers cannot handle the thousands of barley samples and reveal their suitability (or lack of it) for all the world's climates, even if the 'weather' could be faithfully reproduced.

A graphical approach has clarified the degrees of adaptation of different varieties to the S. Australian climate [38]. In S. Australia the rainfall is low and variable, strongly influencing yields. At different sites, over several years, 277 varieties were grown and the yields noted. When $\log_{10}$ (varietal yield) was plotted against $\log_{10}$ (mean yield) of all varieties, at each site, straight lines with varying slopes were obtained (Fig. 12.1). Near horizontal lines indicate relative insensitivity to climate, while lines with steep slopes are poorly adaptable, being very responsive to climatic variations. High-yielding varieties of wide adaptability are needed to accommodate the wide variations in rainfall. Atlas, a N. African type from the U.S.A., gave above average yields at all locations. Others gave below average yields, while English and Scandinavian varieties, adapted to damp maritime climates and fertile soils, gave high yields under the best conditions but very poor yields under adverse conditions. The regression coefficient, t, defining the slope of the line, can be regarded as

448 Barley

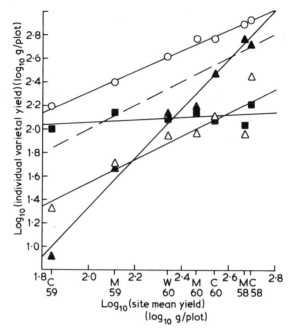

Fig. 12.1 Regression lines relating the yields of four varieties and the mean yield of 277 barley varieties to the mean site yields. Barleys were grown at three different sites in three successive years in S. Australia (after Finlay and Wilkinson [38]). C: Clinton, M: Minlaton, W: Waite Institute. 58, 59, 60 – years 1958, 1959, 1960.
O, Atlas. ■ Bankuti Korai. ▲ Provost. △ BR 1239 ———— Mean of 277 varieties.

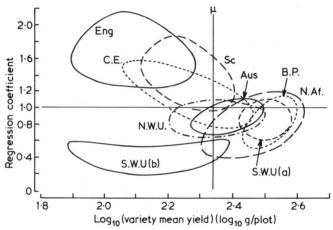

Fig. 12.2 The distributions of regression coefficients plotted against varietal mean yields of barleys from a range of geographical locations, grown in S. Australia (after Finlay and Wilkinson [38]).

a measure of adaptability. 'Stable' varieties have regression coefficients less than 1 (such as Atlas, 0.9); 'unstable' varieties have regression coefficients greater than 1 (such as Provost, 2.13). Scatter diagrams of regression coefficients against varietal mean yields tend to group varieties by areas of origin (Fig. 12.2). Varieties in the bottom, right quadrant yield well and are adaptable. These include cultivars from S. Western U.S.A. (N. African origin), N. Africa (direct), Bolivia and Peru. The data suggest that these are areas where other high yielding, well-adapted varieties may be sought. Such analyses give no guide to grain quality, and may not be applicable to all groups of environments [85].

12.3. Plant selections

The old land varieties were mixtures of near-homozygous plants of very varying characteristics [12,76]. Selection of the best forms led to more uniform stands of plants with enhanced yield and quality. Repeated selection is ineffective on uniform material and so yield and quality 'plateaued'. Selections from land races may still be of immediate value in undeveloped areas. Chevallier, a cultivar with superior malting quality that dominated the English crop for many years was selected in 1824. Selections from Archer, a high yielding, relatively poor quality 'late ripe' land race, gave rise to several important strains of differing characteristics (Table 12.1), as did Spratt – a fan-eared, stiff-strawed type known

Table 12.1 The results of some Irish trials with Archer and Archer selections (from Hunter [76]).
(Trials were usually made at 10 or 12 sites)*

Year of trial	Grain type	Yield			Grain characters	
		cwt	stones/acre	kg/ha	Total nitrogen (%)	Thousand grain weight (g)
1906	Archer (unselected)	23	4	2950	1.51	37.9
	Tystofte Prentice	25	1	3154	1.47	39.5
1907	Archer (mass selected)	24	0	3013	1.55	34.0
	Tystofte Prentice	24	6	3107	1.54	34.5
	English Archer	23	0	2887	1.62	34.3
1908	Irish Archer I	22	1	2777	1.52	36.7
	Tystofte Prentice	22	4	2824	1.49	36.9
1909	Irish Archer I	28	0	3515	1.47	38.4
	Tystofte Prentice	28	3	3562	1.46	39.8
1910	Irish Archer I	20	4	2573	1.55	34.1
	Tystofte Prentice	20	5	2589	1.53	34.3

*(8 stones = 1 cwt 1 cwt/acre = 125.5 kg/ha)

450 Barley

from mediaeval times. Goldthorpe came from a single plant selection made in 1891 [12,76,163]. Selection experiments made in Ireland illustrate the way gains were made. Tystofte Prentice was a relatively uniform Danish selection made from English Archer seed. The unselected 'Archer' grown in Ireland was heterogeneous in ear type, time of ripening, and other characters and, on average, yielded less than T. Prentice. A single-ear selection gave rise to the improved Irish Archer I. English Archer, with supposedly the same seed source as Irish Archer, was also inferior in yield to T. Prentice. The variations in the mean yields with different growing seasons and the observation that in 1906, T. Prentice yielded *less* than unselected Archer at 2 locations out of 12, illustrates some of the difficulties encountered in comparative trials [76]. Varietal superiority is an *average* property in that it is demonstrated at most sites in most seasons. In Sweden, Svanhals, Hannchen and Gull were selected from land races at Svalöf [96]. In the U.S.A. a single seed selection, made on a seed lot from Turkey in 1905, gave rise to Trebi, and selections from Swiss seeds gave rise to Chevron and Peatland.

12.4. Mutation breeding [55,138]

The widespread search for useful barley mutants, spontaneous or induced, was mainly initiated by the workers at Svalöf, in spite of substantial scepticism [141]. It was hoped that mutations induced in the best local varieties would quickly produce improved forms. The detection of mutants among the offspring of treated seed is laborious, many mutations are deleterious, and many otherwise desirable mutations have associated adverse characters which make them unacceptable. Probably less than 1% of all observed mutations are of value to breeders. However, potentially useful mutants with enhanced yield, stiffer straw, enhanced insect and disease resistance, better winter hardiness, different maturation times, improved grain characteristics, altered temperature and light requirements, smooth awns, high grain protein contents, high grain lysine contents and naked grains are known. Some mutant strains have been released for agricultural use while others have served as parents for agricultural varieties [44]. Two Swedish mutants of Bonus are in cultivation. Pallas is an *erectoides* form with an enhanced resistance to lodging, and a high yielding capacity. Mari is a mutant that is 7–10 days earlier ripening but yields less than Bonus in S. Sweden where it was obtained. However in Finland and N. Sweden, Mari outyields Bonus. Phytotron trials show that Mari is relatively insensitive to thermo- and photoperiods, and will head and ripen grain in 8 h photoperiods, when Bonus only makes vegetative growth [54]. Hellas, from the cross Pallas × Herta, has enhanced resistance to straw-breaking after ripening, and lodging resistance and yields are improved. Mari (early) crossed with

Domen (lodging resistant) has given the improved cultivar Kristina. In Japan one mutant having a short culm and another that ripened early were obtained from the same high-quality malting variety. The double-recessive isolated from the cross between the mutants, Gamma No. 8, was a shorter culmed, 4–5 days earlier ripening form, with yield and malting quality equal to those of the parent [70].

The successes have been with mutations giving readily detected phenotypes. Many 'micro-mutations' must escape detection. Some of these could be of value but probably most will be detrimental [44,45]. By collecting 'normal', fertile heads from selfed plants two, three and four generations after mutagenic treatment, and evaluating the progenies it is found that the mean yield of the 'mutated' families has been depressed, but the variance of the yield of the population is increased and some few of the families slightly outyield the best of the controls. In experiments with Haisa II, selection increased the average yield of untreated (control) families by 4.6% (mean of 10), with the best family exceeding the bulk by 6.6%. In the best families from the irradiated stock the increase of yield averaged (of 10), 6.6%, the best family yielding 8.8% more than the bulk. With Wisa, the best 'irradiated' family exceeded the yield of the unselected bulk by 14% [45]. These results are less encouraging than first appears since the best 'mutated' families gave only about a 2% yield advantage over the best 'control' families. Suggestions that cycles of irradiation and selection may improve yields by 10% per cycle seem unrealistic.

The use of mutagens to enhance crossing over in F_1 seeds, so separating closely linked, unwanted characters from genes being transferred in a back-crossing programme seems unusual, yet its feasibility has been demonstrated by separating the 'orange lemma' factor and the tightly linked 'high α-amylase' factor by neutron irradiation [105].

Polyploids seem unlikely to be of commercial importance. Although large, leafy autotetraploid plants may make good fodder, the low grain yields prevent their acceptance.

12.5. Hybridization

Crosses must be made to obtain forms combining desirable characteristics from two or more varieties, and the desired recombinant types must be sought in the progeny. Undoubtedly this form of breeding is the most powerful and the most widely used method of barley improvement [5,15, 23,64,111,163,166]. In the simplest case two barleys are crossed, the heterozygous F_1 seed is harvested and it and subsequent generations, are grown normally and so are mainly self-fertilized. Thus successive generations will be mixtures of more nearly homozygous plants, from which selections are made. It is often implied that after about eight generations the individuals will be homozygous. In fact, if two parents differed by only 100, *unlinked* genes then after seven generations of selfing less than half of the progeny will be homozygous. In practice many more gene differences may be involved and many genes will be linked. Segregation has been observed in barley after more than 18 generations of 'selfing' and, whether it is readily detectable or not, in principle such segregation should go on indefinitely but with a decreasing frequency. Heterozygosity is enhanced by a little outcrossing, the occurrence of 'spontaneous' mutations, and selection in favour of heterozygous plants. The degree of outcrossing depends on the variety, the location, and the season. In some cases none is detected while in male-sterile strains it accounts for all the seed set. In Israel the formation of hybrids from accidental crosses between cultivated *H. vulgare* and wild *H. spontaneum*, indicates that here it occurs readily (Chapter 3). In Saskatoon humid seasons favoured out-crossing which ranged over 5 years from 0.0 to 0.17% for awned varieties and from 0.0 to 0.37% for hooded forms [63]. Elsewhere outcrossing of 0.04%, 0.12% and 0.15% occurred in successive years [145], and in *H. deficiens nudificiens* values of 3.3%, 20.7% and 5.75% were found [128]. Frost or high temperatures can emasculate barley and so facilitate outcrossing [149]. Outcrossing of 20% occurred on one dwarf form [153]. In a 'composite cross' grown in Scotland outcrossing occurred in the range 2.3–4.6% [47]. Thus cultivars are not pure lines, or even collections of truly homozygous plants. Rather they are collections that are sufficiently similar for agricultural and industrial purposes. This heterogeneity explains observed yield responses of 'pure cultivars' to selection [44,45].

Some theoretical aspects of breeding are daunting. For example, to realise all possible gene combinations in the F_2 of a cross between parents differing by only 30 unlinked genes would require that more than the whole land surface of the earth be planted densely at 2 million plants/ha [96]. The impossibility of obtaining 'complete' sets of progeny from a cross is apparent. Breeders must have luck as well as skill and work with the largest possible numbers, and have efficient screening methods, to have a reasonable chance of finding superior strains. In practice a minimum of about 20 000 F_2 plants should be raised from a 'wide' cross between dissimilar parents [64].

Early breeders did not attempt to produce homozygous lines; their objective was to produce improved varieties quickly. No less than four series of hybrids called Plumage–Archer were released [12,76,163]. For the original Plumage–Archer, Beaven (Warminster) made a cross in 1905, and apparently grew 5 F_1 plants, and 218 F_2 plants (24 ear-rows). After selection and evaluation steps the variety was released in 1914, at F_9 [12,163]. Hunter (Ballinacurra) made a cross in 1908 that led to Spratt – Archer. Possibly only one F_1 plant was grown, and 46 F_2 plants.

This variety was released in F_6 [76,163]. Thus both varieties were 'impure', and seed stocks were occasionally changed after reselection, yet they dominated British agriculture for over 30 years. The separate propagation of lines derived from 'normal' ears of Spratt–Archer, about 17 years after the varieties release, showed that 12 of the 41 cultures investigated varied significantly in their growth and grain characteristics. One ripened 7–10 days before the 'parent', and was released as the variety Earl. Another poor quality selection, called Streetly Rogue, ripened 7–10 days *after* the 'parent'. Clearly selection based on the form of the mature ear alone is insufficient to produce uniform plants. In contrast Proctor, derived from the cross Plumage–Archer × Kenia, made by Bell (Cambridge), was not released until 20 years after the cross was made [163]. Kenia, a Scandinavian variety badly adapted to the English climate, has poor malting quality but has stiff straw and a good grain yield. Plumage–Archer has good malting quality. From the F_1 plants 2300 F_2 grains were 'space' planted to allow close examination. The selected plants gave 149 F_3 progenies; there were 48 F_4, and then 10 F_5 progenies so at every stage strong selection was practised to try to secure a type that matched an ideal. In F_6 eight selections were grown in bulk, five were in trial by F_9, and finally one was selected and released as Proctor [13,163]. Proctor displaced the older varieties, being higher yielding, of good malting quality and suitable for combine harvesting. It is now giving way to shorter, more lodging-resistant and yet higher yielding varieties.

Breeding is slow; a cross made now must yield a variety suitable for the agriculture to be found in 15–30 years' time [13,15]. To accelerate breeding programmes two generations of plants may be grown each year alternately at locations in the Northern and Southern hemispheres, e.g. England and New Zealand, North and South America, or in an 'off-season nursery' by growing spring barleys as a winter barley in a mild climate, at a more southerly location. By using special feeding, light and temperature schedules it is possible to get four, or even five to six generations/year in a greenhouse. However most selection steps must be applied to plants grown under agricultural conditions. Crosses are easily made in greenhouses and here plants can be selected for resistance to some adverse conditions such as high soil salinity, low temperatures and some diseases.

12.6. Crossing barley

The objects in crossing barley are to exclude self-pollination and random cross-pollination, to bring sufficient viable pollen from one chosen parent to the stigma of the other, and to ensure that seed is set, developed and harvested. It may be desirable to cross 'both ways' so that each variety is used as a male and a female parent. The 'female' flowers have to be emasculated i.e. the anthers are removed, or are otherwise made non-functional [64,115]. Plants to be crossed must have their pollen and ovules 'ripe' simultaneously, since barley pollen cannot be stored in a viable state. Winter forms must be grown from vernalized grain to head at all. The chances of getting two varieties ripe simultaneously are enhanced by sowing seeds at intervals to give a succession of plants, and by sowing them thinly, to encourage tillering so prolonging the period of heading. By growing at low temperatures, or low light intensities, or by feeding nitrogen fertilizers the vegetative period may be prolonged. Conversely, increasing temperatures and photoperiods favour flowering, though to widely different extents in different varieties. Heads for crossing may be stored for several weeks at low temperatures, and good seed-set still be obtained. Pollen was 57% viable in heads stored 3 weeks at about 2°C (36°F); ovules also survive such storage [112,113].

Spikes are ready to be emasculated when the anthers are light green, and beginning to turn yellow, but before they ripen, become fully yellow, and shed their pollen. At this stage the ear is often in the boot, so the flag leaf must either be slit and moved or unrolled and bent down, or cut away. Replacing the flag leaf may be tedious, but it supplies photosynthate to the developing grain, and protects and supports the ear. The side florets (which may produce pollen even in two-rowed varieties) are removed as are the top and bottom florets of the median rows. From eight to twenty florets are retained on each ear. The ear is held to the light, so that the anthers may be detected in silhouette. In the most usual emasculation method the tips of the palea and lemma are cut off, 'clipped', just above the anthers, which are then removed with fine forceps, or a suction device. Slitting the lemma and removing the anthers through the cut, but leaving the awn in place, results in the production of larger grains, but this technique is tedious [20]. An alternative 'scissor emasculation' involves clipping the spikes just below the level of the tips of the anthers, taking care to cut these. After this treatment mature pollen does not form [107,162]. The pistil can withstand some accidental 'pruning', yet remain operative.

Manual emasculation is tedious, and ways have been sought to circumvent it. Numerous chemicals have been tested as emasculating agents but despite some limited success none have proved to be reliably effective under all conditions. A recent report indicates that a new compound is capable of inducing total male sterility, when used in the greenhouse, with female fertility reduced to 19 to 40% [161]. Excessive heat and cold are known to emasculate florets in the field, but these do not seem to be exploited by breeders. By using male-sterile plants as the 'females' the need for emasculation is circumvented. Attempts have been made to make male-sterile lines self-fertile 'on command' by using sprays of gibberellic acid solution, so making it easy to maintain the lines. Using the first known recessive, male sterile gene *ms*, it was shown that natural cross pollination in the field resulted in 10–45% seed set, while hand pollination achieved 64–90% seed set [126]. The use of male-sterile parents allows massive cross-pollination programmes to be undertaken with minimum labour. Pollen-donor and male-sterile seeds are planted in alternate rows. Only heads from the male-sterile plants, the seeds of which must be crosses, are harvested [126,166].

Emasculated heads are covered with lightweight bags or shielded with cloth 'cages', to keep away stray pollen, and are left for 1–3 days. Normally unfertilized florets will 'gape' as the lodicules force the palea and lemma apart. If this does not happen it is assumed that accidental pollination has occurred, and the closed floret is removed. By keeping the clipped heads cool the period in which gaping occurs, and the stigmata are receptive to pollen, is lengthened. The florets of the heads that are to act as pollen donors are also clipped above the anthers. They are warmed by the sun or in the hand to cause the ripe stamens to extend. Anthers taken from the pollen donor heads (1 or 2) are broken into each emasculated floret, or bunch of pollen donor ears is dusted over the emasculated head. Several heads may be surrounded by a light 'cone' of paper or plastic to limit the spread of the pollen [162]. In the field a group of emasculated plants may be surrounded by a pollen-proof cloth cage, or 'isolator', and pollen-donor ears may be set within the isolator to release pollen over them, or pollen-donor plants may be placed within the cage [86]. Generally at least 30 heads are pollinated in making one cross. The success of attempted pollination is dependent on the weather and the time of day. After pollination the ears are again 'bagged' and labelled. After 2–3 weeks, when the chance of accidental pollination has passed, the bags may be perforated or removed. The F_1 seed may be small, because of the damage done to the 'mother' plant during emasculation. Seed will continue to grow for over a week in ears on peduncles cut from the plant and kept moist [62]. Seed set may be improved e.g. by 'feeding' the ear through the cut flag-leaf with a solution of gibberellic acid [72]. The small seed must be threshed with great care – e.g. by rubbing it from the ears [64]. The F_1 seed should be fully after-ripened, then grown with care, initial germination often being on damp paper. The seedlings should be protected from damage and diseases and be supported during growth. Normally seedlings are grown widely spaced and well supplied with nutrients to encourage tillering and the production of many F_2 grains. By removing a piece of the lemma over the embryo and keeping the embryonic axis wet it is possible to induce vivipary, and get seedlings to form on the ear [114], so shortening the generation cycle.

'Crossing' in greenhouses or growth chambers is convenient, but is also expensive. In the field, diseases, pests and the weather may cause unacceptable losses. A compromise is to grow plants protected in sheltered

456 Barley

bird-proof net enclosures, with poison bait laid for some other pests, using sprays to combat insects and diseases.

12.7. The choice of parents

In producing the earlier hybrid barleys, superior selections of two local, well adapted varieties that were judged to complement each other with regards to desirable characters, such as stiffness of straw, yield, grain quality or disease resistance, were used as parents. This approach is still used. However, 'wider' crosses are often needed. If one parent is ill adapted, as are short-strawed Japanese varieties in England, then it may be better to find another short-strawed parent [16]. In some 'wide' crosses difficulty has been experienced in finding any useful progeny. The object may be to introduce a gene – or a small number of genes – into a successful variety either (1) to remedy some defect – such as susceptibility to a disease, or (2) to create near-isogenic lines (i.e. lines differing in a very limited number of genes linked on short chromosome segments). To achieve this 'back-crossing' is employed [5,23]. The programme of back-crossing and selection is chosen with reference to the dominance, or otherwise, of the gene(s) being transferred, their numbers, and the ease of their detection in the progeny. The F_1 plants are 'back-crossed' to the 'recurrent' parent. The offspring are closely screened, and those that most closely resemble the recurrent parent, but carry the genes to be transferred, are used to make the next backcross. By rigorous selection 'near-isogenic' lines may be produced after six to eight backcrosses, when 97% of the unlinked genes of the donor parent should be eliminated. Sometimes backcrosses must be separated by periods of 'selfing' (self-fertilization), as selection for recessive genes is only possible in homozygous segregants. To combine the characters of several varieties breeders may resort to multiple crosses. For example, the F_1 of the cross A × B may be crossed with C, or the F_1 from the cross D × E. Very elaborate multiple, 'composite', or 'poly-crosses' have been arranged in various ways. For example, a range of varieties may be crossed, in pairs, in all possible combinations, and the plants may be grown in bulk, or the F_1 plants may then be crossed, then the F_2 plants, and so on until theoretically all the genes have had the opportunity to combine. A great many plants must be raised, to have a chance of obtaining a high proportion of the possible recombinations at least initially. By including male-sterile plants in the cross outcrossing, which may be enhanced by management, will continue as the cross is propagated 'in bulk' so giving further chances for genetic recombinations [6,61,166]. Composite crosses have been developed with two aims (1) to investigate the possibility of producing cheaply new, locally adapted varieties from as wide a variety of genotypes as possible and (2) to create readily maintained 'gene banks' in lines

Barley improvement 457

progressively better adapted to local conditions. However, it is too risky to maintain gene stocks exclusively in a bulk cross, as many characters may be lost [160]. In one experiment, in F_8 over 100 aberrant forms, including dwarfs 10–13 cm (4–5 in) high, 'accordion-rachis', shortly curly awns, hoods, forms with the grain-ends exposed, and stiff-strawed forms were detected. The variability released by wide crosses, mixing genes from 'foreign' forms particularly involving certain varieties such as Lyallpur, Everest and Multan, is more simply obtained than with mutation experiments, yet seems not to have been exploited. In California successes have been claimed for composite crosses but they have yet to be adequately evaluated elsewhere. The early composite crosses, (CC), were probably too broadly based. Others have been made using more promising parental forms. In the development of one cross (CC XXI), 6200, mainly six-rowed, spring barley varieties (8–10 seeds each) were planted in alternate rows with male-sterile stocks of locally well-adapted, high-yielding strains. Manual and natural crossing led to the production of 13.2 kg (29 lb) of F_3 seed, on the male–sterile plants [166]. Next year this was increased to 318 kg (700 lb). Eventually about 2300 female and 5200 male stocks were crossed. Natural crossing continued for 10–15 generations until the male-sterile gene disappeared. To exploit this genetic 'pool' in a different climatic area it might be necessary to cross extensively onto locally adapted cultivars. In England composite crosses are being created in attempts to find strains with good 'field-resistance' to diseases incorporating, for example, winter hardiness, short straw, major-gene disease resistance, and male sterility.

In Californian CC XXV, superior local strains were crossed with large-grained, strongly tillering Ethiopian varieties with resistance to many major diseases, including barley yellow dwarf virus, BYDV. In the original CC I, eleven varieties were crossed 32 ways and grown as a mixture, while CC II involved crossing 28 varieties from many areas of the world, in all possible ways [61,154].

12.8. Selection sequences applied to hybrid progenies

Usually when backcrossing is combined with a high selection pressure progeny are soon indistinguishable from the recurrent parent in traits other than those being transferred. However, if the transferred genes are pleiotropic or are associated with closely linked genes for other characters, then the process is complicated. For example, a high yield gene was associated with a closely linked gene for semi-smooth awns in a backcrossing programme [33]. Atlas 46 was derived from Atlas by a backcrossing programme that conferred resistance to mildew and scald. In the absence of disease these varieties gave the same yield. However, when disease was prevalent Atlas 46 outyielded Atlas by up to 17.6%, or as much as 27% at a single location [137].

When two parents are crossed and the F_1 and subsequent progenies are allowed to self-pollinate (self), segregation occurs and potentially each generation contains many more plants than the one before, differing from each other in genotype, and each being more nearly homozygous than its parent. By convention the plants of the F_9 generation are often taken to be 'practically homozygous'. It is impossible to grow sufficient plants in any generation to achieve all possible gene combinations. As a practical measure, the population sizes needed to give good chances of recovering desired genotypes have been estimated [56]. The breeder must find ways of recognizing the segregants that he wants or plants that will give rise to them. Following the inheritance of one or two traits gives little difficulty. However, in breeding plants for agriculture the entire range of 'characters' (from disease resistance, morphology, and growth pattern to grain yield, quality and physiology) must be considered. Selection in early generations may not be successful if segregation of characters occurs in subsequent generations, and the desired phenotype cannot be recovered in a stable, homozygous state. The expression of some characters is influenced by the environment so that genotypic and phenotypic variations are confounded, so confusing the selection process [160]. To have an efficient programme that begins in the early (segregating) generations, plants with as many desirable characters as possible should be selected, concentrating first on those traits that are inherited by a high proportion of a plant's offspring; characters that continue to segregate for many generation should be selected in later generations, when the frequency of segregation is less. Assessment of the 'heritability' of characters has often been attempted, but unfortunately the numerical concept is used in different ways. Furthermore, the heritability of a character is not necessarily the same in different populations or generations [119]. High heritability has been found for heading date, diastatic power, head density, awn roughness, disease resistance, kernel plumpness and plant height. Lower heritability has been attributed to average kernel weight, weight uniformity, fertility, bushel weight and erectness of head. Variable estimates of heritability have been made for diastatic power, barley extract and nitrogen content. The heritability of grain yield is usually low [9,22, 35,39,41,92,95,99,120,129,143]. Despite the low heritability it seems that selection in F_3 can lead to a small increase in yield in subsequent generations. Selecting for grain size is only successful when this is no longer mainly due to heterozygosity, i.e. by F_9 or F_{10} [69]. There are well known correlations between barley characters, so that grain nitrogen content is positively correlated with diastatic power and negatively correlated with extract [97,129,130]. Often it is not known if these correlations are genetic or physiological in origin, so making the outcome of any selection programme unpredictable. When complex groups of characters, such as 'malting quality', are being sought selection may be

Barley improvement

approached in three ways [160]: (i) by using *tandem selection*, the least efficient approach in terms of work and genetic advance, in which one selects until the desired level of one trait is achieved before beginning selection for the next; (ii) *independent culling levels*, the most common approach, where standards are set for the retention of several characteristics, and are applied successively, either to one population, or in succeeding seasons; (iii) *index of selection*, where a single culling is based on a 'score of merit' embracing several traits. The most rapid rate of advance should be achieved with this method.

The management of progenies is directed by the objectives to be achieved and the resources available. Generally the 'Pedigree method', or one of its variants, is used [15,64,93]. In the F_2 and subsequent generations the plants are grown 'spaced', and the breeder regularly inspects the field and scores the plants for tillering, lodging resistance, disease resistance, length of straw, number of grains per head, the numbers of fertile and infertile tillers, and so on. Records are kept, and the behaviour of 'families' is followed through successive generations. At each stage 'fine' plants are noted, unsuitable ones are rejected. Grain from 5% to 20% of the plants will be grown as short 'head rows' in the next generation so that plants from one parent are grown together, and lines from 'superior' families, which are still segregating, will be propagated. The variety eventually released will be the progeny of a group of closely related morphologically indistinguishable plants. Such an approach is laborious, needs a knowledgeable breeder, and fastidious 'book-keeping'. It has produced many of the best varieties. However space-planting unduly favours high tillering varieties that may not yield best in a pure stand. In this, or other sequences, selection may be aided by stressing the population e.g. by inducing a disease outbreak to allow the recognition of resistant plants, or by giving high doses of nitrogenous fertilizers or other treatments to precipitate lodging. By F_6 small-scale 'yield trials' and perhaps malting trials and grain analysis will be in use to help in deciding which families to retain.

An ingenious selection sequence is the 'F_2-progeny method' in which grain of about 10% of the F_2 plants are retained on the basis of F_2-plant height, straw strength, time of ripening and so on. Individual plant progenies are grown and harvested in bulk and the best are repeatedly grown in yield trials in, say, F_4, F_5 and F_6, under near agricultural conditions. Grains from the best families are space-planted and single 'near-homozygous' plant selections are made and are propagated for trial. The method gives evaluations of the yielding capacities of the 'best' families, and the individual selections should outyield the unselected family. It is simpler to use than the pedigree method, but may take longer to produce an agricultural variety [15,93]. Another more laborious but potentially faster approach is the 'pedigree-trial' method. Families from each plant are grown together and single plant selections are made in each generation, beginning in F_2. Superior plants are selected from the best space-planted family in each generation, to provide seed to be grown on, while grain from the rest of the family is grown in bulk at normal 'agricultural' plant densities in a small yield trial, the produce of which is analysed, but is discarded. Thus progressively more reliable estimates of yield (limited in accuracy by the amount of seed grain available) are gained for each pedigree as the genotypes become more uniform and while selection is going on. Plants are grown spaced for selection purposes, while yield trials are made under appropriate, closely spaced agricultural conditions. Inevitably with all selection schemes some unrecognized superior genotypes are discarded.

'Mass' or 'bulk methods' of propagation are much less laborious to operate, and initially require less skill on the breeders' part [15,64,93]. In the simplest case the F_2 seed is grown in bulk in succeeding generations until F_5 to F_8. Single plant selections are then made. This is time-saving as the material evaluated is a mixture of 'near-homozygous' forms. The plants that survive in the mixture are likely to be vigorous, locally well adapted to climate and agricultural practices, and good competitors. However the method is slow, and many potentially useful forms, such as short-strawed or weakly tillering types, may be eliminated. Many variants of this method are used – for example only plants resistant to lodging or a disease outbreak or showing some other desired trait may be propagated – that is 'mass selection' may be applied at some stages. Sieving to select large grains, or selecting fully fertile ears with at least a certain number of grains or having a certain weight are other simple selection techniques. In addition to composite crosses there are other examples where prolonged propagation of a hybrid population in bulk has given rise to useful commercial varieties. For example, early attempts to recover useful varieties from a particular 'wide' cross were unsuccessful, but after 20 years propagation in bulk two commercially useful varieties were isolated [28].

12.9. Competition and 'natural selection' in barley

When barley varieties are grown mixed together, some strains flourish and others disappear. Clearly, in connection with barley grown in bulk either in the course of normal breeding, or as a composite cross, one must know if superior genotypes are likely to survive. When the leafy, taller, and earlier ripening Goldthorpe was grown in alternate rows with Archer selections, Goldthorpe 'outyielded the Archer by 50%, although when grown separately Archer had a yield advantage of 10%' [12]. Comparable results were obtained with other pairs of varieties [12,76], and with near-isogenic two-row and six-row lines [168]. Numerous trials have failed to correlate competitive ability with plant height, time of maturity, seed size, growing habit (erect or prostrate), spring or winter habit, grain yield, hulledness or nakedness, two- and six-rowedness, plant vigour, planting distance or soil fertility as determined in single-variety stands. Heritability of 'competitiveness' is low; diploids are often better competitors than tetraploids. F_1 plants are often poor competitors despite their hybrid vigour [133,134]. Plants of one variety grown from large seeds are better competitors than those from small seeds [68,135]. However, competition does not automatically eliminate superior agricultural types, as seen when 17 mutant or near-isogenic strains of Hannchen were grown mixed together [155]. In a more complex experiment equal numbers of seeds of 12 varieties (of which only two were not morphologically distinct) were mixed and sown [60]. The harvested, mixed seed was re-sown at different locations during 4 to 12 years. Seed samples from each harvest were 'space planted' and the different types were scored. Some varieties rapidly became extinct, and others came to dominate the populations. The dominant varieties were different at different stations. As expected some varieties initially increased at the expense of the poor competitors, but subsequently declined. At some few stations a mixture of forms survived. In every case the survivors were well adapted to the local climate, and were therefore locally high yielding. When a mixture of 4 locally well adapted barley cultivars were grown together for 16 years two varieties, Atlas (88% plants) and Club Mariout (10.5%) survived, although susceptible to leaf diseases, while Hero (0.7%) and Vaughan (0.4%) were nearly extinct. Vaughan had the highest yield in pure stands, out-yielding Atlas by 7% [148]. Neither relative resistance to scald, net-blotch, and mildew, nor heading date nor height explained these results. From this and other trials it seems that disease resistance does not have a high selective advantage. The onset of jointing coincided with the onset of competition between Atlas and Vaughan, before noticeable shading occurred, when Atlas was producing many roots from the crown. The roots of Vaughan developed more slowly [89]. Competition was via the roots for water and not for nutrients or light [29,65]. The original conclusion was that although Vaughan usually outyielded Atlas it had to have some intangible adverse characteristic since it had never occupied more than 5% of the Californian acreage and that the competition experiment backed the farmer's choice. As Vaughan had a thin, readily frayed husk, it was not used for malting and had poor qualities when rolled for feed [59]. Competition for water is unlikely to select for tough husks; the association between farmers' preference for Atlas and its survival in a mixture is probably accidental. Incidentally, Atlas × Vaughan crosses have yielded superior new varieties. When a mixture of 6000 varieties was grown under conditions of severe stress, involving annual late-planting, the yield increased roughly linearly by 57% in 6 years so efficient selection of well-adapted, better yielding genotypes was occurring [118].

Different genotypes survive in different environments [60,139]. When seed from 105 crosses, a complete diallele series, was bulked and grown for some years at nine locations (in Canada, U.S.A., and Norway) plant height, seed size, collar type and various other morphological characters were little changed but black grains declined rapidly at most areas as did two-rowed ears everywhere except Norway [82]. Two-rowed types predominate in N. Europe and are well adapted there. In contrast to some other studies no association was found between smooth awns and yield. When seed samples from all locations were grown in one place for two successive years it was clear that the yields differed. 'Selection' had favoured locally adapted genotypes at each site.

Analysis of competition between different cultivars shows that they compete for the same 'ecological space', the intensity of competition varying with season, location and closeness of planting. Mixtures never outyielded the best parent grown in pure stand, an argument against 'mixed' cultivars [135]. In the one case when it occurred the survival of a 'poor competitor' in a mixture was traced to its carrying barley stripe mosaic virus (BSMV) to which it was more tolerant than the competing variety which became infected when the two were grown together [135]. 'Multi-line' varieties may tend to homogeneity as one component line outgrows the rest [117]. This need not happen. Land races of barley were not uniform, and in some forms of primitive agriculture wheat and barley are grown mixed together, with neither displacing the other. Although mixtures of 'compatible' lines may yield more consistently than pure lines in a variable environment, and show an advantage over the mean parent yield, yields of mixtures never exceed those of the best parents [87,135].

Special breeding techniques may be needed to create suitable multi-line varieties having, for example, numerous sources of disease resistance in different plants with otherwise near-isogenic backgrounds. The changes that occur in bulk-propagated populations from composite crosses have excited much interest because, whether or not the crosses readily give rise to new agricultural varieties, understanding the changes that occur should help many breeding programmes [7,11,61,79,82,155,166]. Only in California have composite crosses been grown for an adequate (20 or more) number of generations for some studies. Some characters e.g. hoods, two-rowed ears, black or naked grains, *deficiens* forms, and fragile ears were rapidly eliminated. There was a trend towards tall, late-heading, wide eared, rough-awned, six-rowed forms with lax, waxy spikes and large blue grains. Forms with colourless auricles, purple lemmae, long rachilla hairs, and long wide heavy kernels were also frequent. The plants of the F_{30} generation of CC II were heterozygous for many characters. Heading time spread over 29 days, the range of lengths of 10 spike intervals was 17–44 mm, height 88.9–132.1 cm (35–52 in), 20–90 kernels occurred on each ear and the thousand-grain weight range was 30.7–82.8g [11]. Selection for kernel weight altered the frequency of occurrence of other characters so genetic linkage or physiological compensation between phenotype characters was occurring. At F_{35} plants were *still* heterozygous [152]. The yield of CC II was initially low, but by F_{16} its yield about equalled that of Atlas and thereafter exceeded it. In the F_{12} generation 356 selections were made, but none outyielded Atlas and only 17 were resistant to scald. In F_{20} 50 plants were selected of which two gave promising lines. In F_{24} 66 plants were selected of which ten gave promising lines. Clearly with the passage of time only well adapted forms survived, of which some were high yielding types. By 1956 19 varieties of barley grown in California had been selected from CC II. While 'non-exploitive selection' is simple, it is protracted and it is not known how well it would work elsewhere. Composite Cross II itself was used as a feed barley, where its heterogeneity was not important. By F_{35} only about 8000 plants of CC II were being propagated each year, too few to maintain the genetic variability of the original stock. The parents of CC II were chosen to be as diverse as possible. By choosing parental forms more suitable for agricultural use, composite crosses giving results in a shorter time are produced. There is a considerable body of opinion that as good, or better, new varieties would be produced in a shorter time by conventional breeding methods.

A suggested explanation for the heterozygosity of F_{19}–F_{35} in CC II is that selection favours heterozygous plants in populations with limited (1–2%) outcrossing [7]. However, in other circumstances F_1 plants are often poor competitors. The preferential loss of some genes confirms that composite crosses are not adequate gene stores [79,160].

12.10. Breeding for quality [14,59,111]

Barley varieties used for forage and grain production should tiller and regrow freely after cutting or grazing to ensure a good yield of grain. Often smooth awned or hooded forms are preferred for direct feeding. Grain to be rolled should have a sound husk but a low husk percentage. For human food, and probably for feeding other monogastric animals, naked barley is preferable. Breeding to increase the oil content of grain has started. High starch contents are selected by choosing 'low-nitrogen' cultivars. Genes for high amylose or high amylopectin (waxy) starches are available (Chapter 4). It is possible to select for barleys with inherently high or low contents of different mineral elements [121], or tolerance to aluminium or acid soils, or response to other minerals. In selenium-deficient areas this element might be supplied to livestock in feed compounded from varieties best able to accumulate it. Conversely, in high-selenium areas plants selecting against this element would be preferred.

For animal feed barley is deficient in protein and for monogastric animals the quality is poor. Efforts are being made to overcome these important defects. Various high protein forms are known [83,100,102]. To screen breeding samples many small samples must be assayed for total nitrogen ($N \times 6.25$ = 'crude protein') and lysine content, or total basic amino acids. Rapid methods for estimating nitrogen include automated chemical techniques and direct nuclear reactions on individual grains that can afterwards be grown. The content of basic amino acids, including lysine, can be estimated by a dye-binding method. Lysine can be estimated specifically in hydrolysates by an automated method utilizing the enzyme lysine decarboxylase, or by column chromatographic amino acid analysis. High lysine is associated with high levels of total soluble amino-acids, which are easily estimated, or low prolamine levels [98,125]. Strains and mutants are known in which the grain nitrogen varies from about 1.2% to 3%, but these values are highly dependent on cultural conditions. In general the nitrogen content of a crop declines with increasing yield (Chapter 7) [101]. To be acceptable high nitrogen lines must yield well, so that they must break the 'usual' yield/grain nitrogen content correlation. Mutants that do this are known [101]. Genes for high-nitrogen and high lysine have yet to be incorporated into acceptable agricultural varieties [102,103]. The economics of using high-lysine barleys must be considered relative to the use of other high quality protein sources, or the use of supplemental lysine. Similarly the continued use of malt depends, in part, on the relative economics of producing enzymes by malting and from microbes. Studies with isogenic lines suggest that two-rowed barleys have inherently greater nitrogen contents than six-rowed forms [159]. Grains with large embryos and thick aleurones would be expected to have higher contents of protein [52]. The first high-lysine gene giving 20–30% more lysine was located in a low-yielding Ethiopian variety (Hiproly) that also carried a gene for high protein content and another for protein adherence to the starch grains in the endosperm [102,103]. Now several high-lysine mutants are known, but usually their grain yield is poor [26]. However, Risø mutant 1508, induced by ethyleneimine, has a grain yield only 10% less than its progenitor, Bomi, a similar nitrogen content but a greatly altered protein and amino acid pattern and an enhanced lysine content (Table 12.2) [77,78]. The non-protein nitrogen content is also enhanced. The Risø 1508 character is due to a single, recessive gene. This gene appears to suppress the formation of some hordeins in the grain [108].

The wide range of quality requirements for malting grain sets complex problems of selection complicated by the numerous positive and negative correlations that occur between grain characteristics and between these and agronomic characters (Chapter 15) [97,130,147]. Maltsters require grain with resistance to grain splitting; a sound husk, a low husk content;

Table 12.2 A comparison of Risø mutant 1508 with Bomi ((a) Ingversen et al. [77,78], (b) Mertz et al. [98]).

	Bomi	Risø 1508
(a) Yield (%)	100	90
(a) Nitrogen (% seed)	1.72	1.75
(a) Lysine (% protein)	3.64–3.82	5.18–5.42
(a) Albumins + Globulins (% protein)	27	46
(a) Prolamine (% protein)	29	9
(a) Glutelin (% protein)	39	39
(b) Protein (% seed)	19.8	19.4
(b) Lysine (% protein)	3.1	5.3
(b) Free amino acids (μm leucine equivalents/100 mg protein)	19.5	68.0

a good rate of water uptake; a high content of carbohydrate (starch) convertible to extract on mashing; moderate nitrogen contents; the ability to germinate rapidly and evenly, and to develop a sufficient enzyme complement in a short time, yet have a low malting loss; a low level of β-glucan or a high level of β-glucanase, resistance to pregermination, yet dormancy that can readily be overcome to allow malting. Prediction equations based on determinations of grain nitrogen and insoluble carbohydrate, and germination tests help. The final evaluation of grain is based on micromalting and pilot malting trials. Micromalting tests can be made on as little as 5 g of grain and are useful in breeding programmes. The 'best' type of grain varies with the kind of malt to be made [31,163–165]. The major criterion of quality in micromalted grain is the yield of hot water extract (HWE). This is dependent on the way each sample is grown, and on its husk and nitrogen contents, so for breeding purposes relative extracts (i.e. the observed HWE *minus* the predicted HWE) may be the selection criterion [163,165]. Thus sub-samples of a particular lot of barley are steeped and germinated for various times, and the extracts obtained are compared with the predicted extract (Fig. 12.3; Chapter 15). To aid the evaluation process, samples with nitrogen contents ranging from 1% to 3% may be obtained by growing plants on gravel irrigated with different nutrient solutions [163].

In a representative breeding programme the malting variety Kinver was crossed with the high-yielding, strong-strawed feed variety Titan. These varieties also have other complementary characters. From 13 F_1 plants 3500 F_2 plants were grown. The plants were exposed to three major fungal diseases, and 139 F_2 plants were reserved. From the seed of each, 10 F_3 plants were subjected to further selection based on chemical and malting tests. Finally a malting-quality line was selected that out-yielded other cultivars by 15–20% [91]. Similar approaches are common –

466   Barley

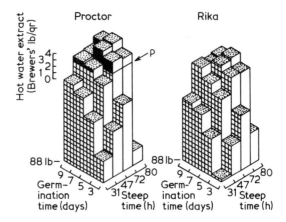

Fig. 12.3 The hot water extracts obtained from two barleys after steeping and germinating for various periods.
The values exceeding the predicted hot water extract are shown in black. For an explanation of the terms used see Chapter 15 (after Whitehouse and Whitmore [165]).

for example in trying to introduce malting qualities from European barleys into strains adapted to the Negev, or S. Australia [31,143]. In Europe, with the highest barley yields in the world, detailed and complex evaluations of breeding stocks are used [90,163]. An attempt has been made to define the inheritance of the rate and extent of germination, which are important to maltsters [36]. Of 28 F_1 'hybrid' barleys only seven had acceptable malting quality; heterosis was of minor significance for malting quality [123].

12.11. Some other objectives in breeding [59,111]

Where barley is grown as a 'smother crop' to depress weed infestations, tall, vigorous, strongly tillering forms with wide, floppy leaves are favoured. In contrast, when the crop is to be used to nurse an undersown crop low tillering and small leaves are favoured, together with early maturity that will allow the undersown legumes or grass to grow away. The life cycle must be adjusted to the climate, temperatures, day length, drought resistance and duration of the growing season. Drought tolerant types sometimes ripen with the ears still in the boot. The plants may need to tolerate or be resistant to prevalent pests and diseases, salinity, extremes of soil pH and excesses of particular minerals, such as aluminium. Cultivars that can germinate in saline soils are easily selected [124]. The straw must be suited for the type of harvesting used. Combine harvesters require short, stiff straw, minimal lodging and no ear shedding or shattering. However, the grain should thresh 'clean', without adherent awns or pieces of rachis. It is advantageous if all ears, whatever tiller or plant they are on, form and ripen together. Cultivars vary in the synchrony of ear emergence, and other traits, and so more synchronous types might usefully be selected [109].

Winter forms vary in the degree of hardiness required. In N. Europe, autumn-sown spring cultivars often survive the winter, while in central continental areas the greatest 'hardiness' may be inadequate to prevent winter killing. In England Maris Otter represents a compromise between malting quality and moderate hardiness. In Continental Europe the much greater hardiness of Tschermak's two-row is essential; this variety is, at best, semi-hardy in many parts of N. America [163]. Hardiness is a complex of properties. Its effectiveness depends on the snow-cover, the rate at which low temperatures are attained, their degree and duration and the likely occurrence of diseases such as 'snow mould'. Sowing dates and depths have a considerable effect on barley winterhardiness. Cold nurseries or artificial freezing chambers are used as aids for selection [19]. It is often easy to recognize 'hardy' and 'tender' strains, but very difficult to recognize degrees of hardiness. Changes in electrical conductivity after freezing may provide a simple measure of degrees of cold damage (Chapter 6).

Resistance to pests or pathogens may be due to major genes, giving 'vertical', race specific gene resistance, or to sets of polygenes which confer 'horizontal', or 'field' resistance (Chapter 9). Major genes are most easily handled, for example, by back-crossing programmes. New genes must continually be found, evaluated, then incorporated into useful varieties, as new pathotypes develop. Field resistance is hard to detect or maintain in breeding programmes that also involve effective major genes so it may be lost. Because of the polygenic nature of field resistance this trait may show transgressive segregation in crosses. A third possible approach would be to set up 'multi-lines' by back-crossing, to give a mixture of similar forms differing in the nature of their resistance to a particular disease. Such an approach should slow down the onset of disease outbreaks (Chapter 9) [74,163]. Genes for disease resistance are located by exposing plant populations to heavy inocula and breeding from tolerant or resistant plants. The methods used must be chosen with care. Thus, if barley ears are inoculated with the spores of loose smut (*Ustilago nuda*), many field-resistant varieties become diseased because their 'pseudo-resistance' is by-passed because their closed-flowering habit normally excludes spores [111].

Many changes have occurred in the mildew susceptibility of British barleys [74,163]. Many major plant genes for resistance and many mildew pathotypes are now known. Older varieties, such as Spratt–Archer, seemed to have been 'in balance' with foliar diseases, perhaps because of field resistance, or because they were multi-lines, or because

468 Barley

only relatively small acreages were grown with little fertilizer and strict adherence to crop rotations. In Proctor (1953–1969) the situation deteriorated a little each year, but it is not clear if this was 'real' or only 'apparent' deterioration, the possible 'real' causes being greater applications of nitrogenous fertilizers and larger spore inocula from the greater barley acreage. A major gene for resistance, *Ml-g*, from a Saarland land race, was introduced into England, in the variety Union. In 1961 it conferred excellent resistance. By 1964 the gene was present in widely grown varieties; pathotypes able to overcome its resistance then became widespread. Another major gene, *Ml – a-6*, originally from *Hordeum spontaneum nigrum* gave complete resistance in breeders trials, but this gave way from 1964 onwards when the gene was distributed in widely grown varieties like Maris Badger (released 1963). Impala (released 1964) carried *Ml – a-6* and *Ml-g*, but pathotypes able to attack it were widespread by 1967. It appears that about 15 years of breeding work with a major gene are 'beaten' by the pathogen within 3 years. Thus, at best, major genes confer only short-term disease control. Attempts are now being made to find ways to exploit 'field' resistance in the hope that it will last better.

12.12. Breeding for higher yields

Yield may be taken as the cash-return/unit area, when special grain qualities and the value of straw must be taken into account, or grain weight/unit area. In the latter case 'naked' grains are at an apparent disadvantage. Attempts to increase yield can be divided into two groups; (i) those that aim to correct specific weaknesses in a particular plant type and (ii) those that aim to produce an inherently higher yielding form. Many attributes such as resistance to disease or lodging, and the ability to resist damage from pesticides and weedicides, may be handled in the first way. The production of widely adaptable varieties that are able to withstand wide climatic fluctuations, is a frequent objective. Transgressive segregation for adaptability occurs [37]. It is simpler to test for adaptability at a range of sites, than wait for numerous seasons to pass at one location. By carrying out selections on a segregating population grown at alternate locations in successive years widely adaptable lines were produced [132]. Most approaches to improved adaptability in plants are empirical.

Selection for grain yield is based on comparative yield trials. Selection by appearance is unreliable. At first it seems that better yielding types might be selected with regard to the 'components of yield', i.e. mean weight/grain, number of grains/ear, and numbers of ears/unit area. The number of ears/unit area is dependent on the seed rate and the number of tillers/plant (Chapters 6 and 7). Surplus tillering capacity may be disadvantageous in which case varieties with 'uniculm' or very limited tillering capacity should be chosen. The regular production of a high proportion of fertile tillers (i.e. few or no wasteful infertile tillers) is a definite advantage. Many successful breeders have selected for a high tillering capacity. However, varieties may achieve high yields by different combinations of yield components, which in any case vary with the cultural conditions [1,21,136]. Attempts to select for high yield by selecting for particular yield components have often been unsuccessful since 'compensation' occurs; for example, in different years or with different seed rates or fertilizer applications, lines may produce different numbers of ears, having more or fewer and lighter or heavier grains [104,146]. Selection for large kernel size often shows negative responses of the other yield components [119]. The two explanations advanced for the reductions in yield found in a population in which selection was for high numbers of kernels per head were based: (i) on the proposed existence of genetic optima for the number of kernels/head and number of heads varying with different environments, and (ii) on an oscillating response of yield components to selection due to their sequential development in time, at the expense of limited resources [3,119]. The oscillatory model suggests that the breeding aim should be to increase the flow of endogenous resources to the grain and raise the genetic 'ceiling' limiting response to excess resources, i.e. increase the storage capacity of the grain.

Newer varieties respond better than older types to added nitrogenous fertilizers, by making more ears per unit area; also they lodge less [136]. Old and new varieties make about the same recoverable yield (grain + straw) in pot experiments but newer varieties produce relatively more grain and less straw. Such observations suggest that short-strawed varieties may be inherently able to yield more grain as well as being more lodging resistant, and being easier to harvest [18,71]. Possibly at the peak 'assimilation time' more assimilate is produced than can be stored in the grain (Chapter 6) [71]. If this is so the search for strains with higher photosynthetic rates may be of little value [88], unless larger 'sinks' for the photosynthate (more or larger grains; or more ears) can be produced.

The need for clearer breeding objectives has led to the construction of 'ideotypes' – morphological models of plants expected to give maximal yields [27,71]. Correct models will vary between locations – for example hoods may be no disadvantage at high altitudes [156]; six-rowed ears may not be universally preferable [163]. In some situations a factor associated with semi-smooth awns gives a substantial yield advantage [33]. In a 'weed-free' environment under favourable conditions an ideal barley type may be fully adapted to the local climate with ultra short, stiff, erect straw, able to produce synchronously two to three bold, fertile tillers, (and no infertile tillers) with many fertile florets per ear. Perhaps the ear should be six-rowed and lax, with all the grains large and carrying large awns or perhaps big, triple awns and large awned glumes to maximize

470 Barley

the photosynthetic area. A large flag-leaf would be desirable. The other leaves should be small and upright. Ear formation should be complete and anthesis should occur early in the growing season; subsequently the flag leaf and ear should remain photosynthetically active for as long as possible. The variety would have to be resistant to pests and diseases, and would have to respond well to fertilizer applications. It should give a relatively good yield at low as well as high fertility levels. Cultivars vary widely in their responses to different fertilizer doses [43]. Establishment after sowing should be rapid, possibly autumn-sowing would be better than spring-sowing, but in this case sufficient tillering should occur to make good thinning caused by winter-killing.

There seems to be a need for backcrossing programmes to transfer potentially useful genes from all sources into the locally most successful, fully adapted genotypes. Only then can the local value of the resulting phenotypes be fully evaluated and logical decisions be made as to the probable best combinations to test further.

12.13. The quantitative evaluation of parents

The choice of parents for a cross is often made intuitively; it is largely luck if a cross 'nicks' well, and superior genotypes are recovered from the progeny [15,163]. In one way of evaluating a pair of varieties as parents a cross is made, the F_1 plants are allowed to 'self' and the characteristics of the bulked, unselected F_2 and F_3 progeny are evaluated. The F_1 generation is not used because of heterosis, which will be less in subsequent generations. Supposedly the 'mean values' obtained from the unselected bulk are a measure of the value of the cross [142]. A more informative approach is to establish a large number of lines from randomly selected F_2 and F_3 (or later generation) plants, evaluate these separately and note the *range* of yields and plant characteristics to test whether useful recombinations or transgressive segregation has occurred. A sophisticated approach involves the use of diallele crosses [15]. For example 28 varieties were crossed in all possible ways and progenies were evaluated in comparison with the parents and lines drawn from mixed bulks in the generations F_2-F_7 [61]. Some cultivars were markedly better parents in crosses than others. Some superior parents were inferior varieties under the conditions of the test. It was concluded that low yielding bulks would rarely give rise to high yielding selections and so might be discarded. Biometrical methods of analysis have been applied to diallele crosses, with the gain of much information [58,67,80,81]. The crosses are made, and the parent lines, F_1 and F_2 bulks are grown in yield trials. The results of the evaluations are grouped in numerical arrays, each array corresponding to one parent and the offspring from all its crosses. Statistical and graphical analysis detects superior parents and parental combinations, contributions of allelic and non-allelic gene interactions, dominance, over dominance, or linkage, to yield. Such analyses suggest that heterosis may always be due to non-allelic interactions, so any high yield that can be gained by producing 'hybrid seed barley' should also be attainable more simply by selecting the appropriate homozygous segregants from a cross [1,2,48,80,170]. The analysis of a 15-parent, 105-cross diallele set, and other crosses, demonstrated the use of this approach [4,142]. However, it is necessary to analyse sub-characters, such as components of yield, before a gain can be attributed to genetic interaction, or a multiplication of characters [163]. Such methods of evaluation are laborious.

A 'predictive' approach for choosing parents with the best chance of creating desired recombinants has been proposed [48,49,51]. Each character is given a numerical vectorial value (partly a subjective process). The parents are chosen on the basis of a vector prediction that their progeny, as an unselected population, will have a mean 'vectorial value', close to that desired, i.e. a hypothetical but realistically practical ideal plant. The mathematical concepts are simplified with the aid of a three-dimensional, spherical model. The assumptions used include 'additivity', that the 'mid-parent' (average) values for each trait is a good prediction of the mean value of this trait in the bulked, unselected progeny, and that by using the components of complex traits, e.g. yield, some of the complexities due to gene interactions are overcome. The approach appears promising, but is complex and partly subjective. Results seem to confirm that overdominance is not a major factor in yield heterosis, and that apparent epistatic effects on yield are frequently due to interactions of the yield components [49]. Probably the optimal ratios between the 'yield components' change with increasing yields [50]. The model suggests rules to use when selecting for yield.

Another proposal is to use 'canonical analysis' as an aid [127,164]. This mathematical approach attempts to take account of all the biometrical data available and to use it in choosing the parents that will give hybrid populations close to the breeding target. It can be used to define the ranges into which the characters of selected plants must fit. The use of a computer is essential, and it is laborious to collect all the essential numerical data. The choice of lines to be propagated may be simplified by giving 'economic value' weightings to the different characters.

12.14. 'Hybrid' barley

The striking heterosis, or 'hybrid vigour', sometimes shown by F_1 plants, and the large gains made with maize *(Zea mays)* has led to attempts to create methods of growing 'hybrid' barley from F_1 seed on the agricultural scale. Heterosis varies in degree, and may be shown by various plant attributes, such as size, tillering capacity or grain size [24,25]. The value of this approach is not universally accepted, since 'hybrid vigour' should be 'fixable' in homozygous plants. However agricultural yield increases in barley of 15–35% and even 50% over the best parent have been claimed when F_1 seed was grown [151,152]. Unfortunately reports have sometimes been based on 'heterosis' relative to mid-parent values, using space-planted seeds, where the extra tillering capacity of the F_1 plant shows to advantage, which it does not do under normal agricultural conditions [66,140]. Thus large-scale agricultural trials are indispensable for evaluating this idea [131,158]. In one set of trials heterosis for forage was from 8 to 31% and for grain yields, 8–37%. Not all crosses show heterosis. Surprisingly heterozygotes between a mutant and its parent are reported to have outyielded the parent [53]. The extent of heterosis varies between years [170].

Male-sterile plants are needed to act as female parents for producing F_1 hybrid seed on a farming scale [32]. Numerous nuclear genes for male sterility, which occur at about 19 loci, are now known. Cytoplasmic male-sterility has been found in crosses with wild barley grasses and in composite cross CC XXI but cannot yet be used, as restorer genes are not known [40,154]. The use of pistilloid mutants as female parents has also been suggested.

Male-sterile plants must be grown near pollen donors to obtain seed set. It is essential to be able to harvest the F_1 seed separately, or to separate it from seed from the self-fertilized pollen-donor plants. In addition the male-sterile and pollen-donor lines must be maintained. The farmer would need to buy fresh F_1 seed from the breeder in each generation since its production would be complex. The cost would be relatively high, and yield would have to be excellent (probably at least 25% above the average in the U.K.) to cover it. Seed from F_1 plants can be suitable for brewing but it is more likely, because of its heterogeneity, to be used as feed [73,152]. The choice of parents can probably only be made by trial [46,66]. A screening method for choosing cereal parents based on the finding that mixtures of mitochondria from 'good' parents have enhanced oxidative phosphorylation, has been proposed, but the findings are not universally repeatable [30,94]. In California, Arizona, and India adequate seed set occurs on wind-pollinated male-sterile plants but elsewhere seed sets are too low at present [66,131,167,169]. To achieve pollination donor plants must exsert their anthers to shed their pollen simultaneously with the gaping of the male-sterile parents. This requirement may be difficult to meet; widely spaced plants, as would be needed, tiller freely and so heads mature over a relatively long period.

The first scheme for producing hybrid barley, i.e. F_1 seed, on an agricultural scale involved linking a recessive male sterile gene *(ms)* to a recessive gene for resistance to a phytocide, such as DDT *(ddt)* [66]. The gene combination $ms \cdot ddt$, closely linked to minimize crossing over, was to be incorporated into the female line by backcrossing. The final backcross, BCF_1, would yield heterozygous progeny $(Ms \cdot Ddt)/(ms \cdot ddt)$. These were to be selfed and the seed (generation BCF_2) would have the constitution shown.

$$1. \frac{Ms\ Ddt}{Ms\ Ddt} : 2. \frac{Ms\ Ddt}{ms\ ddt} : 1. \frac{ms\ ddt}{ms\ ddt} \text{ (resistant; male sterile)}$$

Spraying the BCF_2 rows with the phytocide (DDT) should leave only male-sterile plants alive. Pollen donor (DDT-resistant) and male-sterile rows were to have been interspersed; F_1 seed would have been either harvested separately or otherwise separated from the seed bulk. The male-sterile stock was to be propagated by growing BCF_1 and BCF_2 seeds in alternate rows, and havesting the DDT-sprayed BCF_2 rows; seeds of the constitutions $(Ms \cdot ddt)/(ms \cdot ddt)$ and $(ms \cdot ddt)/(ms \cdot ddt)$ would have been obtained. The scheme failed on three counts: (i) the crossover frequency between the genes for male sterility and phytocide resistance was too high; (ii) DDT is not a reliable phytocide; (iii) the general use of DDT as a phytocide is unacceptable because of the toxic residues it leaves. A second group of schemes, based on the use of balanced tertiary trisomic stocks (BTT) has been proposed, and one has been shown to be practicable (Chapter 11) [116]. Seed of the first 'hybrid' Hembar has been released [169]. The schemes rely on sophisticated genetic stocks; not all require the use of phytocides. The requirements for one scheme are [169]: (a) a trisomic stock, in which the extra interchange chromosome will not be transmitted by the pollen; (b) a suitable recessive male-sterile gene, and (c) an informational gene (colour; albino; shrunken endosperm, lethal, etc) that will allow mixed seed stocks to be sorted. The interchange chromosome must have the dominant genes for 'information' (e.g. green, A) and male sterility ms (fertility, Ms) closely linked to the breakpoint. The recessive genes, ms and a occur twice in the normal diploid chromosome complement, so the BTT, balanced tertiary trisomic, plant is $ms\ a/ms\ a\ Ms\star\ A$. Pollen from such plants is exclusively $ms\ a$, and the eggs are $ms \cdot a$, and $ms \cdot a\ MsA\star$ ($\star$ denoting the extra chromosome) in the ratio 70:30. This unexpected ratio is due to chromosomal irregularities and cell abortion. Rarely $ms\ a/ms\ a$ eggs occur – these give rise to albino plants that die. On selfing the $ms\ a/ms\ a$ plants (70%) give rise to albinos which die. The green, fertile BTT plants $ms\ a/ms\ a\ Ms\ A\star$ (30%) renew and increase the parental form, which is a source of homogeneous $ms\ a$ pollen. A second stock is created by producing a diploid $ms\ A/ms\ a$ stock by backcrossing from the BTT stock. Crossing these stocks $(ms\ A/ms\ a$ with $ms\ a/ms\ a\ Ms\ A\star)$ to give a homogeneous population of green, male-sterile plants $ms\ A/ms\ a$. These serve as female parents in the production of the F_1 seed. By arranging, for example, for the male-sterile plants to have white lemmae and the pollen donor to have

474 Barley

black lemmae it is possible to harvest seed in bulk and separate the F_1 (white) from the selfed pollen donor grain (black) by electronic sorting. The complexities of such a scheme are apparent. In Britain the chance of 'hybrid' barleys coming into general use seems to be small.

12.15. Trial procedures [14,15,64,163]

Evaluation trials are needed throughout breeding programmes. Initially a few plants are grown mixed with other segregants from a cross. If they are to be inspected individually they will be 'space-planted', that is sown at a low seed rate in widely separated rows. Later small bulks will be grown for yield trials. Initially these are limited by the quantities of seed available. Later, when seed stocks are larger, trials will be replicated at each location, and will be carried out at several locations over several years to evaluate the effects of variations in soil and climate. The agricultural techniques used (seed rate, fertilizers, herbicides, etc) should be varied to represent those used locally. Irrigation may be used, but it is hard to apply evenly, and adds to the variability of trial results [10]. The number of trials needed will depend on local conditions. For instance it was concluded that in Minnesota each cultivar should be evaluated with at least three replicates at six locations for 3 years [122].

Arguments concerning the best arrangement of test plots of different varieties spurred on the development of statistics, as the relative merits of regular systematic lattices, latin squares and complete and incomplete random arrangements, and the required numbers of control (check) and test variety plots were debated between Gosset ('Student') and Fisher. Complete and incomplete randomized trials are now usually used, but so are novel arrangements [75]. Early this century Beaven used hand-sown, hand harvested small plots about 120cm × 120cm (4ft × 4ft) of which the central 90cm × 90cm (about 3ft × 3ft) was harvested, often involving eight cultivars with 20 replications, while Hunter employed comparisons between 15 to 35 paired plots [163]. Yield differences of 3% were reliably detected. Small seed-drills and harvesting machines are now used to mimic agricultural practice; this, the cost of labour, and the larger number of comparisons to be made have reduced the number of replications used in trials, with a consequent loss in sensitivity. Frequent replicate plots of control varieties throughout a trial area should be used to correct for variations in soil fertility and microclimate. The modern breeder often cannot detect 5% yield differences at the 5% probability level [163].

The shape of trial plots is dictated by the machinery used to cultivate and harvest them. Barley growing next to paths, in the absence of competition, shows an 'edge effect'; it grows and tillers profusely. Consequently where seed is available 'guard' rows are grown which are discarded at harvest, and the yield of the middle rows (excluding the ends) is noted.

Barley improvement 475

A frequent arrangement is a 'rod-row trial' in which 3–7 rows of seed, about 5.5 m (18 ft) long are planted, and of which the central about 4.9 m (16 ft) of all except the outer 'guard' rows are harvested [15]. Where minimal quantities of spring barley seed are available it is suggested that two rows should be grown between guard rows of winter wheat which, in the absence of vernalization, will not set seed [122]. Fertilizers improve the precision of small-scale yield trials [8]. Among the many considerations are, for example, whether the seeds should be sown in equal numbers, weights or volumes per row, and the source of the seeds [34,144]. Large seeds of a variety often outyield small seeds. Seeds of varieties to be compared should be produced in one season, at one locality and possibly should be graded for size before use in comparative trials (Chapter 7) [84, 135]. The ways in which the results of trials should be recorded and computed are subject to disagreements. With access to computing facilities the results of trials can be stored and submitted to many numerical comparisons [110]. New varieties are also evaluated in state, national, and international trials, being compared against two or more standard varieties. In the United Kingdom comparative trials are organized annually for spring barley and winter barley by the National Institute of Agricultural Botany (N.I.A.B.). Each variety is grown for a number of years on replicated plots at a range of locations. Varieties are scored for agronomic characteristics, disease resistance, time of ripening, yield and grain quality. Preferred varieties are incorporated annually in a recommended list. In western Europe malting varieties are subject to annual, international comparative trials.

12.16. The multiplication of seed [15,57,163]

A breeder's work involves the production of a sufficient supply of pure seed of a new variety, that matches the material that underwent trials. The process resembles the routine multiplication of pure seed except that unwanted segregants are more likely to occur. Departures of seed stock from type, 'degeneration', due to continued segregation, contamination, occasional mutations and cross pollinations, must be rectified. A 'pure' nucleus is established by growing some hundreds of rows from plants that are true to type. Each row is inspected at all stages of the life cycle and off-types are removed – i.e. they are 'rogued'. Seed from the remainder is grown in bulk. Any remaining inhomogeneity necessitates reselection. During multiplication every care is taken to prevent contamination. The plants are grown in a 'clean' field, remote from other fields of barley. Very careful agricultural practice is needed to produce acceptable seed stocks. The breeder, and others, will probably maintain a stock which will be checked and rogued year by year. Sometimes stocks may be 'purified' by reselection then re-released to agriculture. While they must be very similar to the original stocks they cannot be absolutely identical to them. These remarks apply to the production of 'pure' barley cultivars. In many areas the use of multi-lines, composite crosses, hybrid F_1 and other 'impure' seed stocks would be discouraged or prevented.

12.17. Conclusion

In the foreseeable future most improved barley varieties will probably come from the application of well-tried hybridization and selection techniques. Composite crosses between local and 'world collections' of varieties may provide a useful range of ecologically adapted types suitable for use in more conventional breeding programmes. To solve particular problems the use of genetic 'manipulations', such as haploidization followed by diploidization to accelerate the attainment of homozygosity, (Chapter 11), may be valuable. However, there seems to be very little chance of finding a way to achieve a dramatic 'yield breakthrough', as is sometimes expected.

References

1. Aastveit, K. (1961). *Meld. Norg. Landbr. Høisk.*, **40** (pt. 2, report 23), 1–112.
2. Aastveit, K. (1964). *Genetics*, **49**, 159–164.
3. Adams, M.W. and Grafius, J.E. (1971). *Crop Sci.*, **11**, 33–35.
4. Aksel, R. and Johnson, L.P.V. (1961). *Can. J. Genet. Cytol.*, **3**, 242–259.
5. Allard, R.W. (1960). *Principles of Plant Breeding*. New York: Wiley.
6. Allard, R.W. and Hansche, P.E. (1964). *Adv. Agron.*, **16**, 281–325.
7. Allard, R.W., Jain, S.K. and Workman, P.L. (1968). *Adv. Genet.*, **14**, 55–131.
8. Anderson, C.H. (1963). *Can. J. Pl. Sci.*, **43**, 519–521.
9. Baker, R.J., Bendelow, V.M. and Buchannon, K.W. (1968). *Crop Sci.*, **8**, 446–448.
10. Baker, G.A., Huberty, M.R. and Veihmeyer, F.J. (1952). *Agron. J.*, **44**, 267–270.
11. Bal, B.S., Suneson, C.A. and Ramage, R.J. (1959). *Agron, J.*, **51**, 555–557.
12. Beaven, E.S. (1947). *Barley, Fifty Years of Observation and Experiment*. London: Duckworth.
13. Bell, G.D.H. (1951). *J. Inst. Brew.*, **57**, 247–260.
14. Bell, G.D.H. (1969). *Agric. Progr.*, **44**, 24–29.
15. Bell, G.D.H. and Lupton, F.G.H. (1962). In *Barley and Malt*, ed. Cook, A.H. pp. 45–99. London: Academic Press.
16. Bell, G.D.H., Whitehouse, R.N.H. and Jenkins, G. (1969). *Ann. Rep. Pl. Breed. Inst. Camb.*, *1969*, 64–66.
17. Bennett, E. (1965). *Scottish Pl. Breed. Sta. Rec.*, 27–113.
18. Bingham, J. (1971). In *Potential Crop Production, a case study*, ed. Wareing, P.F. and Cooper, J.P. pp. 273–294. London: Heinemann Educational Books.
19. Bingham, J. and Jenkins, G. (1965). *J. Agric. Sci. Camb.*, **65**, 201–205.
20. Bonnett, O.T. (1938). *J. Am. Soc. Agron.*, **30**, 501–506.

21 Bonnett, O.T. and Woodworth, C.M. (1931). *J. Am. Soc. Agron.*, **23**, 311-327.
22 Borthakur, D.N. and Poehlman, J.M. (1970). *Crop Sci.*, **10**, 452-453.
23 Briggs, F.N. and Knowles, P.F. (1967). *Introduction to Plant Breeding.* London: Reinhold Publishing Corp.
24 Carleton, A.E. and Foote, W.H. (1968). *Crop Sci.*, **8**, 554-557.
25 Crook, W.J. and Poehlman, J.M. (1971). *Crop Sci.*, **11**, 818-821.
26 Doll, H., Køie, B. and Eggum, B.O. (1974). *Radiat. Bot.*, **14**, 73-80.
27 Donald, C.M. (1961). *Euphytica*, **17**, 385-403.
28 Dros, J. (1957). *Euphytica*, **6**, 45-48.
29 Edwards, K.J.R. and Allard, R.W. (1963). *Am. Nat.*, **97**, 243-248.
30 Ellis, J.R.S., Brunton, C.J. and Palmer, J.M. (1971). *Nature*, **241**, 45-47.
31 Ephrat, J. and Lachover, D. (1969). *Qual. Pl. Mater, Veg.*, **17**, 221-234.
32 Eslick, R.F. (1970). In *Barley Genetics, II*, ed. Nilan, R.A. pp. 292-297. Washington State University Press.
33 Everson, E.H. and Schaller, C.W. (1955). *Agron. J.*, **47**, 276-280.
34 Faris, D.G. and Guitard, A.A. (1967). *Can. J. Pl. Sci.*, **47**, 219-220.
35 Fick, G.N. and Rasmusson, D.C. (1967). *Crop Sci.*, **7**, 315-317.
36 Finlay, K.W. (1960). *J. Inst. Brew.*, **66**, 51-57; 58-64.
37 Finlay, K.W. (1970). In *Barley Genetics II*, ed. Nilan, R.A. pp. 338-345. Washington State University Press.
38 Finlay, K.W. and Wilkinson, G.N. (1963). *Austr. J. agric. Res.*, **14**, 742-754.
39 Foster, A.E., Peterson, G.A. and Banasik, O.J. (1967). *Crop Sci.*, **7**, 611-613.
40 Foster, A.E. and Schooler, A.B. (1970). In *Barley Genetics II*, ed. Nilan, R.A. pp. 316-318. Washington State University Press.
41 Fowler, C.W. and Rasmusson, D.C. (1969). *Crop Sci.*, **9**, 729-731.
42 Frankel, O.H. and Bennett, E. (1970). *IBP Handbook No. 11 – Genetic Resources in Plants; their exploration and conservation.* Oxford: Blackwell Scientific Publications.
43 Gardener, C.J. and Rathjen, A.J. (1975). *Aust. J. agric. Res.*, **26**, 219-230.
44 Gaul, H. (1964). *Radiat. Bot.*, **4**, 155-232.
45 Gaul, H.P.K., Ulonska, E., Zum Winkel, C. and Braker, G. (1969). In *Induced Mutations in Plants.* pp. 375-398. Vienna: IAEA.
46 Gebrekidan, B. and Rasmusson, D.C. (1970). *Crop Sci.*, **10**, 500-502.
47 Giles, R.J., McConnell, G. and Fyfe, J.L. (1974). *J. agric. Sci., Camb.*, **83**, 447-450.
48 Grafius, J.E. (1959). *Agron. J.*, **51**, 551-554.
49 Grafius, J.E. (1964). *Crop Sci.*, **4**, 241-246.
50 Grafius, J.E. and Okoli, L.B. (1974). *Crop Sci.*, **14**, 353-355.
51 Grafius, J.E. and Wiebe, G.A. (1959). *Agron. J.*, **51**, 560-562.
52 Greer, E.N. (1967). *Ann. Appl. Biol.* **59**, 319-326.
53 Gustafsson, Å. (1953). *Hereditas*, **39**, 1-18.
54 Gustafsson, Å., Dormling, I. and Ekman, G. (1973). *Hereditas*, **75**, 75-82.
55 Hagberg, A., Persson, G. and Wiberg, A. (1963). In *Recent Plant Breeding Research, Svalöf, 1946-1961.* ed. Åkerberg, E., Hagberg, A., Olsson, G., Tedin, O., Myers, W.M. pp. 105-124. London: John Wiley.
56 Hanson, W.D. (1959). *Agron. J.*, **51**, 711-715.
57 Hanson, P. (1973). *Ann. appl. Biol.*, **73**, 111-117.
58 Hanson, W.D. and Robinson, H.F. (1963). *Statistical Genetics and Plant Breeding—Publication 982, National Academy of Sciences.* Washington, D.C.: National Research Council.
59 Harlan, H.V. and Martini, M.L. (1936). *US Dept. Agric; Yearbook of Agriculture*, 303-346.

60 Harlan, H.V. and Martini, M.L. (1938). *J. agric. Res.*, **57**, 189–199.
61 Harlan, H.V., Martini, M.L. and Stevens, M. (1940). *U.S. Dept. Agric. Tech. Bull.*, No. 720.
62 Harlan, H.V. and Pope, M.N. (1926). *J. agric. Res.*, **32**, 669–678.
63 Harrington, J.B. (1932). *Sci. Agric.*, **12**, 470–484.
64 Harrington, J.B. (1952). *FAO Development Paper No. 28 Agriculture, Cereal Breeding procedures*. Rome: FAO.
65 Hartmann, R.W. and Allard, R.W. (1964). *Crop Sci.*, **4**, 424–426.
66 Hayes, J.D. (1968). *Euphytica*, **17**, (Suppl. 1) 87–102.
67 Hayman, B.I. (1954). *Genetics*, **39**, 789–809.
68 Helgason, S.B. and Chebib, F.S. (1963). In *Statistical Genetics and Plant Breeding – Publication 982, National Academy of Sciences*, ed. Hanson, W.D. and Robinson, H.F. pp. 535–545. Washington D.C. National Research Council.
69 Heyland, K.-U. (1963). *Z. Acker-u. PflBau.*, **121**, 130–147.
70 Hirai, S., Osada, T., Hasegawa, K. and Hiraki, M. (1969). *Rep. Res. Lab. Kirin. Brewing Co.*, **12**, 69–77.
71 Holliday, R. and Willey, R.W. (1969). *Agric. Prog.*, **44**, 56–77.
72 Holm, E. (1968). *Hereditas*, **60**, 409–411.
73 Hoskins, P.H. (1968). *Master Brewer's Ass. Am. Tech. Quart.*, **5**, 100–102.
74 Howard, H.W., Johnson, R., Russell, G.E. and Wolfe, M.S. (1969). *Ann. Rep. Pl. Breed. Inst. Camb.*, 6–36.
75 Hoyle, B.J. and Baker, G.A. (1961). *Hilgardia*, **30**, 365–394; 394 A-E.
76 Hunter, H. (1952). *The Barley Crop*. London: Crosby Lockwood.
77 Ingversen, J., Andersen, A.J., Doll, H. and Køie, B. (1973). *Nuclear Techniques for Seed Protein Improvement*. pp. 193–198. Vienna: IAEA.
78 Ingversen, J., Køie, B. and Doll, H. (1973). *Experientia*, **29**, 1151–1152.
79 Jain, S.K. (1970). In *Barley Genetics II*, ed. Nilan, R.A. pp. 422–429. Washington State University Press.
80 Jinks, J.L. (1955). *Heredity*, **9**, 223–238.
81 Jinks, J.L. (1956). *Heredity*, **10**, 1–30.
82 Johnson, L.P.V. and Singh, L.N. (1970). In *Barley Genetics II*, ed. Nilan, R.A. pp. 443–449. Washington State University Press.
83 Kamra, O.P. (1971). *Z. PflZücht.*, **65**, 293–306.
84 Kaufmann, M.L. and Guitard, A.A. (1967). *Can. J. Pl. Sci.*, **47**, 73–78.
85 Knight, R. (1970). *Euphytica*, **19**, 225–235.
86 Köchling, J. (1966). *Züchter*, **36**, 87–89.
87 Lang, R.W., Holmes, J.C. Taylor, B.R. and Waterson, H.A. (1975). *Expl. Husb.*, **8**, 53–59.
88 Lawes, D.A. and Treharne, K.J. (1971). *Euphytica*, **20**, 86–92.
89 Lee, J. (1960). *Evolution*, **14**, 18–28.
90 Lein, A. (1964). *Barley Genetics I, Proc. 1st Barley Genetics Symp. 1963 Wageningen.*, 310–324.
91 Lejeune, A.J., Sisler, W.W., Banasik, C.J. and Harris, R.H. (1951). *J. Inst. Brew.*, **57**, 445–449.
92 Liang, G.H.L., Heyne, E.G. and Walter, T.L. (1966). *Crop Sci.*, **6**, 135–139.
93 Lupton, F.G.H. and Whitehouse, R.N.H. (1957). *Euphytica*, **6**, 169–184.
94 McDaniel, R.G. (1970). In *Barley Genetics II*, ed. Nilan, R.A. pp. 323–337. Washington State University Press.
95 McKenzie, R.I.H. and Lambert, J.W. (1961). *Crop Sci.*, **1**, 246–249.
96 MacKey, J. (1963). In *Recent Plant Breeding Research—Svalof, 1946–1961* ed. Åkerberg, E., Hagberg, A., Olsson, G., Tedin, O. and Myers, W.M. pp. 73–88. London: John Wiley.

97 Meredith, W.O.S., Anderson, J.A. and Hudson, L.E. (1962). In *Barley and Malt*; ed. Cook, A.H. pp. 207–270. London: Academic Press.
98 Mertz, E.T., Misra, P.S. and Jambunathan, R. (1974). *Cereal Chem.*, **51**, 304–307.
99 Metcalfe, D.R., Helgason, S.B. and Hougen, F.W. (1967). *Can. J. Genet. Cytol.*, **9**, 554–564.
100 Mikaelsen, K. (1972). *Hereditas*, **72**, 201–204.
101 Mossé, J., Baudet, J. Demarteau, C. and Monget, C. (1969). *Ann. Physiol. Veg.*, **11**, 51–66.
102 Munck, L. (1970). In *Improving plant protein by nuclear techniques.* pp. 319–330. Vienna: IAEA.
103 Munck, L., Karlsson, K.E. and Hagberg, A. (1970). In *Barley Genetics II*, ed. Nilan, R.A. pp. 544–558. Washington State University Press.
104 Nickell, C.D. and Grafius, J.E. (1969). *Crop Sci.*, **9**, 447–451.
105 Nilan, R.A. (1960). *Sveriges Utsädes-förenings Tidskrift*, **70**, 110–118.
106 Nuttonson, M.Y. (1957). *Barley climate relationships and the use of phenology in ascertaining the thermal and photothemal requirements of barley.* Washington: American Institute of Crop Ecology.
107 Oltmann, W. (1956). *Züchter*, **26**, 315–319.
108 Oram, R.N., Doll, H. and Køie, B. (1975). *Hereditas*, **80**. 53–58.
109 Paroda, R.S. and Hayes, J.D. (1971). *Heredity*, **26**, 157–175.
110 Patterson, H.D. and Talbot, M. (1974). *J. Natn. Inst. agric. Bot.*, **13**, 142–151.
111 Poehlman, J.M. (1959). *Breeding Field Crops.* pp. 23–50; 150–173. New York: Henry Holt.
112 Pope, M.N. (1935). *J. Hered.*, **26**, 411–413.
113 Pope, M.N. (1939). *J. agric. Res.*, **59**, 453–463.
114 Pope, M.N. (1943). *J. Am. Soc. Agron.*, **35**, 161–163.
115 Pope, M.N. (1944). *J. Hered.*, **35**, 99–111.
116 Ramage, R.T. (1965). *Crop Sci.*, **5**, 177–178.
117 Rasmusson, D.C. (1968). *Crop Sci.*, **8**, 600–602.
118 Rasmusson, D.C., Beard, B.H. and Johnson, F.K. (1967). *Crop Sci.*, **7**, 543.
119 Rasmusson, D.C. and Cannell, R.Q. (1970). *Crop Sci.*, **10**, 51–54.
120 Rasmusson, D.C. and Glass, R.L. (1967). *Crop Sci.*, **7**, 185–188.
121 Rasmusson, D.C., Hester, A.J., Fick, G.N. and Byrne, I. (1971). *Crop Sci.*, **11**, 623–626.
122 Rasmusson, D.C. and Lambert, J.W. (1961). *Crop Sci.*, **1**, 259–260; 261–262; 419–420.
123 Rasmusson, D.C., Upadhyaya, B.R. and Glass, R.L. (1966). *Crop Sci.*, **6**, 339–340.
124 Rauser, W.E. (1967). *Can. J. Pl. Sci.*, **47**, 614–616.
125 Rhodes, A.P. (1975). *J. Sci. Fd. Agric.*, **26**, 1703–1710.
126 Riddle, O.C. and Suneson, C.A. (1944). *J. Am. Soc. Agron.*, **36**, 62–65.
127 Riggs, T.J. (1973). *Ann. appl. Biol.*, **74**, 249–258.
128 Robertson, D.W. and Deming, G.W. (1931). *J. Am. Soc. Agron.*, **23**, 402–406.
129 Rutger, J.N., Schaller, C.W., Dickson, A.D. and Williams, J.C. (1966). *Crop Sci.*, **6**, 231–234.
130 Rutger, J.N., Schaller, C.W. and Dickson, A.D. (1967). *Crop Sci.*, **7**, 325–326.
131 Sage, G.C.M. (1968). *Euphytica*, **17**, (Suppl. 1), 103–106.
132 St. Pierre, C.A., Klinck, H.R. and Gauthier, F.N. (1967). *Can J. Pl. Sci.*, **47**, 507–517.

133 Sakai, K.I. (1961). *Symp. Soc. expl. Biol.*, **15**, 245–263.
134 Sakai, K.I. and Iyama, S.Y. (1966). *Jap. J. Breeding*, **16**, 1–9.
135 Sandfær, J. (1970). *Danish Atomic Energy Commission—Risø, Report No. 230.*
136 Sandfær, J., Jørgensen, J.H. and Haahr, V. (1965). *Danish Royal Veterinary & Agricultural College Yearbook*, 153–180.
137 Schaller, C.W. (1951). *Agron. J.*, **43**, 183–188.
138 Scholz, F. (1970). In *Barley Genetics II*, ed. Nilan, R.A. pp. 94–105. Washington State University Press.
139 Schuster, W. and Tugay, M.E. (1971). *Z. PflZücht.*, **65**, 324–344.
140 Severson, D.A. and Rasmusson, D.C. (1968). *Crop Sci.*, **8**, 339–341.
141 Sigurbjörnsson, B. and Micke, A. (1969). *Mutations in Plant Breeding.* pp. 673–698. Vienna: IAEA.
142 Smith, E.L. and Lambert, J.W. (1968). *Crop Sci.*, **8**, 490–493.
143 Sparrow, D.H.B. (1970). In *Barley Genetics II*, ed. Nilan, R.A. pp. 559–574. Washington State University Press.
144 Sprague, H.B. and Farris, N.F. (1931). *J. Am. Soc. Agron.*, **23**, 516–533.
145 Stevenson, F.J. (1928). *J. Am. Soc. Agron.*, **20**, 1193–1196.
146 Stoskopf, N.C. and Reinbergs, E. (1966). *Can. J. Pl. Sci.*, **46**, 513–519.
147 Streeter, J.G. and Pfeifer, R.P. (1966). *Crop Sci.*, **6**, 151–154.
148 Suneson, C.A. (1949). *Agron. J.*, **41**, 459–461.
149 Suneson, C.A. (1953). *Agron. J.*, **45**, 388–389.
150 Suneson, C.A. (1960). *Agron. J.*, **52**, 319–321.
151 Suneson, C.A. (1962). *Crop Sci.*, **2**, 410–411.
152 Suneson, C.A. (1964). *Barley Genetics I, Proc. 1st Internatn. Barley Genetics Symp. 1963. Wageningen*, 303–309.
153 Suneson, C.A. and Cox, E.L. (1964). *Crop Sci.*, **4**, 233–234.
154 Suneson, C.A., Hockett, E.A., Eslick, R.F., Steward, V.R. and Wiebe, G.A. (1970). In *Barley Genetics II*, ed. Nilan, R.A. pp. 32–35. Washington State University Press.
155 Suneson, C.A., Ramage, R.T. and Hoyle, B.J. (1963). *Euphytica*, **12**, 90–92.
156 Suneson, C.A. and Stevens, H. (1957). *Agron. J.*, **49**, 50–52.
157 Suneson, C.A. and Wiebe, G.A. (1942). *J. Am. Soc. Agron.*, **34**, 1052–1056.
158 Upadhyaya, B.R. and Rasmusson, D.C. (1967). *Crop Sci.*, **7**, 644–647.
159 Viuf, B.T. (1969). In *New Approaches to Breeding for Improved Plant Protein.* pp. 23–28. Vienna: IAEA.
160 Walker, J.T. (1969). *Biol. Rev.*, **44**, 207–243.
161 Wang, R.C. and Lund, S. (1975). *Crop Sci.*, **15**, 550–553.
162 Wells, D.G. (1962). *Crop Sci.*, **2**, 177–178.
163 Whitehouse, R.N.H. (1968). *Ann. Rep. Pl. Breeding Inst. Camb.*, 6–29.
164 Whitehouse, R.N.H. (1970). In *Barley Genetics II*, ed. Nilan, R.A. pp. 269–282. Washington State University Press.
165 Whitehouse, R.N.H. and Whitmore, E.T. (1964). In *Barley Genetics I, Proc. 1st Internatn. Barley Genetics Symp. 1963. Wageningen*, 325–334.
166 Wiebe, G.A. (1968). *U.S. Dept. Agric. Handbook No. 338, Barley.* 96–104.
167 Wiebe, G.A. (1969). *Wallerstein Lab. Commun.*, **32**, 83–91.
168 Wiebe, G.A., Petr. F.C. and Stevens, H. (1963). In *Statistical Genetics and Plant Breeding*, ed. Hanson, W.D. and Robinson, H.F. U.S. Natn. Acad. Sci., N.R.C. Publication No. 982, 546–557.
169 Wiebe, G.A. and Ramage, R.T. (1970). In *Barley Genetics II*, ed. Nilan, R.A. pp. 287–291. Washington State University Press.
170 Wienhues, F. (1968). *Euphytica*, **17**, (Suppl. 1) 49–62.

Chapter 13
Some actual and potential uses of barley

13.1. Introduction

In this chapter some uses of grain and straw that are not directly concerned with the production of food and drink are noted. New or more efficient uses of these raw materials are needed to enhance the value of the crop. At present large quantities of straw are burned in the field – a dismal comment on the economics and technical capabilities of 'advanced' societies. More straw is likely to be used, after alkali treatments, in animal feed stuffs (Chapter 14). Still there is a need for industrial uses of this annually renewed material. Handling costs are high, and a prime requirement for a straw-based industry is that producers arrange to maintain a steady supply of top quality, well stored and dried material, probably densely baled for ease of handling.

13.2. Barley grain; a source of starch and protein

Barley grain can be used as a source of 'normal', high amylose, or high amylopectin starches (Chapter 4). It was used as a source of 'fecula' in the past [27]. In most areas other sources are now preferred. In districts where barley is cheap starch may be economically separated from milled barley flour by an alkali treatment, to loosen and dissolve protein and allow the collection of starch by centrifugation. The contaminating 'tailings', (other solids), are separated physically from the starch cake. Acidification of the supernatant liquid precipitates about 80% of the crude protein, which may be collected by centrifugation and might find various uses, including addition to feed stuffs [16]. Marked differences in suitability for processing have been noted for different barley types. Starch yields of 43–63 g/100 g of flour have been obtained. Protein yields varied between 11.5 and 16.6 g/100 g flour. Although sodium hydroxide was the base used in these preliminary trials solutions of ammonia, which is volatile and might be readily recovered and re-used, seem an attractive alternative.

13.3. Minor uses of straw

Straw is the dried stems, leaf-sheaths and the tattered residues of the leaf blades that remain after threshing. It has been used for many purposes. Weaving 'corn dollies', either to maintain a country craft or to magically keep alive and house the spirit of the 'corn-mother' between harvest and spring sowing, is one such purpose. Straws, often bleached or coloured, may be used in various games, for making 'mobiles', and for other ornamental purposes. Straw, often carefully selected was, and is, used in small amounts for making matting, hats, plaited baskets and containers, ropes, and packing material. Drinking 'straws' are now often tubes made of plastic or waxed paper. Straw may be used in building clamps for storing root crops on the farm. Layers of straw are interposed between harvested root-crops (such as turnips or swedes), and the whole mound is covered with a layer of straw about 15 cm (6 in) thick, before being covered and sealed with a layer of soil. Straw is sometimes used as 'backfill' over land-drains.

13.4. Straw in building

On the farm corn ricks were often built on a bed of straw, and were thatched with it. At one time most houses, in the U.K. at least, were thatched with straw – usually a layer more than 30 cm (1 ft) thick was used – but less flammable reeds (*Arundo phragmites*) came to be preferred, and now tiles, slates, or other materials are the usual roof coverings. The grain needed to be combed from the straw, or other forms of threshing had to be gentle, to leave the stems as intact as possible.

Chopped straw was used to bind together many forms of building materials based on dried mud or clay [12]. Walls were formed by pounding firm a mixture of clay and straw between restraining walls of removable shuttering or by building up walls with sun-baked, straw bound bricks ('adobe' work). This type of building is still extensively used in arid areas. Adobe bricks may be 'stabilized' by the incorporation of bitumen. In the U.K. straw-bound walling survives in the 'cob' and 'chalk-mud' walls of the West country. Mud, clay, or crushed chalk, or chalk and clay were well mixed with straw or other fibrous material. Originally the components were trodden together by men and horses. The cob (cleam, clob, clom) was compacted in successive layers, onto a stone base. When high enough it was pared smooth. To protect it from rain it was capped with thatching, slates, stones, or tiles, the capping having a good overhang.

The surfaces might be plastered and limewashed, 'rendered' to a rough cast finish with stones, or be tarred. Such walls, properly maintained, last for hundreds of years [12]. Straw, or even expanded ('popped') grain, perhaps first carbonized, could be used as 'expanders' in making lightweight concrete, or other building blocks. On farms, temporary buildings and enclosures, e.g. to protect sheep, may be made from straw bales and wattles.

Apparently the only frequent use of straw in buildings in the U.K. is as compressed straw slabs or strawboards [19]. Typically the straw is compressed and bonded into rigid slabs, about 256 kg/m^3 (16 lb/ft^3), 5.1 cm (2 in) deep, by 122 cm (4 ft) by 244, 274 or 305 cm (8, 9 or 10 ft). The straw may be 'treated' to diminish fungal attack. The slabs are normally faced and edged with thick paper or thin millboard. They may be given an additional aluminium-foil or other damp-proof, reflective and sealing surface. The insulating power is good, the thermal conductivity 47–55 W/m^2/h/cm/° C (0.6–0.7 B.T.U/ft^2/h/in/° C) being low. The material is light, easily cut and fixed and is useful for insulating e.g. water tanks, roofs and walls. Its weaknesses are its flammability and its susceptibility to deterioration when damp. Its fire-supporting and propagating properties may be diminished by treatment with appropriate paints and plasters. A range of other suggested straw-based materials includes particle-boards, manufactured from milled straw, graded for size and possibly blended with other materials, bonded together with, for example, thermosetting resins.

13.5. Animal bedding, litter, farmyard manure and compost

At one time straw was used to stuff mattresses. A continuing use is as bedding or litter for pets or farm animals, and birds. In a covered yard, fully grown cattle require, per head about 7–9 kg (15–20 lb) of straw each day, amounting to 1.5–2.0 tonne (1.5–2 ton) per beast during the winter [28]. The material, mixed with dung or droppings, warms up, warming the animals. Afterwards it may be stacked and, after rotting, is used as farmyard manure, FYM. The tendency to specialize in either arable farming, with its straw surpluses, or animal rearing, has created problems. The cost of collecting, and carting straw to the animal houses, often situated on other farms, has discouraged its use. Animal housing systems have been developed in which the semi-liquid mixture of urine and faeces ('slurry'), unmixed with straw, is washed, pumped and scraped into holding containers for subsequent disposal. The disposal of slurry creates other problems, not least being the undesirable aspects of spreading the waste on land to be used for forage production (on which animal-farming projects naturally rely), the smell and the need to avoid the pollution of water courses. A simple composting system, in concept resembling the mechanized production of farmyard manure, provides a possible answer [6,17]. In a pilot plant cubicles, lined with plastic, are force-aerated from the bottom. Each is protected from the rain. In the cubicles slurry is pumped and sprayed over straw bales loaded in with the ties removed; liquid draining to the bottom of the heap is returned to a recycling tank. Each cubicle is loaded over a week, for example with 270 l (60 gal) of pig slurry and 33 kg (about 72 lb) of straw daily. The composting is carried out on a 4-week cycle; so a minimum of four cubicles is required. The straw filters off solids and, as it rots, moisture is taken up and evaporated; a maximum temperature of about 66°C (150°F) is reached. After 4 weeks the solid product weighs about one third of the starting materials. An average composition of 79% organic matter, about 3.0% nitrogen, 3.9% phosphorus and 2.2% potassium has been obtained for pig slurry compost. Straw consumption is moderate. For slurries containing 7% solids, straw requirements may be about 0.2 tonne/pig/year and 2 tonne/cow/year. One hundred cows will produce about 660 tonnes (650 tons) of compost annually. The compost may be spread on the fields, or used in domestic gardens or horticulture. The manurial value of the product is not high and, like FYM, it is costly to spread but it usefully increases the organic matter in many soils. It seems eccentric not to mix FYM (or compost) with a complementary, 'artificial' fertilizer, so that the two can be spread in one operation.

13.6. Soil protection, conditioning, or replacement

In some low-rainfall areas of the world, straw is allowed to remain on the soil surface after harvesting, as a mulch to trap moisture and to reduce erosion (Chapter 8). In horticulture and ornamental gardening, limited quantities of straw are used as a basis for composts to dig into the soil, or to provide a mulch around trees and other plants, to suppress weeds, to conserve moisture, or to form a layer to lift the produce clear of the soil, as with marrows, or to minimize the quantities of mud splashed onto the fruit by rain, as with strawberries.

An interesting approach is to grow high-value, intensively cultivated crops on rotted straw, or straw composts. Crops that have been grown in this way can be divided into two groups: (1) higher plants, such as tomatoes or cucumbers, and (2) edible fungi.

Growing glasshouse fruit (vegetable) crops, and possibly some outside crops such as tomatoes, on bales or wads of rotting straw has several advantages. Thus (i) the straw does not carry soil-borne pathogens, which give soil sickness problems; (ii) the rotting straw, which provides a well aerated medium, warms the roots and usefully enriches the air with carbon dioxide; (iii) labour is saved as the rotted straw is easily disposed of as a 'soil-conditioner' at the end of the season and soil cultivation, handling and sterilization is avoided and (iv) the costs of greenhouse heating are reduced [2,3,5,23]. The technique has been used to grow ornamental flower crops (chrysanthemums, stocks), as well as cucumbers, tomatoes, aubergines and sweet peppers under glass, and has been tried in the open with other crops in Central Africa. The straw used must not contain residues of herbicides (such as TBA and picloram). It is placed on an impervious, clean surface, such as a layer of polythene. It may be arranged as whole bales or as wads, initially 20–25 cm (8–10 in) deep. Bales are soaked with water over a period of days (about 1.6 litre/kg on 18 gal/cwt straw). When saturated a composting fertilizer is worked or watered in. Suggested mixtures include (a) 0.45–0.68 kg (1–1.5 lb) of nitro chalk, 0.45 kg (1 lb) of superphosphate, 0.45 kg (1 lb) of potassium nitrate, 113 g (4 oz) of magnesium sulphate, and 85 g (3 oz) of iron sulphate, all per 25.4 kg ($\frac{1}{2}$ cwt) straw, or, (b) 907 g (32 oz) of an ammonium nitrate/lime mixture, 340 g (12 oz) of triple superphosphate, 340 g (12 oz) of magnesium sulphate and 340 g (24 oz) of potassium nitrate per 50.8 kg (1 cwt) of straw. The temperature of the mixture rises and in a few days may reach 55°C (about 131°F); it then falls. Plants are set out in soil next to a stake for support on the surface of the rotting straw, when the temperature has fallen to about 38°C (100°F), which occurs in 3–5 days. The surface soil may be a heap or may be supported in a cylinder of material as used in ring cultures. The straw beds are automatically watered by drip-feeds, and once a week liquid fertilizer is applied. Ventilation in the greenhouse is limited when temperatures are low so that the carbon dioxide from the straw can accumulate in the atmosphere and enhance the photosynthetic rates of the plants. Straw-bale culture is a comparatively recent idea, but it appears to be gaining acceptance.

In Europe straw, in the form of well rotted FYM, is used as the growth medium for the edible mushroom, *Agaricus bisporis*. As now carried out, mushroom cultivation is a specialized and skilful occupation [15,21,29]. Pure fungal cultures are grown on agar, and the mycelium is grown onto dead, sterilized cereal grains, (e.g. wheat), which are used as 'spawn' for inoculating the production beds. The beds are placed in locations with an even humidity and temperature, in special buildings or in caves or disused underground workings. Mushroom beds range from simple troughs to boxes moved from place to place in a highly automated manner. The compost substratum originally preferred was rotted horse manure. This has been increasingly 'extended', or even replaced, by chopped straw or brewers spent grains and manure from other animals. Often 'activators' (additional minerals such as gypsum, ammonium phosphate and mixtures of B-vitamins) are added. The composted material is cooled, packed into the beds, and is inoculated with 'spawn' (fungal mycelium). The mycelium grows into the compost, filling the space available in about 14 days at 25°C (77°F). During growth the accompany-

486 Barley

ing populations of bacteria, fungi and *Actinomycetes* change in a characteristic way. The process must be regulated so that competing organisms and diseases are checked. The compost is then 'cased', by enclosing it in a layer of pasteurized soil to encourage the production of the edible mushrooms (fruiting bodies, sporophores, carpophores). These grow on the surface, and appear at rhythmically occurring intervals called 'flushes' or 'breaks'. They are picked, as available, until the crop becomes uneconomically small – in about 42 days at $14-18°C$ (about $57-65°F$). The lower temperature range, $14-18°C$, favours fruiting while the initial, higher temperature ($25°C$) favours mycelial growth. Afterwards the spent compost is used as a manure and 'soil conditioner' in horticulture. As an economy measure part of it may be retained, stored for a year, and after 'recycling' be mixed with fresh compost. The emptied growing areas and equipment are sterilized before starting a new growth cycle.

No doubt many other edible fungi could be grown on substrata made mainly of straw. For example *Agaricus bitorquis* will grow in pure culture (in contrast to *A. bisporis*) and at a uniform optimal temperature of $23-25°C$ ($73-77°F$). Others – e.g. the Oyster mushroom *(Pleurotus oestreatus)* and the Padi-straw mushroom *(Volvaria volvacea)* are grown on straw composts in various parts of the world, and strains could probably be selected that would grow well on barley straw.

13.7. Some industrial uses of barley

Hydrolysis of straw with dilute acid produces *inter alia* an impure mixture of sugars, notably xylose and glucose, and a spongy residue. The sugars can be used to support the growth of micro-organisms such as *Geotrichum candidum* which are rich in fat (about 20% by weight) and protein, or others which turn the sugars into alcohol [26,31]. The solid residues might be used in the manufacture of fibre board, or be dried and compressed to briquettes to be sold or used in the plant as combustible fuel [20]. Straw was used in the production of power alcohol in Europe, during wartime. By reacting the straw with more concentrated sulphuric or hydrochloric acids the pentosans present are hydrolysed and the free pentoses are dehydrated to give furfural, furfuraldehyde, a substance used, mixed with phenol, in making synthetic resins for mouldings, and as an industrial solvent. Its reduction product, furfuryl alcohol, is also an important industrial chemical, used in the manufacture of resins, adhesives, protective lacquers, and paint strippers. Some of the resins are used as the bonding material in making glass-fibre products. It is also a solvent, and is employed in the chemical, cosmetic, oil-refining, car, textile and synthetic rubber industries. Main sources of the furfural have, in the past, been maize ('corn') cobs and oat hulls, but straw is an obvious alternative. Oat hulls can be converted to furfural in about 10%

Some actual and potential uses of barley

yield. Straw, and polysaccharides from it, have been nitrated to form explosives analogous to nitrocellulose.

Among other suggested uses for straw, some being based on experience with maize cobs, are the production of veneers for chipboards; the use of milled straw as a carrier dust for pesticides, for fur-cleaning and dry-cleaning; as a hard-soap abrasive; as a filler in explosives, plastics, linoleum, roofing materials, glues and floor-tiles; as an abrasive, dependent on its high silica content, for polishing and burnishing, and in grit-blasting; as a component of industrial adsorbents, floor-cleaning compounds, pet-litter, low-density aggregates and soil conditioners. Straw might be converted to active carbon, a useful industrial adsorbent. Puffed, carbonized grains are cheaper extenders than expanded polystyrene. Extracts of brewer's spent grains, and perhaps straw, can be used to make flocculants for waste-water treatment – e.g. from copper refining. Other products are used in altering the surface characteristics of minerals being separated by flotation processes. Spent grains and straw might be used to manufacture textile sizes, binders, glues, and perhaps lubricants [20,25,30].

Many industrial uses might involve extracting the hemicelluloses for further processing leaving a solid residue rich in lignin and crude cellulose. Such residues might be used in board or pulp manufacture. Approached differently pulping operations should be arranged, where possible, to make saleable products from the extractives present in what would otherwise be effluent. It seems that whatever 'primary' use is found for straw the 'waste' materials will need to be processed to useful products, on the same site, to make the economics attractive.

Straw has possibilities as a fuel if it is used near the site of origin, to avoid carriage costs. Straw bales, burnt in specially-designed boilers, may be used to heat water either for heating a farm-house or a radiator heating the air used in grain drying. Burning about four 'average' size, low density bales/day can provide sufficient heat to reduce the moisture content of 300 tons of grain by 4% in 10h [11]. Heating and partly pyrolysing straw, so that about 12% of the dry-matter is lost, produces tar which can bind the compressed material into briquettes, which are suitable for use as a fuel. Such briquettes might be made on site out of industrial straw wastes and residues, then used to provide power for the industrial process itself [20]. Alternatively perhaps untreated residual solids could be burnt directly in suitable furnaces.

13.8. Paper, cardboard and millboard

A major potential use of straw is the production of paper, paper laminates, cardboard and millboard. While straw has been used for making even, fine, bleached paper, and low grade unbleached papers, especially in wartime, in peace time it has largely been displaced by Esparto grass and woodpulp. In part this is due to the technical difficulties encountered in wartime when straw pulp had to be made in unsuitable mills designed for processing other materials. Other problems are organizational and economic. Thus to maintain its quality the straw must be baled and dried at harvest, and be stored dry. Further, the collection and transport must be organized so that a regular supply is available the whole year round. To give an adequate incentive it seems that farmers need a direct interest in the mill's success [33]. The mill must be situated at a site where fuel, chemicals and water are available, and where effluent disposal is a manageable problem. Clearly it should also be near a grain growing area. Where these factors are properly balanced straw mills operate profitably [32,33]. When mixed with other pulps bleached straw pulp gives paper some desirable characters e.g. types that are thin and translucent, with a good 'rattle'. Some papers contain up to 80% straw pulp [10,33]. The yield of pulp from clean straw varies according to the preparation process used and the degree of refinement achieved. Chemical pulps are obtained in 33–48% yield, the lower yield generally being of bleached pulp while semi-chemical pulps are obtained in 45–52% yield (bleached) and 67–75% unbleached. Unbleached mechano-chemical pulps are obtained in yields of 80% or better [33]. Perhaps it is products made from unbleached pulps which offer a substantial outlet for straw [8]. Surprisingly products made directly from hammer-milled straw or cut straw particles bonded together with adhesives seem to find little use [1,14].

In paper making the starting material, e.g. straw, is cleaned to an adequate extent, washed, cut up, compressed and then cooked, usually under alkaline conditions, with varying degrees of agitation. Following other washing, agitation and sometimes bleaching and sieving stages, the suspension of separated fibres in water, the 'stock' or stuff, is formed into a damp pad which, after drying, becomes paper. The paper may be formed on a continuous wire cloth filter in a Fourdrinier machine, or the water may be removed on various cylindrical moulds or board machines. Mill- and card-boards may be made in a similar way, or may be made from superimposed paper laminates. Other natural or synthetic fibres may be blended with the stuff. The damp paper is pressed, steamed, rolled, dried and trimmed [4,7]. Various additives such as dyes, sizes and fillers (e.g. china-clay) may be used. Pulps may be so finely beaten or fibrillated that they become 'jellified', so that the paper has a low porosity and is semi-transparent and 'greaseproof'.

Barley straw contains 'on average' 11.3% water extractable materials, 38.2% alkali-extractable materials, 0.8% waxes, resins and 'chlorophyll', and 49.7% fibre [4]. In other terms the composition is ash, 8–11%, lignin, 15–17%, α-cellulose, 34–37%, and pentosan, 25–27%. Compared to hardwood pulps (< 1%) the ash content is high [13,33]. The fibrous materials important in making the finer pulps are situated in the vascular bundles and sclerenchyma of the stem and leaf-sheaths. In width they range from 9–13 μm, and in length 0.3–1.5mm from the stem, and 0.5–1.7mm from the leaves [22]. Straw pulps contain a range of other cell walls, besides fibres. Selection for highly fibrous straw in breeding programmes might improve its value as a raw material and at the same time increase stem strength.

Pulping is designed to separate intact fibres from the starting material in high yield and to remove soluble materials and some fractions of the hemicelluloses. The severity of the pulping process is matched to the quality of material to be produced. For producing brown paper, cardboards, strawboards, or corrugated packing paper pulping severity is minimal. Often no attempt is made to separate individual fibres, and the unbleached, brown material is not delignified. A strong product can be obtained in high yield. To make fine quality pulps more severe pulping conditions are needed, followed by delignification or bleaching treatments. The fibres obtained from more stringent treatments are partly broken up, and are weaker [7,9,13,14,18,24,33].

Straw for pulping must be free from rotting, and be bright and free from weeds. Any decomposition sharply reduces the yield of fibres. Before pulping it is cut into lengths (about 5cm, 2in) and is broken up in, for example, an edge-runner mill (kollergang) in which a wheel runs over the straw in a circular trough. The broken material is then pulped. In the oldest sytem, used to produce unbleached pulp, it is cooked in rotary digestors at e.g. 3500–42 000 kg/m^2 (5–60 lb/in^2) for 3.5–15 h, with lime (CaO, 8–10% of the straw dry weight). If the pressure is released suddenly the cooked straw can be discharged violently into a blow tank, being well disintegrated in the process. The unbleached product, disaggregated to a minimal extent by mechanical agitation, is obtained in 75–80% yield. It is used to make soft board. To make a harder product, in about 65% yield, other liquors, for example sodium hydroxide or various mixtures containing sodium hydroxide, soda ash, sodium sulphite and lime, are used in the cookers [33].

In 'mechanochemical' pulping systems cooking is in 'hydropulpers' at 98°C (208°F) for 0.5–1.5 h with 10% lime or other substances such as sodium hydroxide and sodium sulphide mixtures. The pulp is washed and sieved and 'knots' (unbroken nodes) are separated and returned to the hydropulper. The yield of unbleached pulp is about 65%. To bleach the pulp two or three exposures to chlorine gas, associated with alkaline washing, or solutions of hypochlorites may be used. Bleached pulp is obtained in about 50% yield. In the caustic soda–chlorine process the main reagents are prepared on site by the electrolysis of brine. Semi-pulp, with an unbleached yield of 60–65%, may be prepared using an initial soak in 2% sodium hydroxide. For making bleached pulp, straw is soaked in 3% sodium hydroxide then is digested, washed, pressed, beaten to open out its structure, and is bleached by exposure to chlorine gas for 20–30 min in the cold, followed by a caustic wash. This process may be repeated. The final, high-quality pulp is obtained in 40–45% yield and may be used, generally mixed with longer fibres from other sources, in making writing, printing or glassine papers.

At least for making unbleached products, straw should be a competitive raw material [13,33]. However, there are problems of effluent disposal, and of recovering reagents from used caustic digestion liquors containing the silicates extracted from the straw. An interesting suggestion is to use potassium hydroxide, or ammonia-based digestion liquors so that the spent solution can be turned into fertilizers enriched with the straw ash. Ammonia has the added advantage that, being volatile, some of it could be stripped from waste-liquors and then re-used. Sulphite liquor has been sold for a range of uses [33].

References

1 Akers, L.E. (1966). *Particle Board & Hardboard.* London: Pergamon Press.
2 Allen, P.G. (1968). *N.A.A.S. Q. Rev.*, **80**, 167–174.
3 Allerton, F.W. (1968). *Tomatoes for Everyone.* pp. 71–80. London: Faber & Faber.
4 Anon. (*ca.* 1877). *Chemistry, Theoretical Practical & Analytical as Applied to Arts & Manufactures, VII*, pp. 513–542. London: W. Mackenzie.
5 Anon. (1964). *Rept. Lee Valley, Expl. Hort. Sta.*, 18–23.
6 Anon. (1975). *Farmer's Weekly*, 18 April, p. 48.
7 Bolam, F. (1965). *Paper Making*, Technical section of the British Paper & Board Makers Association (Inc.), London.
8 Burns, I.G. and Grant, J. (1952). *The World's Paper Trade Review*, *137*, 19 June, 1831–1840.
9 Clapperton, R.H. and Henderson, W. (1947). *Modern Paper-Making*, 3rd edition. Oxford: Basil Blackwell.
10 Cross, C.F., Bevan, E.J. and Briggs, J.F. (1916). *A Textbook of Paper-Making*, (4th edition). London: E & F.N. Spon.
11 Crisford, P. (1976). *Farmer's Weekly*, 5 March, xiv, xvi.
12 Davey, N. (1961). *A History of Building Materials.* London: Phoenix House.
13 F.A.O. (1953). Food and Agricultural Organisation of the United Nations. *Forestry & Products Study, No. 6. Raw Materials for More Paper Pulping Processes and Procedures Recommended for Testing.* Rome (1953).
14 F.A.O. (1958). Food and Agricultural Organisation of the United Nations. *Fibreboard and Particle Board*, Rome.
15 Gaisford, M. (1974). *Farmer's Weekly*, October v, vii.
16 Goering, K.J. and Imsande, J.D. (1960). *J. Agric. Fd Chem.*, **8**, 368–370.
17 Gray, K.R., Biddlestone, A.J. and Clarke, R. (1973). *Process Biochem.*, October, 11–15; 30.
18 Hammond, W.E. (1950). *Nebraska Expl. Sta. Bull.*, **401**.
19 Handisyde, C.C. (1968). *Building Materials—Science and Practice*, (6th edition). London: The Architectural Press.

Some actual and potential uses of barley 491

20 Hansford, R.J., Hughes, R.G., Truman, A.B., Wilson, P.N. and Woods, R.S. (1975). *Report of Working Conference on Straw Utilisation at Oxford, 5th Dec. 1974*, ed. Staniforth, A.R. Oxford: ADAS.
21 Hayes, W.A. (1974). *Process Biochem.*, **21**, December, 24, 28.
22 Lloyd, F.E. (1921). *Pulp and Paper Magazine*, 15 September, 953–954; 20 October, 1071–1075.
23 M.A.F.F. (1973). (Ministry of Agriculture, Fisheries and Food of the U.K.) *Short-term leaflet No. 105* (Amended). *Straw substrates for the production of crops in greenhouses.*
24 Norris, F.H. (1952). *Paper and Paper Making*. London: Oxford University Press.
25 Pomeranz, Y. (1973). *The Industrial Uses of Cereals. (Proc. Symp. 58th Ann. Mtg. Am. Ass. Cereal Chemists, St. Louis, Mo.).*
26 Prescott, S.C. and Dunn, C.G. (1959). *Industrial Microbiology*, (3rd edition). London: McGraw-Hill.
27 Raspail, F.V. (1834). *A New System of Organic Chemistry*—(Transl. Henderson, W. from the French, with notes and additions). London: Sherwood, Gilbert and Piper.
28 Robinson, D.H. (1962). *Fream's Elements of Agriculture*. London: John Murray.
29 Singer, R. (1961). *Mushrooms and Truffles, Botany, Cultivation and Utilization*. London: Leonard Hill (Books) Ltd.
30 Staniforth, A.R. (1975). *Outlook on Agriculture*, **8**, 194–200.
31 Thayson, A.C. and Galloway, M.A. (1928). *Ann. appl. Biol.*, **15**, 392–407.
32 Trow-Smith, R. (1974). *Farmer's Weekly*, October, 75–77.
33 Truman, A.B. (ed. 1974). *Survey of Straw Pulping in Great Britain*. Research Assoc. for the Paper, Board, Printing and Packaging Industries (PIRA), Leatherhead, Surrey, **1** and **2**.

Chapter 14

Barley for animal and human food

14.1. Introduction

A proportion of barley grain is used to make malt which, with or without unmalted grain, is used to make beer, whisky and some other products (Chapters 15 and 16). By-products from these processes, notably malt culms and dust, spent brewery or distillery grains, distillers' solubles, and surplus yeast are used in animal feeding stuffs. To a limited extent malt extracts are used in human food. In Western countries most barley grain is used to feed farm animals – cattle, sheep, goats, pigs, and horses, and poultry. Many of these become food for man. In addition cereals are blended into feeds for game-birds, fish, fur-bearing animals and pets. In some Eastern countries, including Japan, Korea, China and the Himalayan regions, large quantities of barley, often of naked varieties, are used in human food and drink. In the West only small quantities of processed barley grain are so used at present.

Animals diets may include barley grain, barley-derived by-products, hay, straw, or whole-crop silage. In the future grain may, of necessity, be reserved for human consumption. Even if this occurs the straw will remain available as an animal foodstuff.

14.2. The nutritional requirements of animals

Foodstuffs must supply animals' energy requirements. The main energy sources in cereal feeds are the carbohydrates, although proteins and fats may also contribute to a significant extent. In addition animals need a supply of amino acids. Essential amino acids must be supplied in adequate amounts as animals are not able to make them, at least in amounts sufficient for their requirements, from others in their diet. The non-essential amino acids can be inter-converted by the animals. Animals also need supplies of inorganic materials, notably Ca, P, Mg, K, Na, Cl, Fe, Cu, Co, I and Se. Only trace amounts of the last seven are required; indeed in excess they are toxic. Trace quantities of various organic substances are needed, including vitamins or vitamin precursors and certain unsaturated fatty acids. Animal species, including mankind (and birds) differ in their requirements for nutrients depending, for example, on the conditions in which they are 'housed', age, sex, health, degree of activity, the ambient temperature, and whether they are carrying young, lactating or laying eggs. Furthermore individuals apparently differ in their needs for vitamins and essential amino acids. Different species digest foodstuffs with varying degrees of efficiency that is partly dependent on the preparation and presentation of the food, the proportions of the different nutrients present and on substances mixed with it. In addition the compositions of barley and other farm crops are variable. The imprecision of nutritional science is readily explained by these uncertainties [4,74]. The acquisition of complete chemical analyses of the grain or other materials to be used in animal feedingstuffs is both impossible and pointless. In general grades and sometimes nitrogen ('protein') contents are noted and, guided by 'proximate' analyses, rations are designed using adequate 'safety margins' to ensure (in a wasteful way) that the animals' requirements are met. The use of such proximate analyses are dependent on information provided by numerous feeding trials.

Non-ruminants, e.g. human beings, pigs and poultry, only gain from their foodstuffs substances that are initially present, even if in a combined state. In animals with enlarged caeca (horses and rabbits) and still more with ruminants (cattle, sheep and goats), this is not true. The large populations of micro-organisms that inhabit their guts can degrade low-grade foodstuffs containing cellulose and, if simple or complex nitrogenous substances are available, can convert part to a mixture of amino acids. As the micro-organisms 'spill out' from the rumen or caecum the animal digests them, and so receives a balanced supply of nutrients. Ruminants can even be fed on a high-carbohydrate diet, such as barley grains, supplemented with non-protein sources of nitrogen, such as ammonium salts or urea [19]. Consequently the composition of ruminants' diet is less critical than that of non-ruminants.

The abilities of different animals to utilize the components of feedstuffs have been estimated in feeding trials. The digestibility, D, is a measure of the component used by the animal – i.e. not voided in the faeces. Unfortunately various units are in use, and more are being introduced and cannot be fully reviewed here; in Britain for example in considering energy values calories are being replaced by joules [3,4,29,65,100]. Kellner's 'starch equivalent' system of energy units, used for ruminants, is familiar. In terms of 'energy value', digested starch = cellulose = 1; sugars = 0.76; proteins = 0.94, and fats = 1.9–2.4, depending on their nature. Various arbitrary corrections are used with this system. The Scandinavian unit is equivalent to 1 kg 'average' barley, = 0.71 kg (1.56 lb) starch equivalents. Barley generally contains about 81% dry weight (70–71% fresh weight) total digestible nutrients, with a digestible energy of 3.4–3.6 Mcal/kg dry matter and 3.2–3.5 Mcal metabolizable energy/kg dry matter. (One megacalorie, 1 Mcal = 4.184 megajoules, MJ). Metabolizable energy has been calculated from the proximate analysis data for digestible nutrients using empirically derived factors [3]. Thus the metabolizable energy (Mcal/kg) for various materials are digestible crude protein per lb (0.454 kg) feed × 4.3 (coarse fodders) or × 4.5 (others); digestible oil per lb feed × 7.8 (coarse fodder) or × 8.3 (grain); digestible crude fibre per lb feed × 2.9; digestible nitrogen-free extractives per lb feed × 3.7.

Grain crude protein is usually taken as (N × 6.25%). Unfortunately this groups true protein and other nitrogenous materials (non-protein nitrogen, N.P.N; 'amides') together. The food value of protein for non-ruminants depends on its amino acid composition. Barley protein, like that of other cereals, is 'deficient' in lysine, methionine and possibly some other amino acids. Consequently when pig rations are compounded lysine supplements, or complementary feeds rich in lysine must be used. Pig requirements for lysine and methionine increase from 8.9 to 21.2 and from 3.1 to 8.3 g/day respectively, as the pigs' weight increases from 20 to 80 kg [62]. The need for high lysine barleys to reduce the need for expensive supplements is urgent. High-lysine barleys may become important for human consumption. Supplementing wheat flour with 0.25% lysine raises the 'usable protein' content for man by one third; a similar result would probably be obtained with barley flour. To keep a 70 kg (154 lb) man in nitrogen balance takes a daily supply of about 30 g of high-quality protein or 100 g of low quality protein – each supplying 4–6 g of essential amino acids. Among estimates of human nutritional requirements are 3200 Kcal/day for moderately active 70 kg (154 lb) men, 2300 Kcal/day for 58 kg (128 lb) women and peak requirements for calcium (Ca) of 1.4 g/day and for iron (Fe) of 12 mg/day for male adolescents [7]. The essential amino acids for men are lysine, methionine, phenylalanine, threonine, leucine, isoleucine, tryptophan and valine. Young pigs require these, together with arginine and histidine. Poultry also need glycine. Varietal differences in amino acid composition and differences between different lots of grain of one cultivar are well established. The feed values of materials can only be assessed in direct feeding trials. Feeding mice and following weight gain and carcass analyses can serve as a quick check on feed values [74]. Trials using insects and other creatures as the test organisms have been unsatisfactory. The reduction in feed value of overheated grain, caused by the ε-amino group of lysine reacting with other grain constituents, is readily demonstrated. Barley is a superior feed to rye, apparently owing to its smaller content of toxic 5-*n*-alkyl resorcinols, which also occur in significant quantities in wheat bran [13,74]. The strong chelating agent phytic acid interferes with calcium and iron absorption from the gut of mammals, so supplements of these elements must be made to cereal feeds. Barley contains inhibitors of trypsin, chymotrypsin and perhaps other digestive enzymes; the significance of these inhibitors is uncertain (Chapter 4). The feed value of foodstuffs can be reduced by the presence of mycotoxins (Chapter 10). Estimates of digestibilities are sometimes made using *in vitro* digestion techniques, often with rumen liquor or standardized mixtures of enzymes.

Roughages are those foodstuffs, such as hays and straw, which are high in fibre and low in nutrient value. Some roughage is desirable in most ruminant feeds. Concentrates, such as barley grains, have a high utilizable energy content per unit weight. Proximate analyses of foodstuffs normally include moisture content (%), ash content (%), 'crude protein' (N × 6.25, %), ether extractives (*syn*. oil, fat, %), fibre (%), and, by difference, nitrogen free extractives (%), (Tables 14.1, 14.2 and 14.3). Rarely more specific estimates may be made. These fractions have different digestibilities. For example, when adult fowls were fed barley, (composition (%): water, 14.3; crude protein, 11.0, ether extract, 2.1; N-free extractives, 64.8; crude fibre, 5.4; ash, 2.4; sugar, 1.3; starch, 51.8; pentosan, 8.9; cellulose, 4.7; lignin, 1.6), the digestibilities of some of the fractions were: starch and sugars, 100%; pentosan, 25–30%; cellulose and lignin, 0%; crude protein, 75–80%; ether extract 62–76%; N-free extractives, 83–85%; crude fibre, 4–10% [10]. The fractions mentioned 'overlap' and therefore the total given exceeds 100%.

For cattle the food intake is calculated as daily maintenance and production allowances, according to breed, body weight, whether or not pregnant, and by the amount of milk being produced [3,65]. A dairy cow weighing 508 kg (10 cwt) will eat about 15.4 kg (34 lb) dry matter/day. For maintenance this must contain about 3.0 kg (6.5 lb) starch equivalents (S.E.), 0.3 kg (0.65 lb) of digestible crude protein (D.C.P.), 18 g Ca, 25 g P, 7.5 g Mg and 9.0 g Na. In addition, for each gallon (about 4.6 l) of milk produced, the ration must include 1.1 kg (2.5 lb) S.E., 0.23 kg (0.50 lb) D.C.P., 13 g Ca, 8 g P, 3 g Mg and 2.8 g Ca.

Calculating the composition of rations for non-ruminants, e.g. pigs, is more difficult, since allowance must also be made for a correctly balanced supply of amino acids and vitamins A,D,E,B_1,B_2, nicotinic acid, pantothenic acid, pyridoxine, and choline. Vitamin B_{12} must also be supplied; it does not occur in plants. In addition, biotin, folic acid, inositol and *p*-aminobenzoic acid are probably required [29]. Pigs require a diet relatively low in fibre, without roughages. With limited access to foodstuffs a pig of 18 kg (40 lb) may consume 0.91 kg (2 lb) of air-dry food per day, and a 90.7 kg (200 lb) pig, 3.2 kg (7.1 lb). If the food is continuously

Table 14.1 The composition and nutrient value to ruminants of various foods (from M.A.F.F. [65]; Crown copyright: by permission of H.M.S.O.).

Food name	Dry matter content (g/kg)	Metabolizable energy (MJ/kg DM)	Digestible crude protein (g/kg DM)	Analysis of dry matter (g/kg)				
				Crude protein	Ether extract	Crude fibre	N-free extract	Total ash
Barley in flower	250	10.0	46	68	16	316	536	64
Barley (just past milk stage) hay	850	8.8	54	81	22	289	533	74
Barley straw, spring	860	7.3	9	38	21	394	493	53
Barley straw, winter	860	5.8	8	37	16	488	392	66
Barley (whole crop) tower silage	400	9.6	50	95	22	250	570	63
Barley grain	860	13.7	82	108	17	53	795	26
Rye grain	860	14.0	110	133	20	22	802	23
Wheat grain	860	14.0	105	124	19	26	810	21
Fine barley dust	860	13.5	101	136	26	52	750	36
Barley, brewers' grains, fresh	220	10.0	149	205	64	186	500	45
Barley, brewers' grains, ensiled	280	10.0	149	204	64	189	500	43
Barley, brewers' grains, dried	900	10.3	145	204	71	169	512	43
Barley, distillers' grains, fresh	250	11.8	237	320	116	136	396	32
Barley, distillers' grains, dried	900	12.1	214	301	126	110	443	20
Barley, ale and porter grains, fresh	250	10.2	178	240	76	212	428	44
Barley, ale and porter grains, dried	900	10.3	153	219	74	194	477	36
Barley malt culms	900	11.2	222	271	22	156	471	80
Hop leaves and bine (dried)	890	8.2	90	140	39	273	426	121
Hops, spent, fresh	250	6.3	52	172	76	236	456	60
Hops, spent, dried	900	6.4	53	172	77	236	443	72
Yeast, dried	900	11.7	381	443	11	2	441	102
Yeast, wood sugar (dried)	900	12.6	471	523	14	0	381	81

Food names	Gross energy (MJ/kg DM)	$\left[\dfrac{Q}{\dfrac{ME}{GE}}\right]$	Digestible organic matter in dry matter (DOMD, %)	Digestibility coefficients (decimal)				
				Crude protein	Ether extract	Crude fibre	N-free extract	
Barley in flower	17.7	0.56	66	0.68	0.60	0.64	0.75	
Barley (just past milk stage) hay	17.7	0.50	58	0.67	0.42	0.62	0.63	
Barley straw, spring	18.0	0.40	49	0.24	0.33	0.54	0.53	
Barley straw, winter	17.8	0.32	39	0.22	0.29	0.38	0.50	
Barley (whole crop) tower silage	18.0	0.54	62	0.53	0.61	0.53	0.74	
Barley grain	18.3	0.75	86	0.76	0.80	0.56	0.92	
Rye grain	18.4	0.76	87	0.83	0.65	0.53	0.92	
Wheat grain	18.4	0.76	87	0.84	0.63	0.47	0.92	
Fine barley dust	18.4	0.73	83	0.74	0.91	0.24	0.92	
Barley, brewers' grains, fresh	19.6	0.51	59	0.73	0.86	0.39	0.62	
Barley, brewers' grains, ensiled	19.7	0.51	59	0.73	0.86	0.39	0.62	
Barley, brewers' grains, dried	19.8	0.52	60	0.71	0.88	0.48	0.60	
Barley, distillers' grains, fresh	21.6	0.55	65	0.74	0.87	0.47	0.62	
Barley, distillers' grains, dried	21.9	0.55	65	0.71	0.88	0.48	0.62	
Barley, ale and porter grains, fresh	20.2	0.51	59	0.74	0.86	0.39	0.62	
Barley, ale and porter grains, dried	20.1	0.51	60	0.70	0.88	0.48	0.60	
Barley malt culms	18.4	0.61	72	0.82	0.75	0.91	0.73	
Hop leaves and bine (dried)	17.6	0.47	51	0.64	0.72	0.31	0.71	
Hops, spent, fresh	19.6	0.32	35	0.30	0.63	0.17	0.47	
Hops, spent, dried	19.4	0.33	36	0.31	0.65	0.17	0.48	
Yeast, dried	18.3	0.64	75	0.86	0.40	0.00	0.82	
Yeast, wood sugar (dried)	19.2	0.66	81	0.90	0.23	0.00	0.88	

Table 14.2 Proximate analyses of some North American barleys and other materials used in animal feeds (abstracted from Geddes [35,36]; Hill [42]).

Product	Moisture (%)	Protein (%)	Fat (%)	Fibre (%)	N-free extract (%)	Ash (%)
Barley, lightweight	10.2	12.3 (7.6–14.4)	2.3 (1.3–2.8)	8.5 (4–8)	63.7 (62–66)	3.0 (2–3.2)
Barley, hull-less or bald	9.8	11.6	2.0	2.4 (2–3)	72.1	2.1
Barley, bran, nearly all hulls	8.1	5.9	1.3	26.4	51.8	6.4
Barley feed, low grade	9.9	12.3	3.5	14.7	56.2	5.3
Barley feed, high grade	8.0	13.2	3.5	8.4	61.0	4.0
Barley malt	6.6	12.7	2.1	5.4	70.9	2.3
Barley screenings	11.4	11.5	2.8	9.5	60.6	4.2
Brewers' grains, dried 25% protein or over	7.4	26.6	6.8	14.6	41.0	3.6
Brewers' grains, dried, 23–25% protein	6.1	23.8	6.5	14.9	44.9	3.8
Brewers' grains, dried, below 23% protein	7.0	21.6	6.2	16.1	45.2	3.9
Brewers' grains, dried, from Californian barley	8.9	20.0	5.7	18.1	43.6	3.7

Test weights grains, (lb/bu N. American): husked, 39–57; naked, 58,65.
Total digestible nutrients (%): husked, 86–89; naked, 89, 90.

Table 14.3 Analyses of some barleys of the U.S.S.R., and their products (from Myasnikova et al. [75]).

	Water (%)	Crude protein (%)	Fat (%)	Cellulose (%)	Ash (%)	Carbohydrate (%)	Starch (%)	Pentosan (%)	Sugars (%)
Barley grain	14	10.5	2.1	4.5	2.5	66.4	—	—	—
Barley grain	0*	—	—	—	2.4–3.0	—	56–66	9–12	—
Covered grain	0*	13.5	2.4	5.9	2.8	—	60.0	9.0	0.5
Hulls	0*	3.5	0.6	28.0	9.5	—	0	16.5	0.9
Barley flour	0*	12.9	1.8	2.4	2.2	—	77.1	—	—
Pearled barley	0*	9	1.2	1.3	1.2	—	85.0	—	0.5
Crushed barley	0*	11	1.5	2.0	1.5	—	82.0	—	0.4

*Dry matter basis. Other data. Composition of grain: Husks, 9–14%; seed coats 5.5–6.5%; aleurone layers, 11–13%; embryo 2.5–4.0%; starchy endosperm 65–68%. Grain dimensions: length, 7–13 mm; width, 2.8–4.3 mm; thickness 1.3–3.6 mm; d, 1.21–1.27 g/cc; thousand grain weight 22–50 g (dry weight); test weight 530–640 g/litre.

available consumption may be 15–20% greater [29]. Prepared barley grain can comprise 85% of a pig's diet. Pigs thrive on barley preserved wet, with propionic acid. The digestive capabilities of pigs increase with age, so the average digestibility of the organic matter in mixed feeds were found to be: 100 kg (221 lb) pig, 81.6%; 100–180 kg (221–397 lb), 83.7%; 180 kg (397 lb) upwards, 87.5%. At the same time the digestibility, y, of the organic matter in barley was found to be reduced by increasing crude fibre contents, x%, according to the equation $y = 84.93 - 0.42x$ [76]. The crude fibre fraction of barley contained about lignin, 9.0%, cellulose, 82.8%; pentosan, 9.7%.

14.3. Forage and hay

The composition of the plant alters with advancing age (Chapter 6). The vegetative parts of the plant become progressively more fibrous and less nutritious but as the grains form the nutritive value of the whole plant increases (Tables 14.1, 14.4 and 14.5) [99]. During the transition from seedlings to maturity the crude carotene levels fell from 130 to 15 p.p.m., β-carotene declined from 104 to 12 p.p.m., and the crude protein content fell from 15.6 to 5.4% dry matter [98]. The vitamin E, mainly α-tocopherol, has also been determined; plants contained 0.9–2% oil in the dry matter; the tocopherol content of the oils ranged from 1000 to 6960 μg/g [37]. When animals graze young plants they are obtaining a succulent (about 85% moisture), high crude-protein (about 14–18% dry weight) feed. If animals are fed on rapidly growing, newly fertilized barley, there is a risk of nitrate poisoning [105]. There are occasional claims that green barley contains undesirable traces of oestrogenic materials. If whole-crop barley is to be conserved as hay, it must be cut when the yield is sufficient and the nutritive value is suitable. Mature awns, when dry and hard, can damage animals' eyes, mouths and gut-linings. Consequently barley hay should be taken before the awns are silicified and hardened. In N. America hooded, or awnless rather than awned, barleys may be grown for hay-making. Generally, the plants are harvested when the ears are well formed and the grains are milk- or dough-ripe (Table 14.4). Barley hay should be dried rapidly and handled gently to minimize grain or head losses. With field-dried hay quality is dependent on the weather. Hay well force-dried in a hot airflow can be of excellent quality but the economics of drying depend on the costs of equipment and available fuels [28]. The feed value of hay may be enhanced by the inclusion of a proportion of legumes from an undersown crop. In some British trials a highly-fertilized, short-strawed barley gave the best yield of digestible dry matter when compared with other varieties of cereals [17].

Table 14.4a Analyses of Horsford barley (covered, hooded, six-rowed, *H. vulgare*, L., var. *horsfordianum*) cut at different stages of maturity to leave approximately 1 cm (4 in) of stubble (after Sotola [99]).

Date harvested, 1933	Height (in)	Stage		Age (days)	Water (%)	Composition on a water-free basis							
						Ash (%)	Crude protein ($N \times 6.25$) (%)	Crude fibre (%)	Nitrogen-free extract (%)	Fat (%)	Calcium (%)	Phosphorus (%)	Fraction of whole plant (%)
14 June	16	Bloom on some heads		55	84.3	9.7	17.0	20.2	48.3	4.8	0.33	0.46	
21 June	26	Full Bloom		62	82.4	8.0	17.0	22.8	48.6	3.7	0.37	0.35	
28 June	35	Kernels water stage		69	79.5	7.2	9.9	28.4	52.2	2.3	0.23	0.33	
5 July	42–43			76	74.7	6.9	9.2	27.8	53.9	2.3	0.24	0.25	
8 July		Milk		79	*15.7	7.8	8.4	32.3	49.6	1.9	0.26	0.31	
14 July		Medium dough		85	*14.2	7.2	8.0	24.2	58.4	2.3	0.19	0.29	
26 July		Dead ripe		97	*15.0	7.5	7.3	19.3	64.4	1.5	0.14	0.30	
		Whole plants:	Milk			7.0	10.2	29.8	50.9	2.1	0.26	0.30	100
			Dough			7.1	8.3	22.6	60.0	2.0	0.18	0.29	100
			Ripe			7.5	7.0	18.6	65.3	1.7	0.14	0.29	100
		Leaves:	Milk			14.0	8.1	28.8	45.1	3.9	0.47	0.31	22.8
			Dough			16.8	10.1	26.6	42.7	3.9	0.44	0.29	20.7
			Ripe			15.0	8.2	29.8	43.0	4.0	0.51	0.27	20.9
		Stems:	Milk			3.7	3.8	38.9	52.4	1.2	0.12	0.19	36.8
			Dough			4.2	3.1	36.5	55.1	1.2	0.11	0.15	30.2
			Ripe			6.1	2.2	47.9	43.1	0.84	0.06	0.07	15.2
		Heads:	Milk			6.3	13.3	17.1	61.7	1.7	0.11	0.42	40.4
			Dough			5.6	10.2	11.8	70.9	1.5	0.10	0.35	49.3
			Ripe			4.8	9.0	8.8	76.1	1.4	0.09	0.39	63.9

*Samples dried

Table 14.4b Analyses of field-cured Horsford barley hay (after Sotola [99]).

Stage	Dry matter (%)	Ash (%)	Composition (water-free basis)					
			Crude protein ($N \times 6.25$, %)	Crude fibre (%)	N-free extract (%)	Fat (%)	Calcium (%)	Phosphorus (%)
Horsford barley, field-cured hay, 1933								
Milk	84.3	6.6	7.1	27.3	41.7	1.6	0.22	0.26
Dough	85.8	6.2	6.8	20.7	50.0	2.0	0.16	0.25
Ripe	85.0	6.4	6.2	16.4	54.7	1.3	0.12	0.25

Table 14.5 Composition of whole Zephyr barley at different stages of growth (% of dry matter; from MacGregor and Edwards [61]).

	Heading completed	Flowering	Watery kernels	Milky kernels	Mealy ripe (early)	Mealy ripe (late)	Ripe for cutting
Growth stage*	10.5	10.53	10.54	11.1	11.2 (a)	11.2 (b)	11.4
Dry matter	19.1	20.3	26.2	29.0	35.1	38.7	42.2
Fructose	6.0	5.0	4.1	3.1	2.9	3.1	2.2
Glucose	6.0	6.0	4.2	2.9	2.8	2.0	1.1
Galactose	0	0	0	0.16	0.44	0.46	0.58
Xylose	0	0	0	0	0	0	0
Arabinose	0	0	0	0	0	0	0
Sucrose	1.9	1.5	2.3	3.3	2.1	2.0	0.44
† Oligosaccharides	1.6	3.4	6.9	7.6	3.2	3.5	1.5
Fructosans	3.1	3.3	7.2	12.8	12.2	6.6	2.3
Water-soluble carbohydrates (calculated)	19.3	20.0	26.5	32.6	25.5	18.5	8.6
Water-soluble carbohydrates (found)	19.2	20.2	26.7	30.6	25.5	18.1	8.4
Mannitol	0	0	0	0	0	0	0
Starch	0.28	0.25	0.38	1.02	18.5	34.8	41.3
Formic acid	0.09	0.06	0.10	0.06	0	0	0
Acetic acid	0.43	0.71	0.33	0.65	0.64	0.64	0.51
Succinic acid	Trace	Trace	Trace	Trace	Trace	Trace	Trace
Malonic acid	0	0	0.09	0.08	0.07	0.06	0.07
Lactic acid	0	0	0	0	0	0	0
Malic acid	5.6	2.8	1.7	1.3	0.71	0.32	0.15
Citric acid	0.48	0.40	0.54	0.43	0.24	0.09	0.06
pH	5.9	5.9	5.9	5.9	5.9	6.0	6.0

*Growth stage according to Large (1954) – see chapter 6.
† excluding sucrose but including short-chain fructosans

14.4. Silage

Silage is fed mainly to cattle. As a source of silage barley has the advantages of consistent quality and yield, needs little wilting, and is ready to cut (in the U.K.) when there is a pause in harvesting grass, towards the end of July. Yields may approach 18 tonnes dry matter per ha (3–4 tons/acre; Chapter 10) [56]. High-yielding, short-strawed barleys are best for silage production (as with making hay) and, at the expense of more difficult handling, the feed value can be enhanced by undersowing with legumes. In making silage the material should be dry enough (more than 30% dry matter) and should contain sufficient carbohydrates to support an adequate lactic fermentation. Having been chopped and wilted (if necessary) the plant material should be quickly loaded and compacted into a silo or clamp. Under well-sealed conditions additives are not required. Digestibility of the plants is optimal in the early dough stage when the levels of soluble sugar and the moisture content are suitable for ensilage (Tables 14.1, 14.4, 14.5 and 14.6) [30,55,56,61]. Good silage has a feed value about equal to that of the original whole plant. The grains in the silage must be broken to enhance the feed value. However barley silage is poor in crude protein and needs to be supplemented. Analyses show that, in sufficiently dry samples, lactic acid production is favoured

Table 14.6 Composition of whole barley and silages (% of dry matter). Plants of Ymer barley (1049 kg) cut at early milk-ripe stage and ensiled in small towers. No effluent was produced (Edwards *et al.* [30]).

	Barley	*Silage A*	*Silage B*
Dry matter	32.4	32.4	31.6
Ash	12.8	15.3	12.4
Inorganic matter	87.2	84.7	87.6
Crude protein	6.3	6.4	6.3
Ether extract	1.3	1.5	1.6
Crude fibre	24.3	24.8	26.2
Total N	1.0	1.0	1.0
Protein N	0.77	0.39	—
Non-protein N	0.24	0.63	—
Volatile N	—	0.07	—
Water-soluble carbohydrates	25.3	11.7	12.2
Cellulose	25.6	24.6	25.4
Lignin	6.0	6.2	6.5
Acetic acid	—	1.6	1.6
Propionic acid	—	0.16	0.14
Butyric acid	—	Abs.	Abs.
Succinic acid	—	0.09	0.10
Lactic acid	—	3.4	3.1
Ethanol	—	0.53	0.46
pH	5.8	3.9	4.0

during ensilage and undesirable butyric acid fermentation is suppressed (Table 14.6). Generally, crude protein contents fall in the range 6–8%, fibre 20–25%, and starch equivalent values 50–55% dry matter. Some examples of barley silage digestibility figures for cows are dry matter 49–55%, organic matter 62–70%, crude protein, 41–46%; metabolizable energy 2.0–2.2 Mcal/kg; starch equivalents 47–54 [30].

14.5. Barley straw

Barley straw is a useful roughage of limited food value for ruminants. Some varieties produce more digestible straws than others; spring barley straws seem better than winter barley straws [79]. It has a low protein content, low digestibility and a low energy value (Table 14.1). Ground straw is utilized better by cattle than chopped straw which, in turn, is utilized better than long straw. Probably straw fed in a pelleted ration is most acceptable [79]. The broken straw can be mixed with concentrates, pelleted and fed; with long straw the concentrates must be fed separately [88]. To make a significant contribution to the animals' diet straw must comprise a high proportion, so the animals' nutrient intake tends to fall. There are practical limitations to the quantities which should be fed. However, a proportion of straw roughage in the diet is desirable, as it reduces the occurrence of bloat in cattle. Straw can be fed with a urea supplement [26].

Fodder cellulose, prepared from straw by a paper-making process, is more digestible than straw but is uneconomic to produce [18,92]. A more satisfactory way to enhance the digestibility of straw is the on-farm Beckmann process, in which straw is soaked in dilute solutions of sodium hydoxide (e.g. 3h in 8 volumes of 1.5% NaOH or 10 volumes of 0.15% NaOH, for periods of up to 3 days). The straw is then drained, washed to neutrality and fed or ensiled. Some 20–30% of the dry weight is lost, much labour is involved and large volumes of caustic liquid waste are produced. However, the digestibility of the product is increased from about 40% to 60–70%; in other terms it is increased in value from 25 to 60–65 Scandinavian Feed Units/100kg in one trial. The average digestibilities obtained with four lots of material treated by two caustic soaks, then washed and fed to sheep were: dry-matter, 61.3%; organic matter, 65.1%; fibre, 73.4%; N-free extract, 63.7%. The starch equivalent of the treated material averaged 47.8kg/100kg (or lb/lb) dry matter [33]. Because of the associated difficulties Beckmann-type processes are not much used. However, considerable gains in feed value are achieved by spraying chopped straw with low volumes of more concentrated caustic solutions and feeding the product after a period of storage, with or without heat or the application of heat under pressure, and with or without a final application of acid to reduce the alkalinity. Compounded,

Table 14.7 The chemical composition (a) and the digestibility (b) of barley straw treated in various ways and used in feeding trials (from Carmona and Greenhalgh [18]). The treatments were: C – The straw was chopped. M – The straw was milled. CIL – Chopped straw was soaked in NaOH solution (1.5%), then washed; dry weight loss 28%. CIH – chopped straw was soaked in NaOH solution (3%), then washed; dry weight loss 32%. CS – chopped straw sprayed with 16% NaOH; neutralized with propionic acid before feeding. MS – milled straw treated as CS. DM – dry matter. OM – organic matter.

(a)

Treatment		Ash (% DM)	Crude protein (% DM)	Neutral-detergent fibre (% OM)	Acid-detergent fibre (% OM)	Acid-detergent lignin (% OM)	Cellulose (% OM)	Gross energy (Mcal/kg DM)
(C)	Chopped	6.1	5.0	85.6	57.5	7.4	46.0	4.23
(M)	Milled	4.4	3.2	89.7	66.9	8.3	58.5	4.36
(CIL)	Chopped, 1.5% NaOH	3.0	2.4	93.4	74.7	8.2	63.7	4.40
(CIH)	Chopped, 3.0% NaOH							
(CS)	Chopped, sprayed	14.8	5.3	70.0	53.4	6.6	41.1	4.10
(MS)	Milled, sprayed							

(b)

	Treatments						S.E. of differences
	C	M	CIL	CIH	CS	MS	
Digestibility (%)							
Organic matter	45.4	44.9	71.1	69.6	60.8	63.5	± 2.89
Energy	40.4	39.1	65.0	64.4	57.3	60.2	± 3.23
Energy as straw	40.4	39.1	65.0	64.4	53.2	56.5	—
Acid-detergent fibre	49.4	47.9	71.9	73.6	59.0	62.9	—
Cellulose	58.4	57.1	81.5	81.6	73.1	73.8	—
Intake (per day)							
Dry matter (g)	501	696	742	869	950	1048	± 90
Organic matter (g)	471	654	704	841	809	893	± 78
Digestible organic matter (g)	218	293	494	587	492	568	± 54
Gross energy (Mcal)	2.13	2.94	3.21	3.82	3.90	4.30	± 0.37
Energy as straw (Mcal)	2.13	2.94	3.21	3.82	3.56	3.94	—
Intake (per kg $W^{0.75}$ per day)							
Dry matter (g)	26.7	36.2	37.1	44.2	48.4	53.6	± 4.2
Digestible energy (kcal)	46	60	105	125	114	132	—

neutralized and unneutralized, treated straws are accepted by sheep and cattle. With wheat straw treated with sodium hydroxide, then stored for 13–21 days, digestibility increased with increasing quantities of alkali up to 9%. Straw treated with 6% NaOH and stored for 21 days had a digestibility of about 68%, and contained about a third of the original alkali unreacted [104]. Heating, e.g. with steam, causes the alkali to degrade the straw more rapidly. Comparative trials indicate that the sprayed product, whether subsequently stored or steam-heated, is not as digestible as the soaked, but there are no losses and the process is comparatively convenient (Table 14.7) [18,20,63]. In some trials the level of caustic used (16% NaOH) was high, and the need to neutralize the product would have made the treatment uneconomic. Proper control of this process is essential. Feedstuffs manufacturers have started using the process in the U.K. A machine has been devised that applies 5% NaOH to chopped straw and forms it into cubes ('cobs', briquettes) at a high pressure and temperature in a cubing machine. *In vitro* estimates of digestibility were for the straw 40% and, for the treated material about 70%. *In vivo* digestibilities for the treated materials were slightly lower, about 65%. Higher doses of NaOH gave products with higher digestibilities. Treatment of the straw increased the energy value from 25 Scandinavian Units (S.U.)/100 kg to 51 S.U./100 kg (fresh wt.) or 62 S.U./100 kg (dry weight). The product did not need to be neutralized, and could be fed in mixtures. Using a weak base for the treatment, such as ammonia instead of caustic soda, there was a smaller increase in digestibility, but up to 2% N (about 12% 'crude protein') was bound in the product [92]. Alkali-treated straw will probably become a generally accepted constituent of compound ruminant feedstuffs, and may become the most valuable use for straw.

14.6. Barley Grain

Barley grain is rich in starch, and sugar, relatively poor in protein and very low in fat (Tables 14.1, 14.2 and 14.3). Most of the fibre is in the husk, pericarp and testa. The protein is of variable composition, and is not well balanced for non-ruminant nutrition (Tables 14.9–14.11). Feed grain in commerce is priced according to variety, appearance, freedom from dirt and extraneous matter, moisture content, and absence of deterioration, but *not* apparently by its analysis. In N. America grain is often sold by grades, based on grain colour, area of origin and test weights (i.e. kg/hl, lb wt/bu). In vigorously threshed grain there is a supposed correlation between value and increasing test weight. As test weights are lowered by adhering awns – a sign of gentle threshing and good grain handling, it is hard to accept this as a universal criterion of quality [42]. Published analyses provide only very rough guides to grain composition.

Table 14.8 Crude protein (N × 6.25, % dry matter) and amino acid composition (amino acid, g/100g recovered) of some barleys and their separated parts (from Robbins and Pomeranz [94]).

	Husked (covered) barley					Himalaya, a naked barley		
	Whole* kernel	Lemma†	Palea†	Germ*	Dehusked* degermed residue	Whole kernel	Germ	Endosperm
Yield (%)	100.0	7.3	3.1	3.7	85.9	—	—	—
Crude protein	12.4	1.7	2.0	35.0	12.3	17.6	33.2	8.3
Lysine	3.9	6.0	6.1	7.2	3.6	3.2	6.2	2.8
Histidine	2.2	1.5	1.8	3.1	2.2	2.3	3.3	2.0
Ammonia	3.0	3.2	3.4	2.3	3.1	3.5	3.5	3.6
Arginine	4.4	4.9	5.0	9.5	4.4	4.3	9.8	3.9
Aspartic acid	6.8	11.6	11.7	10.6	6.3	6.0	10.3	4.8
Threonine	3.4	5.5	5.5	4.5	3.5	3.2	4.1	2.8
Serine	3.7	5.9	6.1	4.4	3.7	3.9	4.4	3.2
Glutamic acid	26.1	12.8	13.1	14.6	27.0	28.2	15.8	29.5
Proline	11.4	4.9	3.8	3.9	11.8	11.9	4.9	14.0
Cystine/2	1.0	0.1	0.2	0.7	1.1	1.0	0.7	1.2
Glycine	4.2	7.4	7.5	6.7	4.0	3.6	6.3	2.9
Alanine	4.4	7.7	7.9	7.0	4.1	3.8	6.2	3.2
Valine	5.3	7.1	7.1	6.0	5.2	5.2	5.7	5.0
Methionine	2.6	2.0	1.8	2.3	2.6	2.3	2.1	2.9
Isoleucine	3.8	4.5	4.4	3.7	3.8	3.8	3.5	3.7
Leucine	7.1	8.3	8.1	6.9	7.1	6.7	6.5	6.7
Tyrosine	1.9	2.5	2.3	3.0	2.2	1.9	2.7	2.6
Phenylalanine	5.4	4.7	4.5	4.3	5.4	5.3	4.2	5.4

*Average of two varieties, Larker and Piroline
† Average of four varieties including Larker and Piroline

Table 14.9 The yield, crude protein content (N × 6.25, % dry matter) and amino acid composition (amino acid g/100g recovered) of a roller-milled barley sample. The bran fraction contained mainly husk and pericarp; the shorts and tailing flour a mixture of pericarp and aleurone, with some germ and starchy endosperm. The protein content in the flour (65% extraction) is high, due to incomplete separation of starchy endosperm from other materials, and the addition of some tailings flour to adjust the extraction rate (from Robbins and Pomeranz [94]).

	Whole kernel	Flour (65% extraction)	Tailings flour	Shorts	Bran
Yield (%)	100.0	65.0	17.7	11.9	5.4
Crude protein	9.3	9.8	11.3	8.8	3.1
Lysine	4.2	4.1	4.1	4.8	5.0
Histidine	2.4	2.4	2.4	2.1	1.4
Ammonia	3.1	3.1	3.0	2.9	3.5
Arginine	5.3	5.5	5.7	5.9	4.6
Aspartic acid	7.4	7.1	7.5	8.2	8.6
Threonine	3.6	3.6	3.6	3.8	4.2
Serine	4.1	4.0	4.1	4.2	4.7
Glutamic acid	22.6	23.3	22.9	21.2	20.6
Proline	11.4	10.1	9.6	9.2	9.9
Cystine/2	1.1	1.4	1.3	1.1	0.3
Glycine	4.5	4.3	4.7	5.1	5.0
Alanine	4.6	4.4	4.7	5.1	5.0
Valine	5.3	5.2	5.3	5.5	6.1
Methionine	2.5	2.7	2.5	2.5	2.3
Isoleucine	3.6	3.7	3.6	3.7	3.7
Leucine	6.8	7.0	6.8	6.9	7.5
Tyrosine	2.7	3.2	3.0	2.9	2.5
Phenylalanine	4.9	5.0	5.2	5.0	5.1

The composition of barleys grown in one area varies widely (Tables 14.1–14.3).

The vitamin contents of barley grains have not been studied very thoroughly. Analyses often disagree widely. The differences may be real, or may be due to the use of poor analytical techniques. Arbitrarily, some widely discrepant sets of analyses have been ignored.

Ungerminated barley contains no vitamins A, B_{12}, C or D, although the carotenoids and sterols that are present may act as precursors for vitamins A and D respectively. Vitamin E, a mixture of tocopherols occurs in barley oil, a sample of which contained (in mg/g) α-tocopherol, 15.3; γ-tocopherol, 6; ε-tocopherol, 34.2; ζ-tocopherol, 44.5 (Chapter 4) [38]. The other vitamins investigated belong to the B-group. Barley grain contains about 0.1 μg/g of free folates and 0.2 μg/g of bound folates; only formylfolic acid (rhizopteringlutamic acid) was identified. Malt rootlets contained, in addition, folinic acid triglutamate, folinic acid, formylpteroic

Barley

Table 14.10 The protein contents and amino acid composition (g/100g amino acid recovered) of barley, and of malt culms (from Pomeranz and Robbins [84]).

Component	Barley	Commercial culms (average of seven samples)	Laboratory culms (average of five samples)
Protein (N × 6.25) %	12.5	29.6	31.0
Lysine	2.9	5.7	5.4
Histidine	1.8	2.1	2.1
Ammonia	3.3	4.0	4.1
Arginine	4.9	5.5	4.9
Aspartic acid	6.3	18.4	22.0
Threonine	3.3	4.3	3.8
Serine	4.1	4.3	3.9
Glutamic acid	25.5	12.8	13.3
Proline	14.6	7.5	9.0
Cystine/2	0.3	0.2	0.3
Glycine	3.9	4.8	4.3
Alanine	4.0	5.9	5.2
Valine	4.9	5.9	5.2
Methionine	2.3	2.1	1.8
Isoleucine	3.5	4.2	3.7
Leucine	6.4	6.4	5.6
Tyrosine	2.9	2.3	2.1
Phenylalanine	5.2	3.9	3.4

acid (rhizopterin), and unidentified substances, but these could not be assayed because of the high levels (approximately 600 µg/g) of the deoxyribosides of guanine, uracil and thymine [9]. Estimates of choline contents in barley grain (as the chloride), range from 0.96 to 2.2 mg/g dry weight [51,52]. Estimates of thiamine contents in barley grains (in µg/g) include 5.4–7.5; 3.8–9.2; 1.2; 3.0; and 16. The discrepancies underline the difficulties of this type of estimation. More than half of the grains' thiamine is concentrated in the embryo; 8% is in the embryonic axis and 49% is in the scutellum at concentrations of 15 µg/g and 105 µg/g respectively [43]. Riboflavin contents (in µg/g) of 0.8–3.7 have been recorded, with estimates averaging about 1.5 [40,51,52]. Malt culms contain 13.8–15 µg/g [46]. Nicotinic acid (as the amide) occurs in bound forms in cereals not only as the well-known cofactors NAD^+ and $NADP^+$ but also in niacytin [66]. Nicotinic acid in niacytin is unavailable to mammals and to many micro-organisms. Estimates of nicotinic acid in barley grains range from 47 to 147 µg/g. Levels of pantothenic acid (as the calcium salt) in grain are 3.7–4.4 µg/g, and 27 µg/g in malt culms [40,47]. Estimates of vitamin B_6 (pyridoxin, pyridoxal and pyridoxamine) are 11.5 and 2.7–4.4 µg/g [40,47]. The level of vitamin B_6 in malt culms is given as 10 µg/g. Biotin contents of barley grains are 0.05–0.1 µg/g [40].

Table 14.11 Quantities (μg/g tissue) of some minerals in naked Himalaya barley, and some fractions prepared by hand-dissection or pearling (from Liu et al. [59]).

Part 1

Element	Whole kernels	Germ	Sections of degermed grains*		
			Proximal	Central	Distal
P	5 630	12 930	5 280	4 790	5 130
K	5 070	10 900	5 240	4 170	5 020
Mg	1 410	2 940	1 500	1 230	1 370
Ca	406	740	437	302	404
Na	254	—	—	—	—
Fe	36.7	56.5	47.6	27.9	30.6
Zn	23.6	71.3	27.2	14.1	19.9
Mn	18.9	69.4	19.6	14.1	16.5
Cu	15.1	—	—	—	—
Al	4.9	26.5	8.7	3.6	4.0
Mo	1.35	2.32	2.01	1.06	2.21

Part 2

Element	Hand dissected endosperm	Pearling, after abrasion for (sec)			Pearled grain after abrasion for (sec)		
		30	90	270	30	90	270
P	1 120	10 350	10 430	8 150	4 140	3 160	2 210
K	1 440	14 200	12 100	8 780	4 530	3 330	2 830
Mg	78	3 210	3 475	2 280	1 210	779	463
Ca	132	902	729	490	320	258	232
Na	58	800	472	266	161	102	93
Fe	10.0	139	104	76.4	25.7	17.0	13.0
Zn	4.7	78.6	54.7	37.4	20.6	16.9	15.7
Mn	10.0	63.0	39.7	23.2	15.5	13.4	13.3
Cu	3.5	38.0	20.2	11.3	7.1	5.4	6.1
Al	tr.	118	62.5	36.8	1.2	0.8	tr.
Mo	0.41	3.42	2.91	2.56	1.50	1.65	1.11

tr = trace quantity detected

* = Vertically divided parts of the degermed grain.
Proximal – adjacent to the embryo. Distal – apical, furthest from the embryo. On average. germ. 4.3; proximal, 20.9; central, 49.5; Distal, 25.3 (% whole grain).

Inositol occurs in the grain mainly as inositol hexaphosphate, phytic acid. The bulk of the phytate must occur in the aleurone layer, probably in the aleurone grains [21,82]. Estimates of inositol levels (as mg/g) are 0.18 (free) and 1.4–2.9 and 3.2 (bound) [77].

Representative values for the mineral content of barley grains, ash content about 2%, have been given (in mg/100g dry weight) as, K, 580; P, 440; S, 160; Mg, 180; Cl, 120; Ca, 50; Na, 77; Si, 420; Fe, 5; Mn, 2; Cu, 0.5; Ni, 0.02; Mo, 0.04; Co, < 0.005; I, 0.002 [51]. Minerals are variable [72]. The levels of minerals change in growing grains [60]. The distribution of ash, as other grain constituents, is uneven throughout the kernel, a matter of great importance when considering the values of separated grain fractions (Tables 14.8, and 14.11).

Grain for ruminant feeding may be stored dry, or wet under anaerobic conditions, or wet treated with preservatives such as propionic acid (Chapter 10). If storage is adequate the feed value is maintained. Whole grain is not masticated thoroughly and a proportion is wasted, passing out intact in the faeces. Thus grain should be broken before feeding. This may be achieved by rolling dry, rolling into flakes with steam heated rollers after steam or pressure cooking, coarsely milling (cracking or 'grittling') into pieces, or by popping [101]. These treatments increase the digestibility of grain and may improve its acceptability, so that more is eaten. The addition of urea supplements for ruminants may be made by spraying the grain with a saturated solution of urea; this is absorbed quantitatively by the grain; the product's palatability is good [78]. Finer grinding is desirable for grain fed to pigs and poultry. The broken grain may be fed wet, or after cooking, or blended into mixed feeds. As barley grain comprises 85–90% of the diet of some farm ruminants and pigs it is important to achieve efficient utilization and ready acceptance.

Feeding barley to poultry was disliked in the U.K., but in fact chicks can use 65% and adults 75% in the diet [1]. However, barleys from dry areas (e.g. Australia, Lebanon and some parts of the U.S.A.) and steamed barleys are unsuitable for feeding to chickens. Growth is less than expected, and the animals produce sticky, wet droppings which foul the litter. These problems can be overcome by subjecting the barley to a wet treatment, or by supplementing the feed with enzyme preparations from micro-organisms. In some circumstances it may also be advantageous to add enzymes to pig feed [25,54,71]. Probably β-glucan gums are mainly responsible for the adverse effects since it seems to be the β-glucanase of the enzyme mixtures which exerts the corrective effect [14,93]. Probably the enzymes work by reducing the viscosity of the gum.

From the earliest times to the present day barley grain has been used as human food (Chapter 3). The ancient Greeks apparently believed it was the primary food given by the gods and ate a preparation, appropriately called 'alphita', made of roasted and roughly ground grain [6]. Barley is eaten in many forms. Barley flour used alone in baking produces a 'heavy' loaf as barley lacks gluten or other cohesive proteins [24]. 'Flat' barley bread was widely consumed in Europe from before Roman times (gladiators ate so much they were called Hordearii) to close to the present day [5]. In wartime 10–15% of barley flour was added to wheat flour to produce acceptable bread in the U.K. [48]. Breads with 30% barley flour mixed in with strong wheat flour are acceptable, and may produce loaves of unexpectedly large volume. There are marked differences between the baking values of barley varieties. There seem to have been no recent studies on barley bread; the mean analysis of two samples was (%): water, 12.4; nitrogenous substances, 5.9; fat, 0.9; sugar 3.9; carbohydrates, 71; fibre, 5.6 [8]. Because of the high phytate content, barley bread should presumably be made with flour supplemented with calcium and iron as well as vitamins, as is wheat bread [52]. In the East barley, particularly naked barley, is used in human foodstuffs roasted or parched, or as flour-based products, or pearled and in soups. In the West products such as pot (Scotch) and pearl barley and barley groats are prepared by successively grinding away the outer layers of the grain. These products are used mainly in soups, stews and sauces. The milled surface layers are used in compounded animal feeds (Tables 14.8, 14.9 and 14.11). Barley flour may be prepared from pearl barley, in which case it is nearly pure endosperm, or from roller-milled grain (Tables 14.9 and 14.10).

Because of the unequal distribution of vitamins, minerals, proteins, fibre, fat, starch, and other components within the grain, the nutrient values of milled and pearled barley products depend greatly on the proportions of retained aleurone layer and (rarely) germ (Tables 14.8 and 14.9) [57]. Analyses of grains and milling fractions are used as guides to milling operations. They give a good idea of the composition of the grain parts. Barley husk is rich in ash (6% ash containing about 66% silica) [51]. Within the grain the embryo and aleurone layer contain most of the remaining ash [21]. As barley is subjected to successive abrasive operations so the ash content declines (Table 14.11). The micro-organisms associated with the grain, including thermophilic fungi and actinomycetes, are mainly removed with the husk and pericarp [34]. The high thiamine and fat content of the offals (10–20% of materials removed), indicates that aleurone and especially embryo is being removed at this stage when most of the crude fibre, from the husk, has already gone. Pearl barley is relatively poor in vitamins, giving values (mg/100g) of thiamine, 0.12; riboflavin, 0.08, and niacin, 3.1 [32]. The distribution and composition of the ash (mineral components) in naked barley, in its separated parts, in milled and pearled barley and barley flour and by-product fractions have all been studied (Tables 14.8, 14.9 and 14.11) [58,59,83,94]. Reference to the vitamin, mineral and fat figures shows that as the rich surface-layers are removed, so the nutritional value of the main product drops. The outer layers of the grain are rich in protein, and the protein in the aleurone and embryo is proportionately richer in essential amino acids [85]. Pot, or Scotch barley is nutritionally superior to pearl, as well as being cheaper to produce, since it retains more of the outer grain layers.

It is not clear how much barley, or malt and malt extract, is used in western type breakfast cereals. There seem to be no particular problems in roasting, flaking or popping various milled and perhaps cooked barley grain products [50,51,103]. When cereals are puffed – or flaked – phytate is partly destroyed: ranges of 33–90% and 13–50% respectively having been noted [11]. Vitamin B_1 is destroyed in puffing; an analysis for puffed barley gave values of vitamin B_1, 0; riboflavin, 1.1 µg/g; nicotinic acid 76 µg/g; crude protein (N × 5.7), 8.1; Fe, 3.1 mg/100g; total P, 338 mg/100g; phytate-P, 15 mg/100g [11]. Barley flour (farina) contains a level of thiamine (about 1.8 µg/g fresh weight) that does not decline on storage. However more is destroyed on cooking stored material (about 33%) than fresh (about 11%) [95]. Malt flour, a highly diastatic material, is used in making special breads, to which it imparts a characteristic texture and flavour (cf. Chapter 15). Malt extracts, essentially unhopped worts concentrated under reduced pressure to 75–81% solids, SG, 1.415–1.430 may or may not contain diastatic enzymes. They are used to a limited extent in making sweets and confectionery, as a luscious sugary flavouring material containing other nutrients and vitamins. Malt extract is also used as a pleasant-testing carrier to make medicaments, such as vitamin-rich cod-liver oil, acceptable to children.

14.7. By-products for animal feed, derived from barley

By-products derived directly or indirectly from barley, or industries based primarily on barley, include barley pearlings or bran, grain dust, malt culms and dust, brewers' and distillers' spent grains or draff, brewers' spent hops and break, yeast, and distillers' solubles (Tables 14.1–14.3) [103]. The outer pearling fractions though fibrous, are rich in lipid, crude protein, vitamins and ash and have a high value when blended into animal feeds. The composition of pearlings can be varied widely depending on the proportions of husk and inner layers that are included.

Malt culms (sprouts, rootlets, coombs, cummins) are separated from the kilned malt and, together with malt dust, are sold for compounding into animal feeds or may be fed direct to cattle after wetting, as a proportion of their ration. They are fibrous (8.6–11.9% fibre) but are rich in crude protein (25–34%), fat (1.6–2.2%), minerals (6.0–7.1% ash, including phosphate, about 1.8%), and N-free extract (35–44%, or even up to 50%). About 70–80% of the crude protein is true protein. Of the total nitrogen (3.5–4.2%), about one third (1.1–1.5%) is soluble. Asparagine, allantoin, betaine, choline, adenine, tyramine, hordenine and candicine have been noted among the simple nitrogenous substances. Amino acids analyses of sprouts are available (Table 14.10). The pentosan (15.6–18.9%) swells on contact with water, and dried rootlets are hygroscopic. Cellulose (about 10%) and some lignin are also present.

Traces of glucose and fructose are found but sucrose (3.8–7.7%) is the main soluble sugar [41,49,84]. Moisture contents should be less than 6–7%. Sprouts are rich in various vitamins including vitamin E (tocopherols, 8–10 mg/100 g) [84]. These and the high level of nitrogenous materials make them an excellent source of growth factors for many micro-organisms including yeasts and *Eremothecium ashbyii*, a mould that produces riboflavin [86]. Their composition is very variable, depending on the malting processes employed and probably the intensity of kilning. At least lightly kilned rootlets are a rich, and unexploited, source of some hydrolytic enzymes including nucleases, proteases, phosphatases and carbohydrases. It has not been found worthwhile to germinate barley before feeding it to farm animals although vitamin C appears and increases during germination. The vitamin C is destroyed when malt is kilned.

The composition of brewers' and distillers' spent grains and draff depends on the composition of the grist – the proportions and types of malt and unmalted adjuncts used, and the completeness of the starch conversion and the extraction of the soluble materials produced (Tables 14.1 and 14.2, Chapter 16) [49,69,103]. The preferential removal of the soluble sugars and starch, which mainly go to form malt extract, means that the spent grist residues are enriched in fibre, protein, fat and ash. Residues from malt whisky distilleries resemble traditional brewery grains more than residues from grain distilleries. In each case the unmalted adjuncts may be prepared barley, wheat, maize or any of a range of cereals. The spent grains may be sold wet (about 70–80% moisture) or dry (about 10% moisture). While the dry grains may be stored, the wet grains must be used quickly, or treated with a preservative, or dried or ensiled since they deteriorate within hours. Breweries also produce spent hops and hot break and some surplus yeast; these may be disposed of with the grains or separately. Spent grains are valuable as cattle food and may also be fed to sheep. Up to 18 kg (40 lb)/day of wet grains may be fed to fattening cattle and 100 kg (2 cwt) of dried grains may be incorporated in every tonne (ton) of mixed feed. Excessive feeding of spent grains may lead to scouring. The value of draff has often been investigated [31,58,69,91].

Surplus Brewer's yeast may find its way into yeast extract, made from autolysed yeast, for human consumption or it may be used in animal feedstuffs (Table 14.1). Distillers have a residue – stillage – which remains in the stills and contains yeast and residual, non-volatile materials, including husk, residual grain, fat, oligosaccharides, limit dextrins and glycerol, protein and B vitamins, but from which the alcohol has been removed. This nutritious material used to be added to the draff for disposal, or put to waste. Now it is collected, another minor proteinaceous waste (slummage) is added to it, then it is screened, the solids are pressed and the liquid is concentrated, dried and sold as DDS, dried distillers solubles, for use in animal feeds [27,87].

14.8. Non-alcoholic beverages

Modest quantities of non-alcoholic drinks, based on barley and malt, are consumed in various parts of the world. Both barley and malt are roasted and hot water infusions of the ground products are consumed – for example as malzkaffee, 'malt coffee', or mugi-cha the Japanese 'barley-tea'. The roasting causes the heat-dextrinization of the starch, a decline in hemicelluloses, the caramelization of sugars, the formation of melanoidins from the interactions between reducing sugars and amino acids with the consequent development of dark colours and flavour, and an increase in acidity and solubility of the grain products [68]. Cereal extracts, going under a wide variety of titles, are sold as soluble powders to make 'instant' drinks. Apparently these are prepared by making hot water extracts of roasted cereals or roasted malt then recovering the dissolved solids by drying either in rotating, heated drums or in spray-driers. The gross composition of one N. American soluble cereal powder; 'instant Postum' is given as: moisture, 2.4%; crude protein, 6.4%; fat, 0.2%; ash, 7.8%; N-free extract, 83.2%; fibre, 0%. The components of the N-free extract, as parts of the whole powder, were: dextrin, 62.7%; sucrose, 7.4%; glucose and fructose, 6.3%; other carbohydrates, 6.8% [50]. Many chemical components contribute flavour and aroma to roasted cereal products. An early example (1894) was the identification of maltol in the condensate deposited in a 'malt-coffee' roasting kiln [12]. In Japan roasted barley is used in preparing a drink for summer, 'mugi-cha'. Roasting at 160°C (320°F) for 20 minutes produces a good quality product, and causes the release of carbon dioxide as browning and the development of the characteristic odour continues. Reactions occurring in the aleurone are responsible for much of this. Components identified, many of which contribute to flavour, include sulphur compounds, aldehydes, ketones, numerous cyclic nitrogenous compounds, organic acids, lactones, and aromatic substances [22,70,96,102].

'Barley water' is made by soaking pot or pearl barley. The product, often flavoured with lemon, is said to be nutritious and is often given to children and invalids.

Various 'malted' beverages, often forms of 'malted milk', are available. In some of these malt extract, selected for its flavour, is blended with a milk fraction, and the mixture is dried and sold as a soluble powder to be reconstituted with warm water. In other cases powders, containing at least some malt extract, are sold to be made into drinks with warm milk.

14.9. Other potential feeding stuffs

There is a need to improve the supply of high-quality, high-protein feeding stuffs. Stringent flavour, safety and quality standards are set for human foods. Consequently, the two 'novel' products mentioned below are most likely to be used, at least initially, in animal foods. Barley grain as a feed needs to be supplemented with a source of essential amino acids such as high quality protein. It may be justifiable to add buffering salts, an inorganic source of nitrogen, or perhaps urea, to ground barley grain, straw, or alkali-swollen straw and by culturing micro-organisms on the mixture, increase the protein content and enhance its content of essential amino acids at the expense of some carbohydrate. On the pilot scale the filamentous mould *Aspergillus oryzae* has been successfully grown, under non-aseptic conditions, in 1000-litre cultures in modified pig-feed mixing vessels [89]. After 24h culture 9.3kg of 'biomass' (31% crude protein) was receovered from an initial 20kg barley, 1kg urea and 2kg KH_2PO_4 (pH 3.5) inoculated with the fungus. Pigs and rats readily ate the product and grew well. The composition of the crude protein was nutritionally favourable. In other work a thermotolerant Basidiomycete, *Sporotrichum pulverulentum (Phanerochaete chrysosporium)* has been cultivated on cereal flours, brans and cellulose, and other organisms have also been grown in quantity. Various 'mushroom-like' fungi could perhaps be used directly in human food. In one case the spent culture liquid, from which the mycelium had been removed, contained enzymes and other potentially useful substances [44]. While such processes might prove inconvenient to the small farmer, they could be valuable to feed compounders. Brewers and distillers might use this approach to 'upgrade' the value of the daily supply of spent grains.

Table 14.12 The average amino-acid composition of unfractionated, 'cytoplasmic' and 'chloroplastic' barley leaf protein preparations (from Byers [15]).

Amino acids	Protein composition (%)*		
	Unfractionated	Cytoplasmic	Chloroplastic
Aspartic acid	9.57	9.62	9.75
Threonine	5.07	5.41	4.82
Serine	4.40	4.10	4.85
Glutamic acid	11.41	11.94	11.00
Proline	4.68	4.62	4.88
Glycine	5.64	5.38	6.12
Alanine	6.71	6.52	7.05
Valine	6.37	6.50	6.16
Methionine	2.24	2.39	2.28
Isoleucine	4.95	4.74	5.25
Leucine	9.33	8.42	10.43
Tyrosine	4.50	4.92	4.49
Phenylalanine	6.22	5.84	6.97
Lysine	6.61	7.06	5.60
Histidine	2.34	2.66	1.82
Arginine	6.89	7.01	6.29

*g amino acid per 100g recovered amino acids.

It may now be practical to prepare 'plant leaf protein' on a large scale [80, 81]. Barley is not one of the best crops to use for making leaf protein, but the feasibility of so using it is apparent [16]. A green forage, such as whole-crop barley, is harvested and crushed to extract the juice, which contains proteins. This extraction should be carried out under controlled conditions, at as low a temperature as possible, with air excluded, and possibly with additions of anti-oxidants to minimize the oxidation of polyphenolic substances, since the quinonoid products react with the proteins in the extract and reduce their feed value. The residual pulp is still highly nutritious for ruminants, and may be fed directly, or after ensilage. The protein needs to be separated from the extracted juice, probably by filtration of a coagulum produced by a controlled heat treatment. The liquid remaining might be used to support nutritionally useful micro-organisms. Such 'protein', adequately preserved, has a high feed value for chicks, pigs and probably human beings. It is relatively deficient in methionine, but in general its amino acid balance is good (Table 14.12) [15]. Such leaf protein concentrates, LPC, should be good 'supplements' for balancing feeds based on grains.

14.10. The technology of preparing grain for food

For feed, cleaned grain is broken down to an extent depending on what animals are to be fed, and whether the product is to be fed on its own or mixed or pelleted with other foodstuffs, either cooked or raw. Cattle and pigs may be fed grain that has been cracked, crushed, 'crimped', or flaked in roller mills. The mill rolls may be plain or the surfaces may be fluted in various ways, and may rotate at the same or different speeds. In this last case the grain is crushed and sheared [23,53,97]. The moisture content of the grain may be adjusted to about 18% to reduce the level of dust produced and ease the work of the mill. Steam heating to $96-99°C$ ($205-210°F$) for 20 minutes then rolling at the same temperature can produce flaked barley that cattle find palatable [39].

The old stone mills in which grain was fed into a central hole in a disc-shaped rotating stone and moved between this and a stationary lower stone to the periphery, where it emerged ground, have almost disappeared. However, various powered stone mills are still used and analagous metal plate mills (Buhr mills) are in use [97]. In plate mills grain is fed into a central space between a stationary metal disc and a disc rotating at 400-600 r.p.m., or between two contra-rotating discs, giving a differential rotation speed of 3-5000 r.p.m. The plates are faced with abrasive materials, or teeth in the case of pin-mills, which break up the grain as it moves outwards to the edge of the disc. The degree of breakage or coarse-grinding (kibbling) achieved depends on the grain, the plate's surface texture, their relative speeds, the clearance between them and the pressure

Barley for animal and human food

applied to hold them together. Probably the most popular grain-breaking device is the hammer, or percussion mill. Grain is delivered into a cylindrical chamber in which rapidly rotating, loosely mounted steel rods, the 'hammers', or in some cases rigid rods or blades, rotate around a central shaft and smash the grain into small pieces. When the grain size is sufficiently reduced it passes through a metal screen and is carried away by an airstream driven by a centrifugal impellor built into the mill. The milling chamber may have static blades or metal ridges mounted to give irregularity to the inner face of the cylinder, and so help in breaking the grain. The screen, which is usually interchangeable to allow different sizes of perforation to be used, may comprise a part or the whole of the curved wall of the grinding chamber. Recommended sizes of screen perforations are – for young pigs 1.6–3.2 mm (1/16–1/8 in); for fattening pigs, 1.1–4.0 mm (3/72–5/32 in); for sheep 3.2–12.7 mm (1/8–1/2 in), and for cattle, 6.4 mm (1/4 in) [64]. The barley meal is collected from the airstream by gravity, or in a cyclone. Any dust carried with the air leaving the cyclone is trapped in filter-socks.

The prime object of reducing grain size and disrupting its structure is to allow the animal's digestive juices to act on the nutritious internal parts. The effects of milling may be augmented by steaming or pressure cooking. Other ways to achieve grain disruption include (i) hot air expansion, or popping, and (ii) wet heat expansion, or extrusion [101]. Popped barley grain, treated at about 15% moisture, heated at 246°C (475°F) for 30 sec, expanded to 1.5–2 × its original size. The starch was about 37% gelatinized, and lysine recovery was good. Probably an initial moisture content of 25% would be preferable [101]. It is possible to pop (torrify) barley by heating it with microwaves. More drastic treatments yield materials with higher digestibilities. Microscopic examination shows that, although some disruption of cell walls occurs in popping the structure of the endosperm – particularly in the sub-aleurone layer – is well preserved [90].

Animal feeds are frequently mixtures of grain and other materials with mineral and vitamin supplements held together with binders such as molasses, which must be used hot, 55°–60°C (131°–140°F), to reduce its viscosity. Permitted antibiotics or other agents to control diseases, and synthetic oestrogens (stilboestrol and hexoestrol) may be added [53,97]. The reduced, and milled components have to be delivered, in proportions regulated by volume or (better) by weight, into a mixing machine. After leaving the mixer the feed may be formed into pellets, cubes or wafers, which can be stored, or the feed-mash may be directed to the animals, or to a holding vessel. Two of the many types of feed mixers are (i) those in which the components are blended in a horizontal or inclined trough by a sequence of moving slats, and (ii) those in which vertical mixing is achieved by a screw-type auger working in the centre of the mass [23,53,

97]. For forming into pellets, cubes, cobs, crumbs or granules, the mashed feed is heated and shaped by pressure in a die, or by extrusion through a die, being sectioned into lengths by knives on emergence [97]. Sometimes the mix, stirred and subjected to shear, is blended with superheated water, then the dough is discharged from an expander orifice so that the drop in pressure allows the water to vapourize and the mass is swollen. On drying the product, for animal or human consumption, has an 'open' crunchy texture, and the components have been cooked. The hot material is cooled, and sieved before storage. Broken fragments are returned to the feed hopper. The size of the product, about 9–13 mm (3/8–1/2 in) diameter cubes, or about 3 mm (1/8 in) diameter pellets, is chosen to suit the animal to be fed. Sometimes large cubes are broken up, and are fed to the animals as 'crumbles'.

Pot and pearl barleys for human consumption are produced by abrasive and breaking processes, the dust or 'offals' normally being used in animal feed. For feed manufacture husks have been removed experimentally from barley after a treatment with a hot solution of caustic soda. Naked grain requires relatively little preparation other than cleaning and mild scouring before subsequent processing. With hulled grain after cleaning, conditioning (the adjustment of the moisture content) and sometimes bleaching, the husk must be removed, a process variously called blocking, hulling or shelling. The grain to be used must be sound, uncoloured (not having a blue aleurone), plump, and of a uniform size. Where bleaching is permitted (in the U.K. it is not) a bisulphite or sulphur dioxide and steam treatment may be given for 20–30 minutes, followed by storage for 12–24 h [51]. Blocking may be carried out in several types of machine in which impact and abrasion against stone, emery, or other composition-faced surfaces, breaks up and wears away the husk, reducing it to a powder. In batch-type hullers (hollanders) the grain is worked against itself and against the machine surface in the gap between an abrasive-surfaced, rotating drum and an enclosing, slowly contra-rotating perforated metal cage. The machine is filled and emptied automatically [51,53]. Other scouring and hulling machines operate continuously, the grain either working its way along inside a cylinder and being driven against an abrasive surface by paddles mounted on a central rotating shaft or by blades impelling the grain around the periphery of a partial cylinder and into a receiving chamber. Alternatively, the hulling is carried out in a pearling machine, consisting of a rotor, or stack of disc-shaped rotors of stone or metal faced with an abrasive, working within a casing also faced with abrasive. The grain is aspirated by a stream of air, which carries away the powdered offals. With the batch machines the abraded grain and offals are separated after the blocking operation, by sieving and aspiration. The blocked barley may be used directly, or cut up into groats. The grain parts may be subjected to a further series of abrasion and

Barley for animal and human food

aspiration steps (up to 5 more) to make pot (Scotch), and pearl barley and barley flour. It may be necessary to allow the grain to cool in the interval between leaving one machine and entering the next. In some countries, but not in the U.K., talc (a mineral dust) may be added to help polish and 'brighten' the final product. The products are rounded pieces of starchy endosperm with little other material, except from the furrow. In the U.K. pot barley, first pearl and second pearl barleys are on offer. Elsewhere about 12 different size-grades are sold [50,51,53,57].

Pot barley, pearl barley and groats may be flaked. Damped (steamed) and cooked endosperm fragments are flattened between hot rollers. After further drying, if needed, the products of 10–16% moisture content are packaged and used, for example in 'porridges' or breakfast cereals.

Barley flour may be the 'offal' from later pearling operations, or milled pearl or blocked barley. Whereas the first two types are made merely by size reduction and grading, the use of blocked barley involves more complicated sieving and separation steps to reduce the percentage of aleurone and embryo fragments [2,83]. Air-classification of the final flour can separate streams having markedly different compositions.

Small quantities of barley products, and malt extract, are used in breakfast cereals. In addition to the products mentioned it is probably possible to 'puff' pearl barley or pastes of barley and wheat flour. A unique granular breakfast cereal, one of the first to be manufactured, is made from a blend of wheat flour and flour from malted barley. The blended flour is mixed with appropriate proportions of salt, water and yeast. After a period of fermentation, 4.5–5 h at 27°C (80°F) the dough is divided into loaves and baked, for 2 h at 204°C (400°F). The loaves are broken up, the particles are sized, rebaked, for 2 h, 121°C (250°F), then packaged. Powdery residues and 'fines' are blended into subsequent doughs. Considerable starch degradation occurs in this process, and about 45% of the solids are soluble in hot water [50,67].

14.11. Future uses of barley as food

In Western countries small quantities of barley will probably continue to be used as porridges, in soup thickening, in special diets for infants or geriatrics, and for those with allergies to wheat. Barley flour may also be used to 'extend' wheat flour. Ways may be found to improve the efficiency with which barley is utilized in animal feed. It is a mistake to assume that all cereal produced – whole plant or grain – would, if not fed to farm animals, be immediately available for human consumption [45,80]. However the best way to improve the utilization of barley is to use as large a proportion as possible directly for human food. The proportion of the grain employed, the 'extraction rate' in milling terms, is higher in naked grains than in husked varieties. Grain of a naked English barley was used in making malted and unmalted flakes, in milk puddings and as soup thickening and was satisfactory in each case [6]. 'Whole grain' barley flour would only be acceptable if made from naked grains which would only need to be cleaned by scouring before milling [53].

Plant breeders should, at least as a precaution, develop locally adapted, high-yielding naked barleys preferably also with high-protein and high-lysine characteristics, as a standby for augmenting human food supplies. The 'apparent' reduction in yield (9–13%), due to non-adherence of husks is easily compensated by the greater digestibility and easier processing of naked grain. If for some purpose husky material was needed it could be supplemented by chopped or milled straw. Industrial, and farm-scale tests of leaf protein concentrates are overdue. While the direct consumption of microbes, grown on barley with nitrogenous supplements, may not appeal to everyone, some Eastern foodstuffs are traditionally prepared in this general manner (on steamed rice), and there should be little difficulty in producing valuable foodstuffs in this way. Meanwhile by-products and straw should be used to the best effect in feeding farm animals. The adoption of 'dry' caustic treatments of straw is a valuable move in this direction. Perhaps similar caustic or ammonia treatments of spent grains, or even whole or rolled grains, might usefully enhance food value for ruminants.

References

1 Adamson, A.H. (1968). *N.A.A.S. Q. Rev.*, **82**, 86–92.
2 Anon. (1934). *Milling*, **83** (9), 240–241.
3 Anon. (1971). *A.D.A.S. Advisory Paper No. 11. Nutrient Allowances and Composition of Feeding stuffs for Ruminants.* M.A.F.F.
4 A.R.C., Agricultural Research Council (1965). *Nutrient Requirements of Farm Livestock No. 2, Ruminants.* London: A.R.C.
5 Ashley, Sir W. (1928). *The Bread of our Forefathers—an inquiry in economic history.* Oxford: Clarendon Press.
6 Beaven, E.S. (1947). *Barley—Fifty Years of Observation and Experiment.* London: Duckworth.
7 Bender, A.E. (1963). *Chem. and Ind.*, 1668–1675.
8 Blyth, A.W., Blyth, M.W. and Cox, H.E. (1927). *Foods: their composition & analysis.* London: C. Griffin & Co. Ltd.
9 Bolinder, H., Kurz, W. and Lundin, H. (1956). *J. Inst. Brew.*, **62**, 497–504.
10 Bolton, W. (1955). *J. agric. Sci. Comb.*, **46**, 119–122.
11 Booth, R.G., Moran, T. and Pringle, W.J.S. (1945). *J. Soc. chem. Ind., London*, **64**, 302–305.
12 Brand, J. (1894). *Ber. dt. chem. Ges.*, **27**, 806–810.
13 Briggs, D.E. (1974). *Phytochemistry*, **13**, 987–996.
14 Burnett, G.S. (1966). *Brit. Poultry Sci.*, **7**, 55–75.
15 Byers, M. (1971). *J. Sci. Fd. Agric.*, **22**, 242–251.
16 Byers, M. and Sturrock, J.W. (1965). *J. Sci. Fd. Agric.*, **16**, 341–355.
17 Cannell, R.Q. and Jobson, H.T. (1968). *J. agric. Sci., Camb.*, **71**, 337–341.
18 Carmona, J.F. and Greenhalgh, J.F.D. (1972). *J. agric. Sci., Camb.*, **78**, 477–485.

19 Chalupa, W. (1968). *J. Anim. Sci.*, **27**, 207–219.
20 Chandra, S. and Jackson, M.G. (1971). *J. agric. Sci., Camb.*, **77**, 11–17.
21 Clutterbuck, V.J. and Briggs, D.E. (1974). *Phytochemistry*, **13**, 45–54.
22 Collins, E. (1971). *J. agric. Fd Chem.*, **19**, 533–535.
23 Culpin, C. (1969). *Farm Machinery*, (8th edition). London: Crosby Lockwood and Son.
24 Cunningham, D.K., Geddes, W.F. and Anderson, J.A. (1955). *Cereal Chem.*, **32**, 91–106.
25 Daghir, N.J. and Rottensten, K. (1966). *Brit. Poultry Sci.*, **7** (3), 159–163.
26 Davies, G.M., Ositelu, G.S. and Hall, G.R. (1969). *Expl. Husb.*, **18**, 32–37.
27 Duckworth, J. (1955). *J. Sci. Fd. Agric.*, **6**, 177–185; 240–250.
28 Durand, M. (1970). *C. r. Acad. Sci. Paris*, **271D**, 2167–2170.
29 Eden, A. (1973). *A.D.A.S. advisory paper No. 7; Nutrient Requirements of Pigs.* M.A.F.F.
30 Edwards, R.A., Donaldson, E. and MacGregor, A.W. (1968). *J. Sci. Fd. Agric.*, **19**, 656–660.
31 El Hag, G.A. and Miller, T.B. (1972). *J. Sci. Fd. Agric.*, **23**, 247–258.
32 Feldberg, C. (1959). In *The Chemistry & Technology of Cereals as Food & Feed*, ed. Matz, S.A. pp. 619–661. Westpoint, Connecticutt: Avi Publishing Co.
33 Ferguson, W.S. (1943). *J. agric. Sci., Camb.*, **33**, 174–177.
34 Flannigan, B. and Dickie, N.A. (1972). *Trans. Br. mycol. Soc.*, **59**, 377–391.
35 Geddes, W.F. (1944). In *Chemistry & Technology of Food & Food Products* ed. Jacobs, M.B. **1**, pp. 612–632; 783. New York: Wiley Interscience.
36 Geddes, W.F. (1963). *Dominion of Canada Grain Research Lab., Annual Report*, **10**, 84–85.
37 Green, J. (1958). *J. Sci. Fd Agric.*, **9**, 801–812.
38 Green, J., Marcinkiewicz, S. and Watt, P.R. (1955). *J. Sci. Fd. Agric.*, **6**, 274–282.
39 Hale, W.H. (1965). *Feedstuffs*, **37** (27), 40–41.
40 Hall, H.H., Curtis, J.J. and Shekleton, M.C. (1952). *Cereal Chem.*, **29**, 156–160.
41 Hegazi, S.M., Ghali, Y., Foda, M.S. and Youssef, A. (1975). *J. Sci. Fd. Agric.*, **26**, 1077–1081.
42 Hill, D.D. (1933). *J. Am. Soc. Agron.*, **25**, 301–311.
43 Hinton, J.J.C. (1944). *Biochem. J.*, **38**, 214–217.
44 von Hofsten, B. (1976). In *Food from Waste* ed. Birch. G.G., Parker, K.J. and Worgan, J.T. pp. 156–166. London: Applied Science.
45 Holmes, W. (1971). In *Potential Crop Production* ed. Wareing, P.F. and Cooper, J.P. pp. 213–227. London: Heinemann Educational Books.
46 Hopkins, R.H. and Wiener, S. (1944). *J. Inst. Brew.*, **50**, 124–137.
47 Hopkins, R.H., Wiener, S. and Rainbow, C. (1948). *J. Inst. Brew.*, **54**, 264–269.
48 Horder, (Lord), Dodds, C. and Moran, T. (1954). *Bread.* London: Constable.
49 Hough, J.S., Briggs, D.E. and Stevens, R. (1971). *Malting and Brewing Science.* London: Chapman and Hall.
50 Jacobs, M.B. (1944). *The Chemistry & Technology of Food & Food Products.* **I** and **II**. New York: Wiley Interscience.
51 Kent, N.L. (1966). *Technology of cereals—with special reference to wheat.* Oxford: Pergamon Press.
52 Kent-Jones, D.W. and Amos, A.H. (1967). *Modern Cereal Chemistry.* London: Food Trade Press.

53 Kuprits, Ya. N. (1965). *Technology of Grain Processing & Provender Milling*, (Translated from Russian by R. Kondor). Jerusalem: Israel programme of Scientific Translations.
54 Laerdal, O.A., Sunde, M.L., Dickson, A.D. and Phillips, P.H. (1960). *Proc. Ann. Mtg. Am. Soc. Brew. Chem.*, 15–21.
55 Lang, R.W. and Holmes, J.C. (1969). *Expl. Husb.*, **18**, 1–7.
56 Lawes, D.A. and Jones, D.I.H. (1971). *J. agric. Sci., Camb.*, **76**, 479–485.
57 Le Clerc, J.A. and Garby, C.D. (1925). *Ind. Eng. Chem.*, **12**, 451–455.
58 Linton, J.H. (1973). *Master Brewer's Ass. Am. Tech. Q.*, **10**, 161–164.
59 Liu, D.J., Robbins, G.S. and Pomeranz, Y. (1974). *Cereal Chem.*, **51**, 309–316.
60 Liu, D.J., Pomeranz, Y. and Robbins, G.S. (1975). *Cereal Chem.*, **52**, 678–686.
61 MacGregor, A.W. and Edwards, R.A. (1968). *J. Sci. Fd. Agric.*, **19**, 661–666.
62 Madsen, A., Eggum, B., Mortensen, H.P. and Larsen, A.E. (1970). *Yearbook Royal Vet. Agr. Coll. Copenhagen*, pp. 1–11.
63 Maeng, W.J., Mowat, D.M. and Bilanski, W.K. (1971). *Can. J. Anim. Sci.*, **51**, 743–747.
64 M.A.F.F. (1968). *Mechanization Leaflet No. 20 for Farmers & Growers; Farm grinding.*
65 M.A.F.F. (1975). *Technical Bulletin No. 33, Energy allowances and Feeding Systems for Ruminants.* London: H.M.S.O.
66 Mason, J.B. and Kodicek, E. (1970). *Biochem. J.*, **120**, 509–513.
67 Matz, S.A. (1959). In *The Chemistry & Technology of Cereals as Food & Feed*. ed. Matz, S.A. Westport, Connecticutt: Avi Publishing.
68 Milić, B.L., Grujić-Injac, B., Piletić, M.V., Jajšić, S. and Kolarov, L.A. (1975). *J. agric. Fd. Chem.*, **23**, 960–963.
69 Miller, T.B. (1969). *J. Sci. Fd. Agric.*, **20**, 477–481.
70 Mizunuma, Y., Shimizu, Y., Matsoto, S. and Okada, I. (1970). *Eiyo To Shokuryo*, **23** (4). 281–5 (Japan)—*via* abstract.
71 Moran, E.T. and McGinnis, J. (1968). *Poultry Sci.*, **47**, 152–158.
72 Morgan, D.E. (1968). *J. Sci. Fd. Agric.*, **19**, 393–395.
73 Morgan, D.E. (1972). *J. Natn. Inst. agric. Bot.*, **12**, 471–476.
74 Munck, L. (1972). *Hereditas*, **72**, 1–128.
75 Myasnikova, A.V., Rall', Yu.S., Trisvyatskii, L.A. and Shatilov, I.S. (1965). *Handbook of Food Products – Grain and its Products* (translated by S. Nemchonok). Jerusalem: Israel Programme for Scientific Translations.
76 Nordfeldt, S. (1954). *Ann. R. Agric. Coll. Sweden*, **21**, 1–29.
77 Norris, F.W. and Darbre, A. (1956). *Analyst*, **81**, 394–400.
78 Ørskov, E.R., Smart, R. and Mehrez, A.Z. (1974). *J. agric. Sci., Camb.*, **83**, 299–302.
79 Palmer, F.G. (1976). *ADAS Q. Rev.*, **21**, 220–234.
80 Pirie, N.W. (1969). *Food Resources—Conventional & Novel*. Harmondsworth: Penguin.
81 Pirie, N.W. (ed. 1971). *I.B.P. Handbook 20 – Leaf Protein, its agronomy, preparation, quality and use*. Oxford: Blackwell Scientific Publications.
82 Pomeranz, Y. (1973). *Cereal Chem.*, **50**, 504–511.
83 Pomeranz, Y., Ke, M. and Ward, A.B. (1971). *Cereal Chem.*, **48**, 47–58.
84 Pomeranz, Y. and Robbins, G.S. (1971-May). *Brewers Digest*, **46**, 58–64.
85 Postel, W. (1957). *Z. PflZücht*, **37**, 113–136.
86 Prescott, S.C. and Dunn, C.G. (1959). *Industrial Microbiology* (3rd edition). London: McGraw-Hill.

87 Pyke, M. (1971). In *Potential Crop Production*, ed. Wareing, P.F. and Cooper, J.P. pp. 202–212. London: Heinemann Educational.
88 Raven, A.M., Forbes, T.J. and Irwin, J.H.D. (1969). *J. agric. Sci. Camb.*, **73**, 355–363.
89 Reade, A.E., Smith, R.H. and Palmer, R.M. (1972). *Biochem. J.*, **127**, 32 P.
90 Reeve, R.M. and Walker, H.G. (1969). *Cereal Chem.*, **46**, 227–241.
91 Reveron, A.E., Topps, J.H., Miller, T.B. and Pratt, G. (1971). *J. Sci. Fd. Agric.*, **22**, 60–64.
92 Rexen, F. and Moller, M. (1974-Feb.). *Feedstuffs*, 46–47.
93 Rickes, E.L., Ham, A.E., Moscatelli, E.A. and Ott, W.H. (1962). *Arch. Biochem. Biophys.*, **96**, 371–375.
94 Robbins, G.S. and Pomeranz, Y. (1972). *Cereal Chem.*, **49**, 240–246.
95 Robinson, A.D., Hiltz, M.C., Campbell, R. and Levinson, A. (1945). *Can. J. Res.*, **23 F**, 1–8.
96 Shimizu, Y., Matsuto, S., Mizunuma, Y. and Okada, I. (1970). *Agric. biol. Chem.*, **34**, 437–441.
97 Simmons, N.O. (1963). *Feed Milling*, (2nd edition) Leonard-Hall Books Ltd: London.
98 Smith, A.M. and Robb, W. (1943). *J. agric. Sci.*, **33**, 119–121.
99 Sotola, J. (1937). *J. agric. Res.*, **54**, 399–415.
100 Swan, H. and Lewis, D. (eds. 1974). *University of Nottingham, Nutrition Conference for Feed Manufacturers, 1973*. London: Butterworth.
101 Walker, H.G., Lai, B., Rockwell, W.C. and Kohler, G.O. (1971). *Cereal Chem.*, **47**, 513–521.
102 Wang, P.S., Kato, H. and Fujimaki, M. (1970). *Agric. biol. Chem.*, **34**, 561–567.
103 Watson, S.J. (1953). *J. Soc. Chem. Ind.*, **62**, 95–97.
104 Wilson, R.K. and Pidgen, W.J. (1964). *Can. J. Anim. Sci.*, **44**, 122–123.
105 Wright, M.J. and Davison, K.L. (1964). *Adv. Agron.*, **16**, 197–217.

Chapter 15

Malting

15.1. Introduction

In malting, cereal grains (usually barley) are germinated for a limited period and are then dried and lightly cooked (kilned). At present wheat, sorghum and a little rye seem to be the only other cereals used. The product, from which the dried rootlets (culms, coombes, cummins, sprouts) have been separated is called malt. Malt is an 'intermediate' in several manufacturing processes. Malt types differ, depending on their intended use e.g. the manufacture of malt flours, of diastatic or non-diastatic malt extracts, of cereal syrups, of various types of distilled spirits, beers, and of malt vinegar. The malt culms and the residues from these manufacturing processes are used in animal feedstuffs (Chapter 14).

For simplicity, a preliminary outline of the malting process is given first [18]. Selected barley is 'steeped', usually by immersion in water, for a period chosen to achieve a particular moisture level. The water is drained from the grain, which is usually moved to a new location where it germinates. Conditions are regulated to keep the grain cool (generally below 18° C, 64° F) and to minimize water losses. As the grain germinates the 'acrospire' (coleoptile, spire, blade) grows beneath the husk and pericarp, while the 'chit' (coleorhiza, root-sheath) appears at the base of the grain, and is split by the emerging rootlets. At intervals the grain is mixed and turned to provide more uniform growth opportunities, and to prevent the roots matting together. As the embryo grows the endosperm undergoes 'modification', that is its structure is altered by hydrolytic enzymes, which accumulate in the tissue (Chapter 5). The malting process is regulated by the initial choice of barley, the duration of growth, the temperature, the grain moisture content, changes in the steeping schedule, and by the use of additives. When 'modification' is sufficient it is stopped by kilning the 'green' malt, that is by drying and cooking it

Malting 527

in a current of hot, dry air. Kilning can change the enzyme content of the finished malt from causing a slight enhancement of the levels found in the green malt to none (i.e. total destruction). At the same time the colour and flavour of the malt is altered. The dry, brittle culms are then separated and the finished malt is stored. Dry malt is stable on storage. Unlike barley, it is readily crushed. It contains relatively large quantities of soluble sugars and nitrogenous substances and, if it has been kilned at low temperatures, it contains high levels of hydrolytic enzymes. When crushed malt is mixed with warm water, e.g. as malt flour in baking, or as the 'grist' in brewing or distilling, the enzymes catalyse the partial or complete hydrolysis of the starch, other polysaccharides, protein and nucleic acids accessible to them, whether from the malt or from materials mixed with it, yielding sugars and other simple soluble substances (Chapter 16). Malt also confers colour, aroma and flavour to the product. The nature of the dissolved solids formed during mashing, the 'extract', varies with the malt, and details of the mashing procedure. The extract obtainable from a malt is determined in various ways and the quantities may be expressed in several different units. The solution of extract, 'wort', may be dried to a syrup or a powder (to be sold as 'malt extracts') or may be processed further and fermented by yeasts to give alcoholic products. The alcohol may in turn be oxidized to acetic acid by micro-organisms to give malt vinegar. Malt may be milled to give a flour that is used in baking.

The conversion of barley into malt incurs losses of dry matter. These losses are: (i) materials dissolved or dust washed away in the steep; (ii) the respiratory loss of water and carbon dioxide, and (iii) the culms. The culms have a monetary value (less than of the malt on a weight basis), but the steep effluent is a liability as its high biological oxygen demand (B.O.D.) must be reduced before it can be allowed into any waterway. The maltster aims to reduce malting losses to a minimum, while producing good malt in as short a time and in as economical a manner as possible. Understanding malting is complicated by the use of numerous units of weight (e.g. 1 quarter, Qr, barley = 448 lb; 1 Qr malt = 336 lb; 1 lb = 0.4536 kg) and volume (various bushels) besides metric units. Also the starting materials and the products have various moisture contents (typically 9–26% barley, 1.5–7% malt). By convention moisture contents are usually expressed on a percentage wet weight (fresh weight, 'as is') basis [9,18,21,42,43,48].

15.2. The selection and acceptance of malting barley

Malting barleys are selected under differing sets of circumstances: (i) in breeding programmes when decisions on which lines to propagate must be made using small grain samples (Chapter 12); (ii) in purchasing bulk lots of grain (Chapter 5); (iii) in deciding which lots of grain, held in store, are fit to be malted. Usually the maltster will have to assess the value of a barley lot to be purchased by investigating a small sample in some haste, so rapid methods of analysis may be used. Afterwards the bulk of grain must be found to be of equal – or superior – quality to the sample, before it is accepted (Chapter 5) [18,19,28,48]. Barley of a suitable variety will be purchased from a region known to produce acceptable grain. Small-grained six-rowed barleys, with nitrogen contents of 2.0–2.6%, may be preferred for highly enzymic malts; malts from plump, two-rowed barleys of moderate nitrogen content, e.g. 1.6%, are demanded in many ale breweries [16,18,21]. Californian (N. African) large-grained, thick hulled six-rowed varieties have a rather low enzyme-forming potential. At purchase the barley is inspected. The sample must only contain grain of the stated cultivar, and not a mixture of varieties; only a minimal quantity of other cereal grains, impurities and trash is tolerated. Grains should be short and plump, rather than 'pinched', long and thin, or unduly large, smooth-husked and 'coarse', and should fall in some chosen range of thousand corn weight or distribution of sizes ('assortment') as determined by sieving. The test weight (lb/bu or kg/hl of packed grain) may be determined. The grain should contain few bruised, or broken corns, or grains split along the ventral furrow which go mouldy during malting. Pre-germinated grains are also unacceptable. The husk should appear 'bright', yellow and 'ripe', and not greenish, except with pigmented varieties, when the pigmentation should be correct. Grains should not be 'weathered' (stained and discoloured) due to damp storage or microbial attack in the field (Chapter 9). There should be no signs of insect attack or heat damage (Chapter 10). Heat damage is easily overlooked. The husk should have a 'thin' appearance (a term *not* related to its physical thickness) and be slightly wrinkled. The grain should smell and taste 'sound'. When cut transversely samples should show a chalky white opaque ('mealy' or 'floury') endosperm rather than hard, greyish translucent 'flinty' or 'steely' fractures – which tend to indicate nitrogenous and immature grain. Barleys having a viability (germinative capacity) of at least 96% are preferred for malting, although standards vary with the choice available (Chapter 5). The moisture content is estimated to determine the quantity of dry grain being bought, the likely storage characteristics of the sample, the degree of urgency in drying it, and to attempt to discover whether or not it has been partly dried already. The nitrogen content is also measured, since this has a positive correlation with the enzyme content of the finished malt, but a negative correlation with its yield of extract (Table 15.1). Numerous relationships occur between analyses of barley samples and of lightly kilned malts derived from them, when the malting has been carried out in a standard manner. For example, barley extract and starch content both correlate with malt extract; nitrogen content correlates with diastatic power (Table 15.1) [28].

Standardized sets of analytical methods are in use. Sometimes other criteria, e.g. grain swelling or the autolytic release of substances from ground grain, are evaluated [4,9,11,16,18–21,28,34,48]. The recognition of the regularity principle of grain composition (Chapter 6), the positive correlation between barley grain size and malt extract and the negative correlation between barley nitrogen and malt extract led Bishop to propose extract prediction equations. The inverse correlation between extract and nitrogen content exists because extract is derived mainly from the carbohydrates of malt which decline as the proportion of nitrogenous materials increases (Table 15.1). Prediction equations either rely on a knowledge of the variety used (Equation 1) or on a special determination of 'insoluble carbohydrate', similar to the feed analyst's 'fibre' (Equation 2). 'Insoluble carbohydrate' is mainly an indirect measure of husk content; the husk contributes practically nothing to malt extract. Husk percentage declines with increasing grain size in samples of one variety. The equations given predict the extract, as determined by the Institute of Brewing, or I.o.B. Method, to be obtained from well-modified, lightly kilned malts, prepared from sound barleys. The varietal 'constants' are really slightly variable, and are determined experimentally.

$$T.E. = A - 11.0\,N + 0.22\,G \quad (1)$$
$$T.E. = 138.2 - 9.5\,N - 3.0\,I \quad (2)$$

where:

$T.E.$ = hot water extract (lb/Qr, dry basis). A = 'varietal constant' for the type of barley – generally from 109 to 115 for British two-rowed varieties. G = The dry weight of 1000 grains (thousand corn wt., TCW). N = Nitrogen content of the barley (% dry weight). I = Insoluble carbohydrate (% dry weight), remaining after specified acidic and alkaline extractions.

Other equations allow for the effects of ungerminated corns; yet others discard the use of G, since this determination is tedious and imprecise (Chapter 5) [18]. Computer established correlations, based on numerous samples, indicate that extract does not decline exactly linearly with increasing grain nitrogen, and the prediction equations should take account of grain nitrogen, husk fineness, assortment on a 2.8 mm screen, germination capacity, and germination pattern [41]. Occasionally 'direct' estimates of barley extract are made. The grain is ground and mashed with an external supply of enzymes, and the extract is calculated from the specific gravity of the solution obtained. Barley extracts correlate with malt extracts, at least within one laboratory, and when a single malting method is used.

Table 15.1 Mean analyses of twelve Canadian barley varieties grown at different sites and arranged in order of nitrogen contents, and data on some derived malts (Results expressed as % dry matter; from Anderson and Ayre [1], Ayre *et al.* [5], Meredith and Anderson [27]; by permission of the National Research Council of Canada).

Total nitrogen (%)	Salt-soluble			Alcohol-soluble protein nitrogen (%)	Insoluble protein nitrogen (%)	Starch (%)	Crude cellulose-lignin (%)	Barley Extract (%)
	Non-protein nitrogen (%)	Protein nitrogen (%)	Total (%)					
1.540	0.292	0.305	0.598	0.480	0.462	59.2	10.0	80.8
1.738	0.342	0.320	0.662	0.582	0.494	57.7	10.8	78.2
1.932	0.333	0.312	0.646	0.699	0.587	56.1	11.0	77.5
2.278	0.338	0.348	0.686	0.893	0.699	54.2	10.9	76.8
2.294	0.349	0.462	0.812	0.893	0.589	55.1	9.6	77.9
2.333	0.359	0.398	0.757	0.921	0.655	53.6	10.2	74.7
2.357	0.342	0.399	0.741	0.962	0.654	54.8	9.8	76.5
2.381	0.352	0.445	0.797	0.932	0.652	52.4	10.6	74.6
2.526	0.361	0.411	0.772	1.030	0.724	53.3	9.9	75.5
2.668	0.395	0.396	0.791	1.182	0.695	52.7	10.0	74.4
2.674	0.349	0.393	0.742	1.208	0.723	51.7	10.4	73.6
2.687	0.394	0.412	0.807	1.065	0.815	52.7	9.9	74.2

Table 15.1 (Contd.)

Barley			Malting data			Malt			
Total nitrogen (%)	1000-kernel weight (g)	Plump barley (%)	Steeping time (h)	Malting loss (%)	Sprouts (%)	Extract (%)	Wort nitrogen (%)	Diastatic power (°L)
1.540	34.8	92.2	99	6.8	1.6	78.0	0.61	63
1.738	34.8	84.7	83	6.5	1.7	76.3	0.65	85
1.932	32.9	79.3	74	7.0	1.6	75.5	0.78	100
2.278	33.5	89.5	77	6.9	2.0	74.3	0.82	121
2.294	33.5	86.6	84	7.2	2.0	74.7	0.73	116
2.333	30.6	65.0	75	7.2	2.3	73.0	0.75	105
2.357	28.0	84.0	77	7.0	2.0	73.9	0.78	117
2.381	32.5	50.3	65	6.6	1.8	72.8	0.87	122
2.526	28.0	69.7	81	7.4	1.9	73.3	0.87	133
2.668	32.5	82.9	75	7.1	2.0	71.9	0.79	140
2.674	31.8	78.0	74	8.0	2.4	70.5	0.78	149
2.687	30.8	67.3	76	7.4	1.7	71.6	0.81	133

After a grain lot has been accepted it is precleaned and dried to 11–13% moisture, possibly mixed with an approved insecticide, then stored cool (Chapter 10). The 'traditional' maltster keeps grain lots of different varieties, germinative capacities and nitrogen contents separate. To achieve acceptable bulks by mixing different lots of grain flexible grain handling and storage facilities are essential. Often only two or three different varieties are purchased to simplify the problems. The simplest method of building up a stock of barley for making a 'standard malt' is to purchase samples of a limited number of varieties, each sample having less than a chosen nitrogen content and more than a particular germinative capacity and, following drying, store them in a mixed bulk.

Malting barley should not be harvested before it is ripe [37]. In regions such as N. Europe, where dormancy occurs, the grain needs to be 'after-ripened'; that is its dormancy must be reduced to a low level before it can be malted. A period of post-harvest maturation may improve the malting quality of barley even though its germinative energy remains high and unaltered [48a]. The old method of maturing barley was to leave it unthreshed, in a stack in the field. With combine-harvested grain the procedure is to 'sweat' it (warm and dry it slowly and carefully to about 12% moisture) then to store it uncooled in bulk at initially about 25°C (77°F) for 1–2 weeks before cooling to below 15°C (59°F). Dormancy, (both 'profound' dormancy and 'water sensitivity', Chapter 5), varies with variety and especially with the season. Normally barley is malted only when dormancy has declined. Small-scale trials can indicate whether or not a large bulk of barley is ready for malting [24].

Small-scale malting was in limited use by 1900. In the 'stocking malting' technique about 0.5–1 kg (1.1–2.2 lb) of grain is sewn into a loosely woven cloth bag (nylon net is suitable) and is steeped, germinated and kilned with the bag embedded in a large bulk of malting grain, a technique which provides a 'typical malting environment' for the sample. This useful technique is inflexible. Small-scale malting is carried out on samples ranging from about 40 g (micromalting) to 20 kg (about 44 lb, pilot-scale malting) [18,28]. The equipment varies from simple glass or plastic tubes, jars, bottles or shelves held in attemperated, humidified chambers to complex 'pneumatic' containers with automatic turning devices. Small-scale malting may also be used to work out optimal malting sequences [18,35]. Malting losses and rootlet production may also be determined. Sufficient malt may be produced to allow extensive analyses, to assess such wort features as extract, soluble nitrogen, tannin content, viscosity, fermentability and the ability to support yeast growth. Often there are 'scale-up' problems in taking a malting process from a small-scale system into commercial plant. Temperature control and handling are much more precise on the small scale. Furthermore small-scale kilning is somewhat dissimilar to the production process.

Malting 533

15.3. Barley handling

Before it is steeped barley is cleaned and graded (Chapter 10). Grain of less than 2.2 or 2.3 mm width is rejected, and is used in animal feed. As grain of differing sizes, but from one lot, varies in nitrogen content and rates of water uptake and modification the best practice is to divide each lot into batches of similar width and malt these separately under optimal conditions. However as malting factories work according to preset, 'generally acceptable' schedules, which are varied as little as possible, processing separate grades is often neither practicable nor worthwhile. Some slight mechanical damage to the grain caused by the conveyors or by deliberate abrasion in special machines can help overcome dormancy, improve malting performance, and facilitate the response to exogenous gibberellic acid (Chapter 5) [7,31]. However excessive abrasion, or mechanical dehulling, causes problems such as leaking of sugars from the grain, with a consequent heavy growth of contaminating microorganisms [36,49].

15.4. Steeping

In steeping, weighed quantities of barley are discharged into a 'steep' tank, containing water. Often the water is vigorously aerated and as the grain water mixture is agitated dust, chaff and light grains float to the surface and, with some water let in from below, 'weir over' into an outlet pipe. These 'skimmings' are collected by sieving and are used in animal feed. Steep temperatures are often 'controlled' only by the initial temperatures of the water and grain, and the thermal inertia of this mixture. In many areas steep temperatures of about $10-16°$C ($50-60°$F) are common. Steeping may continue, with water changes, for 50–70 h. In a few places steep water is warmed and steeps are being insulated to help regulate this process. High steep temperatures can cause subsequent germination problems. As the grain takes up moisture it swells, ultimately by about $\times 1.3-1.4$. The moisture content at the end of steeping depends on the nature of the grain and the intended type of malt. In order to produce a poorly modified traditional pale lager, malt from two-rowed European barley requires steeping to about 41–43% moisture, while to make a better modified pale ale malt the preferred traditional moisture content would be about 43–45%. For a high-nitrogen barley (about 1.8% N) destined for a distillery or vinegar factory 46–49% moisture might be preferred. Higher water contents result in faster modification, but greater malting losses. As steeping proceeds the carbon dioxide content of the liquor rises and the oxygen content falls to a negligible value. As fermentation continues ethanol accumulates in the grain and steep liquor and, after a prolonged anaerobic period, may harm it. Excessive aeration, designed to minimize this fermentation should be avoided otherwise some grains may chit under water, and their moisture contents will become excessive [12]. Oversteeped grains grow too fast ('bolt'), and waste occurs. Phosphates, and organic substances are leached from the grain, the microbial population of the steep increases, and products of microbial metabolism, including acetic acid, accumulate. The acetic acid may be formed from the ethanol that leaks from the grain. The dissolved materials and the skimmings normally represent a loss of 0.5–1.5% (dry weight) of the original barley. Some of the dissolved materials check subsequent growth and may induce a state resembling 'water sensitivity' (Chapter 5). To minimize these effects the steep-liquor may be replaced by fresh water 1 to 4 times. More changes are used in prolonged steeps or when the barley contains a high proportion of damaged grains, since these decompose and the water becomes foul. Steep water is expensive to obtain and discharge, the disposal charges being elevated by the need for purification. In traditional 'Bavarian' practice grain was steeped in running water, while in 'Bohemia' the grain was rested in the air for up to 24 h between immersions [14]. Such 'air rests', used with induced downward aeration to remove carbon dioxide, are now used to enable water-sensitive grain to be malted. Water-sensitive grain is steeped to a moisture content of 33–37% in 6–16 h and is then 'air rested' for 12–24 h before re-immersion for a further period of, say, 24 h to achieve the desired final moisture content. During the air-rest period the grain takes up the film of surface-moisture so that the surface dries; it also undergoes some change which permits subsequent germination. In 'flush-steeping' the air-rest allows a temperature increase in the grain mass. The 'flushing' immersions, which cover the grain for only a few minutes, cool the grain and renew the film of surface moisture. The excellent aeration and increase in grain temperature result in rapid and vigorous germination, but large volumes of steep liquor are used.

Steeps are sometimes aerated by sparging compressed air through nozzles or perforated aeration rings mounted in the base of the tank. Alternatively, or in addition, 'fountains' or 'geysers' are used; in this device a tube extends from above the water surface to near the bottom of the grain-water mixture; compressed air is blown up into the tube and the grain-air-water stream is carried upwards and is discharged above the water surface, ensuring good mixing and aeration [18]. The top of the tube may carry hollow, rotating arms that distribute the grain-water mixture to the sides of the steep. Aeration maintains the viability of the grain and increases the vigour and evenness of its subsequent germination. Another aeration method is to 'pump-over' the grain-water mixture either from the base of the steep back into the top, or from one steep into the next. The grain-water mixture may be discharged over a perforated cone which carries away some of the moisture while spreading the grain before it

Malting 535

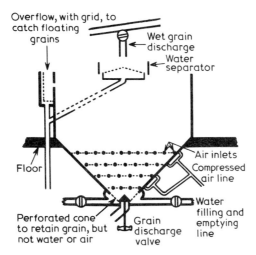

Fig. 15.1 Schematic section through a self discharging conical-bottomed steep tank with air-inlets for aeration and a device to separate some of the water and spread the grain when a grain-water mixture is 'pumped over' (after Vermeylen [48]).

falls into the tank (Fig. 15.1). Limited aeration and considerable mixing is achieved. Circulation of water through the steep, via an external circuit containing a water purification and oxygenation device, has also been used.

A logical way of applying aerated water to the grain is by spraying. Experience shows that grain needs at least one immersion, presumably to wash away inhibitory substances. Subsequently intermittent spraying may be used to renew the film of surface moisture and this, combined with forced ventilation of the grain mass, favours rapid and uniform germination. By spraying on just sufficient moisture at intervals to renew the film of surface moisture around each grain, a process termed 'incremental spray-steeping', little effluent is produced. Specialized 'steeps' are now used for spray-steeping (Fig. 15.2).

Substances may be added to steep liquor to check rootlet growth, to overcome dormancy or water sensitivity, to control micro-organisms, or to 'clean' the grain (Chapter 5). Lime water (calcium hydroxide), sodium or potassium hydroxide and sodium carbonate solutions have been used for short periods. The stronger alkalis are more effective in extracting 'testinic acid', a mixture of substances including polyphenols and protein, from the grain but are more likely to damage it. In the U.S.A. a lime-water wash has been followed by acidification to pH 2.5 with sulphuric acid for 3–5 h followed by a drain, and continued steeping in water [48a]. Sulphurous acid, chlorinated water and hypochlorites have been used to check microbial growth, but the use of sulphurous acid seems to have

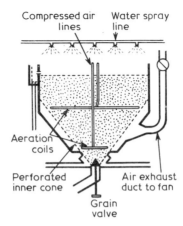

Fig. 15.2 A conical-bottomed, self-emptying vessel designed to allow normal steeping, or spray steeping and downward aeration of the grain during air-rests, or the initial stages of germination.

lapsed and the other substances are generally avoided because of the danger of imparting antiseptic 'phenolic' taints to the malt. Dilute formaldehyde sometimes breaks dormancy and is so efficient at checking micro-organisms that steep liquors may re-used. Formaldehyde kills *Fusarium* moulds on barley, avoiding consequent problems in beer made with infected malt. *Fusaria* can also induce overmodification. Other micro-organisms can prevent germination, or impart off-flavours. There is also the risk that they will produce toxins. The inclusion of formaldehyde in the steep checks embryo growth and can reduce malting losses; when used in a resteep formaldehyde reduces the tannin contents of worts from treated malts [18]. Occasionally hydrogen peroxide (0.1–1%) is added to steep liquors to help overcome dormancy. This, like gibberellic acid and bromates, can be used more economically if sprayed onto the grain at steep out, or at the start of an air-rest period. In some continuous malting plants the grain is first washed while immersed in water. Subsequently other water is added by spraying. Containers in which grain can be vigorously agitated and washed in batches before transfer to conventional steeps, have also been used.

Steeps (cisterns, tanks) are normally metal vessels, cylindrical or rectangular in cross-section, with hopper bottoms to allow emptying by gravity. Water can be withdrawn separately and air may be sucked down through the air-resting grain. Other devices, to allow aeration, are normally fitted (Fig. 15.1). Recently 'spray steeps' have been built in which gas may be withdrawn through the greater part of the hopper bottom, and water or solutions of additives may be sprayed over the surface of the grain bed (Fig. 15.2). Liquor drains to the bottom and may be collected and recirculated, or discarded. As the heat output is small and rootlet growth is minimal in the initial stages of germination, mixing and turning the grain is not essential, so these modified steeps provide adequate conditions and a cheap way of enlarging germination space. However the grain must be transferred to the germination compartments before it is much grown, otherwise the roots may mat, and the grains will bind together. Also during the transfer conveyors may damage the chitted grain. The division between the steeping and germination stages need not be sharp. 'Multiple steeping' processes, various sequences of grain immersions followed by airrests, can greatly speed malting and reduce malting losses [39,40]. Water may be sprayed onto the grain ('sprinkling') during germination, to make good losses caused by evaporation. It is difficult to ensure that all grains in conventional steep tanks are treated equally. To reduce this problem, flat-bed steeps have been developed (Fig. 15.3) [2]. The same units can hold the germinating grain, and water may be added back for 'resteeping'.

Grain may be discharged from steeps either pumped while mixed with water, 'wet casting', or conveyed after a period of draining, 'dry casting'. In wet casting the grain–water mixtures are easy to direct. However the grain arrives wet at the germination site, and so initial germination is delayed. Wet-casting is unsuitable for chitted grains because excessive water uptake occurs. Dry casting avoids these problems and solutions of additives, such as potassium bromate or gibberellic acid, may be sprayed uniformly and economically onto the grain as it is being moved. When applied in this way only about a quarter to a fifth as much additive is needed to obtain a given effect as if it is dissolved in the last immersion steep, when presumably about 80% goes to waste in the effluent. Physically bruising grain in the steep slightly speeds water uptake, and reduces subsequent germination. The addition of gibberellic acid to such bruised grain hastens subsequent malting to a greater extent than normal [8].

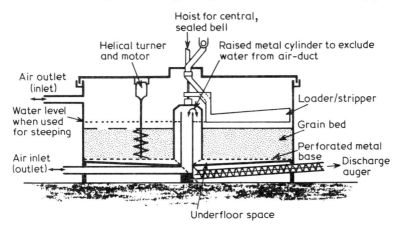

Fig. 15.3 A Boby cylindrical flat-bed steeping, resteeping and germination unit (after various sources).

538 Barley

15.5. Germination equipment [18,21,42,43,48]

In traditional malting, grain was germinated on smooth, impermeable floors. Several such floors were arranged one above another in a building. They were used only in winter-time, as the low air temperature was needed to control the germination process; the temperature on each floor was regulated by opening or closing shuttered windows. In modernized floor maltings which are used throughout the year, the side windows are sealed and thermally insulated, and attemperated air of high humidity is circulated. The grain, when cast from the steep, is heaped into a rectangular 'couch' up to 76 cm (2.5 ft) thick to encourage a rise in temperature, which hastens the uptake of the film of surface water. As the grain begins to grow the 'piece' is thinned, by spreading it over a greater area, to reduce the temperature. The grain is turned and mixed once or more daily, either manually or by machines, and is gradually moved along the malting floor, being 'thinned' or 'thickened' to lower or increase the temperature as is necessary. Thus a series of pieces move in sequence down a malting floor. Ultimately each piece is loaded onto a kiln. The grain on the floor grows in an atmosphere that is saturated with moisture and rich in carbon dioxide. Each piece is normally about 10 cm (4 in) deep initially. Generally temperatures of 13–17°C (about 55–63°F) are used, but at the end of the process, when rootlets are beginning to wither and respiration is declining, the piece may be thickened and a temperature of 20°–25°C (68–77°F) is reached. The temperature of the piece is often 5–7°C (9–12.6°F) above that of the air. The temperature may differ by 0.5–1.5°C (about 1–3°F) between the top and bottom of the grain layer. Germination periods vary, but are usually 6–10 days.

Floor malting is expensive in space and manpower. To handle germinating grain in large quantities 'pneumatic' malting equipment is used, in which the temperature of a large bulk of grain is controlled by the forced passage of cooled air, as nearly as possible saturated with water vapour. Mechanical arrangements are made for gently mixing and turning the germinating grain. Such plant is equipped for mechanical loading and emptying. As the air is warmed by passage through the grain it inevitably picks up moisture, and the grain tends to dry. To compensate the grain may be 'sprinkled', or sprayed, with carefully regulated volumes of water which are applied during the turning process. The airstream also provides oxygen to and carries carbon dioxide away from the grain. Malting temperatures are usually 13–17°C (about 55–63°F) often rising over 20°C (68°F) on the last day of germination. The airflow is adjusted to change the air in the grain about × 3 per min., and the temperature differential across the grain between the air inlet and leaving the grain is normally held below 3°C (5.4°F). The airstream is fan driven, and is conditioned by passage through water sprays adjusted to a suitable temperature. The air entering the spray chamber is partly fresh and partly recirculated from the stream leaving the grain. High carbon dioxide concentrations are never achieved in the air-stream, due to the gases high solubility in the spray-water, so this does not check growth. Several types of pneumatic equipment are in use and very many more have been proposed [3,18].

In box, compartment or Saladin maltings the grain bed, about 1 m (about 3.3 ft) deep, is supported on a perforated metal deck and is restrained by side- (and usually end-) walls (Fig. 15.4). Capacities are very variable – 10 000–40 000 kg (22 000–88 000 lb) of barley would be usual. The tops of the side-walls carry rails which support mechanical turning, levelling or stripping (unloading) devices. Steeped grain may be delivered from overhead spouts. When the full charge of steeped grain has been loaded the turners pass through and level it. At intervals the grain is turned to 'lighten the piece', separate matted roots and reduce the resistance to the airflow. Turners are designed to treat the grain gently to avoid bruising, tearing the roots, or other damage. They are of several designs: a carriage may carry a row of rotating, vertical helical metal screws, or buckets and chains which lift the grain and then drop it behind the turner as it moves forward. The turner carriage is sometimes fitted with spray-bars to allow the grain to be sprinkled. Boxes may be stripped (unloaded) by turners being adjusted to lift the grain into a conveyor placed beside the box, or by 'bulldozing' the grain from an open end into a conveyor, or by a pneumatic lifting system.

Malting drums generally fall into two categories. In the older type the horizontal cylindrical metal container is nearly full of grain and the attemperating air passes, in one direction or the other, between a central duct and a series of ducts or a perforated wall around the sides [18].

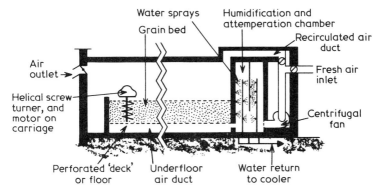

Fig. 15.4 A section through a malting germination box (compartment, or Saladin box), in which the humidified and attemperated air is driven up through the grain mass.

540 Barley

A sliding valve seals the ducts that happen to be above the level of the grain. Sprinkler bars are incorporated to allow the addition of water during turning, which is achieved by intermittently rotating each drum around its long axis. In 'decked' or 'box' drums the grain bed is formed on a flat, perforated deck. The result is an 'enclosed box', except that the side walls curve and turning is achieved by occasional rotation (Fig. 15.5). A drum may be loaded by spouting the grain in either at one end, or through a row of side doors which are opened for the purpose. The bed is formed by rotating the drum. As the grain forms an inclined surface it is necessary to turn the drum until the perforated deck is below the grain, and parallel to the inclined surface. The drum is then turned back so that the grain surface and deck are parallel and horizontal. This ensures an even airflow through the bed. Drums never seem to be thermally insulated. Drums may be stripped by augers, or may be rotated so that when a row of doors is opened the grain falls into a hopper, and then into a conveyor system. Often these doors are in a row along the length of the drum. Sometimes they are arranged round the middle; as the drum rotates the grain in the centre drops out and fixed helical baffles, twisted in opposite senses, guide the grain from the ends of the cylindrical space towards the middle, and out of the doors.

Less usual malting equipments allow the germinating grain to be covered with water, for resteeping (e.g. Fig. 15.3). Boxes and drums that may be used for steeping and kilning as well as germination also exist [18]. In the 'Popp' system the deep bed of grain, formed on a perforated floor in a vertical metal cylinder, is turned by the violent release of compressed air below the bed of grain. In the 'Wanderhaufen' or 'moving piece' system the germinating grain is moved in steps along the length of an extended box or 'street'. Grain pieces are loaded at one end and the finished pieces are stripped from the other end and are transferred into

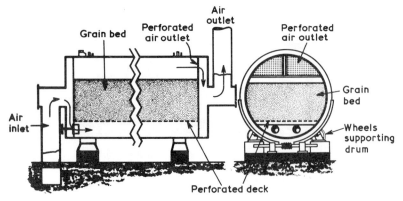

Fig. 15.5 A box- or decked-drum, in which the air passes up through the perforated deck and the bed of germinated grain.

Malting 541

kilns. This system depends on the use of special turners which minimize the mixing between successive pieces when moving them along one step at a time, each time the grain is turned. The different sections of the street can be individually attemperated. In the continuous Domalt and Solek malting plants a metered stream of grain is continuously washed, formed into beds on moving supports, and is spray steeped, turned and kilned as it is carried through the appropriate sections of the plant.

15.6. Kilns and kilning [9,16,18,43,48]

Rarely malt is used 'raw' or 'green', for example in potato distilleries, where unflavoured alcohol is produced. Such green malt has a high enzyme content, but it cannot be stored at ambient temperatures, and may impart an unpleasant flavour to products such as beer. In Belgium small quantities of 'wind-malt' for making special beers are prepared by allowing green malt, spread in thin layers in lofts, to air dry. Usually malts are kilned, that is they are dried and partly cooked in a current of warm air.

The consequences of kilning are: (i) A dry product that may be stored without difficulty, and from which the roots may readily be separated ($\leqslant 7\%$ moisture) and which is 'friable' and is therefore easily crushed in the milling operation. Moisture contents of 1.5–6.0% are common. (ii) The enzymic composition of the grain is modified. If the initial stages of kilning are carried out warm, with minimal evaporation (stewing) enzyme development and endosperm breakdown will continue for a short period. High temperatures reduce enzyme activities in the finished malts. At air temperatures of 40° C (104° F) or less enzyme destruction is minimal. (iii) As a consequence of chemical changes the character of the malt can be altered. The raw grain flavour is removed and desirable flavours and colours develop. Kilning may last from 18 h to 4 days and the air temperatures vary in the range 30° C (86° F) to 105° C (221° F) depending on the type of kiln being used, the type of malt being made, and the stage of the kilning process.

When malt is dried in a constant airflow at a fixed temperature, moisture is removed at a steady rate during the initial 'free drying' state when moisture content of the grain is in the range 46% to about 23%. The rate of drying then falls. Initially this is due to the length of time taken for the moisture to reach the surface of the grain, but at some grain moisture that has been estimated as about 12% (higher for less well modified malts; and up to 17% for raw barley) the difficulty of water removal increases still more, since the remaining water is bound in the grain (but see Chapter 10) [18,47,48]. By lightly crushing green malt, before the kiln is loaded, the rate of evaporation can be accelerated. Since air is costly to heat and drive through the grain bed its drying capacity should

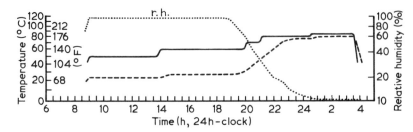

Fig. 15.6 The temperature of the air above and below the malt and the relative humidity of the air-off during a sequence for kilning a pale malt on a single-floor kiln (after Schuster and Weinfurtner [43]). ———, Temperature under the malt, ----. Temperature over the malt. ...... Relative humidity.

be used to the maximum. It should emerge from the grain with a relative humidity (r.h.) of about 95%. Total saturation is not desirable since condensation in the bed of grain or kiln must be avoided. As the rate of evaporation falls so does the r.h. of the emerging air. Initially the airflow through the grain is reduced to allow more time for it to pick up moisture. Also the temperature of the air is raised to increase the rate of drying. There are limits to the extent to which the airflow can be reduced, yet the r.h. of the emerging air continues to fall. Gradually a proportion of the hot air is recirculated with freshly heated air until at the end of kilning, when little evaporation is occurring, as much as 75% of the air may be recirculated. Recirculation saves much heat, and reduces fuel costs. During the initial stages the 'air-on' temperature may be increased from say 45°C (113°F) to 60°C (140°F). When the malt is 'hand-dry', by convention that is 5–8% moisture, the temperature is raised to a 'curing' value of 80–85°C (176°–185°F) for light malts, or up to as much as 105°C (221°F) for dark malts. The air emerging from the grain ('air-off') has a lower temperature than the underfloor, 'air-on', temperature due to the cooling caused by the evaporation of water from the malt (Fig. 15.6). As kilning proceeds 'air-off' temperatures approach 'air-on' temperatures, a sign that drying is slowing down.

Kilning is sometimes carried out in special drums, or in 'vertical kilns' where hot air passes sideways through a vertical column of malt restrained by perforated walls. Kilns usually carry a horizontal bed of grain on a perforated floor and hot air is either forced from below from a fan-pressurized chamber, or it is 'induced' by either natural draught, or a fan in the exit of the kiln chamber. Traditional British kilns have one floor and are mostly modernized to permit mechanical levelling of the loaded green malt, turning during kilning, and mechanical stripping at the end of the kilning cycle (Fig. 15.7). Ducting has sometimes been added to allow air recirculation. In Continental Europe and N. America older kilns may have one, two or three floors (Fig. 15.8). Hot gases from

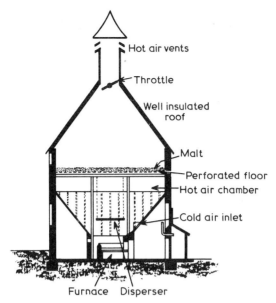

Fig. 15.7 An old type of traditional British malt kiln, before modernization. Modernization would include the installation of a thermostatically controlled furnace, provision for a fan-driven airstream and recirculation of the gases from above the malt, and the provision of mechanical grain loading, turning and stripping equipment.

the lowest floor, where curing is carried out, are diluted with fresh air and are used to dry malts with higher moisture contents, on the upper floor(s). When the lowest floor is stripped of the cured malt the malts from the upper floors are dropped one stage, and green malt is loaded onto the top floor. Such kilns overcome the need for air recirculation, and can have a high thermal efficiency.

In Britain kilns are usually 'directly' heated, i.e. the products of combustion pass through the malt. The flavours so conferred may be accentuated deliberately; for example, peat is burnt and the smoke is passed through malts used in Scotch whisky manufacture and some special brewing malts were treated with woodsmoke. 'Indirect' heating is less thermally efficient; the furnace heats air, water, or steam, which warms a heat-exchanger which in turn heats the air. Indirect heating keeps undesirable combustion products from the malt. Fuels used include coal, coke, anthracite, natural gas, and various oils. Solid fuels are tested to ensure that only those with a low arsenic content are used. Oils should have a limited sulphur content, since in some circumstances mottled 'magpie malt' can be produced by the presence of sulphur trioxide in the combustion gases. Apart from the appearance magpie malt can be of good quality. In some areas, e.g. N. America, where very pale malts are

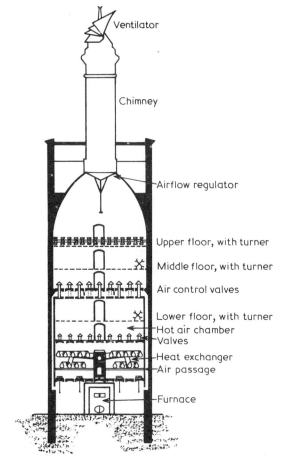

Fig. 15.8 A traditional Continental European three-floor kiln with indirect air heating and mechanical turning. Modernization would include building in a fan to achieve a steady air-flow.

required for brewing, sulphur dioxide is fed into the kiln to retard colour development and perhaps bleach the malt while its moisture content is over 40%. The pH of mashes made with 'sulphured' malts are unusually low. The sulphur dioxide treatment leads to increased levels of soluble nitrogen in the extract.

In the older kilns the malt is regularly turned, to try to ensure that all the grains are treated approximately equally. However in deep loading, or compartment (Winkler or Müger) kilns the deep bed of grain (about 1 metre, 3.3 ft) is not turned. A very rapid airflow, containing the combustion gases, is driven up through the grain bed and a drying zone spreads up from the bottom (Fig. 15.9). These kilns are compact, well insulated and purpose built to allow air-recirculation, rapid loading and

Malting 545

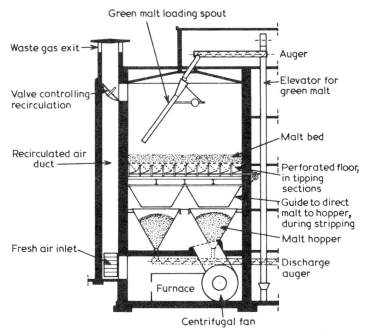

Fig. 15.9 A modern deep-loading, self stripping kiln, designed to allow air-recirculation (after Vermeylen [48]). The malt is not turned during kilning. Stripping (unloading) is achieved by tipping the sections of the louvred kiln floor, which allows the malt to fall into the hoppers above the discharge auger.

stripping. Stripping is usually achieved by tipping the floor either as one unit or in several sections, so that the malt falls into a hopper leading to a conveyor. The rapid 'turn around' ensures that the kiln never completely cools, saving fuel which would otherwise be used in heating the kiln structure. Efficiency can be increased further by working kilns in pairs, so that hot, dry 'curing' air at a low flow-rate from one provides part of the air needed for another working in the drying stage. One is always used for drying, the other is always used for curing, and both can be designed for optimal use.

The kilned malt is rapidly cooled, beaten and screened to separate the rootlets. Rootlets are separated from malt because of their unpleasant, bitter flavour. They are collected and packed in waterproof sacks. The cool malt is conveyed to a dry store. Speed is important since malt and rootlets are hygroscopic. Many believe that malt should be stored for at least some weeks before use. The reasons for this are unclear.

15.7. Malt analyses

Many chemical studies have been carried out on malting grain (Chapter 5, Table 15.2) [15,16,18,34]. Analyses used to evaluate malt and their

Table 15.2 Chemical analyses of 'average' European two-rowed barleys and their malts (from Harris [15]).

Fraction	Proportions (% dry weight)	
	Barley	Malt
Starch	63–65	58–60
Sucrose	1–2	3–5
Reducing sugars	0.1–0.2	3–4
Other sugars	1	2
Soluble gums	1–1.5	2–4
Hemicelluloses	8–10	6–8
Cellulose	4–5	5
Lipids	2–3	2–3
Crude 'protein' (N × 6.25)	8–11	8–11
Salt soluble protein ⎰ albumin	0.5	2
⎱ globulin	3	—
Hordein – 'protein'	3–4	2
Glutelin – 'protein'	3–4	3–4
Amino acids and peptides	0.5	1–2
Nucleic acids	0.2–0.3	0.2–0.3
Minerals	2	2.2
Other substances	5–6	6–7

importance are generally unfamiliar, and so some are outlined here [4,9,11,18,20,34,48]. If malt is 'slack' – is wetter than it should be – it may deteriorate, the purchaser is buying unwanted water, and the user may have processing difficulties. Off flavours and smells should be absent. An unmixed batch should contain corns of approximately uniform colour and uniformly grown acrospires usually to about 0.7–0.8 of the grain length. Excessively grown ('overshot') or ungerminated grains are undesirable. A normal malt corn should have a near-white, floury endosperm that is well modified, i.e. it is readily crushed to a flour, and with no hard end. No corns with steely endosperms should be present. Resistance to crushing may be determined by a machine, or by chewing. Malt should have the flavour characteristic of its type. Sorting on sieves and estimates of weight per unit volume (kg/hl, lb/bu) and thousand corn weight are sometimes used. A crude 'floater' or 'sinker' test is used to distinguished well-modified malt, which floats on water or a salt solution ($d = 1-1.2 \, g/ml$) from raw barley, or undermodified malt, which are denser and sink. The test is imprecise because air-bubbles are apt to be trapped in the husk.

Extracts of malt should have a low viscosity. This is not an adequate criterion of modification in ale malts but it seems to serve with lager malts (Table 15.3). High wort viscosity leads to operational difficulties. Cold water extracts (C.W.E.) of malt, prepared with water at $0°C$ or with

Table 15.3 Changes in the composition of barley malted at two temperatures (from Piratzky [32]).

Days	Total N (%)	Extract (%) Graf[†]	Extract (%) Coarse[‡]	Soluble N (%) of Total N. Graf[†]	Soluble N (%) of Total N. Coarse[‡]	Congress wort Maltose* (%)	Congress wort Formol N (% soluble N)	pH	Relative viscosity, water = 1000	Boiled 5 h p.p.t. (mg)	Boiled 5 h protein (% in p.p.t.)	Trichloracetic acid p.p.t. (mg)	Trichloracetic acid p.p.t. protein (% in p.p.t.)
\multicolumn{14}{l}{Germination temperature 13–15° C (55°–59° F)}
0	1.82	77.8	—	15.9	—	—	—	—	—	—	—	—	—
1	1.86	78.4	—	17.2	—	—	—	—	—	—	—	—	—
2	1.83	78.3	—	19.8	—	—	—	—	—	—	—	—	—
3	1.81	80.2	73.1	32.3	26.2	63.0	17.9	6.09	1340	58.0	33.1	82.6	21.9
4	1.76	80.7	75.7	36.7	33.5	64.4	15.9	5.97	1202	53.4	47.5	71.4	38.8
5	1.79	80.1	77.6	38.9	34.3	63.4	19.3	5.96	1121	47.4	64.4	65.1	45.0
7	1.80	80.7	78.0	40.6	35.4	66.1	20.1	5.95	1081	36.9	96.3	57.3	59.4
9	1.79	80.8	78.3	41.9	37.0	68.3	19.4	5.95	1068	39.1	96.3	57.9	63.7
11	1.79	80.9	78.5	41.7	39.6	70.2	20.1	5.93	1062	37.5	90.7	58.9	56.9
16	1.85	80.1	78.2	45.5	38.6	71.1	19.9	5.92	1063	37.8	90.0	57.5	65.6
\multicolumn{14}{l}{Germination temperature 18–21° C (64.5°–70° F)}
0	1.82	77.8	—	15.9	—	—	—	—	—	—	—	—	—
1	1.86	77.4	—	16.2	—	—	—	—	—	—	—	—	—
2	1.90	78.8	—	17.1	—	—	—	—	—	—	—	—	—
3	1.83	79.2	73.0	30.9	25.9	62.1	17.4	6.19	1316	67.1	35.0	46.7	37.5
4	1.84	79.4	76.5	33.9	29.4	69.4	16.7	6.10	1149	50.3	52.5	45.4	57.5
5	1.70	79.4	76.6	36.0	27.3	66.1	17.4	6.04	1098	43.0	66.9	51.8	61.2
7	1.79	78.7	76.4	33.0	28.3	65.2	18.2	6.04	1082	35.5	83.7	54.4	54.9
8	1.79	79.8	77.2	33.7	30.5	68.7	17.6	5.99	1069	34.7	88.8	53.0	58.8

* 'Apparent' maltose – indirect estimate
† 'Graf' – mashes made with added enzyme
‡ Grist coarsely ground.

dilute solutions of mercuric chloride or ammonia to check enzyme activity, may be prepared. The specific gravity of the resulting solution is a measure of the 'preformed' soluble materials, mainly carbohydrates present in the malt. The result is expressed as a percentage of the malt solids dissolved. For malts from two-rowed barleys a C.W.E. of 18% is taken to indicate undermodification; 18–20% is adequate; 20–22% is well modified and more than 22% indicates overmodification. Values of 15–18% are acceptable in six-row malts. The pH of cold water extracts changes during malting (see below) [17]. The yield of cold water extract rises and falls, (Table 15.3).

Enzymic activities, diastatic power and 'α-amylase', are often determined. Most of the methods used lack specificity, yet the results can be useful guides to processing. The total nitrogen content of the malt (T.N., % dry matter) and the nitrogen that comes into solution under standard extraction conditions, the total soluble nitrogen (T.S.N.), are regularly determined. In many cases it is usual to multiply the nitrogen values by 6.25 and call the result 'protein'. The T.S.N. may be divided into subclasses in various ways. For example after standard boiling the nitrogen precipitated is the coagulable nitrogen, while that variable part remaining in solution is the 'permanently soluble nitrogen', (P.S.N. sometimes taken as 0.94 of the T.S.N.). The nitrogenous materials may be fractionated e.g. by the Lundin precipitation system, or by gel exclusion chromatography. Formol titrations or determinations of total amino acids or sometimes individual amino acids may be used. The ratios T.S.N. × 100/T.N. (%), the Kolbach index, and P.S.N. × 100/T.N. (%) the 'nitrogen index of modification' have been used as indications of malt quality. Within any one malting system they work well, but additives may alter these ratios, (bromate depresses it, while sulphur dioxide increases it) without necessarily altering other malt characters. This illustrates the fact that different measurements used as indications of 'degrees of modification' often correlate badly when a wide range of malts is considered. In traditional ale malts (from two-rowed barleys) nitrogen indices of modification of < 33% indicate undermodification and > 37% overmodification. With six-rowed malts an index of 30% indicates good modification; for two-rowed malts 36% is a 'comparable' figure [16].

Generally it is the quantity and quality of the *extract* that malt provides, either when mashed alone or mixed with other cereal preparations, that are the criteria of overwhelming importance. By 'extract' we refer to the solids brought into solution by a 'mashing' procedure. Ground malt is mixed with warm water, and after a period the solution is separated from the insoluble residual solids (Chapter 16). Brewers call this solution 'wort'; others give it different names. The dissolved solids are the extract. The quantity obtained from a given weight of malt is determined by; (i) the preformed soluble materials in the malt (i.e. the cold water extract);

Malting

(ii) the levels of enzymes that convert insoluble to soluble substances in the malt; (iii) the levels of potentially extractable materials that may be so converted; (iv) the degree of grinding received by the malt; (v) the temperature programme of the mash; (vi) the ratio of liquid to malt used in the extraction process; and (vii) the presence of salts or other materials in the mashing liquor. The quantity of extract in solution is calculated from its specific gravity at a fixed temperature, making various assumptions [18]. For *one type* of wort the extract may be found using a very precise refractometer calibrated against comparable worts of known specific gravity. In British ale brewing most of the mashing process is at one temperature, and so the analytical determination of hot water extract (H.W.E.) by the Institute of Brewing (I.o.B.) method involves a mash at 65°C (149°F). In the American Society of Brewing Chemists (A.S.B.C.) and European Brewery Convention (E.B.C.) methods the temperature of the mash is progressively increased in a predetermined way to resemble more closely the changes that occur in decoction or other mashing systems. In each case the malt is milled in a closely specified way. Extract may be expressed as the percentage of malt solids that come into solution (misleadingly based on the specific gravities of sucrose solutions). The I.o.B. method gives extract in Brewer's pounds per standard quarter of malt, of 336 lb (i.e. lb/Qr). The weight of a barrel of wort (36 imp. gal; 1.6365 hl) in lb, less the weight of a barrel of water, (360 lb), both at 15.5°C (60°F) is the extract in that barrel, in brewer's pounds. When 50 g of malt are mashed and the volume is made up to 515 ml (at a fixed temperature) than the SG of the supernatant liquid relates to the malt extract as:

$$E = 3.565 \, (SG - 1000) - 4.8$$

Where E = extract (lb/Qr) and SG = specific gravity (relative to water as 1000) [18]. This system is simple, and no assumptions are made about relationships between specific gravity and the weight of materials in solution. Commercial two-row ale malts vary in extract usually between 96 and 103 lb/Qr. In ASBC analyses two-row malts usually give extracts in the range 75–80%. Six-row malts may give extracts as low as 70%. In all these cases the extract is corrected to a dry malt basis. Poorly modified malts give low extracts when coarsely ground, compared to when they are finely ground so the difference in extract between finely and coarsely ground samples is a measure of modification. In some special techniques extra enzymes may be added to the experimental mash, e.g. in 'Graf' mashes, to obtain an estimate of maximum potential extract. The time taken for the starch to disappear from the mash, the misnamed 'saccharification' time, gives an indication of whether or not the malt contains sufficient α-amylase. Excessive saccharification times are unacceptable. The filtration time should not be excessive, or the 'wort' is

550 Barley

too viscous. The filtered wort should be 'bright' (clear), and not opalescent or turbid. The mash should have no unpleasant smells or off-flavours. The wort may also be analysed for total soluble nitrogen or its fractions, polyphenols, pentosans or other fractions. Its viscosity may be determined directly, as may the degree to which its contents can be fermented by yeasts. In the latter case values of about 65–78% 'fermentability' are obtained by a conventional calculation if the wort is first boiled to inactivate residual enzymes. Lower values indicate that the malt gives a poorly fermentable wort. Higher fermentabilities are desirable in distilleries, but would be unusual in beer breweries. If unboiled 'worts' are fermented, as employed in distilleries, higher-fermentabilities of 80–85% are encountered, due to the continued degradation of initially non-fermentable dextrins to sugars by enzymes carried from the mash.

15.8. Changes that occur in the malting grain

Using traditional methods, with slow growing two-row European barleys, it takes 8–10 days germination time to make a well modified British ale malt, and 6–8 days germination time to make a pale, less well modified lager malt (Fig. 15.10). It is now usual, by using higher germination temperatures, additives, and newer steeping techniques, to make acceptable ale malts, from two-rowed barleys, with 3.5–5 days germination. In comparison, it takes 5–7 days germination to make a good malt from high-nitrogen six-rowed N. American barleys. Many types of malting schedule are now used, with and without additives. As malting barley germinates, grows and respires, carbon dioxide and water are lost and the grain alters in its chemical and physical composition (Chapter 5; Table 15.2). Respiration is associated with heat output; if this is excessive it may be difficult to cool a large mass of germinating grain. Additions of potassium bromate depress respiration and heat outputs. Good aeration in the steeps, and the additives hydrogen peroxide and gibberellic acid help to break dormancy, and accelerate germination. Grain high in nitrogen takes up water more slowly, but eventually respires faster than grain having less nitrogen, modifies faster, produces more enzymes but sustains a higher malting loss. Grain with a high moisture content may germinate slowly, but then it grows and modifies fast, and develops a good extract at the expense of a high malting loss. However, if it is grown too wet the extract declines, the extract substances being used in growth and respiration. Being more wet it is more expensive to dry on the kiln. Oversteeping may reduce the initial respiration rate, but subsequently the respiration intensity is enhanced (Chapter 5) [10]. Maximal respiration rates (carbon dioxide output) are usually achieved by germination days 3–6, and intensities correlate well with eventual malting losses [28]. When grain is germinated warm (e.g. 22° C; 71.6° F, compared to about

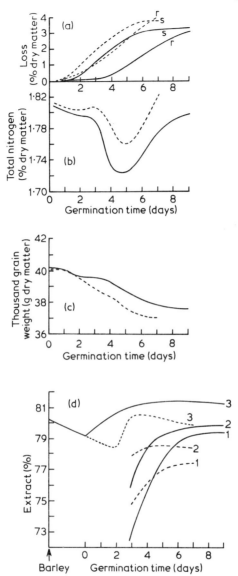

Fig. 15.10 The dry matter losses, the total nitrogen content, the thousand grain weight and the extracts obtainable (in three different ways) from malts grown for different periods of time at low temperatures of 13°–17° C (∼ 55°–63° F) and high temperatures of 19°–22° C (∼ 66°–72° F) (after Piratzky and Rehberg [33]). Grain germinated at 13°–17° C ———; grain germinated at 19°–22° C --------.
(a) Malting losses, r = respiration losses; s = rootlets. (d) Extracts were obtained with different types of mash. (1) Made with coarsely ground malt; (2) made with finely ground malt; (3) 'Graf' mashes, made with additions of enzymes to supplement those provided by the malt itself.

14°C; 57.2°F) the alterations in the grain are initially faster, but while the malting losses are enhanced the final enzyme content, extract and soluble nitrogen levels, are depressed (Fig. 15.10; Table 15.3) [44]. Suddenly reducing the temperature, e.g. from 18°C (64.4°F) to 13°C (55.4°F), part way through germination can produce malt with an elevated α-amylase content in a shorter time [29,45].

By tradition maltsters follow the progress of modification by noting the malt's physical state. When most of the acrospires are 0.7–0.8 the length of the grain, and the whole endosperm is easily 'rubbed out' between the fingers, malt is considered ready to kiln. As individual grains grow and modify at different rates, it is important to inspect at least 100 corns. The different changes constituting growth and 'modification' do not always alter in parallel (Chapter 5). Respiration rises to a maximum then falls. The proteolytic and diastatic activities in the malt rise, but not in parallel. In the latter case this is mainly due to the increased solubility of β-amylase and the *de novo* synthesis of α-amylase. Numerous other enzymes alter in amount. Normally enzymes are at least partly inactivated during kilning; the degree of inactivation differs. Thus α-glucosidase is readily destroyed, β-amylase less so, and α-amylase is relatively resistant to thermal inactivation. As the enzymes accumulate they catalyse the partial degradation of the endosperm cell walls, some storage proteins (mainly the hordein fraction) and a little starch. Thus the composition of the endosperm alters and the cold water extract, after a fall during steeping and the first day of germination (caused by leaching and respiration) rises as sugars accumulate in the grain. The pH of extracts of grain falls during steeping, e.g. from about 5.8 to about 5.2 but rises again during germination to as much as 6.0, only to fall to about 5.7 on the kiln [17]. The buffering substances extracted from the grain, which include phosphates and amino acids, also undergo alterations in amount. Some estimates of the mineral contents of malts, and the original barleys are available [22,23].

When grain is dried after differing periods of germination it is found that the hardness of the series of 'malts' declines and that the hot water extract increases to a maximum value, and then it may decline. The difference in extract obtained from finely and coarsely ground samples decreases with increasing germination time, e.g. from 8.2%, or 10.2 lb/Qr, to 0.6%, or 0.7 lb/Qr for a fully modified malt. The viscosity of extracts declines to a nearly constant value, (Table 15.3) and the soluble 'pentosans' (furfural-yielding substances) rise to a constant value. Extracts from undermodified malts tend to be opalescent or turbid. Nitrogenous substances move from the starchy endosperm into the embryo, and some are lost with the rootlets. However, the total weight of the finished malt declines with continuing germination time, so the nitrogen (%) content of the malt at any time is due to conflicting trends, removal by the rootlets, and the concentration due to declining dry weight (Fig. 15.10). The soluble nitrogen obtainable from malt tends to increase to a constant value with increasing germination time.

Malting losses are highest when overmodified malts, containing high levels of soluble sugars and amino acids, are made to allow the production of dark colours and rich flavours during kilning. Conversely they are lowest when 'short-grown' undermodified malts are being prepared. Steep losses vary with barley cleanliness and usually are 0.5–1.5% (dry weight) of the original sample. Respiratory losses range from 3–8% and losses on rootlets 2.5–4.5%, giving commercial losses in dry weight of 6–14% [18,21,48]. Large losses of starch occur during malting. These are substantially greater than the differences between the percentage starch contents of malts and the barleys from which they were made. Consequently a reduction of malting losses greatly reduces losses of potential extract. Rootlet losses and a major part of respiratory losses are due to the embryo. Beyond a certain time the embryo is not necessary to ensure modification. Reliable ways of reducing malting losses were developed when it was found that gibberellic acid, (GA_3, 0.025–0.50 mg/kg grain) dissolved in water, and applied to the grain at steep-out would break dormancy, ensure more rapid and even germination and accelerate the modification process, shortening germination by 2–3 days. As enzyme production and modification are enhanced more than embryo growth, malt having a peak extract is obtained sooner, in slightly higher yield, and with an extract enhanced by up to 2%. If the GA_3 is used in moderation top quality malts, indistinguishable from the best traditional products, are obtained. If excess GA_3 is applied malts are overmodified, excessive levels of enzymes, cold water extract and soluble nitrogenous materials are formed, and dark malts giving highly fermentable worts are produced at the expense of high malting losses [18]. Many attempts have been made to accelerate the penetration of gibberellic acid into the grain. The induction of rapid chitting by spraying a solution of gibberellic acid in dilute hydrogen peroxide is one possibility. Another approach, undergoing commercial trials, involves 'abrading' barley.

Gibberellic acid enhances grain respiration, and the extra heat produced by treated grain is hard to dissipate in some malting plants. Potassium bromate reduces respiration and rootlet growth, and is often applied in conjunction with gibberellic acid, being sprayed onto the grain (100–200 mg/kg barley). The only residue is potassium bromide [25]. Treated malts have characteristically short, thick rootlets. Good quality malt is produced in high yield. Many other treatments have been devised to reduce malting losses, with or without the use of gibberellic acid. Processes that have had some success, at least experimentally, include: (i) Holding the grain in a closed container when the fall in oxygen and build up in carbon dioxide checks respiration and growth; Kropff malting. (ii) Spraying solutions or otherwise applying growth-retarding chemicals such as coumarin, organic solvents, ammonia, sulphur dioxide, acetic acid, various inorganic acids, used steep liquors, copper sulphate, potassium bromate, and formaldehyde. Technical difficulties or other objections are associated with the use of several of these substances [38]. (iii) Grain growth may be checked by 'resteeping', that is covering germinating grain with water and obtaining anaerobic conditions for various periods. When successful the roots die and their contents leach out. The process is more certain if warm water is used in the resteep. The resteep liquor becomes foul as micro-organisms grow on the root leachate. This problem can be overcome by the addition of small quantities of formaldehyde to the resteep liquor, which also makes the process more reliable. (iv) An initial 'hot steep' at 40° C (104° F) followed by an application of GA_3 and normal 'germination' produces malt in a high yield. (v) Other physical treatments of the grain such as freezing the embryo, or drying, de-rooting and rewetting the partly germinated malt seem to be impractical. By continuously 'bumping' the grain, by moving it and ensuring that it falls from a height and by applying gibberellic acid it is possible experimentally to get malt with a loss of only about 3.5% [46]. The power requirement is uneconomically high for commercial malting, but the technique indicates how low malting losses might be. Customers may specify that certain additives may not be used in the preparation of their malts.

During kilning enzymes survive high curing temperatures best if the malt is relatively dry. Under these conditions colour and flavour development is minimal. On the other hand, if kilning temperatures are elevated while the malt is still 'green' and damp, colour and flavour development and enzyme destruction are favoured. Gum-degrading enzymes are liable to substantial inactivation on kilning. The greater heat lability of β-amylase compared to α-amylase can lead, when malts are strongly cured, to products with a normal extract, but giving dextrinous worts with low fermentabilities. During kilning the cold water extract and levels of some reducing sugars, soluble pentosans, and particularly sucrose, may increase in amount, while β-glucan may decline. During curing the starch may, to a slight extent, undergo 'pyrodextrinization', be involved in melanoidin forming reactions and become slightly less susceptible to enzymic degradation [26]. Hydroxymethyl furfural is formed during curing. Pyrodextrinization is significant in roasted malts and cereals (cf. Chapter 14). Colour and aroma are developed mainly as a result of 'Maillard' reactions occurring between the reducing groups of sugars and amino groups, mainly of amino acids. Some addition compounds between amino acids and fructose are produced during kilning, and may be regarded as intermediates in Maillard reactions [50]. These reactions occur most readily when the reactants are present in high concentrations (i.e. the grain has a high cold water extract), the grain is wet, and the

Malting 555

temperature is high. To achieve high colours and flavours, therefore, malt is overmodified on the floor or 'stewed' on the kiln (heated with limited ventilation and drying) and is then cured while relatively moist. In the preparation of Munich dark malts overmodification is achieved by loading the malt into a bed about 50 cm (20 in) deep and covering it so that the temperature rises to about $50°C$ ($122°F$) and only falls as the build up of carbon dioxide and depletion of oxygen limit respiration, Melanoidins are polymeric, colloidal materials, coloured brown, and with strong reducing properties. They contribute to beer stability and they and the small molecular weight substances also produced in the Maillard reaction contribute to the glorious flavours found in some malts and malt extracts. Small contributions to malt colour may be made by caramelized sugars and oxidized polyphenols [6]. Among the numerous minor components that may contribute to malt aroma and flavour are a series of aldehydes, ketones, alcohols, amines, and miscellaneous other substances, including sulphur-containing compounds and nitrogenous bases (Chapter 14). Presumably roasted malts contain many of the substances found in roasted barley. Peated distillery malts, for Scotch whisky manufacture, take up many substances from the peat smoke and contain various alkanes, alkenes, aldehydes, alcohols, esters, fatty acids, aromatic and phenolic substances, including phenol and cresols.

Distillers' malts, made by germinating barley cool and wet (46–48% moisture) to obtain maximal enzyme levels are lightly kilned to about 6% moisture so that minimal enzyme inactivation occurs. Worts of high fermentability are obtained when the malts are mashed. If the distillery uses unmalted 'adjuncts' then malts of very high diastatic power, prepared from selected high-nitrogen, six-rowed barleys, are preferred. For brewing purposes a very wide range of malts may be prepared. In some cases (e.g. black malts) the intense colour and characteristic flavours and aromas are developed by a roasting process at temperatures of $200-230°C$ ($390-446°F$); great care is needed to avoid carbonization. Enzymes are totally inactivated by such roasting procedures. In making caramel malts the wet, green malt (or rewet malt) is heated to $60-75°C$ ($140-167°F$) for 30–45 min. Massive starch and protein breakdown occurs within each grain, and the endosperm liquefies. The product readily colours on drying and curing, and on cooling the interior sets to a caramel-like mass, with a fine flavour and aroma. The special malts mentioned are used in small quantities to give characteristic colours, flavours and aromas to particular products. Acid malts, coated with lactic acid during their preparation, are used to adjust the mash-tun pH in brewing. Chit malts, germinated for a very short period, are used for cheapness and to convey 'raw grain' characters to beer while circumventing local laws forbidding the direct use of raw grains.

Sulphur dioxide reaching the green malt on the kiln in the drying

Table 15.4 Some I.o.B. analyses of top quality brewing malts (from (a) Hind [16], (b) Northam [30]). At present (1976) average nitrogen contents are substantially higher and many 'standard malts' have hot water extracts of e.g. 98 lb/Qr (dry basis).

Barley type	Pale ale malt two-rowed barley		Mild ale malt two-rowed barley		Californian coast six-rowed	Californian mariout six-rowed	Wisconsin-38 lager six-rowed
	(a)	(b)	(a)	(b)	(a)	(a)	(a)
Extract (lb/Qr, dry)	102.0	103.7	99.8	100–103.2	93.7	92.2	86.4
Moisture (%)	1.5	1.7	2.6	1.8–2.0	1.9	2.2	3.6
Colour (1 in cell)	4.5	—	8.0	—	6.0	4.0	2.0
Colour (°EBC)	—	5.0	—	6.7	—	—	—
Cold water extract (%)	18.0	18.0	17.8	17–19	16.8	16.9	17.3
Diastatic power (°L)	36	35	31	33–48	31	50	170
Total N (T.N., %)	1.34	1.35	1.47	1.44–1.56	1.62	1.62	2.37
Permanently soluble N (P.S.N., %)	0.51	—	0.51	—	0.49	0.48	0.68
Total soluble N (T.S.N., %)	—	0.49	—	0.50–0.58	—	—	—
P.S.N./T.N. (%)	38.0	—	34.9	—	30.1	29.4	28.8
Thousand corn weight (g)	35.3	—	34.4	—	37.3	39.4	24.7

Malting 557

Table 15.5 The composition of American malts, analysed by the A.S.B.C. (1958) methods (from Kneen and Dickson [21]).

Type of barley	Mid-Western six-rowed	Californian six-rowed	Western two-rowed
Kernel weight (mg)	30	39	37
Moisture (%)	4.2	4.4	4.2
Extract (%, fine grind)	76.8	77.0	80.5
Extract (%, coarse grind)	74.8	75.0	79.0
Extract difference (%)	2.0	2.0	1.5
Colour ($\frac{1}{2}$ in cell, °L, S.52)	1.6	1.4	1.2
Total protein (%)	12.3	11.0	10.5
Soluble protein (% total)	40	35	38
Diastatic power (°L)	135	65	90
α-Amylase (20° units)	38	25	30
Acrospire growth $\frac{1}{2}-\frac{3}{4}$ (%)	92	85	91
Assortment:			
on 7/64 in (0.28 cm) screen	32	68	85
on 6/64 in (0.24 cm) screen	53	26	10
on 5/64 in (0.20 cm) screen	14	6	1
through	1	0	0

gas-stream enhances the levels of extract and soluble nitrogen and reduces the colour. Applying sulphur dioxide or sulphurous acid to the germinating grain checks malting losses, and has similar effects on wort composition as supplying sulphur dioxide on the kiln [13,38].

Chemical analyses of two-rowed barleys, and malts derived from them, are shown in Table 15.2. In six-rowed barleys and malts the husk contents (and so the lignin, cellulose and hemicellulose levels) are higher, while starch contents are less by about 5–8%. Technical analyses of various malts are shown in Tables 15.4 and 15.5.

References

1. Anderson, J.A. and Ayre, C.A. (1938). *Can. J. Res.*, **16 C**, 377–390.
2. Anon. (1969). *Internatn. Brewer's J.*, **105**, Sept., 72–75.
3. Anon. (1976). *Brewers' Guard.*, Jan., 31–33.
4. A.S.B.C. (American Society of Brewing Chemists; 1958). *Methods of Analysis of the American Society of Brewing Chemists.* Madison, Wisconsin: ASBC.
5. Ayre, C.A., Sallans, H.R. and Anderson, J.A. (1940). *Can. J. Res.*, **18 C**, 169–177.
6. Bathgate, G.N. (1973-Apr.). *Brewer's Digest*, 60–65.
7. Baxter, E.D., Booer, C.D. and Palmer, G.H. (1974). *J. Inst. Brewing*, **80**, 549–555.
8. Bloch, F. and Morgan, A.I. (1967). *Cereal Chem.*, **44**, 61–69.
9. Clerck, J. de. (1957). *A Textbook of Brewing*, (Transl. Barton-Wright, K.) **1**. London: Chapman and Hall.

10. Day, T.C. (1896). *Trans. Proc. bot. Soc. Edinb.*, **60**, 492–501.
11. E.B.C. (European Brewery Convention; 1963). *Analytica—E.B.C.* (2nd edition). Amsterdam: Elsevier.
12. Ekström, D., Cederquist, B. and Sandegren, E. (1959). *Proc. Eur. Brew. Conv., Rome*, 11–26.
13. Graff, A.R. (1972). *Master Brewer's Ass. Am. Technical Q.*, **9**, 18–24.
14. Graham, C. (1874). *J. Soc. Arts.*, Feb. 6, 221–226.
15. Harris, G. (1962). In *Barley and Malt; Biology, Biochemistry and Technology.* ed. Cook, A.H. pp. 431–582; 583–694. London: Academic Press.
16. Hind, H.L. (1943). *Brewing Science and Practice*, **1**—(Brewing Materials.). London: Chapman and Hall.
17. Hopkins, R.H. and Kelly, H.E. (1929). *J. Inst. Brewing*, **35**, 402–409.
18. Hough, J.S., Briggs, D.E. and Stevens, R. (1971). *Malting and Brewing Science* (revised 1975). London: Chapman and Hall.
19. Hudson, J.R. (1960). *Development of Brewing Analysis, A Historical Review.* London: Institute of Brewing.
20. I.o.B. (Institute of Brewing Analysis Committee 1971). *J. Inst. Brewing*, **77**, 181–226.
21. Kneen, E. and Dickson, A.D. (1967). In *Kirk-Othmer's Enclyopaedia of Chemical Technology* (2nd edition). **12**, 861–886.
22. Kolbach, P. and Rinke, W. (1963). *Monats. Brau.*, **16**, 11–16.
23. Liu, D.J., Pomeranz, Y. and Robbins, G.S. (1975). *Cereal Chem.*, **52**, 678–686.
24. Macey, A. and Stowell, K.C. (1959). *Proc. Eur. Brew. Conv., Rome*, 105–112.
25. Macey, A. and Stowell, K.C. (1961). *Proc. Eur. Brew. Conv., Vienna*, 85–98.
26. MacWilliam, I.C. (1972). *J. Inst. Brewing*, **78**, 76–81.
27. Meredith, W.O.S. and Anderson, J.A. (1938). *Can. J. Res.*, **16 C**, 497–509.
28. Meredith, W.O.S., Anderson, J.A. and Hudson, L.E. (1962). In *Barley and Malt; Biology; Biochemistry; Technology*, ed. Cook, A.H. pp. 206–270. London: Academic Press.
29. Narziss, L. (1970). *Wallerstein Labs. Commun.*, **33**, Aug., 77–85.
30. Northam, P.C. (1965). *Brewer's Guard.*, **94**, 97–110.
31. Northam, P.C. and Button, A.H. (1973). *Proc. Eur. Brew. Conv., Salzburg*, 99–110.
32. Piratzky, W. (1936). *Woch. Brau.*, **53**, 105–108.
33. Piratzky, W. and Rehberg, R. (1935). *Woch. Brau.*, **52**, 89–93; 101–104.
34. Pollock, J.R.A. (1962). In *Barley and Malt; Biology; Biochemistry; Technology*, ed. Cook, A.H. pp. 303–398; 399–430. London: Academic Press.
35. Pollock, J.R.A. (1971). *J. Inst. Brewing*, **77**, 57–70.
36. Pomeranz, Y. and Burger, W.C. (1973). *Proc. A.M. Am. Soc. Brew. Chem.*, 83–85.
37. Pomeranz, Y., Standridge, N.N. and Shands, H.L. (1971). *Crop. Sci.*, **11**, 85–88.
38. Ponton, I.D. and Briggs, D.E. (1969). *J. Inst. Brewing*, **75**, 383–391.
39. Pool, A.A. (1964). *J. Inst. Brewing*, **70**, 221–225.
40. Pool, A.A. (1967). *Brewer's Guild J.*, Apr., 188–196.
41. Reiner, L. (1975). *Brauwissenschaft*, **28**, 363–364.
42. Schuster, K. (1962). In *Barley and Malt; Biology, Biochemistry, Technology*, ed. Cook, A.H. pp. 271–302. London: Academic Press.
43. Schuster, K. and Weinfurtner, F. (1963). *Die Bierbrauerei I. Die Technologie der Malzbereitung.* Stuttgart. Ferdinand Enke Verlag.

44 Shands, H.L., Dickson, A.D., Dickson, J.G. and Burkhart, B.A. (1941). *Cereal Chem.*, **18**, 370–394.
45 Shands, H.L., Dickson, A.D. and Dickson, J.G. (1942). *Cereal Chem.*, **19**, 471–480.
46 Sipi, M. and Briggs, D.E. (1968). *J. Inst. Brewing*, **74**, 444–447.
47 St. Johnston, J.H. (1954). *J. Inst. Brewing*, **60**, 318–340.
48 Vermeylen, J. (1962). *Traité de la Fabrication du Malt et de la Bière*, **I, II**, Association Royale des Anciens Elèves de l'Institut Superieur des Fermentations. Gand., 750; 751–1624.
48a Witt, P.R. (1959). In *The Chemistry and Technology of Cereals as Food and Feed*. ed. Matz, S.A. pp. 475–509. Westport, Conn: Avi Publishing.
49 Witt, P.R. (1970). *Brewer's Digest*. Feb., 48–57.
50 Yoshida, T., Horie, Y. and Kuroiwa, Y. (1972). *Proc. 12th Conf. Inst. Brewing*, (*Austral. and N.Z. Section, Perth*), 135–146.

Chapter 16

Some uses of barley malt

16.1. Introduction

Malts are used in the manufacture of other products. Malt flour, prepared by milling and sifting to remove the husks, is used to a small extent in baking. However for most purposes malt is mashed, that is the milled product is mixed with warm water and sometimes with other starchy materials. After a reaction period, 'the stand' the sugary liquid, 'wort' or 'wash', is usually separated from the residual solids and is processed further. Ultimate products include malt extracts (powders and syrups), diastase, beer, whisky, gin, vodka, alcohol, and malt vinegar. In traditional beer brewing, and in the production of Scotch malt whisky, only malt is used in the mash. In most other cases malt enzymes are used to 'convert' starch from other sources – e.g. barley, maize, wheat or potatoes, so that the malt provides enzymes, a proportion of the extract, and a range of nitrogenous and other yeast nutrients. In products where malt 'character' is unimportant, e.g. in preparing alcohol or vodka, it is sometimes economic to use green malt or enzymes of microbial origin to supplement the malt or even to replace it. Economic and other factors decide whether alcohol from fermentations or from petrochemicals is used for a particular purpose. Generally only alcohol from fermentations may legally be used in 'potable liquors'.

Malt has been used as a source of nutrients for a range of unusual fermentations, e.g. the production of butanol, glycerol, and acetone [46]. These processes have fallen out of use, but changing conditions may once again make them economic. Extracts of rootlets, ('malt sprouts') have often been used to supply vitamins and nitrogenous nutrients for micro-organisms [46]. It is feasible to use acid-hydrolysed barley grits as microbial nutrients, e.g. in lactic acid fermentations. The uses of malts are

Some uses of barley malt 561

outlined first to introduce the reasons for using different types, and for mashing them in different ways.

Malt extracts and syrups are prepared by concentrating worts by evaporation. These products possess a range of colours, flavours, enzyme complements and fermentabilities (i.e. their carbohydrate compositions vary). Enzyme rich powders, prepared from such extracts, are called diastase, a term also applied to the natural mixture of starch degrading enzymes. In brewing beer a clear wort from the mash tun is boiled with hops, then is clarified, cooled and aerated before being fermented by yeast. The product, after the removal of most or all of the yeast, is clarified, and packaged. The hop boiling process flavours the wort, sterilizes it, precipitates a protein sludge and inactivates residual enzymes so that its composition is 'fixed'; as the higher dextrins are not fermentable by yeast they survive in the finished beer. In the distilling and vinegar brewing industries the 'wort' (known by various names) is not boiled, and indeed in N. America it is not even separated from the residual solids of the mash, so that during the fermentative stage surviving enzymes continue to degrade the dextrins to a level such that the products are fermentable by yeast and more alcohol is produced. The failure to boil gives problems with microbial contamination from the malt during subsequent processing [11]. Special care is taken to favour enzyme survival, and to maximize alcohol production. In products such as beer and malt whisky care is taken to obtain the correct flavours. In the production of 'neutral spirit' for gin, vodka or other purposes, the rectification (fractional distillation) is so discriminating that few off-flavoured, minor products are carried into the final material. In the production of malt vinegar the wort is fermented by yeast, and the alcohol is subsequently oxidized to acetic acid by bacteria.

Thus mashing must produce worts of the composition needed to give a particular product. The nature of malt and the changes occurring during mashing are so complex that there is little likelihood of their being specified in detail. The choices and uses of raw materials are guided in a semi-empirical way by experience and as much scientific quality control as is practicable. In *broad* terms some chemical aspects of mashing can be described.

16.2. Mashing

When malt is extracted with ice-cold water 15–22% of its dry matter is dissolved. About half of this dissolved material is fermentable by yeast. When mashes are made at progressively higher temperatures more of the malt solids come into solution, forming the extract. The quantities of extract obtained increase with mashing time up to a maximum usually reached in about 1 h, generally 75–82% of the malt solids, at suitable temperatures, up to about 65°C (149°F). The fermentabilities of the wort solids also change with time, but not in parallel with the total extract. Thus the same total extract is obtained at 65°C (149°F) or 70°C (158°F), but the fermentability is lower at 70°C (Fig. 16.1) [24,69]. These alterations are chiefly due to the changing quantities and relative proportions of different carbohydrates in the extract. Analogous differences occur with the nitrogenous and other components of the wort solids [10,22,24,65].

When ground malt, with or without unmalted adjuncts (the grist), is mixed with warm water the water penetrates the solids and dissolves a variety of materials, some of which diffuse into the liquid between the solid particles. At the temperatures used in mashing hydrolytic enzymes act to degrade insoluble substances, notably starch, and the degradation products also dissolve. Substances in solution are also degraded further.

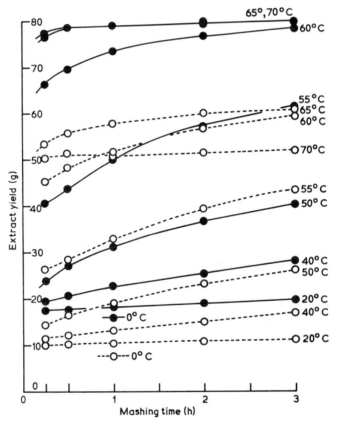

Fig. 16.1 The increases in total extract and fermentable extract with time in mashes made at various constant temperatures (from Windisch *et al.* [69]).
●——● total extract. ○----○ Fermentable extract. Temperature scale equivalents. 0°C, 32°F; 20°C, 68°F; 40°C, 104°F; 50°C, 122°F; 55°C, 131°F; 60°C, 140°F; 65°C, 149°F; 70°C, 158°F.

Some uses of barley malt

At the temperatures needed to degrade or 'convert' starch sufficiently to bring it into solution, $60°–65°C$ ($140°–149°F$), enzymes act rapidly, but they are progressively inactivated. Many enzymes are involved in wort production, and these vary in their susceptibilities to heat inactivation. Consequently malts prepared and mashed at different temperatures give particular worts of differing compositions. Traditional British ale brewers choose well modified, highly cured malts with comparatively small enzyme contents that, in combination with infusion mashing at about $65°C$ ($149°F$), yield worts that, after the hop-boil, contain a mixture of solids of which about 70–75% are fermentable, and have a particular colour, a chosen level of soluble nitrogen, and so on. By using malts of differing degrees of modification and enzyme content, with or without unmalted adjuncts it is possible to achieve very similar worts by altering mashing conditions [22,24]. In decoction and temperature-programmed mashing systems the initial mash temperature is low, and is progressively increased to about $75°C$ ($167°F$) during the mashing period. In decoction mashing some part of the mash is also boiled. Such temperature programmes allow heat-labile enzymes to act before they are denatured, and also permit the more stable enzymes to act rapidly at elevated temperatures, before they too are destroyed.

Traditionally malt is milled and is broken up so that the husk is preserved as completely as possible to form a filter bed, and the wort can be run easily from the residual solids. Usually prepared 'adjuncts' are mixed with the milled malt just before mashing begins. By breaking up the malt finely and ensuring complete mixing rapid solution of the extract is possible, but the separation of the liquid from the residual solids becomes more difficult until, when the entire grist is reduced to a fine flour, the usual simple filtration devices are inadequate and it becomes necessary to use centrifuges, or rotary vacuum filters [30].

The ratio of ground grist to water influences the yield of extract and its composition. The details vary a little, but are due to enzymes, such as proteases and particular carbohydrases, surviving better in the 'thick' mashes due to the partial protection given by their substrates against heat inactivation. Thus the production of hexoses and soluble nitrogenous substances are favoured. At the same time the rate at which total extract increases, is reduced as the high concentrations of sugars occurring in thick mashes partly inhibit the hydrolytic enzymes that degrade the starch (product inhibition of enzyme action). Usually about 90–92% of the solids of brewer's wort are carbohydrates, 5–6% are 'crude' protein' ($N \times 6.25$), 1.5–2% are ash, and the remainder is a mixture of trace components, such as lipids, phenolic substances, vitamins, and organic acids [22,24,37]. During mashing most ($>98\%$) of the starch is degraded and dissolved, 'converted', but only 30–40% of the malt nitrogenous substances come into solution. When unmalted cereals, which are deficient in enzymes are included in the grist proportionately fewer nitrogenous substances dissolve, and more care is needed to ensure that adequate starch conversion occurs.

In infusion mashing initially insoluble polysaccharides, other than starch, probably contribute less than 2% to the wort carbohydrates, although in decoction mashing this figure may reach 6% [22]. A comparison of the 'potentially extractable' carbohydrates of malt with those of wort show how greatly starch contributes to extract (Table 16.1) [23]. The carbohydrates found in wort include arabinose, xylose, ribose, fructose, glucose, sucrose, maltose, maltotriose, maltotetraose, other higher α (1 $\longrightarrow$ 4)-linked glucans, with some α (1 $\longrightarrow$ 6) branch points (dextrins), isomaltose, panose (6^2-α-glucosyl-maltose), iso-panose (6-α-maltosyl-glucose), maltulose, maltulotriose, nigerose, 'glucodifructose' (a mixture of 1- and 6-kestoses), higher fructosans, pentosans, and β-glucans. Some of these materials (e.g. the pentoses) are of sporadic occurrence only. Traces of glycoproteins may also be present (Table 16.1, 16.2) [3,8,13,22,37,38,49]. The pentoses occur mainly in decoction worts. Maltulose is formed by the epimerization of maltose during wort boiling. β-Glucans are important, although present in relatively small amounts, as their high viscosity retards or even prevents adequate wort separation after mashing, and sometimes gelatinous precipitates of β-glucans separate from strong beers [2,15,24]. In brewery worts about 80% of the dextrins (branched and unbranched) contain 4–20 glucose residues per molecule and the remaining 20% over 20 hexose residues per

Table 16.1 A comparison of the major chromatographic fractions of the wort carbohydrates with the 'potentially extractable' carbohydrates of malt* (from Harris, et al. [23]).

Malt carbohydrates (monosaccharide equivalents as % wort solids)		Carbohydrates of wort (monosaccharide equivalents, % of total wort carbohydrate)	
Starch	85.5	Dextrins †	22.2
Glucosan	2.5	Maltotetraose	6.1
Fructosan	1.4		
Maltotriose	0.58	Maltotriose	14.0
Maltose	0.99	Maltose ‡	41.1
Sucrose	5.1	Sucrose	5.5
Glucose	1.7	Glucose	8.9
Fructose	0.71	Fructose	
Total	98.48	Total	97.8

*Extract of malt, laboratory 104.6 lb/Qr; Barley 102.2 lb/Qr.
Recovery of potentially extractable hexose apparently 99.4%.
† Include pentosans.
‡ Includes trace of iso-maltose.

Table 16.2 The levels of the main carbohydrates found in commercial syrups, compared with wort from an all-malt mash (from MacWilliam [38], Roberts and Rainbow [49]). Carbohydrates, expressed as % total carbohydrates. Nitrogen, expressed as % syrup solids, except for wort.

	Glucose and fructose	Sucrose	Maltose	Maltotriose	Dextrins	Total nitrogen (%)
Barley syrup	7–27	0–5.5	46–60	7.5–17.3	5.1–27	0.62–0.80
Green malt syrup	5.6–7.5	4.6–7.7	48.4–55.1	8.9–18.5	22.6–23.1	0.64–0.70
Malt extract	12.5–33.5	0–3.8	18.1–54.0	5.8–9.3	20.7–39.4	0.7–1
All malt brewing wort	8–12 (G, 8 to 11; F, 1 to 4).	1–5	42–50	13–16	21–30	(80–95 mg/100 ml)

molecule [14]. In vinegar or distillers 'worts' these dextrins are hydrolysed by enzymes occurring in the wort and the products are utilized by yeast. In the mash hydrolytic enzymes must have access to the starch. In malt the partially degraded endosperm cell walls allow the penetration of enzymes to the starch within. In raw cereal powders (flours), which are sometimes used as adjuncts, the fine grinding used in their preparation breaks up the endosperm cells and exposes the starch. In cooked or steamed and rolled (flaked) preparations the heating (and physical treatment) disrupts the cell walls and swells the starch grains. In the case of flaked cereals the flattened shape is favourable for enzyme penetration. The partial swelling (gelatinization) of starch makes it more susceptible to degradation by enzymes. Such unmalted 'adjuncts' are used primarily as cheap sources of extract [43]. Cereal adjuncts, used as malt replacements in mashing, have effects on the wort composition. They act as 'nitrogen-diluents', yielding worts containing less soluble nitrogen, and the use of wheat and, to a lesser extent, barley flour yields beers having unusually good head retentions (foam stabilities). Mashes containing barley flours and flakes tend to yield viscous worts due to the presence of extra quantities of high molecular-weight β-glucans. These may be degraded by adding microbial β-glucanases to the mash. Barley adjuncts include whole, wet-rolled grains, whole-grain flakes, torrified (i.e. heated and popped) grains, pearl-barley flakes, flour and air-classified flour. The use of microbial enzymes to 'convert' unmalted adjuncts has been extensively investigated [44]. Using raw barley, which contains the enzyme β-amylase, with additions of the bacterial enzyme pullulanase, which breaks $\alpha(1\longrightarrow 6)$ linkages and so is a debranching enzyme, consideable starch degradation can be achieved in a mash [12]. When adjuncts are 'cooked' to gelatinize the starch sulphur-containing substances with characteristic smells, such as dimethyl sulphide, are often formed [50].

At low temperatures starch granules are relatively resistant to enzymic degradation. However at about 60° C (140° F) the large grains in barley and malt starches swell appreciably, the first step in 'gelatinization', and their susceptibility to enzymatic attack increases [4,66]. Swelling is faster and more extensive at higher temperatures and amylose is leached into solution until the granules break up and, in the absence of liquefying enzymes, the mass sets to a gel on cooling. Malt starch is more susceptible than barley starch to enzymic degradation. The small starch grains are resistant to swelling, and may resist enzyme attack even after boiling. These tend to remain in the spent grains after mashing [4,66]. The major aspects of starch degradation occurring in mashes can be explained by the combined action of α-amylase and β-amylase. However other enzymes play some part, and their activities are essential in obtaining good fermentabilities in vinegar and distillery worts. α-Amylase is an *endo*-$\alpha(1\longrightarrow 4)$-glucanase, that can attack starch grains and catalyse the hydrolysis of links at random, but at reduced rates near the chain ends or branch-points. The enzyme is stabilized by calcium ions. By degrading granular starch the enzyme brings it into solution. β-Amylase is an exo-α(1 ⟶ 4)-glucanase that hydrolyses the penultimate links of the non-reducing ends of starch grains, with the release of maltose. The enzyme cannot attack starch grains or by-pass α(1 ⟶ 6)-branch points, and so the ultimate products of its action on soluble starch are maltose and a β-limit-dextrin. β-Amylase is more readily inactivated by heat than α-amylase. During the metabolism of wort most yeasts utilize only the simpler sugars up to, and including maltotriose. When starch is degraded by α-amylase the products are only 16–20% fermentable by brewing yeasts. The corresponding values for β-amylase are 70%, and for a mixture of the two enzymes about 80%. The conversion process in beer breweries is terminated by the hop boil which fixes the fermentability, usually in the range 70–75%. In malt whisky distilleries the minimum acceptable fermentability of the unboiled extract is 86% [11]. The 'extra' fermentability is due to debranching enzymes hydrolysing α(1 ⟶ 6) bonds in the oligosaccharides. Acting together the amylases degrade starch more completely than either acting alone or in sequence. Relatively small amounts of α-amylase facilitate the breakdown of granular starch by β-amylase, by 'dextrinizing' the granules and bringing the starch into solution, where it is exposed to attack by β-amylase, and by breaking the starch chains within the α(1 ⟶ 6)-branch points, so exposing more non-reducing chain ends to the action of the second enzyme. In the temperature range 63°–67° C (about 145°–153° F) β-amylase is inactivated significantly faster at the higher temperatures. The effect of α-amylase is much less marked. Consequently less fermentable worts are producing by mashing at the higher temperatures (Fig. 16.1). Higher levels of α-amylase can partially 'compensate' and increase fermentabilities in worts from higher temperature mashes by bringing the starch into solution faster so that the surviving β-amylase can 'saccharify' the dextrins, before being inactivated. Initial high levels of β-amylase also favour high wort fermentabilities [24]. In infusion mashes extract yield is maximal at about 65°–68° C (149°–154° F), but the maximum yield of fermentable carbohydrate is obtained at about 65° C (149° F). The maximal production of reducing sugars occurs at about 60°–63° C (140°–145° F), and the most fermentable wort is obtained (in poor yield) at 57°–59° C (135°–138° F). In vinegar brewery- or distillery-worts practically all the dextrins are finally fermented by yeast. In part this is due to the use of 'highly attenuating' yeast strains, which produce extracellular amylases. By initially mashing at lower temperatures with malts rich in enzymes and by partly separating the enzyme-rich extract before remashing the extracted malt residue at a higher temperature enzyme survival in the wort is enhanced. Limit dextrinase, α(1 ⟶ 6)-glucosidase activity in the wort continues to act in the presence of yeast, and ensures that ultimately the degradation of the dextrins is complete. Limit dextrinase activity does not correlate closely with potential wort alcohol yields, which may be determined directly for quality control purposes [1,11,61]. In some circumstances enzymes (enzyme-rich malt flour; microbial amyloglucosidase or pullulanase) may be added to the wort to achieve maximal fermentability. As mash tempertures increase so the pH falls; conversely as wort is cooled its pH rises. At mashing temperatures, about 65°C (149°F), starch is most rapidly dextrinized at pH 5.3–5.9 (i.e. its ability to give a blue colour with iodine is destroyed), and attenuation of worts by yeast is maximal when the mash is made at pH 5.4–5.7. The pH optimum of β-amylase, about 5.1, is slightly below that of α-amylase, about pH 5.3, which is unstable below pH 4.9.

Worts contain many nitrogenous substances, including the 'common' amino acids and proline, peptides, proteins, ammonia, choline, betaine, various amines including methylamine and hordenine, purine and pyrimidine bases, ribonucleosides, deoxyribonucleosides and various vitamins [7,22,37]. At higher mashing temperatures the ratio of nucleosides to free bases increases as the nucleosidase activity is heat-labile. Nucleic acids and the phosphorylated nucleotides seem to be completely degraded during mashing. A brewers' wort of SG 1040 (relative to water, SG 1000) may contain 650–750 mg nitrogen/l, of which 140–190 mg may be α-amino-nitrogen, and 155–225 mg will be in compounds, 'protein', of molecular weight exceeding 5000 [49]. Purines, pyrimidines and their derivatives contain about 10%, and peptides and polypeptides contain about 30–40% of the soluble nitrogen [37]. The higher molecular weight substances are important in connection with the colloidal instability of beer and malt vinegar and other desirable attributes of beers, such as foam-stability and some aspects of 'character'. Peptidases and carboxypeptidases are active in mashing but they are more heat-labile than the amylases [22,41]. In consequence the apparent temperature optimum for the formation of soluble nitrogen is distinctly higher at shorter mashing times, and maximal yields of soluble nitrogen are obtained from mashes made at 50°–60°C (122°–140°F). The optimal mash pH values for the largest yields of different nitrogenous substances differ for the different fractions. Only 30–40% of malt nitrogen comes into solution during mashing. Greater quantities are dissolved in thicker mashes and by mashing for longer periods at low temperatures. Yeasts need a supply of simple nitrogenous substances to support their growth. In addition to any individual flavour characteristics, such substances may alter the production of flavour components by yeast. Thus valine tends to suppress diacetyl formation. Yeast only utilizes simple nitrogenous compounds. Unusually high levels of nitrogenous substances may cause wasteful, excessive yeast growth, which causes carbohydrate to be diverted from making alcohol to yeast dry matter, and can support the growth of undesirable contaminating micro-organisms. Levels of wort nitrogen are controlled by the choice of low-nitrogen malts, mashing conditions, and the use of unmalted adjuncts. Usually unmalted adjuncts contribute little soluble nitrogen to worts, and raw barley flour contains a protease inhibitor that *reduces* the quantity of nitrogen dissolved [33].

Vitamins present in wort include choline, folic acid, *myo*-inositol, nicotinic acid, pantothenic acid, pyridoxine, riboflavin and thiamine. Some of these are essential for healthy yeast growth [55]. For anaerobic growth yeast also requires certain unsaturated fatty acids and sterols which may be limiting in some worts. Worts contain small variable quantities of free fatty acids, and other lipids, which in excess have adverse effects on flavour, and on head (foam) stability in beer. Worts tend to be rich in lipids when 'rapid' techniques are used for separating them from the mash. Some lipid may be removed by Kieselguhr filtration during clarification. Worts also contain many phenols, some of which contribute flavours to the final products. Barley phenols tend to be astringent, while those from peat smoke are volatile and contribute to the flavour of Scotch malt whisky, which is made with 'peated' malt. Unfortunately polyphenols tend to polymerize and become 'tannins' which react with, and cross-link, the proteins which are present. In distilled products no proteins are present. In beers and malt vinegars protein-polyphenol complexes can separate from solution as hazes. The turbidity caused by the haze particles detracts from the product's appearance. Polyphenols may be reduced in amount by the addition of formaldehyde to mashes, or by the use of adsorbents. Many different precautions in processing may be used to reduce any tendency to haze formation in the product. For example, the degradation of proteins or protein-tannin complexes in beer is encouraged by adding proteolytic enzymes, such as papain, and tannin formation is minimized by the rigid exclusion of air [24].

In addition to fatty acids worts contain other organic acids, salts, inorganic phosphate, phytic acid (*myo*-inositol hexaphosphate) and probably lower phosphate esters of inositol. The acids, amino acids, and especially the phosphates contribute to the buffering capacity of the mash. Among the numerous minor wort components various sulphur compounds have attracted interest mainly for their eventual contributions to aroma and flavour.

In the simplest infusion mashing systems the 'stand', the period of about 1–3 h in which the mash is left to allow starch conversion is at about $65°C$ ($149°F$), this being a suitable temperature for producing British ale wort. However in other situations where the activities of heat-labile enzymes are needed, a range of combinations of rising mashing temperatures are used. The acidity of mashes are often adjusted, usually to within the range pH 5.3–5.6. A cold water extract of malt, made with

570 Barley

distilled water at $0°C$ ($32°F$) has a pH of about 6.3. When the malt is mashed at about $65°C$ ($149°F$) with water containing 'temporary hardness' ($CaCO_3$), with distilled water, and with permanently hard water (containing $CaSO_4$) the wort pH values, when cooled to room temperature, may be 6.0, 5.8 and 5.65 respectively. At mashing temperatures, $65°C$ ($149°F$) the pH values will be about 0.35 units less, so that by mashing with permanently hard water a satisfactory final mash pH is obtained. Further, the calcium ions favour the survival of α-amylase in the mash and, in brewing beer, a better 'break' formation (separation of coagulated materials), at the end of the hop boil, and better yeast flocculation. Mash pH is sometimes adjusted by the addition of hydrochloric, sulphuric, or lactic acids. This last may be produced on site from wort, by fermentation with *Lactobacilli*, and is used where food regulations prohibit the use of mineral or 'chemically produced' acids. Within limits reducing the pH of the mash optimizes extract yield, increases proteolysis, gives a more favourable wort pH for yeast growth, and reduces the extraction of polyphenols. When wort is being separated from the mash, and the spent goods are being washed either by sparging or by remashing with hot liquor, the lack of buffering substances allows the pH to rise. In consequence the extract in the weak final wort, the 'last runnings', is relatively rich in ash, silicates, polyphenols, high molecular weight nitrogenous substances, and dispersed polysaccharides [52]. These components may lead to product instability and poor flavours. Various procedures are used to reduce this adverse balance, including treatment with active charcoal to adsorb some proteins and polyphenols. The spent grains (draff) from mashes and sometimes surplus yeast, are normally used in animal feedingstuffs (Chapter 14). A recent suggestion is to use liquor pressed from spent grains as an unobjectionable antifoaming agent in fermentation vessels. Possibly the antifoaming activity is due to lipids [48].

16.3. Some aspects of yeast metabolism

Yeasts used in the fermentation industries are small, unicellular fungi; they are the subjects of several comprehensive publications [51]. The strains used for different industrial processes vary a little, notably in their ability to utilize particular carbohydrates. Under anaerobic conditions hexoses and oligosaccharides that the yeast is able to utilize are mainly degraded by the glycolytic pathway to yield ethanol (ethyl alcohol) and carbon dioxide. In breweries it is sometimes economic to collect and condense the carbon dioxide not retained in the beer for subsequent use. Numerous other substances are also produced in minor amounts. Brewing yeasts, *Saccharomyces cerevisiae* and related strains, do not utilize pentoses, but ferment sucrose, hexoses, maltose and maltotriose, in that order of preference. Maltulose and maltulotriose are also slowly utilized,

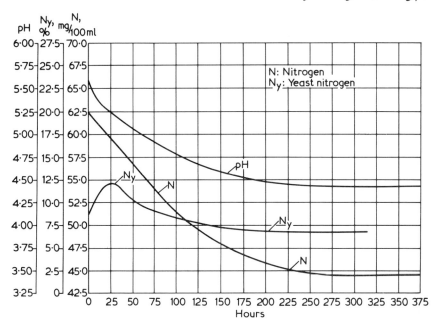

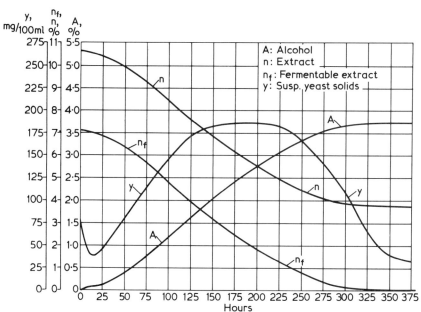

Fig. 16.2 Changes occurring in the fermentation of a wort by yeast (from Trolle [62]). During the initial period of rapid yeast growth the temperature rises, and is normally checked by cooling the fermenter. At the end of the fermentation yeast separation may be encouraged by further cooling.

572 Barley

but isomaltose, nigerose, maltotetraose, higher dextrins, β-glucans and pentosans are not [8,39]. Thus maltotriose and most simpler sugars are grouped as 'fermentable' sugars, while more complex saccharides are grouped as 'unfermentable dextrins'. In contrast other yeasts, such as *Saccharomyces diastaticus*, *Schizosaccharomyces pombé*, and *Brettanomyces bruxellensis*, rapidly degrade higher dextrins and 'attenuate' worts highly. They also utilize the panoses and some isomaltose. They are therefore useful to distillers, but are undesirable contaminants to brewers. *Brettanomyces bruxellensis* utilizes maltotriose and maltotetraose *before* simpler sugars. In contrast *Saccharomyces uvarum* utilizes maltose, but not maltotriose or higher oligomers. Brewing yeasts remove amino acids from worts at varying rates and to differing final extents. Proline is scarcely utilized [24].

Fermentations are complex, and are largely outside the scope of this work (Fig. 16.2) [11,24,62]. In brewing beer alcohol production is often halted by the yeasts clumping together (flocculating) and separating from the fermenting liquid, either into the 'head' or foam caused by the escaping carbon dioxide at the top of the liquid (top fermentation yeast-strains) or as a precipitate at the bottom, (bottom fermentation yeast strains). Various steps may be taken to regulate the effectively fermenting yeast population in suspension – by stirring ('rousing') to increase it, or by 'skimming' the head to remove yeast. Yeast separation may be encouraged by adding 'finings' and cooling the beer. Finally the beer may be centrifuged or filtered. To complete a fermentation the yeast must be kept in suspension, and must survive the levels of alcohol produced. The yeast strain chosen, the level and nature of any microbiological contamination, the wort concentration and composition and the temperature programme followed by the fermentation all influence the flavour of the final product. While in brewing beers the fermentations are cooled, i.e. they are 'attemperated' to various extents, attemperation is not used in whisky fermentations. Yeast death and autolysis can give poor flavours to beers. Over one hundred minor flavour and aroma substances, found in beers, wines, vinegars and distilled beverages, are produced by yeasts, including higher alcohols, esters, acids and carbonyl compounds, including vicinyl diketones [59].

16.4. Malt extracts and barley syrups

Barley syrups and malt extracts have very variable compositions (Table 16.2) [6,19,38,42,49]. In the preparation of syrups raw barley grains are milled and may be subjected to a one or two stage hydrolytic process using acids or enzymes (from microbes, or kilned or green malt) or both. Extracts prepared using acid hydrolysis are relatively rich in salts. Ash contents range from 1.6 to 3.8% of the solids. With malt extracts the malt is mashed and the wort is separated in one of the ways usual for brewing or distilling. The sweet 'worts' are then concentrated in double – or triple – effect vacuum evaporators to give syrups or 'extracts' containing about 75–80% dry materials, SG 1400–1440. The materials may be handled warm to reduce their viscosities so that they will flow and can be moved more readily. Without dilution the only spoilage organisms likely to occur in syrups are osmophilic yeasts. Rarely the syrups may be dried further, in vacuum-pan evaporators or spray driers, to give powders of 95–98% dry solids.

The extracts and syrups find uses in home brewing, in commercial breweries (as wort extenders and as primings), in baking special breads and biscuits, as yeast foods, in making breakfast cereals, in malted drinks, in confectionary, or as carriers, for example of cod- or halibut-liver oils. Enzyme-rich extracts were used to remove the starch size from textiles in laundries and during manufacture, but heat-stable bacterial amylases are now preferred. 'Diastase rich' malt extracts are prepared from highly enzymic malts by mashing at low temperatures, e.g. less than 49° C (120° F) collecting the first wort, then remashing the solid residues at 68° C (about 154° F). The combined worts are concentrated at the lowest possible temperatures. Such extracts tend to be pale and are highly fermentable. In contrast extracts for use in foodstuffs are often dark, being made from malts rich in melanoidins, and are richly flavoured. Extracts for use in sweets and cakes may need to be totally devoid of enzyme activity, particularly α-amylase. To achieve this the extract is heated to at least 80° C, sometimes with prior acidification, to destroy the enzyme. Extracts are available with colours ranging from 3° to 500° EBC (dark beers are 45°–100° EBC) and with diastatic powers from 0 to 350° L. Other specified properties may include flavour characteristics, the extract content, the percentage solids, the pH, the refractive index, the sugar content, the spectrum of carbohydrates present, the nitrogen content, the arsenic content, and the degree of microbiological purity.

16.5. Brewing beer

Recent brewing practice has been extensively reviewed [10,24,54,65]. In the past two main groups of brewing techniques could be distinguished, the British ale brewing systems and the Continental European lager brewing systems. It is useful to begin by considering these, although the distinctions between them have become blurred, and new techniques are now in use.

In the British system, well-modified and cured malt is coarsely ground in simple roller mills, and the grist is mixed with warm water to give an initial temperature of 63°–67° C (about 145°–153° F). The mash is held in a large mash-tun, a circular vessel with a slotted false bottom that supports the mash bed and allows the wort to be withdrawn from below. Rotating spray-arms (sparge-arms) are positioned over the mash for spraying (sparging) hot water onto the top of mash. The hot water percolates downwards and washes out residual wort solids. The mash may be about 1.8–2.4 m (6–8 ft) deep. An infusion mash is usually regarded as a single-temperature process but in fact the temperature may be raised, say to 67°–70°C (153°–158°F) by underletting, that is by pumping hot water beneath the mash, and mixing it in with mechanical mixing equipment. Then, during wort separation the 'sparge' water is applied at about 75°C (167°F), so the effective mash temperature is gradually raised. 'Run-off' (wort collection) is continued until the wort specific gravity (SG) falls to about 1003–1005 from an initial value as high as 1070–1100. The drained, spent grains are discharged by mechanical means. The separated sweet wort is boiled with hops, the spent hops and 'hot break' (trub, the precipitated coagulum which forms), are separated and the cooled, aerated wort is fermented with a 'top yeast', a strain or mixture of strains of *Saccharomyces cerevisiae*. The inoculation, or 'pitching rate' is usually in the range 0.09–0.3 kg/hl (0.3–1 lb/barrel). Care is taken to keep air (oxygen) away from the beer after fermentation. The yeast is separated and after various treatments the beer is packaged in casks, kegs, or bottles. In the oldest process 'priming' sugars were added to the beer, together with a secondary yeast which slowly metabolized them and kept the beer 'in condition' i.e. anaerobic, and charged with carbon dioxide. The yeast formed a layer at the bottom of the container, and the beer had to be dispensed without disturbing this layer. It is now more usual to have the beer nearly sterile and to store it under a slight top-pressure of carbon dioxide.

In traditional lager brewing less well-modified, lightly kilned malts are milled in complex (e.g. six-roll) mills designed to break up the grain to an exactly predetermined extent while leaving the husks as intact as possible. This is achieved by crushing and sieving, and finally remixing the differently treated fractions. The mash is made 'thinner' (with a smaller grist to water ratio) than is used in infusion mashing, to allow it to be moved by pumping. The temperature is initially at about 35°–40°C (95°–104°F). After about 1 h a proportion of the stirred mash is withdrawn and is heated in a separate container, the mash copper. The temperature is increased, then held at about 65°C (149°F) to allow starch conversion, and then it is brought to boiling to break up the grist and gelatinize any residual starch. The hot slurry is subsequently pumped back into the stirred main mash, raising the temperature to about 52°C (126°F). The withdrawal, heating and return of a proportion of the mash is a 'decoction'. Second and third decoctions raise the temperature of the main mash to about 65°C (149°F) and 76°C (169°F) respectively. The initial low temperature allows heat-labile enzymes, e.g. proteases and phytase, to act, the intermediate temperature ensures complete starch conversion, and the final high temperature ensures complete solution of wort substances, and considerable enzyme inactivation. Residual gelatinized starch in decoctions is readily converted when returned to the main mash. Enzymes present in the materials boiled in the decoctions are destroyed. The classical three-decoction process is frequently replaced by more rapid one- or two-decoction systems. At the finish the mash is pumped to a shallow vessel with a slotted bottom, the lauter tun, or to a mash filter, and the wort is collected. After a boil with hops, usually seedless hops in contrast to British ale brewing practice, the cooled aerated wort is fermented at a low temperature with a 'bottom' yeast, such as *Saccharomyces carlsbergensis*. Finally the beer is stored for some weeks ('lagered') at near $0°C$ ($32°F$). It is then filtered and kegged, bottled or canned.

These processes have been modified in various ways. For example where the law permits their use cooked or uncooked unmalted adjuncts – cereal flours or flakes or washed and rolled barley – may be added directly to the mash, sometimes with microbial enzymes [49]. Raw cereal preparations, often maize or rice 'grits', may be 'cooked' in a special vessel – usually with a small quantity of microbial α-amylase or highly diastatic malt to ensure partial starch degradation, 'liquefaction'. After heating to $100°C$ ($212°F$) to disrupt most of the starch granules the 'grits' mash is pumped, with mixing, into an all-malt mash to complete the 'saccharification' process. Green malt has been used successfully in brewing beer experimentally, but there are often unpleasant flavours in the final products. Lightly-dried malts seem a more practicable alternative. Extract and supplementary enzymes may be supplied by dissolving diastatic malt extracts in the mash, or extract only by dissolving barley syrups, sucrose, or invert sugar in the copper used for boiling the hops.

The milling process has been modified, for example by damping the surface of the malt or by actually milling it wet to minimize the degree to which the husk is broken up and reduce dust formation. The intact husk 'opens up' the mash and allows the wort to drain through, and separate from it. By using warm water in a wet-mill, milling and 'mashing in' are achieved simultaneously. Numerous newer types of brewing equipment are now in use, to speed the mashing process, or to hasten wort separation. By having an adequately stirred thin mash, in a vessel equipped with heaters, a mash can be warmed directly without the complexities of underlets or decoctions. Such temperature-programmed mashes can yield extracts of normal quality and in excellent yield [26,34]. The 'sweet wort', separated from the mash, is boiled with hops. Micro-organisms are killed, and enzymes are destroyed. Minor chemical changes occur during the boil; sugars are isomerized, some melanoidins are formed, calcium phosphates precipitate, and the pH falls. In addition a coagulum ('trub' or 'hot-break', mainly of proteins and polyphenols) forms and separates, removing some of the potential haze-forming substances. Polyphenols come from the malt and from the hops. Substances, 'copper finings', may be added to help the break to form. At the same time evaporation increases the concentration of the wort. However the main purpose of the copper-boil is to extract flavour substances from the hops [24]. 'Hops' are the female inflorescences of the specially-grown plant, *Humulus lupulus*. These hop cones are bundles of green bracts, each of which carries lupulin glands at the base, containing essential oils and resins. The hop oil contains over 100 components. While some of the hop oil is boiled off, the residual materials do add to the beer character. The resins are a complex mixture of substances that are partly extracted and isomerized in the boil to give, among other substances, the bitter *iso-α-*acids. Fresh hops usually contain 2.5–9% of the important resin α acids and hop rates of 0.2–0.9 kg/hl (0.02–0.09 lb/gal) wort may be used, depending on the hops and the beer being made. In addition minor amounts of oxidized hop substances occur in beer. Because of the inefficiency of the extraction process and the undesirable characteristics of the essential oils of some hops which are, however, rich in resins it is sometimes economic to separate the resins from the hops, isomerize them chemically and then add them directly to the beer to adjust the level of bitterness.

After cooling and aeration the wort is inoculated, 'pitched', with yeast to start the fermentation. Whereas all fermentations used to be carried out in batches in open or closed vessels of various shapes some breweries now ferment wort continuously [24]. Wort is pumped either through a 'cascade' of stirred tanks or up through a cylindrical tower. As it meets the yeast retained and multiplying in the vessel it is fermented and emerges as beer, carrying with it some of the yeast which may, or may not, be returned to the fermenter after separation. At least some continuous fermenters require considerable skill if they are to be operated successfully. Alternatively by running fresh wort into a small quantity that is already fermenting rapidly in a 'batch' vessel the overall fermentation rate is enhanced. Very large, enclosed, free-standing well-insulated cylindrical fermentation vessels with conical bottoms and external attemperators are coming into widespread use.

Beers are not stable indefinitely. On storage 'non-biological' hazes, formed by associations between polyphenolic 'tannins' and proteins, may separate. Also flavour changes occur during storage; indeed, desirable changes are encouraged when freshly fermented 'green beer' undergoes maturation during storage before being sold. Microbiological contaminants may cause unpleasant flavours. 'Biological haze' in finished beer is caused by micro-organisms increasing in numbers until they cause visible turbidity. A mystifying problem that occasionally occurs is

Some uses of barley malt

'gushing' or 'overfoaming'. When a container of such a beer is opened a rush of carbon dioxide-driven foam pours out. In some cases, but apparently not all, this defect is caused by brewing with malts heavily infected with fungi, *Fusaria* or *Aspergilli* [20]. Fortunately the mycotoxins ochratoxin A and citrinin, formed on mould-infested barley, seem to disappear during the malting and brewing processes [16].

Beers vary in many respects, the intensity of hop and other flavours, the degree of sweetness, the colour, the original specific gravity (OG) of the wort before fermentation and the final alcohol content. The last two are not absolutely linked, since brewing worts vary in their fermentabilities [24,45]. Near beers, according to a British definition, have OG values of 1004–1016 and alcohol contents of less than 2%. Normal beers have OG's in the range 1030–1080, and alcohol contents in the range 2.5–8.4%. In the past OG values in excess of 1100 were common. The 'quality' of a beer is essentially the combination of a preference expressed by customers for a particular type and the absence of definite faults, such as the presence of micro-organisms or the absence of specified amounts of dissolved carbon dioxide. The characteristic flavour of beer is not so much due to the ethyl alcohol as to the presence and relative proportions of hundreds of minor components, whose individual contributions are far from clear. In addition to hop bitter substances, carbohydrates, amino acids, peptides, proteins, ammonia, amines, possibly other nitrogenous bases, various sulphur compounds, aldehydes, esters, various alcohols, organic acids, minerals and phenolic substances contribute to beer flavour [17,18,21,71].

In general, flavour is controlled during beer production by operating according to experience and by recognizing and correcting the causes of particular off-flavours. For example, excessive aeration or too rapid fermentation rates may cause yeast to produce unacceptable levels of esters, particularly ethyl acetate. In other circumstances the β-diketones diacetyl and 2,3-pentanedione accumulate and impart unpleasant flavours to the product. 'Stale' flavours may be due to the degradation products of unsaturated fatty acids.

16.6. Malt vinegar [5,17,18,25,46,67]

Vinegars (Fr. vin aigre – soured or acidified wine) are prepared by double 'fermentation' processes. In the first stage a sugary solution is fermented anaerobically by yeasts to give a solution of alcohol. The sugars may originate from grapes, other fruit, or malt. In the second stage the alcohol in the 'wine', 'cider' or 'beer' is oxidized to acetic acid by bacteria. Distillation plays no part in the production of malt vinegar, but 'spirit' vinegars are prepared from rectified alcohol, fortified with nutrients to support bacterial growth and 'distilled vinegars' are themselves distilled before use. Generally ethanol and acetic acid derived from petrochemicals are not used.

The malt-vinegar brewer may use an all malt grist, or malt with a proportion of raw cereal. This last may be cooked or uncooked. The malt is often a high nitrogen, highly enzymic, lightly kilned sample. The 'wort' may be recovered in many sorts of equipment. The most usual method has been to prepare an infusion mash at 63°–65° C (145°–149° F) with the inclusion of 'weak wort' in the liquor (see below). After settling the first 'wort' is collected, SG 1040–1060, or even 1080. Wort may be collected from below through slotted plates which form a false bottom and from the top of the mash from the pool of turbid liquid which forms on the surface. The spent grains may be sparged by spraying on liquor from above, as in beer-brewing, or may be 'remashed' by mixing in fresh hot water. Remashing may be repeated more than once. The last, weak solution of extractives is used as the mashing liquor in making a fresh mash. The cooled, unboiled wort is inoculated with a large quantity of yeast (2 lb/barrel; about 5.5 g/l) and is fermented at a temperature of 25°–30° C (77°–86° F). The yeast used may be a flocculent strain of *Saccharomyces cerevisiae* mixed with highly attenuating *Saccharomyces diastaticus*. Strains may be selected for their ability to give a superior flavour to the final product. In the fermenting 'beer' the final SG, after 2–4 days, is 998–1000, and the alcohol content is 4–8%. The final pH may be 3.4 or less. No residual dextrins remain, partly as the result of the continued action of malt enzymes, and partly from including *Saccharomyces diastaticus* in the fermentation. A very high degree of wort attenuation can also be achieved by adding amyloglucosidase, a fungal enzyme which degrades dextrins to glucose. Fermentations may be carried out in batches or continuously. After the removal of the yeast the 'beer' ('liquor', 'wash', 'mash') may be stored for a period, or may be directed to the acetifiers at once. If stored then preservatives, up to 1% sodium chloride and 70 p.p.m. sulphur dioxide, may be added.

Originally 'acetification' was carried out by acetic acid bacteria embedded in a zoogleal mat floating on the surface of the liquid, supported by a light wooden grating. Oxygen from the air and ethanol in the liquid reached the organisms by diffusion. However for optimal acetification, higher temperatures, better mixing, and better oxygenation are required. At present acetification is carried out in two main types of equipment. In the older 'trickling' or 'quick' acetifyers the 'sump' of a wooden vat is charged with beer. This is pumped to the top of the vessel and is sprayed in. The vessel, above the sump, is filled with a porous packing, e.g. of beechwood shavings or birch-twigs, which is covered with a bacterial slime. The alcohol in the descending liquid is oxidized to acetic acid by the bacteria. The liquid drains back to the sump, from which it is recirculated. The large quantities of air needed are admitted through downward-sloping pipes inserted through the sides of the vinegar generator. As the oxidation process is exothemic the generator is warm, and the air is drawn through by convection. Each charge is recirculated, sometimes with cooling applied on the external circulation pipe, until after 4–5 days acetification is complete. Newer acetifiers are tall vessels, cylindrical in form, and are made of stainless steel. The contents are vigorously stirred and aerated, and may be chilled by using external cooling jackets. They may operate at temperatures up to 40°C (104°F). Deep culture acetifiers used 'batchwise' achieve 90–95% conversion of ethanol to acetic acid, compared to 98% or better for continuously operated vessels, and 65–70% for the old 'trickling' acetifiers.

The acetic acid bacteria, e.g. *Acetobacter aceti*; *A. rancens*, are notable for their variable morphology, the difficulties in attempting to classify them and their readiness to alter their metabolic patterns. Generally selected pure cultures are used in the manufacture of vinegar [56]. In deep cultures a failure of the aeration system lasting for 15 sec results in the death of the bacteria, presumably due to the rapid accumulation of acetaldehyde, the intermediate in ethanol oxidation. The loss of a batch of vinegar is costly, so 'back-up' aeration facilities with automatic switching gear are always connected to acetifiers.

When acetification is complete, and after the separation of the bacteria, malt vinegar is allowed to mature. Traditionally this is achieved by storing for up to 6 weeks in wooden vats containing chips of beechwood, or other materials. During this time colloidal substances precipitate, the product becomes less liable to throw a haze, and minor chemical changes occur which improve flavour. The final product may be treated with adsorbents to reduce its haze potential, and may be filtered and pasteurized before packing. The colour intensity may be adjusted by the addition of a caramel.

Malt vinegars contain 4–8% acetic acid, and have pH values of 2.8–3.2 (at 5% total acid) compared to the pH of 2.46 for 5% pure acetic acid [5]. The characteristic flavour and character is due to some of the minor components which include ethanol, other alcohols, esters, acetoin, diacetyl, various carbohydrates and amino acids, proteins, gums, minerals, polyphenols, various vitamins, glycerol and lactic acid [27,67].

As with beer-brewing the by-products of malt vinegar manufacture, the spent grains, yeast and bacterial sludges, are often used in making animal feeding-stuffs. The vinegar itself is used as a table condiment, and in the manufacture of a wide range of preserves, pickles and sauces. At present vinegar is not a source of acetic acid for industrial purposes. Besides the tendency to 'throw a haze' the major 'fault' that may occur in vinegars is the occurrence of 'vinegar eels', *Anguillula aceti*. These tiny eelworms, (nematodes), live swimming freely in the vinegar, and may be so numerous that the liquid appears to shimmer. They reduce the acetic acid concentration of the final product.

16.7. Distilled 'potable spirits'

Industrial ethanol is now normally made from petrochemicals. However, in the past industrial spirit has been made by distilling fermented, sugary solutions and locally it may again become economic to prepare it in this way. Although 'neutral spirits' such as vodka, in which minor 'congeners' are removed as completely as possible, are prepared by distilling fermented liquors, these are not necessarily derived from malt mashes, although in the past they generally were. Almost any starch-rich material may be converted, using microbial enzymes or small quantities of highly diastatic malt. Malt is often added with microbial enzymes to ensure an adequate supply of yeast nutrients for the fermentation step. Other sugary liquids – molasses, or solutions of beet or cane sugar – are also fermented. The 'neutral spirits' may be flavoured to produce gin, or similar products. However, in the preparation of the various types of whisky (whiskey, Irish; Gaelic *uisge beathe*, water of life), in which minor components are retained in the distilled product, the natures of the components in the mash have large effects on the final quality [63]. While the production of spirits is technically simple, it is a fact that tiny variations in raw materials and processing techniques lead to marked flavour differences; many products appear to be unique. For the present purpose distinctions are made between Scotch malt whisky, Scotch grain whisky, Irish whiskey and N. American whiskeys such as bourbon and rye [5,9,25,35,46,58,63]. However whisk(e)y is made in many other parts of the world. For Scotch malt whisky the mash comprises peated barley malt only. The malt must be capable of giving a good 'spirit yield' [11]. The initial mash is made at about $63°C$ (approximately $145°F$). After about 1 h the wort is separated, and the solids are remashed at about $75°C$ ($167°F$). The liquid from this mash is added to the first. The solid residues are remashed a third time, at about $85°C$ ($185°F$) but this time the weak 'wort' is used as the first mashing liquor of the next mash. In Scotch grain distilleries the grist is 88–98% maize, which is cooked under pressure at $143°C$ ($290°F$), with a little bacterial α-amylase or malt to liquefy the starch. After cooling the maize starch is converted with highly diastatic malts, prepared from six-rowed barleys, (DP 180–200°L), or even with green malt, at about $65°C$ ($149°F$). Remashing is employed, the first two worts going to be fermented, the third being used in the subsequent malt mash, and the fourth being used in the next maize boil. For Irish whiskey the grist is about 40% malt and 60% raw barley, dried to about 4% moisture then milled, between stones or in a hammer mill, to a fine flour. About

2% oats is sometimes included also. The grist is mashed about four times. In the mash tuns drainage is from the bottoms and the sides.

The slightly cloudy, unboiled worts (wash) are pitched (inoculated) with surplus brewers' yeast, or specially grown distillers' yeast or a mixture of the two. The suspended solids increase the rate of yeast growth and fermentation, and alter the proportions of the minor fermentation products [40]. Fermentation is violent and is complete in about 72 h; temperatures are not allowed to rise over about 32 °C (90 °F). Other micro-organism 'contaminants' seem to add to the 'character' of the final product. Sometimes antibiotics, e.g. penicillin, may be added to control unwanted organisms [36]. Alcohol levels finally reached in the wash (9–10% in Scotland, about 14% in Ireland) are sufficient to kill the yeasts. The alcoholic washes, containing the yeast, are then distilled – in pot stills in Irish and Scotch malt whisk(e)y manufacture, in Coffey stills when making grain whisky.

In N. America the practice is sharply different in that mashes are made with grists containing for example at least 51% maize (bourbon whiskey), or at least 51% rye (rye whiskey). A typical bourbon grist might contain 70% maize, 15% rye and 15% malt. The malts used may be made from barley, wheat, or rye. The raw grains are cooked under pressure, cooled and 'doughed in' with the malt. After mashing yeast is added to the whole sweet mash, and after fermentation the whiskey is distilled from this mixture in a continuous still; the solids are not separated first.

Pot stills vary in detail. These simple batch-stills are normally made of copper, or stainless steel. Each consists of a large body which contains the fermented wash or other liquor to be distilled, and various tubes which convey the vapours to the condenser. Still bodies are either heated directly, or by steam heat exchangers, and the contents are kept in suspension by using equipment such as 'rummagers' which, by means of loops of chain dragged across the bottom of the still, prevent material caking onto the heating surface. Direct heating causes more pyrolysis, and pyrolytic products such as furfural impart particular flavours to the product. Vapour from the heated wash rises up a variously-shaped, parallel-sided or bulbous, wide, vertical tube, then along a horizontal tubular 'lyne-arm', and down a condenser. The dimensions of the vertical tube, and the lyne-arm, whether or not the lyne-arm is cooled and if condensate from it is returned to the body of the still all influence the degree to which the vapours are rectified, and the minor flavour substances, the 'congeners', are collected. The condensers used to be tubular copper helices, or 'worms', cooled in tubs of water, but efficient multi-surface condensers are now used. 'Scotch' and 'Irish' both undergo at least two distillation steps. For example, in the wash still the fermented liquor is distilled. The product, having about three times the alcohol concentration of the wash, is distilled again. However, the first and last fractions from each distillation, the fore-shots and feints, are returned to an earlier distillation stage. The rate of distillation and where the cuts are made between the foreshots and feints and the liquor going forward to be processed further all influence the flavour of the final product. Scotch has an alcohol content of 67–70% at this stage, Irish, 83–88%.

Scotch grain whisky, N. American whiskey, and neutral spirits for gin, vodka or industrial uses are produced in various continuous stills based on the 'patent' still of Aeneas Coffey. Such stills consist of two main units, the analyser column, in which the alcohol and other volatiles are stripped from the wash, and the rectifying column (a reflux fractionating column) from which alcohol and fractions of selected composition can be withdrawn. Many fractions, being rich in alcohol as well as unwanted congeners, are returned to the analyser column. The incoming stream of wash is preheated, and used as a source of cooling liquid, by passing through pipes in the rectifying column, before delivery to the analyser. The wash stream, preheated by passage through the rectifier, is discharged at the top of the analyser column, together with returned 'feints'. As the liquid moves down the column, being spread in thin layers on perforated metal plates, it is boiled and stripped of its volatiles by an ascending stream of steam. Spent wash emerges from the bottom. The hot vapours from the analyser rise up the rectifier, and are cooled and refluxed on perforated plates fitted with bubble caps. Progressively more volatile components are condensed at increasing heights on the column, as the temperature declines.

Freshly distilled whiskys are colourless, and are raw and harsh flavoured. To become mellow they must be matured by storage in wooden containers for periods of three to twelve years. The choice of wood, charred white oak in the case of American whiskeys, used sherry casks in the case of Scotch whisky, has an effect on the flavour. During maturation the ethanol concentration tends to fall, materials are extracted from the wood, the colour changes, and esters are formed [64]. Eventually the spirits are diluted to the final alcohol content, caramel is added to give colour, and the product is bottled. Whiskeys such as Irish may be sold 'straight' but Scotch whisky is generally sold as a blend of malt and spirit whisky, although 'straight malts' are sold. Whiskys are normally about 70% proof (30% underproof; about 40% ethanol; pure ethanol is 175.35% proof. Proof spirit (U.K.) has an SG of 0.92308 at 51°F, about 10.5°C). The characteristic flavours are due to the presence of wood extractives, many dozens – perhaps hundreds – of minor fermentation products and, in Scotch, peat smoke volatiles [28,29,60,68]. Some analyses are shown in Table 16.3 but they are relatively crude, and do not define the characteristic flavours of individual products.

'Neutral spirit', nearly pure aqueous ethyl alcohol, may be derived

Table 16.3 Some average analyses of Scotch whiskys. (All results expressed g/100 l at 100° proof, except proof rating itself (degrees proof, U.S.) and colour; (from Schoeneman and Dyer [53]).

	Proof (U.S.A.)	Total acids	Fixed acids	Esters	Fusel oil	Aldehydes	Solids	Total colour	Org. sol. colour	Ethyl acetate	n-propanol	Iso-butanol	Iso-amyl alcohol
Scotch grain whisky, raw	136.4	0.8	0.4	9.6	35.4	1.6	4.0	—	—	9.4	25.2	25.4	0.5
Scotch grain whisky, matured	128.9	9.4	2.1	11.0	33.7	2.8	31.0	1.9	1.6	12.0	15.9	22.0	5.3
Scotch malt whisky, raw	130.9	6.3	0.4	29.1	152.7	4.6	2.2	—	—	24.7	21.5	49.7	100.7
Scotch malt whisky, matured	120.3	42.8	16.9	36.3	161.9	9.2	246.1	11.1	9.3	33.5	18.3	56.2	107.4
Scotch blended whisky	86.2	19.5	5.4	18.4	81.2	5.9	107.2	7.2	4.4	14.6	17.8	37.2	38.0

from numerous sources. Formaldehyde may be added to mashes to control microbial contaminants. Rarely neutral spirit is used for industrial purposes, when it is 'denatured' to make it unpalatable, so that its price need not include a luxury tax. The denaturants include dyes, methanol, pyridine and mixtures of many other substances [63]. The addition of methanol is objectionable since addicts who partially purify 'meths' (methylated spirits), and then drink it, fail to remove this slow poison, their nervous systems are damaged and they become blind and progressively more stupid, before death intervenes.

Vodka, in western countries at least, seems to be neutral spirits containing high levels of ethanol, but from which minor components, the 'congeners', have been removed as thoroughly as possible, for example by re-rectification and passage through columns of activated charcoal.

Gin is 'neutral spirit' flavoured with the essential oils of various seeds, fruits, etc., called 'botanicals' [25,31,57]. The flavour may be conferred by extracts ('essences') of the botanicals, but the best gins are said to be made by re-rectifying the spirit with the botanicals; these are either added to the body of the still, or are held in a perforated container placed in the still head. The generally-used 'botanical' is juniper berries (*Juniperus communis*; French genièvre, hence 'geneva', gin), but cinnamon, corriander, orange and lemon peels, angelica, aniseed, nutmeg, liquorice, orris, almonds and other 'secret' constituents are used. Sugar is added to some unusual gins. The terpenoid components of the essential oils, such as α-pinene, myrcene, sabinene and linalool are not stable, so the characters of gins alter during storage [25,31,57].

The by-products of distilling, the spent grains or 'draff' and the residues from the still, may be concentrated and dried and are used in animal feeds (Chapter 14) or go to waste, or may be digested anaerobically [32,47].

References

1 Artis, W.G. and Bawden, R.F. (1945). *Cereal Chem.*, **22**, 22–40.
2 Barrett, J., Bathgate, G.N. and Clapperton, J.F. (1975). *J. Inst. Brewing*, **81**, 31–36.
3 Bathgate, G.N. (1969). *Chem. and Ind.*, April, 520–521.
4 Bathgate, G.N., Clapperton, J.F. and Palmer, G.H. (1973). *Proc. Eur. Brew. Conv., Salzburg*, 183–196.
5 Bunker, H.J. (1972). In *Quality Control in the Food Industry*, **3**, ed. Herschdoerfer, S.M. pp. 81–149. London: Academic Press.
6 Burbridge, E. and Hough, J.S. (1970). *Process Biochem.*, **5** (4), 19–22.
7 Charalambous, G., Bruckner, K.J., Hardwick, W.A. and Weatherby, T.J. (1974). *Master Brewers Ass. Am. Tech. Q.*, **11**, 193–196.
8 Clapperton, J.F. and MacWilliam, I.C. (1971). *J. Inst. Brewing*, **77**, 519–522.
9 Court, R.E. and Bowers, V.H. (1970). *Process Biochem*, **5**(10) 17–20.
10 De Clerck, J. (1957, 1958). *A Textbook of Brewing*, (Transl. Barton-Wright, K.). **I** and **II**, London: Chapman and Hall.

11 Dolan, T.C.S. (1976). *J. Inst. Brewing*, **82**, 177–181.
12 Enevoldsen, B.S. (1970). *J. Inst. Brewing*, **76**, 546–552.
13 Enevoldsen, B.S. and Bathgate, G.N. (1969). *J. Inst. Brewing*, **75**, 433–443.
14 Enevoldsen, B.S. and Schmidt, F. (1974). *J. Inst. Brewing*, **80**, 520–533.
15 Erdal, K. and Gjertsen, P. (1971). *Proc. Eur. Brew. Conv., Estoril*, 49–57.
16 Gjertsen, P., Myken, F., Krogh, P. and Hald, B. (1973). *Proc. Eur. Brew. Conv. Salzburg*, 373–380.
17 Greenshields, R.N. (1974). *J. Sci. Fd. Agric.*, **25**, 1307–1312.
18 Greenshields, R.N. (1974). In *Mini-symposium—Fermentation and Biotechnology—10th Anniversary Symp. Inst. Food Science and Technology* (U.K.) 10–15.
19 Griffin, O.T., Collier, J.A. and Shields, P.D. (1968). *J. Inst. Brewing*, **74**, 154–163.
20 Gyllang, H. and Martinson, E. (1976). *J. Inst. Brewing*, **82**, 182–183.
21 Harold, F.V., Hildebrand, R.P., Morieson, A.S. and Murray, P.J. (1961). *J. Inst. Brewing*, **67**, 161–172.
22 Harris, G. (1962). In *Barley & Malt*, ed. Cook, A.H. pp. 583–694. London: Academic Press.
23 Harris, G., Hall, R.D. and MacWilliam, I.C. (1955). *Proc. Eur. Brew. Conv., Baden-Baden*, 26–36.
24 Hough, J.S., Briggs, D.E. and Stevens, R. (1971). *Malting and Brewing Science*, (Revised 1975). London: Chapman and Hall.
25 Hough, J.S., Young, T.W. and Lewis, M.J. (1975). In *Materials and Technology*, **VII**, *Vegetable Food Products*. pp. 787–884. Longman: De Bussy.
26 Hudson, J.R. (1973). *Proc. Eur. Brew. Conv., Salzburg*. p. 157–169.
27 Jones, D.D. and Greenshields, R.N. (1971). *J. Inst. Brewing*, **77**, 160–163.
28 Kahn, J.H. (1969). *J. Ass. Off. Anal. Chem.*, **52**, 1166–1178.
29 Kahn, J.H., Shipley, P.A., La Roe, E.G. and Conner, H.A. (1969). *J. Fd Sci.*, **34**, 587–591.
30 Kieninger, H. (1969). *Proc. Eur. Brew. Conv., Interlaken*, 139–149.
31 Laatsch, H.U. and Sattelberg, K. (1968). *Process Biochem.*, **10** (3), 28–31; 35.
32 Lines, G. (1935). *J. Inst. Brewing*, **81**, 6.
33 Linko, M., Eklund, E. and Enari, T.-M. (1965). *Proc. Eur. Brew. Conv., Stockholm*. 105–120.
34 Lüers, H., Krauss, G., Hartmann, O. and Vogt, H. (1934). *Woch. Brau.*, **51**, 361–365.
35 Lyons, T.P. (1974, Dec.). *Brewer*, **60**, 634–637.
36 MacKenzie, K.G. and Kenny, M.C. (1965). *J. Inst. Brewing*, **71**, 160–165.
37 MacWilliam, I.C. (1968). *J. Inst. Brewing*, **74**, 38–54.
38 MacWilliam, I.C. (1971). *J. Inst. Brewing*, **77**, 295–299.
39 MacWilliam, I.C. and Phillips, A.W. (1959). *Chem. and Ind.*, 364.
40 Merritt, N.R. (1967). *J. Inst. Brewing*, **73**, 484–488.
41 Mikola, J., Pietilä, K. and Enari, T.-M. (1971). *Proc. Eur. Brew. Conv. Estoril*, 21–28.
42 Maule, A.P. and Greenshields, R.N. (1970). *Process Biochem.*, **5** (2), 39–44.
43 Maule, A.P. and Greenshields, R.N. (1971). *Process Biochem.*, **6** (7), 28–31.
44 Nielsen, E. Bjerl. (1971). *Proc. Eur. Brew. Conv., Estoril*, 149–170.
45 Prechtl, C. (1972). *Master Brewers Ass. Am. Tech. Q.*, **9** (4), 200–204.
46 Prescott, S.C. and Dunn, C.G. (1959). *Industrial Microbiology*, (3rd edition). London: McGraw-Hill.
47 Rae, I.J. (1966). *Process Biochem.*, **1** (8), 407–411.
48 Roberts, R.T. (1976). *J. Inst. Brewing*, **82**, 96.

49 Roberts, R.H. and Rainbow, C. (1971). *Master Brewer's Ass. Am. Tech. Q.*, **8** (1), 1–6.
50 Ronkainen, P. (1973). *J. Inst. Brewing*, **79**, 200–202.
51 Rose, A.H. and Harrison, J.S. (*eds.*) (1969, 1970). *The Yeasts*, **I, II & III**. London: Academic Press.
52 Schild, E. (1936). *Woch. Brau*, **53**, 345–350; 353–357.
53 Schoeneman, R.L. and Dyer, R.H. (1973). *J. Ass. Off. Anal. Chem.*, **56**, 1–10.
54 Schuster, K. and Weinfurtner, F. (1963). *Die Bierbrauerei*. Stuttgart: Ferdinand Enke Verlag.
55 Scriban, R. (1969). *Brasserie*, 489–503.
56 Shimwell, J.L. (1954). *J. Inst. Brewing*, **60**, 136–141.
57 Simpson, A.C. (1966). *Process Biochem.*, **1** (7), 355–358; 365.
58 Simpson, A.C. (1968). *Process Biochem.*, **3** (1), 9–12.
59 Suomalainen, H. (1971). *J. Inst. Brewing*, **77**, 164–177.
60 Suomalainen, H. and Nykänen, L. (1970). *Process Biochem.*, **5** (7), 13–18.
61 Thorne, C.B., Emerson, R.L., Olson, W.J. and Peterson, W.H. (1945). *Ind. Eng. Chem.*, **37**, 1142–1144.
62 Trolle, B. (1951). *Proc. Eur. Brew. Conv.*, Brighton, 127–140.
63 Valaer, P. (1944). In *Chemistry & Technology of Food and Food Products*, ed. Jacobs, M.B. **2**, pp. 764–803. New York: Interscience.
64 Valaer, P. and Frazier, W.H. (1936). *Ind. Eng. Chem.*, **28**, 92–105.
65 Vermeylen, J. (1962). *Traité de la Fabrication du Malt et de la Bière*. **I & II**. Assoc. Royale des Anciens Élèves de l'Institut Supérieur des Fermentations, Gand.
66 Vine, H.C.A. (1913). *J. Inst. Brewing*, **19**, 413–448.
67 White, J. (1971). *Process Biochem.*, **6** (5), 21–25; 50.
68 Williams, A.A. and Tucknott, O.G. (1972). *J. Sci. Fd. Agric.*, **23**, 1–7.
69 Windisch, W., Kolbach, P. and Schild, E. (1932). *Woch. Brau.*, **49**, 289–295; 298–303.

Index

Abscisic acid
    content in grain, 239
    effect on coleoptile growth, 259
    effect on leaf unrolling, 259
    effect on root uptake processes, 258–9
    enzyme production, 210
    stomatal closure, 246, 249–50
    synthesis, 117
Acaropsis spp., 410
Acarus spp., 410
Acarus siro
    See Flour mite
Acetaldehyde, 181
Acetic acid
    effect on dormancy, 190
    malt vinegar production, 578–9
    penetration of grain, 181
    use in malting, 554
Acetobacter aceti, 579
Acetobacter rancens, 579
Acetylcholine, 308
Acid dichromate, 5
Acids
    permeation of grain, 181
Acrospire
    See Coleoptile
Actinomyces, 399
Adenosine diphosphate glucose (ADPG), 99
S-Adenosyl methionine, 147
Adobe, 482
Agaricus bisporis
    See Mushroom
Agaricus bitorquis, 486
Agmatine, 151
Agriotes spp.
    See Wireworm
Agropyron mosaic virus, 350
Agropyron repens
    See Couchgrass
Agrostemma githago
    See Corncockle

Agrostis spp.
    See Bent grasses
Alanine, 140, 146
Albumin, 162
Alcohol (ethyl)
    production from malt, 560, 580–4
    production from straw, 486
    See also Ethanol
Aldehydes, 108, 151
Aldrin, 349
Ale
    See Beer
Aleurone layer
    chemical composition, 201
    development, 50
    germination, 12–13
    metabolism, 211–3
    morphology, 7–8
    respiration, 213
    variety differentiation, 78
Alkaloids, 262
n-Alkanes, 108
Alkanols, 108
5-n-Alkyl resorcinols, 108
Allelopathic substances, 262
Allium vineale
    See Wild Onion
Alopecurus myosuroides
    See Black grass
Alpha-chloralose, 414
Alternaria spp., 401, 403
Alternaria amstelodami, 401
Alternaria candidus, 401
Alternaria chevalieri, 401
Alternaria ochraceus, 401
Alternaria repens, 401
Alternaria ruber, 401
Alternaria tenuis, 400
Althous haemorrhoidalis
    See Wireworms
Aluminium phosphide, 411

588   Index

America
  early barley cultivation, 84–85
Amines, 147–54
Amino acids
  animal feed requirements, 492, 494
  barley content analysis, 508–10
  breeding for high levels, 464
  content in dry matter, 231
  content in germinating grain, 207
  content in grain, 236–7
  embryo culture, 207–8
  metabolism, 136–47
  release from roots, 244–5
  sulphur, 146–7
  yields, 313
1-(3-Aminopropyl)-pyrroline, 151
Aminotriazole, 341
Ammonia
  application, 329
  starch recovery, 482
  use in malting, 548
  waste treatment, 490
Ammonium chloride, 306
Ammonium ions, 137
Ammonium propionate, 370
Ammonium sulphate
  effect on soil pH, 306
  effect on yields, 300
δ-Aminolaevulinic acid (ALA), 126
Amylase, 79
α-Amylase
  content in barley, 100–1
  content in grain, 239
  content in malting grain, 552
  degradation, 215
  formation, 209–13
  wort production, 566–7
β-Amylase
  content in barley, 100–1
  content in grain, 239
  content in malting grain, 552
  enhancement by sulphur, 300
  formation, 213
  wort production, 566–7
Amylopectin
  See Starch
Amylose
  See Starch
Anagasta kuehniella
  See Mediterranean flour moth
Aneuploids, 432–3
Angoumois grain moth, 406, 409
Anguillula aceti, 579
Anguina tritici
  See Dagger nematodes
Animal bedding, 483
Animal feed
  barley by-products, 514–6

barley quality, 464
  energy values, 493–5
  fertilizer effects, 305
  fodder cellulose, 505
  forage, 305, 500–3
  grain, 392, 507–12
  grain preparation, 518–20
  hay, 370, 371, 500–3
  malt vinegar by-products, 579
  mash draff, 570, 584
  nutritional requirements, 492–500
  silage, 372, 504–5
  straw, 505–7
  upgrading protein content, 516–8
Anther
  development, 41
  filament, 41
  morphology, 41–3
  See also Stamen
Anthesis
  description, 46
  effect of light, 268
  timing, 46
Anthocyanins, 129
Anthracnose, 364
Antibiotics, 190
Antu, 414
Apex, 39
Aphids, 346
Apical meristem, 6
Apical primordia, 17
Arabinose
  content in barley, 89
  content in germinating grain, 202
Araecerus fasciculatus
  See Coffee bean weevil
'Archer' barley *cv.* (cultivar)
  selection, 449–50, 460
Aromatic substances, 131–6
Aromatic rings, 131, 136
Arsenicals, 414
Ascochyta hordei
  See Ascochyta leaf spot
Ascochyta leaf spot, 364
Ascorbate, 107
Ascorbic acid, 89, 96
Ash
  content in awns, 58
  content in dry matter, 231
  content in grain, 232–3
  straw burning, 334
Asparagine, 207
Aspartate, 43
Aspartic acid, 142, 146
Aspergillosis, 370
Aspergillus spp.
  growth in stored grain, 393, 400, 403
Aspergillus flavus, 401, 403

Index 589

Aspergillus glaucus, 399, 401
Aspergillus halophilicus, 399, 401
Aspergillus oryzae
  contamination of stored grain, 404
  upgrading protein content of feed, 517
Aspergillus strictus, 399, 401
Aster yellow viruses, 350
'Atlas' barley
  crossing, 457
  selection, 461
  yield in S. Australia, 447–8
'Atlas 46' barley
  crossing, 457
ATP, 157
Aubry test, 183
Auricles
  development, 15
  morphology, 22
'Aurore Bretagne' barley
  germination, 189
Australian spider beetle, 406, 409–10
Auxin
  effect on growth, 259, 262
  enzyme production, 210
Avena spp.
  See Oats
Awners, 331
Awns
  appearance, 17
  development, 41
  effect on yield, 276–7, 279
  morphology, 58
  varietal differences, 58–62
  water loss, 247

Backfill, 482
Bacteria
  growth in stored grain, 399
Bacterial disease, 351–2
Balers, 334–5
Bales
  handling, 371
'Bankuti Korai' barley
  yield in S. Australia, 448
Barban, 341
Barberry, 356
Barium carbonate, 414
Barium chloride, 184
Barley (used for whole plant/crop only; otherwise, see separate parts)
  analyses, 498–9
  biochemistry, 89–173
  breeding improvements, 445–80
  by-products for animal feed, 514–6
  classification, 67, 76–81
  crossing, 419–22
  cultivated
    classification, 77

  origins, 81–85
  energy value, 494
  food value to ruminants, 496–7
  genetics, 419–44
  growth, 222–91
  metabolic pathways, 102–7
  morphology
    reproductive parts, 39–75
    vegetative phase, 1–38
  yield
    world, xiii-xiv
Barley leaf stripe disease, 363
Barley stripe mosaic virus (BSMV), 349
  resistance, 462
Barley tea, 516
Barley water, 516
Barley yellow dwarf virus (BYDV), 349–50
  resistance, 457
  transmission, 345, 346, 350
Beer
  bitter substances, 576
  brewing, 573–7
  fermentation, 572
  flavour, 403, 576–7
  mashing temperature, 563
  production, 561
  stability, 568, 576–7
  starch degradation, 567
  steeping, 533
Beetles, 407–8
Bell reaper, 330
Benazolin, 342
Benodanil, 357
Benomyl, 361
Bent grasses, 340
Benzene hexachloride BHC, 349
Benzyl alcohol, 346
Benzylaminopurine
  effect on germination, 193
  enzyme production, 210
Betaine
  changes during water stress, 250
  metabolism, 147
'Betzes' barley
  carbohydrate contents, 237
Beverages
  See Drinks, beer, whisk(e)y, gin, barley tea etc.
γ-BHC, 407, 410, 412
Bifurcose, 96
Bins, 391–3
Biochemistry, 89–173
Biotin, 160, 510
Bird control, 345
Black bindweed, 340
Black chaff disease, 350
Black grass, 340
Black loose smut disease, 359

Black point disease, 362
Black rat, 412
Black stem rust, 356
Blade
   See Coleoptile
Blight
   See Fungal infections
Blissus leucopterus
   See Cinch bug
Blocking, 520
'Bomi' barley
   chemical composition, 464–5
'Bonus'
   mutations, 450
   stem dimensions, 35
Boot, 17
Boron
   deficiency effects, 253
   uptake, 251–2
Botanicals for gin, 584
Bourbon, 581
'BR 1239' barley
   yield in S. Australia, 448
Bread, 512–3
Breeding
   adapted varieties, 446
   choice of parents, 456–7, 470–1
   crossing, 453–6
   disease-resistance, 365, 464
   for quality, 463–6
   for yield, 468–70
   hybridization, 451–3, 471–4
   mutations, 450–1
   natural selection, 460–3
   objectives, 463–70
   optimization for photosynthesis, 280
   plant selection, 449–50
   selection sequences, 457–60
   trials, 474–5
Brettanomyces bruxellensis, 572
'Breust. Granat' barley
   root system, 27
Brewing, 573–7
Broad-nosed grain weevil, 407
Bromate, 548
Bromegrass mosaic virus, 350
Brown house moth, 409
Brown rat, 412
Brown rust, 357
Bud primordium, 6
Building blocks, 483

C-l compounds, 125
Calandra granaria
   See Grain weevil
Calcium
   deficiency effects, 253
   uptake, 251, 258
Calcium carbonate, 300
Calcium hydroxide, 535
Calcium hypochlorite, 190
Calcium nitrate, 306
Calcium propionate, 373
Calixin, 353
Calyptra
   See Root cap
Calyptrogen
   See Root cap
Canandra oryzae
   See Rice weevil
Candicine, 150, 151
Candida spp., 393
Carbohydrates
   animal feed requirements, 492
   barley content, 89–102
   changes during germination, 202–3
   content in dry matter, 231–2
   content in grain, 235–9
   correlation with malt extract, 529
   grain competition, 278–9
   malt, 564–6
   release from roots, 244–5
   wort, 564–6
Carbon dioxide
   effect on root growth, 243–4
   effect on yield, 274
   non-photosynthetic fixation, 103
   output by germinating grain, 195–9
   photorespiration, 124
   photosynthetic fixation, 121–5
Carbon disulphide, 411
Carbon pathway photosynthesis, 123–4
Carbon tetrachloride, 411
Carboxin, 360
Cardboard, 487–90
'Carlsberg' barley
   yield, 297
Carotenoids, 117
Caryopsis, 1
Casparian strip, 32
Catechin, 130
Caterpillars, 346
Cattle
   food intake, 495, 496–7
   grain feed, 512
Caulophilus latinasus
   See Broad-nosed grain weevil
Cellobiose, 97
Cells
   aleurone, 7–8
   antipodal, 44, 47–48
   cross, 4
   division in embryo, 50–51
   division in roots, 29–30
   epidermal, 3, 22
   hair-, 46

meristematic, 29
sheaf, 5–6
sheaths, 24
silica, 3
synergid, 44
tube, 4
Cellulose, 202
Cephalosporium gramineum
    See Cephalosporium leaf stripe
Cephalosporium leaf stripe, 364
Cephus pygmaeus
    See Sawfly
Cercobin, 361
Cercosporella herpotrichoides
    See Eyespot
    Cereal enanismo, 350
Cerealia, 77
Cereals
    archeological evidence of origins, 83–85
    breakfast, 514, 521
    companion crop yield, 299
Ceresan 100, 351
Ceresan M, 351
Chaetomium spp., 403
Chaff
    See Husk
Chalazal tissue
    See Pigment strand
Chalk, 300
Chalk-mud walls, 482
Chamberlain system, 315, 361
Charlock, 340
'Chevron' barley
    selection, 450
Chickweed, 307
China
    early barley cultivation, 84
Chipcote, 25, 351
Chipcote 75, 351
Chit
    See Coleorhiza
Chlordane, 345
Chloridoid types, 81
Chlorocholine chloride
    effect on stem thickness, 308
    gibberellin inhibition, 212
    greening inhibition, 260
2-Chloroethyl phosphonic acid, 211
Chloroglyphus spp., 410
4-Chloro-2-methyl-phenoxyacetic acid, 342
Chlorophacinone, 414
Chlorophyll
    a, 123
    b, 123
    -deficient mutants, 438–9
    in ovary wall, 44
    synthesis, 125–8, 260
Chloropicrin,
    insect control, 411
    nematode control, 344
Chloroplasts, 112, 113
Chlorops pumilionis
    See Gout fly
Chlorosis, 353
Choline
    barley content, 147, 510
    in water stress, 250
    metabolism, 147
Chromosomes
    abnormal, 427–30
    behaviour, 422–7
    morphology, 425–6
    telocentric, 433–4
Chyletus spp., 410
Cinch bug, 345
Cirsium arvense
    See Thistle
Cirsium vulgare
    See Thistle
Cladosporium spp., 400, 403
Clamps, 482
Classifications, 76–81
    botanical, 76
    isozyme patterns, 79
    keys, 76, 77
Claviceps purpurea
    See Ergot
Claws
    See Auricles
Cleavers, 340
Climate
    adaptation, 446–9
    effect on dormancy, 188
    effect on yield, 222, 309–11
Clipping, 313
'Club Mariout' barley
    selection, 461
Cob walls, 482
Coenzyme A, 160
Coffee bean weevil, 407
Coldewe test, 183
Coleoptiles
    effect of light on growth, 270
    germination, 9–12
    growth, 14, 259
    morphology, 6, 21–22, 25
Coleorhiza
    germination, 9, 14
    morphology, 7
Collar, 66
Colletotrichum graminicolum
    See Anthracnose
Combine
    See Harvesting, combine
'Composite Cross I'
    selection, 457

'Composite Cross II'
  selection, 457, 462–3
'Composite Cross XXI'
  crossing, 472
'Composite Cross XXV'
  selection, 457
Compost, 483–4
Confused flour beetle, 406
'Conquest' barley
  chemical composition, 238–9
Contarinia tritici
  See Midges
Conveyors, 376–7
Coolers, 392–3
Coombes
  See Rootlets
Copper deficiency, 253
Copper sulphate
  effect on yield, 300
  penetration of grain, 181
  slug control, 344
  weed control, 340
Corn
  See Grain
Corn dollies, 482
Corn moth, 409
Corncockle, 339
Cortex roots, 30–32
Corticium solani
  See Sharp eyespot
Couchgrasses, 340
Coulters, 324–6
Coumarin
  barley content, 128
  effect on dormancy, 190
  effect on germination, 193
  use in malting, 554
Coumaroyl-agmatine, 151
Coumatetralyl, 414
Covered smut, 358–9
Crithe Döll, 77
Crossing
  back-, 456, 457
  barley classification, 77
  choosing parents, 456–7, 470–1
  complex characters, 440–1
  composite, 456–7
  diallele, 470
  for improved strains, 445–6
  hybridization, 451–3, 471–4
  inheritance, 419–22, 457–60
  mutations, 439, 450–1
  objectives, 453
  ploidy levels, 431–3
  selection sequences, 457–60
  techniques, 453–6
Cryptolestes ferrugineus

  See Rust red grain beetle
Cryptophagus saginatus
  See Fungus beetle
Culm
  lengths, 19
  lodging, 33–5
  morphology, 17–25
  See also Rootlets
  See also Stems
Cultivators, 327
Cummins
  See Rootlets
Cutworm, 346
Cyanide
  fate in barley, 147
  resistance by grain, 239
Cytochrome oxidase, 107
Cytokinins
  barley content, 117
  leaf senescence, regulation, 261
Cytology, 422–34

Dagger nematode, 344
Dalapon, 341
Damage, 175, 333, 533
Darnel, 339
Day period
  effect on grain moisture, 228
  effect on growth, 268–70
  induction of mutations, 72
Dazomet
  nematode control, 344
  weed control, 340–1
D-D, 343
DDT
  harm to photosynthesis, 125
  insect control, 346, 412
  resistance, 472–3
Decortication
  germination enhancement, 191–2
  with chemicals, 176, 180, 184, 191
Dehydroascorbic acid, 89
Deoxyribonucleic acid (DNA)
  biochemistry, 155–6
  pollen content, 47
Deoxyribose, 89
Depleted cell layer
  See Endosperm, starchy
Deuterium oxide, 179
Dextrinase, 568
Dew traps, 321
Dibbling irons, 327
1, 2-Dibromo-3-chloropropane (DBCP), 344
2, 4-Dichlorophenoxyacetic acid (2, 4-D), 308, 342

Index 593

Dichloroprop, 342
Dichlorovos, 410, 412
Dieldrin, 349
Diethyl sulphate, 435
Dimethoate, 346
Dimeton methyl, 346
m-Dinitrobenzene, 184
3, 5-Dinitro-*o*-cresol (DNOC), 342
Dinoseb, 342
Diploids
   characteristics, 430–1
   classification, 77
Disease
   rodent transmission, 413
   resistance
      barley classification, 78
      breeding for, 365, 464, 465
DNA
   See Deoxyribonucleic acid
Docks, 340
Domalt equipment, 541
'Domen' barley
   dormancy, 188
   mutations, 451
Dormancy
   grain evaluation, 186–95
   hypothesis, 194–5
   pretreatment for reducing, 184
   types, 187
   water sensitivity, 183
Downy mildew, 354
Draff, 570, 584
   animal feed, 515, 570, 584
Drainage machinery, 320–2
Driers, 370–1, 386–9
Drilling, 298–9
Drills, 327–9
   combine, 328
Drinks, non-alcoholic, 516
Drosophila spp.
   See Flies
Drought
   effect on growth, 248
   resistance, 466
Dry matter
   alteration during growth, 239
   composition, 200, 231
   loss on steeping, 179, 527, 551, 553
   loss on storage, 373
Drying
   grain, 313
   hay, 370–1
Dwarf rust, 357

Ear
   abnormalities from herbicides, 342
   branching, 72
   classification of barleys, 78
   development, 17, 39–52
   photosynthesis, grain yield, 277–9
   growth stages, 229
   morphology, 65–72
   ridges, 39
   tweaked, 66
'Early Beardless' barley
   yield, 312
Eckhardt test, 184
Eelworm, 343
Egg cell, 44, 47
Egypt
   barley origins, 81–2, 84, 85
Electricity
   effect on germination, 192
Electron transport chain, 103, 106–7
Emasculation, 454–5
Embryo
   abnormal development, 51
   chemical composition, 201
   contents, 7
   culture *in vitro*, 207–8, 209
   development, 51
   feeding studies, 146
   germination, 9–14
   morphology, 1, 6–7
   release of enzymes, 209–15
   reserves from endosperm, 12, 208–15
   respiration, 198–9
   sac, 44, 47
   starch content, 13–14
Endodermis, 30
Endosperm
   breakdown on germination, 12–14
   chemical composition, 201–2
   development, 48–49
   evaluation of quality, 176–7
   mobilization of reserves, 208–15
   morphology, 1, 7–9
   nitrogenous contents, 206
   respiration, 198–9
   See also Aleurone layer, modification
   specific gravity, 8–9
   starch contents, 201–2
   starchy, 7–9, 50
   types, 8–9
Endrosis sarcitrella
   See White shouldered house moth
'English Archer' barley
   See 'Archer' barley
Entoleter, 405
Enzymes
   barley content, 97–102, 163
   debranching, 100
   degradation, 215
   grain content, 239

malting grain, 548, 555
release from embryo, 209–15
release into endosperm, 209
wort production, 562–3, 566
Ephestia elutella
   See Warehouse moth
Epicarp, 4
Epidermis, 3
Epoxides, 435
Ergot, 357–8, 403
Ergotism, 404
Erysiphe graminis
   See Powdery mildew
Estolide, 108
Ethanol, 582, 584
   effect on dormancy, 190–1
   formation, 195–6, 199
   penetration of grain, 181, 533
   See also Alcohol, ethyl
Ethanolamine, 147
Ethoxycaffeine, 435, 436
Ethyl acetate, 181
Ethyl alcohol, 582, 584
   See also Ethanol
Ethylene
   effect on root growth, 245
   effect on top growth, 262
   enzyme production, 210
Ethylene dibromide (EDB)
   insect control, 411
   nematode control, 344
Ethylene dichloride, 411
Ethylene glycol, 181
Ethylene oxide, 411
Ethyleneimines, 435
Ethylmethanesulphonate, 435
Europe
   early barley cultivation, 84
European grain moth, 409
'Everest' barley
   crossing, 457
Eyespot, 361

Fallows, 293, 300, 315
False clothes moth, 409
Farinator, 176
Farmer's lung, 370, 403
Farmyard manure (FYM)
   application, 329–30
   effect on nitrogen content, 304
   effect on yield, 300–1
   formation, 483–4
   mushroom cultivation, 485–6
Fats
   animal feed requirements, 492
   content in grain, 239
Fatty acids
   animal feed requirements, 493

barley content, 108–13
degradation, 112–3
in stored grain, 402
α-oxidation, 112–3
synthesis, 109–13
wort content, 569
Feints, 582
Fenitrothion, 412
Fermentation
   brewing and distilling, 570–2, 574, 576
   silage, 372–3
Fertilization (reproduction), 46
Fertlizers
   application, 305–7, 328, 329
   composition, 306
   dose rates, 306–7
   effect on chemical composition, 236
   effect on water stress, 247–8
   effect on yield, 297, 299–307, 469
   in straw for horticultural use, 485
Festucoid types, 80–81
Flail, 331
Flavin adenine dinucleotide (FAD), 160
Flavin mononucleotide (FMN), 160
Flavonoids, 130
Flavonols, 108
Flies, 409
Florets
   development, 41
   effect of temperature, 264
   fertility, 67
   maturity, 46
   supernumerary, 41
Flour, 512–4, 521–2
Flour mite, 406, 410
Flowering
   See Anthesis
Flowers
   growing in straw bales, 485
Fluoroacetamide, 414
Folic acid, 160
Foot-rot, 362
Forage
   animal feed value, 300–3
   yield, 311–3
Fore-shots, 582
Formaldehyde
   effect on water sensitivity, 190
   use in steeping, 536, 554
Formalin, 181
Formic acid
   addition to grain, 393
   addition to silage, 373
Freezing effects, 190
Frit fly, 348
Frost
   effect on yield, 309
   resistance, 265–6

Fructosans, 96–97
Fructose
  content in barley, 89
  content in germinating grain, 202
  effect of cold, 265
Fruit
  growing in straw bales, 484–5
Fucose, 89
Fuel
  straw use, 487
Fungal disease, 351–64
  blight
    control in malting, 536
    detection in grain, 177
    in growing crop, 350, 362
    in stored grain, 400, 401, 403
Fungi
  beer spoilage, 576–7
  field, 400
  growth in stored grain, 399–404
  toxin-formation, 403–4
Fungicides
  application in field, 329
  application to stored grain, 394
Fungus bettles, 409
Furfural, 486
Fusarium spp.
  See Fungal diseases
Fusarium graminearum, 403

Galactolipids, 113
Galactose, 89
Galacturonic acid, 89
Galium oparine
  See Cleaver
'Gamma No. 8' barley
  mutations, 451
Gene-banks, 446, 456–7
Gene-buffering, 446
Genetics, 420–44
Germination
  changes in malting grains, 550–2
  chemical effects, 193–4
  equipment for malting, 538–41
  growth stages, 228
  moisture effects, 190–1
  morphological changes, 9–14
  physical treatment effects, 191–3
  pregermination effects, 192
  storage effects, 188–90
  testing for germinability, 182–6
  variety effects, 188
  water uptake, 178
Gibberella spp.
  See Fungal diseases
Gibberella saubinetti, 403
Gibberellic acid
  barley content, 118
  effect on embryo, 210
  effect on growth, 262
  effect on leaf unrolling, 259
  effect on morphology, 72
  effect on mutants, 438
  embryo culture, 208
  enzyme production, 211–3
  germination inducement, 192, 193
  penetration of grain, 181
  self-fertilization, 454
  stomatal opening, 246
  use in malting, 537, 550, 553–4
Gibberellins
  biosynthesis, 119
  content in grain, 239
  effect of red light, 259
  effect on frost-resistance, 263
  effect on growth, 262
  enzyme production, 210
  metabolism, 118
  production, 211–3
Gin, 582, 584
'Glacier' barley
  growth from different seed size, 296
Globulin, 162
β-Glucan, 97
β-Glucanase, 213
Glucans, 202
Gluconeogenesis, 92–93
Glucosamine, 89
N-Acetyl-D-glucosamine, 162
Glucose
  barley content, 89, 98
  content in germinating grains, 202, 204
  effect of cold, 265
α-Glucosidase
  barley content, 101
  formation, 213
Glucuronic acid, 89
Glume
  development, 39
  lemma-like, 56–57
  varietal differences, 56–58
  See also Husk
Glutamate, 43
Glutamic acid, 137–9
Glutamine
  anther content, 43
  embryo culture, 207
  synthesis, 137
Glutathione, 163
Glutelin, 162
Glycerol
  metabolism, 102
  penetration of grain, 181
Glycollic acid, 124–5
Glycolysis, 102–7
Glycophagus spp., 410

596 *Index*

Glycoproteins, 162
C-Glycosylflavones, 130
Glyoxylate cycle, 105
'Golden Promise' barley
 yield, 314
'Goldthorpe' barley
 selection, 450, 460
Goosegrass, 340
Gout fly, 347
Grading, 177, 378–82
Graf mashes, 549
Grain
 abnormal formation, 51, 396
 amylopectin starch source, 481
 amylose starch source, 481
 asymmetry, 54
 brewers' spent, 515
 carting, 233–4
 characterization, 55
 chemical composition while germinating, 202–7
 chemical composition while growing, 232–9
 chemical composition while quiescent, 199–202
 cleaning, 377–82, 533
 colour, 55, 176
 damage evaluation, 185–6
 deterioration on storage, 394–9, 402
 development, 47–52
 drying, 382–9
  calculations, 385
 dust explosions, 375
 endosperm evaluation, 176–7
 fecula source, 481
 feed value, 292, 507–14
 formation in supernumerary florets, 41
 gas-exchange during germination, 195–9
 genetic purity evaluation, 177
 germination
  changes on malting, 550–2
  chemical composition, 202–7
  equipment for malting, 538–41
  evaluation, 182–6
  gas-exchange, 195–9
  malting, 526
  morphological changes, 9–14
 grading, 177, 378–82
 growth stages, 228–9
 handling, 375–7
 heat output, 195
 human food, 512–4
 length, 227
 longevity, 394–9
 micro-organism presence determination, 177
 milling, 518–9
 moisture content, 228, 231
 moisture content at harvest, 313
 moisture content determination, 177
 moisture content during malting, 533
 moisture content in storage, 382–6
 morphology, 1–14
 nitrogen content determination, 177–8
 outer layer evaluation, 176
 permeability, 180–2
 prices, 292
 protein yield, 481
 quality, 174–215
  rain effects, 313
  wild onion effects, 339, 340
 reception for storage, 375
 respiration, 195–9, 395
 sampling with small numbers, 174–6
 selection for malting, 528
 shapes, 53–55
 sizes, 55, 200
  effect on yield, 222–3
  nitrogen content, 304–5
 starch source yield, 481
 storage, 389–414
 technology of preparation for food, 518–21
 temperature effects on development, 47–48
 tests, 174–6
 variants, 52–65
 variety differentiation, 78–79
 vernalization, 266–8
 viability curves, 396–8
 vigour evaluation, 186
 water penetration, 9–10, 178–82
 water vapour pressure, 382–4
 weighing, 377
 weight, 176, 200, 227–8
 weight loss, 225, 404
Grain weevil, 406, 407
Gramine, 147, 154
Gramineae
 Chloridoid sub-family, 81
 classification, 79–81
 Festucoid sub-family, 80
 Panicoid sub-family, 81
Grasses, 79–81
Grasshoppers, 345
Gravity perception, 29, 243, 259
Grazing, 313
Greenbug, 345–6
Grist
 animal feed, 515
 mashing, 562–3
Growth period, 225, 228
Growth rate, 225
'Gull' barley
 selection, 450
Gums, 97–98
Guttation fluid, 246

Hairs
    germination of root, 9
    morphology, 3
    ovary, 44
    variety differences, 78
'Haisa II' barley
    mutations, 451
Halo spot, 364
'Hannchen' barley
    chemical composition, 233
    growth, 224
    growth rate, 226
    plant density effects, 223
    root growth, 241
    selection, 450, 461
Hansenula spp., 393
Haplodiplosis equestris
    See Midges
Haploids
    characteristics, 431–2
    chromosome adjustment, 77
Hardening to cold, 265–8
Hardiness, 467
Hare control, 345
Harrow, 326
Harvesting
    ancient, 84
    combine, 313, 331–4
    damage to grain, 176, 333, 533
    machinery, 330–4
    malting barley, 532
    objectives, 313–4
    optimal period, 370
    silage, 372
    whole plant, 335–7
Hay
    animal feed value, 500–3
    drying, 370–1
    moisture content, 370–1
    storage, 369–71
'Hellas' barley
    mutations, 450
Helminthosporic acid, 210
Helminthosporium spp., 400, 401
Helminthosporium sativum, 362, 365
Helminthosporium teres
    See Net blotch
Helminthosporol, 210
Hemicellulose, 97
Hentriacontan, 108
Herbicides
    application, 329, 341
    effect on barley, 342
    effect on yields, 307–8
    induction of morphological changes, 72, 342
    selective, 311–2
    types, 340–3
Heritability, 458–9

Herniarin, 193
'Hero' barley
    selection, 461
'Herta' barley
    mutations, 450
Heterodera avenae
    See Eelworm
Heterosis, 471–2
Hessian fly, 346
Hexaploids, 431
'Himalaya' barley
    composition, 508, 511
    dimensions, 22
'Hiproly'
    lysine content, 464
Hofmannophila pseudospretella
    See Brown house moth
Holocellulose, 97
Hoods
    development, 41
    morphology, 41
    varietal differences, 62
Hoppers, 329
Hops
    beer brewing, 576
    food value, 496–7
Hordatines, 131
Hordein, 162
Hordenine, 147, 150, 151, 154
Hordeum, 77
Hordeum agriocrithon, 83
H.bulbosum, 432
H.deficiens nudificiens, 452
H.langunculiforme, 83
H.paradoxon, 83
H.sativum, 78
H.spontaneum
    ancient harvesting, 330
    crossing, 452
    cytology, 422
    dispersal unit, 67
    distribution, 82–85
    dormancy, 188
    rachis morphology, 67
    steeping effects, 191
H.spontaneum nigrum, 468
H.vulgare
    analysis, 501–2
    classification, 78
    crossing, 432, 452
Hordeum mosaic virus
Hordothionin, 162
Hormones, 262
'Horsford' barley
    analyses, 501–2
Hot water extract
    determinations, 549
    malting grain quality evaluation, 465

House mouse, 412
Hullers, 520
Human food
　barley processing, 520–1
　grain, 512–4
　malt extracts, 573
　nutritional requirements, 494
　upgrading protein content, 516–8
Humidity, relative, 382–4
Hummelers, 331
Humulus lupulus, 576
'Hunter' barley
　crossing, 452
Husk
　chemical composition, 200–1
　colour, 55
　morphology, 1–3
　varietal differences, 52–53
Hyaline layer, 5
Hybridization, 451–3
　incompatibility between species, 80
Hydrazoic acid, 435
Hydrochloric acid, 373
Hydrogen cyanide, 411
Hydrogen peroxide
　effect on germination, 190
　penetration of grain, 181
　reduction of dormancy, 184
　use in malting, 536, 550
Hydroperoxide isomerase, 113
25-Hydroxy-hentriacontan 14, 16-dione, 108
Hypodermis, 4

'Impala'
　mildew resistance, 468
India
　early barley cultivation, 84
Indian meal moth, 409
Indoles, 152–4
Inflorescence
　appearance, 18
　ear development, 39–52
'Ingrid' barley
　fatty acid content, 109
myo-Inositol, 89, 96, 511
Insecticides, 345–9
　application in field, 329, 345–9
　application to stored grain, 395, 405, 410–2
Insects
　control, 345–9
　occurrence in stored grain, 404–10
Internodes
　lengths, 66
　morphology, 19–21
　See also Stems
Invertase, 91
Iodide, 252
Iodine, 181–2

'Irish Archer I' barley
　selection, 449–50
Iron deficiency, 253
Irradiation
　control of insects in stored grain, 405
　effect on coleoptile growth, 259
　effect on germination, 195
　mutations, 435
Irrigation
　cause of soil salination, 250
　effect on yields, 309–11
　history, 320–1
　machinery, 320–2
　problems, 311
Isoleucine, 140, 146
Isozymes, 79

'Jackson' barley
　yield, 312
Japan
　early barley cultivation, 84

Karyotypes, 77
ent-Kaurene, 118
'Kenia' barley
　crossing, 453
Kernel
　See Grain
Kernel smudge, 362
Kestose, 96
Keto-acids, 137
Ketones, 108
Khapra beetle, 399, 406, 408
Kilns, 541–6
Kinetin
　effect on coleoptile growth, 259
　effect on germination, 193
　effect on leaf senescence, 261
　effect on leaf unrolling, 259
　enzyme production, 210
　stomatal opening, 246
'Kinver' barley
　crossing, 465
'Kristina' barley
　mutations, 451

Lactic acid
　in silage, 373
　silage content, 373
　use in malting, 555
Lactobacilli spp., 402
Lager
　brewing, 574–5
　steeping, 533
Laminaribiose, 97–98
Lamina of leaf, 279
Land-planes, 321
Leaf beetles, 347

Index 599

Leaf-blade, 24
Leaf-blotch, 314, 354
Leaf miners, 347
Leaf-sheath, 15 *et seq*
Leaf rust, 357
Leaf scald, 354
Leaf spot, 354
Leatherjackets, 344, 346
Leaves
  abnormalities from herbicides, 342
  area index (LAI), 273, 275
  bases, 15
  blade midrib, 24
  distichous arrangement, 23
  effects of mildew, 354
  effect of photosynthesis on yield, 273–6, 279
  effect of slugs, 344
  embryonic, 6
  flag, 17
  greening, 259–60
  growth, 15–25, 226
  light interception capacity, 273–5
  protein for feed, 517–8
  senescence, 18, 260–1
  unrolling, 259–60
  veins, 23–24, 25
  water loss, 246–7
  water potential, 246
Lectin, 162
Legumes, 299, 302, 315
Lema melanopa
  See Leaf beetle, 347
Lemma
  appendage, 41, 59–62
  bases, 62
  development, 40
  morphology, 1–3
  triple awned, 59
  variety differences, 59–62, 78
'Lenta' barley
  seeding rate and yield, 275
Leptohylemyia coarctata
  See Wheat-bulb fly
Lesser grain borer, 406, 407
Leuco-anthocyanins, 130
Light
  effect on germination, 193
  effect on growth, 268–70
  induction of mutations, 72
Lignans, 128, 131
Lignin
  content in barley, 3, 128, 131
  content in dry matter, 231
Ligule
  development, 15
  morphology, 22
Lime
  application, 329
  correction of soil acidity, 300
Limestone, 300
Limothrips cerealium
  See Thrips
Lindane, 349
Linoleic acid, 109
Lipase, 109
Lipids
  content in barley, 107–21
  content in germinating grain, 204
  content in worts, 569
Lipoxygenase, 113
Lithium hydroxide, 181
Lodging
  causes, 34
  effect on yield, 302, 308, 469
  stems, 19, 33–35
Lodicules
  morphology, 3
  variety differentiation, 78
Lolium termulentum
  See Darnel
Longidorus spp.
  See Dagger nematode
Loose smut, 180, 359
Lucel, 353
'Lyallpur' barley
  crossing, 457
Lysine
  animal feed requirements, 494
  breeding for high levels, 464
  stored grain, 403
  synthesis, 140, 146

Machinery
  damage to grain, 185–6
  grain cleaning, 377
  harvesting and threshing, 330–4
  irrigation and drainage, 320–3
  post-sowing, 329–30
  sowing, 327–9
  straw treatment, 334–5
  tillage, 322–7
Macro-elements, nutrients, 250–1
Magnesium
  deficiency effects, 253
  effect on yields. 300
  uptake, 251
Magnetism, 193
Malathion, 412
Maleic hydrazide, 308
Malt
  acid, 555
  analysis, 545–50, 556
  beer brewing, 573–7
  caramel, 555
  coffee, 516
  colour, 554–5
  crushing, 527

culm food value, 514–5
distillers', 555
enzyme activity, 548, 555
extracts
  food value, 514, 515
  pH, 552
  production, 561–70
  quantity obtained, 548–50
  use, 572–3
  viscosity, 546–8, 552
flour, 514
green, 541
magpie, 543
mashing, 560–70
nitrogen content, 548, 552–3
nutrient source, 560
rootlets, 510, 514–515, 545
uses, 561–86
vinegar, 577–80
wind-, 541
Malting, 526–59
  aeration, 534–5
  air-rests, 191, 534
  barley
    breeding for quality, 464–6
    changes on germination, 9–14, 550–2
    fertilizer application, 302
    harvesting, 532
    prices, 292
    requirements for grain germination, 182
    selection, 527–32
  Bavarian, 534
  Bohemian, 534
  casting, 537
  drums, 539–40
  flush steeping, 534
  germination equipment, 538–41
  grain cleaning, 377, 533, 635–6
  kilning, 526–7, 541–5, 554–7
  Kropff, 553
  micromalting trials, 182, 465
  multiple steeping, 537
  pneumatic equipment, 538–9
  process, 526–7, 532
  See also Grain
  small-scale, 532
  spray steeping, 535
  steeping, 533–7
  water uptake, 178, 179–80, 533, 538
Malto-oligosaccharides, 101
Maltose
  content in barley, 101
  content in germinating grain, 202, 204
Maltotetraose, 101
Maltotriose, 101
Manganese deficiency, 253
Mannitol, 247
Mannose, 89

'Mari' barley
  mutations, 450
'Maris Badger' barley
  mildew resistance, 468
'Maris Concord' barley
  yield, 296
'Maris Mink' barley
  chemical composition, 236
'Maris Otter' barley
  hardiness, 467
  yield, 296
'Maris Puma' barley
  yield, 296
Mashing, 560–70
Maturation, 19
Mayetiola destructor
  See Hessian fly
'Maythorpe'
  yield, 297
Mayweed, 342
MCPA
  See 4-chloro-2-methyl phenoxyacetic acid
Mealiness
  effect of water, 178–9
  grain evaluation, 176
  grain morphology, 8–9
  malting grain, 528
Mecoprop, 342
Mediterranean flour moth, 409
Meiosis, 423–5
Menanoplus spp.
  See Grasshoppers
Mercuric chloride
  in malt analysis, 548
  penetration of grain, 181
Mercuric cyanide, 181
Mercuric nitrate, 181
Mercuric sulphate, 181
Mesopotamia
  barley cultivation, 84
Metaldehyde, 344
Methabenzthiazuron, 341
Metham-sodium
  nematode control, 344
  weed control, 340–1
Methyl bromide
  insect control, 411
  nematode control, 343
Methyl isothiocyanate (MIC), 344
S-Methyl methionine, 147
N-Methyl tyramine, 147, 151
Mice control, 345, 412–4
Microelements, nutrients
  animal feed requirements, 493
  barley contents, 511–2
  effect on yield, 300
  uptake, 251–2
Micromalting, 182, 465

Micromonospora spp., 403
Micro-organisms
    contamination of stored grain, 394, 399–404
    control in malting, 536
    detection in grain, 177
    disease, 349–364
    effect on dormancy, 190
    effect on root uptake processes, 257–8
    spread by insects, 404
Micropyle
    development, 44
    morphology, 4, 9
    pollen tube entrance, 46
'Midas'
    growing conditions, 294
Middle East
    barley origins, 82–4
Midges, 346
Mildew
    effect on yield, 314
    resistance, 352–3, 467–8
Milfaron, 353
Millboard, 487–90
Milling, 518–9, 563
Millipedes, 345
Milstem, 353
Minerals
    barley content, 511–2
    deficiency effects, 252–6, 263, 309
    effect on growth, 262–4
    grain content, 233–5
    requirements, 250–6
    uptake by roots, 256–9
    uptake during growth, 232
'Missouri' barley
    yield, 312
Mites
    occurrence in growing crop, 345
    occurrence in stored grain, 404–6, 410
Mitosis, 422
Modification, 526–7
    See also Endosperm breakdown
Mole-drainer, 322
Molluscs, 344
Molybdenum deficiency, 253
Monosaccharides, 89–96
Monosomes, 432
Morfamquat, 342
Morphology
    effect on yield, 279
    reproductive parts, 39–75
    vegetative phase, 1–38
Moths, 409
Mould, 395, 399
    See also Fungi
Mow-burn, 402
Mowers, 335–6

Mucor spp., 401
Müger kiln, 544–5
Mulching, 293, 484
'Multan' barley
    crossing, 457
Mus musculus
    See House mouse
Mushrooms
    growing in straw compost, 485
Mutagenesis, 352, 434–9
Mutations
    albina, 438, 439
    alboviridis, 438
    bracteatum, 58
    calcaroides, 61, 62
    characteristics, 436–7
    chromosomes, 425, 426, 434–9
    divided lemma, 60
    eceriferum, 438
    erectoides, 437–8
    frequency, 436
    in-breeding, 450–1
    maculata, 438
    occurrence in stored grain, 395–6
    rachilla, 64
    striata, 439
    tigrina, 438
    tillering, 17
    viridis, 438
    xantha, 438, 439
Mycetea hirta
    See Fungus beetle
Mycorrhiza, 30–31
Mycotoxicosis, 403, 404
Myleran, 436

Nabataeans, 321
NAD+, NADH, NADP+, NADPH
    See Pyridine nucleotides
Nebularine, 435
Nemapogon granella
    See Corn moth
Nematodes
    control, 343–4
    infection of stored grain, 399
'Nepal' barley
    crossing, 421
Nephrosis, 404
Net blotch, 364
Nicotinic acid, 510
'Nissani' barley
    growth from different seed size, 296
Nitrate
    biochemistry in barley, 137
    uptake, 250
Nitric acid, 181
Nitrogen
    breeding for high levels, 464

content in different grain sizes, 304–5
content in dry matter, 231
content in germinating grain, 204–5
content in grain, 233–4
content in malt, 548, 552–3
content in wort, 568–9
correlation with malting extract, 529
deficiency effects, 252, 256
effect on dormancy, 190
effect on growth, 264
effect on root growth, 242
effect on yield, 302–5
grain content determination, 177–8
uptake during growth, 232
Nitrogen mustard, 435
Nitrous oxide, 435
Nodes
See Stems
Norbormide, 414
North America
climate adaptation of barley, 447
Norway rat, 412
Nucellar tissue, 5
Nucellus, 44
Nuclei, polar, 44, 47
Nucleic acids, 155–61
Nucleotide cofactors, 157–61
Nullisomes, 432
Nutrients
deficiency effects, 309
effect on root growth, 242–3
effect on yield, 299–307
embryo culture, 207–8

Oat blue dwarf, virus, 350
Oats
classification, 79
wild
increase, 340
'Old Scotch Common' barley
dormancy, 188
Ophiobolus graminis
See Take-All fungus
Organic acids
barley content, 103
Organic solvents, 181
'Orge Maroc'
germination, 187
'Orge du Prophete'
glumes, 56
Ornithogalum spp., 357
Oryzaephlius surinamensis
See Saw-toothed grain beetle
Oscinella
See Frit fly
Oscinis frit
See Frit fly

Ovary
development, 43–45
wall disintegration, 50
Ovule
formation, 43, 424–5
maturity, 45–46
Oxidases, 107
Oxidative hexose monophosphate pathway, 94–95
Oxidative phosphorylation, 106–7
Oxygen
effect on germination, 193, 194
effect on root growth, 243–4
uptake by germinating grains, 195–9
Oyster mushroom, 486

Padi-straw mushroom, 486
Palea
development, 40
morphology, 1–3
varietal differences, 62–65
'Pallas' barley
mutations, 450
Panicoid types, 81
Pantothenic acid, 510
Paper
straw production, 487–90
Papillae, 39
Paraquat, 341
Parching, 330
Parenchyma, 3, 32
Paris Green, 344
Particle-boards, 483
Pasteur effect, 102
Pearl barley
food value, 513
production, 520–1
starch content, 201–2
Pearlings, 514
'Peatland' barley
selection, 450
Pectin, 96
Peduncle, 17–18
Pelletization, 520
Penicillium spp.
growth in stored grain, 393, 400, 401, 403
Penicillium cyclopium, 400
Penicillium pupurogenum, 403
Penicillium rubrum, 403
Pencillium urticae, 404
Pencillium viridicatum, 404
Pentosans
content in barley, 98
content in germinating grain, 202
Pentose phosphate pathway, 94–95, 102–7
Peptidases, 163, 215
Peptides, 147, 163

Pericarp
  development, 50
  fungi occurrence, 400
  morphology, 1, 3–4
  water conduction 178
Pericycle, 30
Perisperm
  See Nucellar tissue
Peroxidase
  degradation, 154
  formation, 213
Persicaria, 340
Pesticides, 345–9
Pests
  effect on growth, 262
  in stored grain, 404–10
Phenolic acids, 128
Phenolic substances, 128–31
Phenols
  penetration of grain, 181
  wort contents, 569
Phenotypes, 72
Phenylalanine
  ammonia lyase, 128
  synthesis, 136–7
Phenylphosphonic acid diamide, 308
Phosphate
  barley biochemistry, 157
  effect on yield, 301–2
  uptake, 258
Phosphatase, 213
Phospholipids, 113
Phosphoric acid, 373
Phosphorus
  content in dry matter, 231
  deficiency effect, 252, 257
  effect on root growth, 242
  uptake during growth, 232, 250–1
Phosphorylase, 99–100
Phosphorylcholine, 147
Photoperiod
  See Day period
Photophosphorylation, 122
Photorespiration, 121–5
Photosynthesis
  biochemistry, 121–5
  carbon pathway, 123
  effect on yield, 272–80
  electron transport pathway, 122
  initiation, 259–60
Phylloquinone, 120–1
Phytochrome
  barley content, 128
  effect of light, 270
Phytophthora macrospora
  See Downy mildew
Picloram, 342
Pigment strand, 4, 5, 51
Pigs, 495, 500
Pipecolic acid, 140, 146
Pistil, 40
Plant competition, 262–3
Plant density
  effect on tillers, 263
  effect on yield, 222, 271, 275, 298–9
Planting
  effect of timing and depth on hardiness, 266
Plastid quinones, 121
Plastoquinone, 120
Pleurotus oestreatus
  See Oyster mushroom
Plodia interpunctella
  See Indian meal moth
Plough
  ancient, 323
  chisel, 325–6
  coulter-, 322
  disc, 325
  modern, 322–7
  scratch, 322, 327
  tractor-, 324–5
Ploughing
  danger of soil compaction, 321
  depth, 293–4, 340
  effect on nitrogen fixation, 305
'Plumage–Archer' barley
  crossing, 452–3
  effect of fertilizer, 225
  nitrogen content, 234
  water culture, 222
  yield, 224
Pollen
  development, 41–42
  DNA content, 47
  effect of temperature on development, 47–48
  germination, 46
  maturity, 45–46
  morphology, 46
  tube development, 46–47
Pollination, 46
Polyamines, 148–9
Polyethylene glycol, 247
Polygonum convolvulus
  See Black bindweed
Polygonum persicaria
  See Persicaria
Polysaccharides,
  barley content, 90
  effect on frost-resistance, 265
Popp equipment, 540
Porphyrin, 125–8
Postum, 516
Pot barley
  food value, 513–4

production, 520–1
Potassium
  deficiency effects, 253, 255–6
  effect on root growth, 242
  effect on yield, 301–2
  uptake during growth, 232, 251, 258
Potassium bromate, 550, 553
Potassium carbonate, 181
Potassium chromate, 181
Potassium hydroxide
  penetration of grain, 181
  use in steeping, 535
  waste disposal, 490
Potassium nitrate, 181
Potassium sulphate, 181
Poultry
  food intake, 495
  grain feed, 512
Powdery mildew, 351–3
Pratylenchus spp.
  See Root lesion nematode
'Proctor' barley
  amino acid composition, 236
  crossing, 453
  effect of fertilizer, 225
  growing conditions, 294
  mildew-resistance, 468
  yield, 296
Proline
  content in anther, 43
  content in germinating grain, 204, 206
  effect of water stress, 250
Propagation
  F_1-progeny method, 459
  mass methods, 460
  pedigree method, 459
  pedigree trial method, 459–60
Propionic acid
  addition to grain, 393
  addition to silage, 373
  reduction of moulding, 370
  silage acidity, 372
Protease, 163, 207
Protease inhibitors
  barley content, 162, 163, 495
Proteins
  animal feed requirements, 492, 494
  barley analysis, 508–10
  breeding for high levels, 464
  content in aleurone layer, 8
  content in barley, 161–3
  content in dry matter, 231
  content in germinating grain, 207
  content in grain, 235, 403
  leaf production, 517–8
  micro-organism culture, 517
  patterns, 439
  yields, 313

'Provost'
  yield in S. Australia, 448–9
Pseudomonas atrofaciens, 350
Pseudostems, 15
Ptinus tectus
  See Australian spider beetle
Puccinia anomala
  See Brown rust
Puccinia glumarum
  See Yellow rust
Puccinia graminis
  See Stem rust
Puccinia hordei
  See Brown rust
Puccinia striformis
  See Yellow rust
Pullularia pullulanis, 400
Pulping, 489–90
Pulvinus, 19, 22
Purines, 156–7
Putrescine, 151
Pyrethrin, 411
Pyridine nucleotides, 103, 107, 157–61
Pyridoxal phosphate, 160
Pyrimidines, 155–7
Pyrodextrinization, 554
1-Pyrroline, 151
Pyruvate, 102
Pythium spp.
  See Root rot

Quality
  breeding for, 463–6
  effect of grain size, 223
  effect of lodging, 33
  effect on subsequent yield, 295
  grain germination, 174–215
  rainfall effects, 313
Quinones, 120–1

Rabbit control, 345
Rachilla
  development, 40
  morphology, 3
  varietal differences, 65
Rachis
  accordion, 67
  development, 39
  morphology, 65–67
Radiation
  See Light
Raffinose
  content in barley, 91
  content in germinating grain, 202
Rainfall
  effect on yield, 309–11

Rat control, 345, 412–4
Rattus norvegicus
  See Brown rat
Rattus rattus
  See Black rat
Reapers, 330
Red rust, 356
Red squill, 414
'Research' barley
  seeding rate and yield, 275
Respiration
  cyanide-resistant, 107
  effect of light, 270
  enzyme production, 213
  in germinating grains, 195–9
  in growing grain, 239
  in leaves, 261
  in malting grain, 550–2
  in silage, 372
  in stored grain, 395
Rhizome, 15
Rhizopertha dominica
  See Lesser grain borer
Riboflavin, 160, 510
Ribonucleic acid, 155–6
Ribose, 89
Ribulose diphosphate carboxylase, 124
Rice weevil, 406, 407
Ricks, 330
Ridges, 39
'Risø' 1508 barley
  chemical composition, 236, 464
RNA
  See Ribonucleic acid
Rodent control, 412–4
'Roger' barley
  root growth under water stress, 248
Roller field, 326
Root lesion nematode, 344
Root rot, 362
Rootlets
  composition, 510, 514
  germination, 9,25
  malting process, 526
Roots
  adventitious
    growth, 227, 242
    morphology, 15, 25–26
  branching patterns, 243
  cap, 6, 29, 243
  coronal
    see Adventitious
  depth, 26–27
  effect of eyespot, 360
  effect of light, 270
  effect of nematodes, 343
  effect of O_2 deprivation, 32
  effect of temperature, 265
  effect of soil density, 244
  effect of soil moisture, 242, 245–6
  effect of soil nutirents, 242
  effect of water stress, 247–8
  growth, 227, 240–5
  lateral, 32
  lengths, 242
  microflora, 30–31
  morphology, 25–32
  nodal
    see Adventitious
  number in embryo, 7
  parameters for growth, 240
  pressure, 31
  regeneration, 29
  removal effects of growth, 262
  respiration, 199
  seminal, 25–26
    growth, 242
  sheath
    see Coleorhiza
  uptake of nutrients, 256–9
  uptake of water, 246
  weight, 242
Rotation system, 315, 340
Roughage, 495
Rumex spp.
  See Docks/Sorrels
Rust, 314
Rust red flour beetle, 406, 408
Rust red grain beetle, 406, 408
Rye
  classification, 79
  food value, 496–7
Rye whiskey, 581
Ryegrass
  as undersown crop, 315
  control of fungus, 361
Rynchosporium secalis
  See Leaf blotch

Saccharomyces carlsbergensis, 575
Saccharomyces cerevisiae, 570, 574, 578
Saccharomyces diastaticus, 572, 578
Saccharomyces uvarum, 572
Sacks, 390
Saladin box, 539
Salt
  See Sodium chloride
Salts
  penetration into grain, 181
Sample divider, 175
Sampling, 174–6
Sawfly, 347
Saw-toothed grain beetle, 406, 408
Scab, 362
Scald, 354
Schizosaccharomyces pombe, 572

Schönfeld test, 182–3
Schönjahn test, 183
Sclerophthora macrospora
　See Downy mildew
Scotch barley, 513–4
Screens, 378
Scutellum, 7
Secale cereale
　See Rye
Seed
　broadcasting, 327
　choice, 294–9
　dressings, 294
　drilling, 327–9
　effect of density on yield, 298–9
　effect of size on yield, 295, 469
　effect of sowing date on yield, 295–8
　multiplication, 475–6
　quality evaluation, 295
　setting, 327
Seedling
　growth cessation, 226
　growth stages, 228
　vernalization, 266–8
Selenium
　deficiency, 463
　effect on crossing over, 421
　uptake, 252
Selenium salts
　viability testing, 184
Slenophoma donacis
　See Halo spot
Senescence, 18, 260–1
Separators, 379–82
Septoria, 314
Septoria passerinii
　See Leaf spot
Serin, 113
Sewerage effluent, 311
Sharp eyespot, 360
Shikimate pathway, 136–7
Ship rat, 412
Shoots
　growth rate, 225
Silage
　dry matter content, 372
　food value, 504–5
　lactic acid content, 373
　respiration, 372
　storage, 372–5
Silica gel, 386
Silicates, 305
Silicon, 252
Silos, 373–4, 391–3
Silver nitrate
　penetration of grain, 181
　prevention of steep damage, 191
Simazine, 308

Sinapsis arvensis
　See Charlock
Siteroptes graminum
　See Mite
Sitophilus granarius
　See Grain weevil
Sitophilus oryzae
　See Rice weevil
Sitophilus zeamays, 407
Sitotroga cerealella
　See Angoumois grain moth
Slugs, 344
Snow-rot, 364
Sodium, 251, 258
Sodium bromate, 181
Sodium bicarbonate, 181
Sodium carbonate, 181
Sodium chloride
　effect on growth, 250, 311
　effect on yield, 305
　penetration of grain, 181
　prevention of steep damage, 191
Sodium fluoroacetate, 414
Sodium hydroxide
　penetration of grain, 181
　starch recovery, 481
　use in steeping, 535
Sodium hypochlorite, 176
Sodium metabisulphite, 373
Sodium nitrate, 181
Sodium trichloroacetate, 341
Soil
　acidity effects, 253, 300
　compaction, 294, 321
　density effects on root growth, 244
　drainage, 293
　minerals uptake, 250–6
　moisture conservation, 309–10
　moisture effects, 309–11
　moisture effects on root growth, 242, 245
　nutrient effects on root growth, 242
　pH
　　effect of fertilizers, 306
　　effect on growth, 252
　ploughing depth, 293–4
　preparation for seed, 293–4
　salination, 250, 311
　temperature effects, 265
　treatment with straw, 484–6
　waterlogged, 309
Solek equipment, 541
Sorrel, 340
South Australia
　climate adaptation of barley, 447
Spermidine, 151
Spermine, 151
Spikelet
　contents, 44–45

effect of water stress, 249
Spire
  See Coleoptile
Sporotrichum pulverulentum, 517
Spot blotch, 362
Sparging, 574
'Spratt' barley
  selection, 449–50
'Spratt-Archer' barley
  crossing, 452–3
  mildew resistance, 467–8
  yield, 297
Sprayers, 329
Sprouts
  See Rootlets
Squalene, 116
Stamens
  carpelloid, 52
  development, 40
  See also Anthers
Starch
  appearance in embryo, 13–14
  content in barley, 98–100, 201–2
  content in endosperm, 8
  content in germinating grain, 202
  content in growing grain, 238–9
  degradation during mashing, 563–4, 566–8
  deposition in endosperm, 50
  transitory, 14
Statoliths
  See Gravity perception
Steeliness
  grain evaluation, 176
  grain morphology, 8–9
  hordein content, 304
  nitrogen content, 304
Steeping, 533–7
  changes occurring, 554
  effect on dormancy, 190–1
Steeps, 536–7
Stellaria media
  See Chickweed
Stem rust, 356
Stemphylium spp., 400
Stems
  apex, 17
  branch
    See Tillers
  dimensions, 34–35
  effect of eyespot, 360–1
  growth rate, 225
  growth stages, 228–9
  main
    morphology, 15
  morphology on growth, 14–21
  rhizomatous, 15
  shooting, 17
  strength, 34
  thickness control, 308
  variety effects, 19
Sterility
  chromosome abnormalities, 427–30
  in animals from toxins, 403, 404
  in smooth-awned barley, 309
  male, 440–1, 454–5
Sterols, 108, 117
Stillage, 515
Stills, 581–2
Stomata
  morphology, 24–25
  water loss, 246–7
Storage
  barley hay, 369–71
  barley heads for crossing, 454
  effect on dormancy period, 188
  grain, 389–414
  grain for animal feed, 512
  losses, 369
  silage, 372–5
  straw, 372
Stores
  flat-bed, 390–1
  See also Bins, Silos
Straw
  animal feed, 505–7
  baling, 334–5
  burning, 334
  dry matter, 232
  formation, 19–20
  moisture content, 371
  morphology, 20–21
  post-harvest treatment, 334–5
  prices, 292
  storage, 372
  use
    animal bedding, 483
    building, 482–3
    farmyard manure, 483–4
    industrial 486–90
    minor, 482
    paper industry, 487–90
    soil treatment, 484–6
Strawboards, 483
Straws, 482
'Streetly Rogue' barley
  crossing, 453
Streptomyceteceae spp., 402
Stripe rust, 356
Strychnine, 414
Styles, 40, 43
Sub-aleurone region
  See Endosperm, starchy
Subsoiler, 322
Sucrose
  content in anther, 43

content in barley, 91
content in germinating grain, 202, 204
content in stored grain, 402–3
effect of cold, 265
embryo culture, 207
Sugars
content in barley, 89–102
content in dry matter, 231
content in grain, 237–8
gibberellin inhibition, 212
nucleotide, 91
penetration of grain, 181
yeast utilization, 570, 572
Sulphate
biochemistry in barley, 146
effect on yield, 300
uptake, 251
Sulpholipids, 113
Sulphoquinovose, 90
Sulphur
content in dry matter, 231
deficiency effects, 253
Sulphur dioxide
damage to crops, 308
use in malting, 535, 544, 548, 554, 555–7
Sulphur dust, 356, 357
Sulphuric acid
addition to silage, 373
grain decortication, 176, 180, 184, 191
weed control, 340
Sulphurous acid, 535–6, 557
Suspensor, 51
'Svanhals' barley
selection, 450
Swathing, 313–4
Synergids, 47
Syrups, 572–3

Take-All fungus
control, 361
effect on yield, 315
Tarsonemus spp., 410
2, 3, 6-TBA, 342
Tagmen
See Testa
Tellurium salts, 184
Temperature
effect on crossing, 420–1
effect on flowering for crossing studies, 454
effect on grain development, 47–8
effect on grain germination, 182, 187, 189
effect on grain survival 394
effect on growth, 264–8, 269
effect on mashing, 561–3, 569–70
effect on respiration, 198
effect on water uptake, 179
germination of malting grains, 538, 547

grain drying, 384–6
induction of mutations, 72
kilning malted grains, 541–2
mashing, 549
stored grain, 392
Tenebrio molitor
See Yellow mealworm
'Tennessee' barley
cold damage, 266
Terbutryne, 341
Terpenes, 116, 120
Testa
chemical composition, 201
morphology, 1, 4
Testinic acid
barley biochemistry, 131
extraction during malting, 535
Tetraploids
characteristics, 430–1
chromosome adjustment, 77
Tetrazolium salts, 184
Thallium sulphate, 414
Thatching, 482
Thiamine, 510
Thiamine pyrophosphate, 160
Thiourea, 395
Thistles, 340
Threshing, 313
ancient, 330–1
machinery, 330–4
Thrips, 346
Tillage
machinery, 322–7
objectives, 293–4
Tillers
effect of plant density, 263
effect of temperature, 264
effect on yield, 468–9
growth rate, 225
growth with adventitious roots, 26
morphology, 15–18
See also Stems
Tinea granella
See Corn moth
Tipula paludosa
See Leatherjacket
'Titan' barley
crossing, 465
Tocopherol, 120–1
Tocopherol quinone, 120–1
Tocotrienol, 120–1
Toxaphene, 345
Toxoptera graminum
See Greenbug
Tractors, 294, 308
Transamination, 137
Transglucosylase, 100
Translocation, 428–30

Index 609

Transpiration, 245–6
'Trebi' barley
  dormancy, 188
  selection, 450
Trefoil
  as undersown crop, 299, 315
  control of fungus, 361
Triallate, 341
Tribolium castaneum
  See Rust red flour beetle
Tribolium confusum
  See Confused flour bettle
Tricarboxylic acid cycle, 102–7
Trichloroacetic acid, 181
Trichothecium spp., 401
Triglycerides, 108–16
Trisomics, 432–3
Triticeae, 76, 79
Triticum spp.
  See Wheat
Triticum vulgare
  chromosomes, 431
Trogoderma grananium
  See Khapra beetle
Tryptophan, 137
'Tschermak's barley'
  hardiness, 467
'Tunis' barley
  growth from different seed sizes, 296
Turners, 539
Typhula incarnata
  See Snow-rot
Tyramine, 128, 151
Tyrosine
  barley content, 128
  synthesis, 136–7
'Tystofte Prentice' barley
  selection, 449–50

Ubiquinone, 120
Ultraviolet radiation, 193
Umbelliferone, 128, 134, 193
'Union' barley
  mildew-resistance, 468
Unloaders, 374–5
Uridine diphosphate glucose (UDPG), 91, 98
Urea, 181, 493
Ustilago hordei
  See Covered smut
Ustilago nigra
  See Black loose smut
Ustilago nuda
  See Loose smut

Valine, 140, 146
Varieties of barley
  Abyssinian
    dormancy, 188
    grain size, 55
    origins, 85
  adaptation to climate, 446–9
  awned
    crossing, 419–20, 421
  black-grained
    crossing, 420
  breeding for quality, 463–6
  breeding for yield, 468–70
  calcaroides, 62
  Canadian
    analysis, 530–1
  Chinese
    awns, 59
    naked, 53
  classification, 72–3
  compositum, 71, 72
  definition, 77
  differences in culm length, 19
  differences, in dormancy, 188
  differences in dry matter, 372
  differences in fertility, 51
  differences in grain form, 52–65
  differences in grain survival, 394
  differences in lateral floret development, 39
  differences in leaf dimensions, 22–23
  differences in nitrogen content, 304
  differences in node number, 21
  differences in resistance to moisture, 249
  differences in response to temperature regulation, 264–5
  differences in root depth, 28
  differences in root number, 7
  differences in soil moisture utilization, 245
  differences in soil pH tolerance, 252
  differences in spikelet fertility, 45
  differences in stomata number, 24–25
  differences in striations, 9
  differences in tillering, 17, 468–9
  differences in water utilization, 310
  differences in wax compositions, 108
  differentiation, 78
  disease-resistant, 365, 464, 465
  Dwarf
    grain size, 55
    node number, 21
  eighteen-rowed mutation, 72
  erectoides
    stem structure, 21, 35
  Ethiopian
    See Abyssinian
  evaluation, 294–9, 446, 474–5
  four-rowed, 68, 72
  German
    grain size, 55
  hardy, 467

'Himalaya'
  analysis, 508
  hoods, 62
  mineral content, 511
  naked, 53
hooded
  crossing, 419–20, 421
  mutations, 62, 439
husked
  analysis, 508
hybridization, 451–3
Indian
  lodging, 34
  rooting systems, 28
intermedium, 77
irregulare, 71, 72
Japanese
  naked, 53
latisquamose lodicules, 78
mildew-resistant, 352–3, 467–8
mutation breeding, 450–1
naked, 53
  human food, 522
natural selection, 460–3
Near Eastern
  dormancy, 188
North African
  dormancy, 188
North American
  analysis, 498
parvisquamose lodicules, 78
purity, 446
Russian
  analysis, 499
selection, 449–50, 456–60
six-rowed
  analysis, 556–7
  crossing, 77, 420, 421–2
  definition, 68
  for malting, 528
  germination time, 550
  grain form, 54
  grain size, 68
  growth in North America, 447
  hexastichon, 68, 78
  intermedium, 68, 78
  irregulare, 72, 78
  lateral florets, 39
  nipponicum, 70
  nitrogen content of malt, 548
  nodes, 66–67
  origins, 81, 83–85
  pallidum, 77
  parallelum, 78
  rachilla, 65
  spikelet fertility, 45
  tetrastichum, 78
  trifurcatum, 70

vulgare, 68, 78
spring
  yields, 296–8
staggered growth, 294
sub-calcaroides, 62
ten-rowed mutations, 58, 72
tolerance to greenbug, 345
trials, 474–5
two-rowed
  analysis, 546, 556–7
  crossing, 77, 420, 421–3
  deficiens, 68
  definition, 68
  distichon, 68, 77, 78
  erectum, 68, 78
  for malting, 528
  germination time, 550
  glumes, 56–58
  grain form, 53
  growth in North America, 447
  growth in North Europe, 462
  lateral florets, 39, 68
  nitrogen content of malt, 54–8
  nutans, 68, 78
  origins, 81, 84–85
  spikelet fertility, 45
  zeocrithon, 68
winter
  yield, 296–8
white-grained
  crossing, 420
Vascular bundles
  coleoptile, 6
  morphology, 3, 20–21, 24
'Vaughan' barley
  selection, 461
Vegetables
  growing on straw bales, 484–5
Veins, 23
'Velvon' barley
  sterility, 309
Ventilation, 390–2
Ventral furrow, 4
Vernalization, 266–8
Viablity tests, 182–5
Vigour
  grain evaluation, 186
Vinegar, 577–80
Vinegar eels, 579–80
Virus diseases, 349–50
Viscosity
  malt extracts, 546
Vitamin B_6, 160
Vitamin E, 403
Vitamins
  animal feed requirements, 493
  barley content, 161, 509–11
  wort content, 569

Vitavax, 360
Vodka, 580, 584
Volvaria volvacea
   See Padi-straw mushroom

Wanderhaufen equipment, 540–1
Warehouse moth, 406, 409
Warfarin, 414
Water
  culture, 222
  effect on dormancy, 190
  effect on yield, 309–11
  grain content, 177
  penetration into grain, 178–82, 533
  stress, 247–50
  uptake during growth, 232, 245–7
  vapour pressure, 382–4
Wax
  biochemistry, 108
  cuticular, 108
  leaf morphology, 25
  prevention of water loss, 247
  testa, 108
  variety differences, 108
Weather
  See Climate
Weeds
  control, 339–43
    chemical, 307, 329, 340
    economics, 343
    nonchemical, 293–4, 340
  effect on yield, 307
Weevils, 406–7
Weighers, 377
Wheat
  classification, 79
  food value, 496–7
Wheat-bulb fly, 348
Wheat gall nematode, 344
Wheat mosaic virus, 350
Wheat striate mosaic virus, 350
Whiskey, 580–1
Whisk(e)y
  analysis, 583
  flavour, 555
  kilning malt for, 543
  maturation, 582
  production, 580
  starch degradation in mash, 567
White-shouldered house moth, 409
Wild oats, 341
Wild onion, 339, 340
Windrowers, 332
Winkler kilns, 544–5
Winnowing, 331
Winterkilling, 308–9, 467
Wireworm, 348–9
'Wisa' barley

  mutations, 451
Wort
  beer brewing, 576
  carbohydrate content, 564–6
  definition, 548
  fatty acid content, 569
  fermentabilities, 567
  lipid content, 569
  malt vinegar production, 578
  nitrogenous substances, 550, 568–9
  pH, 568, 569–70
  phenol content, 569
  production, 561–70
  viscosity, 566
  vitamin content, 569

Xanthomonas translucens, 350
Xanthophylls, 117
Xanthoxin, 246
Xylobiose, 97
Xylose
  content in barley, 89
  content in germinating grain, 202

Yeast
  food value, 496–7
  growth in stored grain, 399
  metabolism, 570–2
  nitrogen requirement, 568–9
  surplus Brewer's, 515
Yellow mealworm, 409
Yellow phosphorus, 414
Yellow rust, 356
Yields
  actual *vs* potential, 314–5
  breeding for, 468–70
  definitions, 292
  effect of brown rust, 357
  effect of CO_2, 274
  effect of climate, 222
  effect of different agricultural practices, 292–319
  effect of eyespot, 361
  effect of fertilizer, 297, 299–307, 469
  effect of frost, 309
  effect of grain size, 222–3
  effect of harvesting, 313–4, 333
  effect of herbicides, 343
  effect of leaf blotch, 354
  effect of leaf stripe, 363
  effect of lodging, 33, 302, 308, 469
  effect of mechanical damage, 308–9
  effect of minerals, 251
  effect of nematocides, 343
  effect of nutrients, 299–307
  effect of osmotic stress, 247–9
  effect of other crops, 299
  effect of photosynthesis, 272–80

effect of plant density, 222, 271, 275
effect of powdery mildew, 351
effect of rabbits, 345
effect of seed density, 298–9
effect of seed quality, 295
effect of seed size, 295, 469
effect of sowing date, 295–8
effect of temperature, 264
effect of variety, 294–8, 468–70
effect of virus, 349
effect of water supply, 309–11
effect of weeds, 307
effect of winterkilling, 308–9
factors affecting, 270–80
forage barley, 311–3

heritability, 458
in S. Australia, 447–9
losses
    insects in stored grain, 404
    rodents, 412–3
trials, 474–5
world, xiii-xiv
'Ymer'
    analysis, 504

'Zephyr' barley
    analysis, 503
    growing conditions, 294
Zinc, 253
Zinc phosphide, 414

633.16 Briggs, Dennis
BRI   Edward.

   Barley

Kirtley Library
Columbia College
Columbia, Missouri 65216